1856272

OCT 26 2011

PIPE STRESS ENGINEERING

by
Liang-Chuan (L.C.) Peng and
Tsen-Loong (Alvin) Peng
Peng Engineering, Houston, Texas, USA

© 2009 by ASME, Three Park Avenue, New York, NY 10016, USA (www.asme.org)

All rights reserved. Printed in the United States of America. Except as permitted under the United States Copyright Act of 1976, no part of this publication may be reproduced or distributed in any form or by any means, or stored in a database or retrieval system, without the prior written permission of the publisher.

INFORMATION CONTAINED IN THIS WORK HAS BEEN OBTAINED BY THE AMERICAN SOCIETY OF MECHANICAL ENGINEERS FROM SOURCES BELIEVED TO BE RELIABLE. HOWEVER, NEITHER ASME NOR ITS AUTHORS OR EDITORS GUARANTEE THE ACCURACY OR COMPLETENESS OF ANY INFORMATION PUBLISHED IN THIS WORK. NEITHER ASME NOR ITS AUTHORS AND EDITORS SHALL BE RESPONSIBLE FOR ANY ERRORS, OMISSIONS, OR DAMAGES ARISING OUT OF THE USE OF THIS INFORMATION. THE WORK IS PUBLISHED WITH THE UNDERSTANDING THAT ASME AND ITS AUTHORS AND EDITORS ARE SUPPLYING INFORMATION BUT ARE NOT ATTEMPTING TO RENDER ENGINEERING OR OTHER PROFESSIONAL SERVICES. IF SUCH ENGINEERING OR PROFESSIONAL SERVICES ARE REQUIRED, THE ASSISTANCE OF AN APPROPRIATE PROFESSIONAL SHOULD BE SOUGHT.

ASME *shall not be responsible for statements or opinions advanced in papers or . . . printed in its publications* (B7.1.3). Statement from the Bylaws.

For authorization to photocopy material for internal or personal use under those circumstances not falling within the fair use provisions of the Copyright Act, contact the Copyright Clearance Center (CCC), 222 Rosewood Drive, Danvers, MA 01923, tel: 978-750-8400, www.copyright.com.

This book has been cataloged with the Library of Congress.
ISBN: 978-0-7918-0285-4
ASME Order No. 802854

CONTENTS

Acknowledgments xi
Preface.. xiii
Nomenclature xv

Chapter 1
Introduction 1
1.1 Scope of Pipe Stress Analysis.. 2
1.2 Piping Components and Connecting Equipment... 4
1.3 Modes of Failure.. 9
    1.3.1 Static Stress Rupture.. 9
    1.3.2 Fatigue Failure.. 12
    1.3.3 Creep Rupture 14
    1.3.4 Stability Failure.. 17
    1.3.5 Miscellaneous Modes of Failure 18
1.4 Piping Codes 19
1.5 Industry Practice 22
    1.5.1 Load Cases.. 23
    1.5.2 Local Support Stresses 24
    1.5.3 Local Thermal Stresses 24
    1.5.4 Pressure Effect on Flexibility 25
    1.5.5 Stress Intensification for Sustained Loads 25
    1.5.6 Support Friction 25
    1.5.7 Guide and Stop Gaps 26
    1.5.8 Anchor and Restraint Stiffness 26
    1.5.9 Small Piping 26
1.6 Design Specification 26
    1.6.1 Owner's Design Specification 26
    1.6.2 Project Specification 28
1.7 Plant Walk-down 30

Chapter 2
Strength of Materials Basics... 33
2.1 Tensile Strength 33
    2.1.1 Modulus of Elasticity 34
    2.1.2 Proportional Limit 34
    2.1.3 Yield Strength, S_y 35
    2.1.4 Ultimate Strength, S_u 35
    2.1.5 Stresses at Skewed Plane 35
    2.1.6 Maximum Shear Stress, $S_{s,max}$ 36
    2.1.7 Principal Stresses 36
2.2 Elastic Relationship of Stress and Strain 36
    2.2.1 Poisson's Ratio 37
    2.2.2 Shear Strain and Modulus of Rigidity... 38

2.3 Static Equilibrium 38
    2.3.1 Free-Body Diagram 39
    2.3.2 Static Equilibrium 40
2.4 Stresses due to Moments 40
    2.4.1 Stresses due to Bending Moments 40
    2.4.2 Moment of Inertia 42
    2.4.3 Polar Moment of Inertia 42
    2.4.4 Moment of Inertia for Circular Cross-Sections 43
    2.4.5 Stresses due to Torsion Moment 43
2.5 Stresses in Pipes 45
    2.5.1 Stresses due to Internal Pressure 45
    2.5.2 Stresses due to Forces and Moments 47
2.6 Evaluation of Multi-Dimensional Stresses 49
    2.6.1 General Two-Dimensional Stress Field 49
    2.6.2 Mohr's Circle for Combined Stresses 51
    2.6.3 Theories of Failure 52
    2.6.4 Stress Intensity (Tresca Stress) 52
    2.6.5 Effective Stress (von Mises Stress) 53
2.7 Basic Beam Formulas 53
    2.7.1 Guided Cantilever 55
2.8 Analysis of Piping Assembly 55
    2.8.1 Finite Element 56
    2.8.2 Data Points and Node Points 57
    2.8.3 Piping Assembly 58

Chapter 3
Thermal Expansion and Piping Flexibility 61
3.1 Thermal Expansion Force and Stress 61
    3.1.1 Ideal Anchor Evaluation 61
    3.1.2 The Real Anchor 62
3.2 Methods of Providing Flexibility 63
    3.2.1 Estimating Leg Length Required 63
    3.2.2 Inherent Flexibility 65
    3.2.3 Caution Regarding Quick Check Formulas 65
    3.2.4 Wall Thickness and Thermal Expansion Stress 66
3.3 Self-Limiting Stress 66
    3.3.1 Elastic Equivalent Stress 67
3.4 Stress Intensification and Flexibility Factors 67
    3.4.1 Ovalization of Curved Pipes 68
    3.4.2 Code SIFs 69
3.5 Allowable Thermal Expansion Stress Range 71
3.6 Cold Spring 76
    3.6.1 Cold Spring Gap 77
    3.6.2 Location of Cold Spring Gap 78
    3.6.3 Cold Spring Procedure 78
    3.6.4 Multi-Branched System 79
    3.6.5 Analysis of Cold Sprung Piping System 79
3.7 Pressure Effects on Piping Flexibility 80
    3.7.1 Pressure Elongation 80
    3.7.2 Potential Twisting at Bends 81

	3.7.3 Pressure Elongation Is Self-Limiting Load	82
	3.7.4 Pressure Effect on Bend Flexibility and SIFs	82
3.8	General Procedure of Piping Flexibility Analysis	83
	3.8.1 Operating Modes	83
	3.8.2 Anchor Movements	84
	3.8.3 Assignments of Operating Values	84
	3.8.4 Handling of Piping Components	85
	3.8.5 The Analysis	86
3.9	Problems With Excessive Flexibility	86
	3.9.1 Problems Associated With Excessive Flexibility	88
3.10	Field Proven Systems	88

Chapter 4
Code Stress Requirements 91

4.1	"Design" Chapter of the Piping Codes	91
4.2	Loadings to be Considered	92
	4.2.1 Pressure	93
	4.2.2 Temperature	93
	4.2.3 Weight Effects	95
	4.2.4 Wind Load	95
	4.2.5 Earthquake	96
	4.2.6 Dynamic Fluid Loads	96
	4.2.7 Harmonic Anchor Displacement Loads	98
	4.2.8 Passive Loads	98
4.3	Basic Allowable Stresses	98
	4.3.1 Bases for Establishing Allowable Stresses	98
	4.3.2 Code Allowable Stress Tables	99
	4.3.3 Weld Strength Reduction Factor	100
4.4	Pressure Design	101
	4.4.1 Straight Pipe	102
	4.4.2 Curved Segment of Pipe	104
	4.4.3 Miter Bends	106
	4.4.4 Branch Connections	109
	4.4.5 Pressure Design for Other Components	113
4.5	Stresses of Piping Components	113
	4.5.1 Calculations of Component Stresses	113
	4.5.2 Sustained Stresses	117
	4.5.3 Occasional Stresses	119
	4.5.4 Thermal Expansion and Displacement Stress Range	121
	4.5.5 Code Stress Compliance Report	124
4.6	Class 1 Nuclear Piping	125

Chapter 5
Discontinuity Stresses 133

5.1	Differential Equation of the Beam Deflection Curve	133
5.2	Infinite Beam on Elastic Foundation With Concentrated Load	135
5.3	Semi-Infinite Beam on Elastic Foundation	138
5.4	Application of Beam on Elastic Foundation to Cylindrical Shells	139
5.5	Effective Widths	141
5.6	Choking Model	142

5.7	Stresses at Junctions Between Dissimilar Materials	143
	5.7.1 Uniform Pipes With Similar Modulus of Elasticity	144
	5.7.2 A Pipe Connected to a Rigid Section	146
5.8	Vessel Shell Rotation	147

Chapter 6
Pipe Supports and Restraints ... 151

6.1	Device Terminology and Basic Functions	151
6.2	Support Spacing	157
6.3	Analysis of Piping Systems Resting on Supports	159
6.4	Variable Spring and Constant Effort Supports	162
	6.4.1 Variable Spring Hanger Selection Procedure	163
	6.4.2 Constant-Effort Supports	165
	6.4.3 Spring Support Types and Installations	166
	6.4.4 Setting of Loads — Hot Balance and Cold Balance	168
6.5	Support of Long Risers	170
	6.5.1 Support Schemes	170
	6.5.2 Support Loads	171
	6.5.3 Analysis Method	172
6.6	Significance of Support Friction	172
	6.6.1 Effects of Support Friction	172
	6.6.2 Method of Including Friction in the Analysis	174
	6.6.3 Application of Friction Force	176
	6.6.4 Methods of Reducing Friction Force	176
6.7	Support of Large Pipes	178
	6.7.1 Saddle Supports Using Roark's Formula	179
	6.7.2 Ring Girder Supports	180
	6.7.3 Saddle Supports by Zick's Method	185
	6.7.4 Support Types	190
6.8	Pipe Stresses at Integral Support Attachments	194
	6.8.1 Power Boiler Formulas for Lug Stresses	194
	6.8.2 Kellogg's Choking Model	198
	6.8.3 WRC-107 Stress Evaluation	202
6.9	Treatment of Support Stiffness and Displacement	205

Chapter 7
Flexible Connections ... 209

7.1	Basic Flexible Joint Elements and Analytical Tools	211
	7.1.1 Generic Flexible Connections	211
	7.1.2 Bellow Elements	213
7.2	Using Catalog Data	218
	7.2.1 Background of Catalog Data	218
	7.2.2 Using the Catalog	219
	7.2.3 Calculating Operational Movements	223
	7.2.4 Cold Spring of Expansion Joint	224
7.3	Applications of Bellow Expansion Joints	225
	7.3.1 Application of Axial Deformation	225
	7.3.2 Lateral Movement and Angular Rotation	230
	7.3.3 Hinges and Gimbals	232
7.4	Slip Joints	235

7.5	Flexible Hoses	237
	7.5.1 Types of Metallic Hoses	237
	7.5.2 Application and Analysis of Flexible Hoses	238
	7.5.3 Analysis of Hose Assembly	241
7.6	Examples of Improper Installation of Expansion Joints	241
	7.6.1 Direction of Anchor Force	241
	7.6.2 Tie-Rods and Limit Rods	242
	7.6.3 Improperly Installed Anchors	244

Chapter 8
Interface with Stationary Equipment 247

8.1	Flange Leakage Concern	247
	8.1.1 Standard Flange Design Procedure	248
	8.1.2 Unofficial Position of B31.3	250
	8.1.3 Equivalent Pressure Method	251
	8.1.4 Class 2 Nuclear Piping Rules	254
8.2	Sensitive Valves	255
8.3	Pressure Vessel Connections	256
	8.3.1 Loadings Imposed to Piping from Vessel	257
	8.3.2 Vessel Shell Flexibility	258
	8.3.3 Allowable Piping Load at Vessel Connections	264
	8.3.4 Heat Exchanger Connections	265
8.4	Power Boiler and Process Heater Connections	265
8.5	Air-Cooled Heat Exchanger Connections	268
8.6	Low-Type Tank Connections	270
	8.6.1 Displacement and Rotation of Tank Connection	270
	8.6.2 Stiffness Coefficients of Tank Nozzle Connection	272
	8.6.3 Allowable Piping Loads at Tank Connections	274
	8.6.4 Practical Considerations of Tank Piping	280

Chapter 9
Interface with Rotating Equipment 285

9.1	Brief Background of Allowable Piping Load on Rotating Equipment	286
	9.1.1 When Nobody Knew What to Do	286
	9.1.2 First Official Set of Allowable Piping Loads	287
	9.1.3 Factors Behind the Low Allowable Piping Load	287
9.2	Evaluation of Piping Load on Rotating Equipment	288
	9.2.1 Effect of Piping Loads	288
	9.2.2 Movements of Nozzle Connection Point	289
	9.2.3 Analysis Approach	290
	9.2.4 Selecting the Spring Hangers to Minimize the Weight Load	292
	9.2.5 Multi-Unit Installation	293
	9.2.6 Fit-up the Connection	294
9.3	Steam Power Turbine	295
9.4	Mechanical Drive Steam Turbines	296
	9.4.1 Allowable Loads at Individual Connection	296
	9.4.2 Allowable for Combined Resultant Loads	297
	9.4.3 Basic Piping Layout Strategy	298
9.5	Centrifugal Pumps	300
	9.5.1 Characteristics Related to Piping Interface	300

viii   Contents

		9.5.2 Basic Piping Support Schemes	302
		9.5.3 Non-API Pumps	303
		9.5.4 API Standard 610 Pumps	304
	9.6	Centrifugal Compressors	308
	9.7	Reciprocating Compressors and Pumps	313
		9.7.1 Pulsating Flow	313
		9.7.2 Pulsation Pressure	315
		9.7.3 Pulsation Dampener for Reciprocating Pumps	316
		9.7.4 Some Notes on Piping Connected to Reciprocating Machine	319
	9.8	Problems Associated With Some Techniques Used in Reducing Piping Loads	321
		9.8.1 Excessive Flexibility	321
		9.8.2 Improper Expansion Joint Installations	322
		9.8.3 Theoretical Restraints	322
	9.9	Example Procedure for Designing Rotation Equipment Piping	324

Chapter 10
Transportation Pipeline and Buried Piping 329

10.1	Governing Codes and General Design Requirements	330
	10.1.1 B31.4 Liquid Petroleum Pipeline	331
	10.1.2 B31.8 Gas Transmission Pipeline	333
10.2	Behavior of Long Pipeline	335
	10.2.1 Pressure Elongation	335
	10.2.2 Anchor Force	335
	10.2.3 Potential Movement of Free Ends	336
	10.2.4 Movement of Restrained Ends	337
	10.2.5 Stresses at Fully Restrained Section	337
10.3	Pipeline Bends	339
10.4	Basic Elements of Soil Mechanics	340
	10.4.1 Types of Soils	340
	10.4.2 Friction Angle	340
	10.4.3 Shearing Stress	341
	10.4.4 Soil Resistance Against Axial Pipe Movement	341
	10.4.5 Lateral Soil Force	343
	10.4.6 Soil-Pipe Interaction	344
10.5	Example Calculations of Basic Pipeline Behaviors	346
	10.5.1 Basic Calculations	346
	10.5.2 Soil-Pipe Interaction	347
10.6	Simulation of Soil Resistance	348
10.7	Behavior of Large Bends	349
10.8	Construction of Analytical Model	351
10.9	Anchor and Drag Anchor	353

Chapter 11
Special Thermal Problems 357

11.1	Thermal Bowing	357
	11.1.1 Displacement and Stress Produced by Thermal Bowing	357
	11.1.2 Internal Thermal Stresses Generated by Bowing Temperature	359
	11.1.3 Occurrences of Thermal Bowing	360
	11.1.4 The Problem Created by a Tiny Line	364
11.2	Refractory Lined Pipe	365

		11.2.1 Equivalent Modulus of Elasticity	365
		11.2.2 Hot-Cold Pipe Junction	366
11.3	Un-Insulated Flange Connections		369
11.4	Unmatched Small Branch Connections		369
11.5	Socket-Welded Connections		370

Chapter 12
Dynamic Analysis — Part 1: SDOF Systems and Basics 373

12.1	Impact and Dynamic Load Factor		373
12.2	SDOF Structures		375
	12.2.1	Working Formula for SDOF Systems	376
	12.2.2	Un-Damped SDOF Systems	376
	12.2.3	Damped SDOF Systems	382
	12.2.4	Summary of the Characteristics of SDOF Vibration	386
12.3	Damping		387
12.4	Sonic Velocity Versus Flow Velocity		389
	12.4.1	Sonic Velocity	390
	12.4.2	Flow Velocity	392
12.5	Shaking Forces due to Fluid Flow		395
12.6	Safety Valve Relieving Forces		397
	12.6.1	Open Discharge System	397
	12.6.2	Closed Discharge System	401
12.7	Steam Turbine Trip Load		403

Chapter 13
Dynamic Analysis — Part 2: MDOF Systems and Applications 409

13.1	Lumped-Mass Multi-Degree of Freedom Systems		409
	13.1.1	Mass Lumping	410
	13.1.2	Free Vibration and Modal Superposition	412
13.2	Piping Subject to Ground Motion		413
	13.2.1	Response Spectra Method	415
	13.2.2	Combination of Response Spectra Analysis Results	417
	13.2.3	Comparison of Modal Combination Methods	419
	13.2.4	Puzzles of Absolute Closely Spaced Modal Combination	421
	13.2.5	Compensation for the Higher Modes Truncated	422
	13.2.6	Design Response Spectra	423
13.3	Account for Uncertainties		426
13.4	Steady-State Vibration and Harmonic Analysis		428
	13.4.1	Basic Vibration Patterns	428
	13.4.2	Allowable Vibration Displacement and Velocity	429
	13.4.3	Formulation of Harmonic Analysis	437
	13.4.4	Evaluation of Vibration Stress	444
13.5	Time-History Analysis		446
	13.5.1	Treatment of Damping	446
	13.5.2	Integration Schemes	448
	13.5.3	Time Step, Stability, and Accuracy	450
	13.5.4	Example Time-History Analysis	451

Appendix A	459
Appendix B	461
Appendix C	462
Appendix D	464
Appendix E	466
Appendix F	474
INDEX	477

ACKNOWLEDGMENTS

This book is essentially the summary of the knowledge accumulated by the authors through 40 years of practice as piping mechanical engineers. I, the senior author, would like to use this opportunity to express my appreciation and gratitude to many friends, colleagues, and supervisors for providing those learning opportunities and environments. First, I would like to thank Ron Hollmeier of Pioneer Service & Engineering in Chicago for offering me my first pipe stress job developing a computer program for pipe stress analysis in 1967. This job allowed me to stay in the United States and led to a long, interesting career. I am grateful to Bechtel's Bill Doble and Joe Gilchrist, who did not hesitate to send me to English classes and put me on interesting jobs such as the Trans-Alaskan pipeline and Black Mesa coal slurry pipeline projects. My most memorable work was done at Nuclear Services Corporation in San Jose. Working as part of a team that included Bob Keever, Randy Broman, Doug Munson and myself, and with help from Professor Gram Powell of University of California-Berkeley and valuable inputs from Mel Pedell and Dane Shave of Stone and Webster, we created the NUPIPE pipe stress software, which became a very powerful tool in the design of nuclear piping. I am greatly indebted to Don Mckeehan and Ed Bissaillon of M. W. Kellogg for their encouragement and implementation of the SIMFLEX software. As a result of the authors' long association with M. W. Kellogg, this book is noticeably influenced by Kellogg's philosophy and approaches mentioned in the second edition of the Kellogg's book — *Design of Piping Systems* (1956, John Wiley and Sons). Suggestions from Ray Chao and David Osage of Exxon Research were very helpful during the development of the PENGS program. The estimated 1000 engineers, who came to my training classes conducted in a dozen countries, have greatly widened my perspective on piping mechanical work. The authors are very grateful to ASME Press for valuable comments and excellent editing. We would also like to thank the twin sisters Lina and Linda for reading the manuscript. The fact that these two non-technical sisters have read through the entire draft of the manuscript has greatly encouraged us.

Liang-Chuan (L.C.) Peng
Tsen-Loong (Alvin) Peng

PREFACE

Pipe stress analysis calculates the stress in a piping system subject to normal operating loads such as pressure, weight, and thermal expansion, and occasional loads such as wind, earthquake, and water hammer. Because all piping systems are connected to equipment such as vessels, tanks, pumps, turbines, and compressors, the piping stress analysis also involves evaluation of the effect of the piping forces and moments to the connecting equipment. As the piping stress is controlled by the arrangement of the supports and restraints, the scope of piping stress includes also pipe supports. The whole scope of this work is generally referred to as piping mechanical.

Before the advent of the electronic computer, pipe stress analysis was handled by very specialized engineers. Only large corporations and specialized firms had the personnel to do the job. It normally took a specialist to use the calculator non-stop for a couple of weeks just to analyze the flexibility of a moderately complex piping system to absorb the thermal expansion of the pipe. Because only very few of these engineers knew how to analyze piping stress, most engineers treated it as some type of a mysterious subject. Engineers saw there were expansion loops, offsets, and special supports such as spring hangers and constant effort supports, but did not really know why they were there. The limited scope of pipe stress analysis dealing with the piping flexibility for absorbing thermal expansion was called piping flexibility analysis.

With the arrival of the electronic computer in the 1970s, and especially the personal computer in the 1980s, suddenly everybody knew how to analyze pipe stress. This has generated even more mystery about the field. Nowadays, we occasionally see an electrical engineer, although discouraged, conducting the analysis just as proficiently as a mechanical or a structural engineer. This partly stems from the fact that colleges and universities normally do not offer any course on pipe stress. This leaves the knowledge and skill of pipe stress and piping engineering to be learned by self-study and actual practice. Practitioners who obtain the best computer program and comprehend the manual most will do the better job.

With the rapid advancement in computer technology, a piping flexibility analysis nowadays takes only few minutes via an appropriate computer software. Therefore, the task of the stress engineer has been shifted from the traditional stress calculation to stress engineering. The emphasis is not on how to calculate the stress, but rather on how to utilize the analysis tool to design a better plant. However, just because it is easy to get the stress calculated, engineers often depend too much on the computer and forget about the fundamentals and engineering common sense. Without the fundamentals and common sense, one may not even be aware of the unreasonable results produced by the computer, to say the least about good engineering. This book emphasizes engineering common sense as well as the basic principles. The following are some examples of piping problems that might have been solved by just good engineering common sense:

- A plant operated smoothly for the first 10 years, and then experienced a leakage at the main process piping about every 4 months after a major revamping. Experts were consulted, sophisticated analyses were performed, and expensive modifications were made to no avail. Had the engineers used the basic thermal stress common sense, the problem would have been solved with very little effort (see Section 11.2.2).
- The pressure thrust force is very critical at a bellow expansion joint. Anchors are often needed at bellow expansion joint installations to resist the pressure thrust force, but

- sometimes an anchor placed at the wrong location may just be the cause of the problem. The wrong anchor had been contributing to severe vibration at some rotating equipment. Had we known that the pressure thrust force at an expansion joint is generally much higher than the force that can be tolerated by the rotating equipment, the problem would not have occurred (Chapters 7 and 9).
- A small steam purge line in a large process pipe had caused the plant piping to twist wildly, breaking many connecting leads. Several major re-routing of the piping systems were made, but the problem persisted. Had the involved engineers had some idea about thermal bowing, the problem could have been easily corrected (see Section 11.1.4).

Providing the knowledge for solving the problems such as the ones listed above is the primary goal of the book. Chapter 1 summarizes the scopes and requirements related to piping mechanical activities, and the subsequent chapters discuss how to deal with them. Chapter 1 contains some of the authors' inside views, which we hope will help the readers progress more confidently and comfortably into the piping mechanical field.

Nowadays, making a calculation with a computer is so fast that we often hear about the "what-if" approach in engineering. What all this "what-if" approach accomplishes is making numerous random trials and the wish that one of these trials will hit the mark sooner or later. The problem is that after a few trials, most people lose the ability to make sense of the trials. The more they try, the more they get confused. In contrast, this book puts emphasis on the "what, why, and how" to guide the readers into this 3-W approach — that is, to be aware of the problem, understand the cause of the problem, and to solve the problem or prevent it from happening.

The authors will try to explain all the necessary tasks of pipe stress engineering with basic fundamentals. Although only a few very fundamental equations are introduced, theoretical backgrounds will be covered in as much detail as possible. The book is titled *Pipe Stress Engineering* to distinguish it from a regular pipe stress analysis book, which normally lacks the coverage of the engineering aspects. Although this book is intended for piping mechanical engineers, it is also a suitable reference book for piping designers, plant engineers, and civil-mechanical engineers. The book can be used as the textbook for a one- or two-semester elective course given at the senior or graduate level.

Nomenclature

(All are in consistent units. Special usages and non-consistent units are noted in the main text. Abbreviations are listed at the end)

A	Cross-section area
A	Thickness allowance including corrosion, thread, etc.
A	Flow area
A_b	Total net cross-section area of flange bolts
A_c	Corrosion allowance of pipe thickness
A_m	Manufacturing under-tolerance of pipe thickness
A_p	Pressure area encircled by the bolt circle of flange
A_T	Nozzle throat area or valve orifice area
a	Bellow effective pressure thrust area
a	Sonic speed
a	Tank nozzle radius
B	Flange inside diameter
B	Tank bottom plate thermal expansion factor
b	Flange effective gasket width
b	Thickness of support saddle, in pipe axial direction
b	Width of the beam cross-section
b	Width of a small element or strip
C	Cold spring factor
C	Flange bolt circle diameter
C	Thermal bowing local stress factor, defined in Figure 11.2
$[C]$	Damping matrix of the structural system
C_1, C_2	*i*ntegration constants to be determined by boundary conditions
C_2, K_2	ASME Class-1 nuclear piping stress indices for displacement loading
C_{ED}	End coefficient with respect to vibration displacement, defined in Eq. (13.43)
C_{EV}	End coefficient with respect to vibration velocity, defined in Eq. (13.52)
C_K	Kármán force coefficient on vortex shedding force
C_w	Ratio of total pipe weight, including pipe metal weight, content, and insulation, to pipe metal weight
C_X	Vibration stress amplification factor due to concentrated weight
c	Soil cohesion stress
c	Viscous damping
c_p	Specific heat of gas under constant pressure condition
c_v	specific heat of gas under constant volume condition
D	Diameter of pipe or circular cross-section
D	Diameter of vessel shell or run pipe
$\{D\}$	Displacement vector in global coordinates, for both element and overall structures
$\{D'\}$	Element displacement vector in local coordinates
D_c	Combined equivalent diameter of all the nozzles at rotating equipment
D_e	Equivalent diameter of an individual nozzle at rotating equipment

DLF	Dynamic load factor
d	Displacement
d	Diameter
d	Diameter of branch pipe
d_e	Elastic displacement limit of an elastic-plastic restraint
d_p	Pitch diameter of bellow
E	Modulus of elasticity
E	Longitudinal joint efficiency, normally expressed with allowable stress as (SE)
E_e	Equivalent modulus of elasticity for composite pipe
e	Strain
e_{com}	Combined total equivalent axial bellow deformation per convolution
e_x	Axial bellow deformation per convolution
e_y	Equivalent axial bellow deformation per convolution due to lateral displacement y
e_θ	Equivalent axial bellow deformation per convolution due to rotation θ
F	Force
F	Gas pipeline design factor (defined in Table 10.2)
F	Pipeline anchor force, or potential expansion force
$\{F\}$	Force vector in global coordinates, for both element and overall structures
$\{F'\}$	Element force vector in local coordinates
$\|F\|$	Absolute value of force F
F_1	Force equivalent to total impulse function $= \rho A V^2 + PA$
$\{F_C\}$	Vector of cosine components of the harmonic force
F_K	Kármán force or vortex shedding force
F_{max}	Net maximum peak shaking force of a given piping leg
F_n	Normal force
F_{nS}	Pressure force at point n due to standing pressure wave
F_{nT}	Pressure force at point n due to traveling pressure wave
F_P	Tank nozzle pressure force defined in Eq. (8.30)
F_R	Radial force
F_R	Resultant force
F_s	Shear force
$\{F_S\}$	Vector of sine components of the harmonic force
$F(t)$	Force as function of time
FRX	Rotation spring constant about x axis for a generic flexible joint
FRY	Rotation spring constant about y axis for a generic flexible joint
FRZ	Rotation spring constant about z axis for a generic flexible joint
FTX	Translation spring constant in x axis for a generic flexible joint
FTY	Translation spring constant in y- axis for a generic flexible joint
FTZ	Translation spring constant in z- axis for a generic flexible joint
f	Line force per unit length of active line in an attachment
f	Natural frequency of the structural system, or frequency of a vibration
f	Pipeline friction force per unit length of pipe
f	Stress range factor for calculating allowable expansion stress range
f	Support friction force
$f_1(\beta x)$	Beam on elastic foundation function defined by Eq. (5.17a)
$f_2(\beta x)$	Beam on elastic foundation function defined by Eq. (5.17b)
$f_3(\beta x)$	Beam on elastic foundation function defined by Eq. (5.17c)
$f_4(\beta x)$	Beam on elastic foundation function defined by Eq. (5.17d)
f_w	Working (nominal) axial spring constant per bellow convolution
G	Flange gasket load reaction circle diameter
G	Shear modulus of elasticity, modulus of rigidity

g_1	Flange hub thickness at back of flange
g_x, g_y, g_z	Cold spring gap in x, y, z directions, respectively
H	Depth of soil cover, from top of pipe to soil surface
H	Height of the beam cross-section
H	Ring force per unit circumferential breadth
h	Flexibility characteristic of piping component, defined in Table 3.1
h	Safety valve height, defined in Fig. 12.17
h	Soil depth at an arbitrary point
I	Moment of inertia of a pipe or beam cross-section
$[I]_D$	Identity matrix, or unit diagonal matrix, with 1 on diagonal and 0 at elsewhere
I_p	Polar moment of inertia
i	Stress intensification factor
i	Imaginary number, square root of -1
i_C	Stress intensification factor for circumferential stress
i_L	Stress intensification factor for longitudinal stress
K	Bulk modulus of the liquid $= -dp/(dv/v)$
K	Stiffness or spring constant of support structure
$[K]$	Stiffness matrix in global coordinates, for both element and overall structures
$[K']$	Element stiffness matrix in local coordinates
K_A	Coefficient of active lateral soil pressure
K_h	Spring constant per unit pipe length for horizontal soil resistance
K_L	EJMA bellow lateral spring rate $= K_V$
K_N	Napier constant for steam flow
K_R	EJMA bellow rotational spring rate $= K_{R\theta}$
K_{Ry}	Bellow rotational spring rate due to lateral end deflection
$K_{R\theta}$	Bellow rotational spring rate due to end rotation, when free lateral deflection is allowed
K_V	Bellow lateral spring rate due to lateral end deflection. End moment also created
K_v	Spring constant per unit pipe length for vertical soil resistance
K_x	Bellow axial spring rate
k	Flexibility factor of piping component
k	Ratio of specific heats $= c_p/c_v$
k	Spring constant and directional stiffness
k	Spring constant of foundation per unit length of beam
L	Distance from the center of tank nozzle to tank bottom
L	Length
L	Pipeline active length
L	Support spacing
$[L]$	Transformation matrix between local coordinates and global coordinates
LMP	Larson-Miller parameter defined in Eq. (1.2)
ℓ	Elongation
ℓ	Support lug length
ℓ	Length of element
M	Mass
M	Moment
$[M]$	Mass matrix of the structural system
M_{mass}	Concentrated mass
M_C	Circumferential moment
M_L	Longitudinal moment
M_R	Resultant moment
M_t	Torsion moment

M_w	Molecular weight
M_y	Moment produced by bellow lateral displacement without free end rotation
M_θ	Moment produced by bellow rotation with free lateral displacement
$M_{(\theta+\Delta)}$	Moment produced by bellow rotation without free lateral displacement
m	Flange gasket factor
m	Mass per unit length of pipe
$\dot{m}$	Mass flow rate
N	Circumferential force per unit length of pipe
N	Number of bellow convolutions
N	Number of operating cycles
N_S	Strouhal number on vortex shedding frequency; $N_S = fD/V$
n	Number of stiffening rings at each support saddle
P	Pressure
P_d	Flange design pressure
P_e	Equivalent pressure due to force and moment acting on a flange
P_S	Pressure amplitude of a standing wave
P_T	Pressure amplitude of a traveling wave
p	Pressure
p^*	Critical pressure at sonic velocity state
Q	Pipeline end resistance force
Q	Total support load
Q	Volumetric flow rate
$[Q]$	Influence or relation matrix relating ground motion to every part of the structure
q	Distributed external force per unit length of beam
q	Pitch of bellow convolution
q_a	Deformed pitch at centerline of bellow
q_c	Compressed pitch at pitch diameter of bellow
q_e	Extended pitch at pitch diameter of bellow
q_h	Horizontal soil resistance force per unit length of pipe
q_v	Vertical soil resistance force per unit length of pipe
R	Bend radius
R	Gas constant = $\mathcal{R}/M_w$
R	Radius of curvature
R	Radius of circular cross-section
R	Radius of vessel shell
R	Reaction force
$\mathcal{R}$	Universal gas constant
R_a	Acceleration response spectra
R_{aj}	Acceleration response spectra of jth independent support motion
R_C	Radius of the crown on a vessel head
R_d	Displacement response spectra
R_v	Velocity response spectra
r	Radius of an arbitrary circular ring
r	Radius of pipe cross-section
r	Radius of round attachment
r	Rotation
r_m	Mean radius of pipe cross-section
S	Basic allowable stress for pipeline
S	Stress
S_A	Basic allowable thermal expansion stress range
S_b	Flange bolt stress

S_c	Allowable stress of pipe material at cold ambient condition
S_E	Expansion stress
S_{EB}	Benchmark expansion stress range
S_{el}	Endurance strength of pipe material
S_{elA}	Allowable endurance strength
S_F	Flange gasket stress due to axial force
$\{S_g\}$	Vector of independent support motion components
S_h	Allowable stress of pipe material at hot operating condition
S_h	Pressure hoop stress
S_{hp}	Hoop pressure stress
S_i	Stress intensity (=twice of the maximum shear stress)
S_{lp}	Longitudinal pressure stress
S_M	Flange gasket stress due to bending moment
SMYS	Specified Minimum Yield Strength
S_P	Flange gasket stress due to pressure
S_{PW}	Longitudinal pipe stress due to pressure and weight
S_T	Local thermal stress due to thermal bowing
S_u	Ultimate strength
S_y	Yield strength
S_{yc}	Yield strength at cold condition
S_{yhx}	The lesser of yield strength at hot condition and 160% of the stress producing 0.01% creep in 1000 hours at the operating temperature
T	Absolute temperature, K (=273 + °C) or R (=460 + °F)
T	Period of vibration
T	Temperature
T	Vessel thickness or run pipe thickness
T^*	Critical absolute temperature at sonic velocity state
t	Thickness of bellow
t	Thickness of branch pipe
t	Thickness of pipe, generic
t	Thickness of tank shell at nozzle location
t	Time
t_d	Time duration of an impulse loading
t_o	Effective valve opening time
V	Bellow lateral force
V	Velocity
V	Volume
V	Tangential shear force per unit circumferential breadth
v	Shear force
v	Specific volume
W	Total flange bolt load
W	Total shear force at pipe cross-section
W	Weight load in force unit
W	Weight of the free body, or weight load
W	Weld strength reduction factor
W_p	Weight of pipe per unit length of pipe
W_s	Weight of soil cover per unit length of pipe
w	Weight per unit length
$\{X\}$	Displacement vector of the structural system
X_A	Distance between the top of nozzle and tank bottom plate
X_B	Distance between the bottom of nozzle and tank bottom plate

X_C	Distance between the center of nozzle and tank bottom plate
$\{X\}_g$	Ground motion displacement vector
x	Axial displacement of beam or bellow
x, y, z	Coordinates in x, y, z directions, respectively
Y_C	Allowable coefficient, Fig. 8.25, for circumferential moment on tank nozzle
Y_F	Allowable coefficient, Fig. 8.25, for axial force on tank nozzle
Y_L	Allowable coefficient, Fig. 8.25, for longitudinal moment on tank nozzle
y	Adjustment coefficient (see Table 4.1) for pipe thickness calculation
y	Flange gasket seating stress
y	Local lateral displacement of beam or bellow
y	Pipeline end axial movement
y	Radial displacement of pipe shell
Z	Section modulus of a pipe or beam cross-section
Z_p	Polar section modulus
ZPA	Zero period acceleration of the response spectra curve
z	Compressibility of real gas $= pv/(RT)$

Greek symbols

α	Constant defining participation of mass in damping, see Eq. (13.76)
α	Vibration allowable stress reduction factor, 1.3 for carbon and low alloy steels and 1.0 for austenitic stainless and high alloy steels
α	Thermal expansion rate, as expansion per unit length per unit temperature
β	Angle from top of pipe to edge of saddle
β	Angle of branch intersection
β	Branch/run diameter ratio
β	Characteristic parameter of beam on elastic foundation, defined by Eq. (5.26)
β	Constant defining participation of stiffness in damping, see Eq. (13.76)
β	Frequency ratio of the applied frequency to the natural frequency of the system
γ	Rotational deformation
Δ	Clearance between tube and tube sheet hole
Δ	Deflection
$\Delta..$	Difference of ..
ΔA	Amplitude decay due to step-by-step time-history analysis
ΔT	Period elongation due to step-by-step time-history analysis
Δt	Integration time step for time-history analysis
ζ	Damping ratio, the ratio of damping to critical damping
η	Nozzle flow efficiency
θ	Angle of circular wedge
θ	Angle of miter
θ	Angle of the inclining plane
θ	Rotation
θ	Support saddle angle
λ	Tank geometrical parameter defined in Eq. (8.30)
μ	Friction coefficient
ν	Poisson ratio
ξ	Coordinate or coefficient of normal mode space, i.e., $\{X\} = [\Phi]\{\xi\}$
$\{\xi_C\}$	Cosine component of modal coordinate vector of harmonic response
$\{\xi_S\}$	Sine component of modal coordinate vector of harmonic response
π	3.141592
ρ	Angle of the maximum bending moment location at a support ring
ρ	Density, mass per unit volume

Σ	Summation of
τ	Shear stress
υ	Flow velocity
$\{\Phi\}$	Eigenvector, or natural vibration shape
$[\Phi]$	Eigenvector matrix with eigenvectors as columns
ϕ	Angle from top of pipe
ϕ	Bellow rotation per convolution
ϕ	Soil internal friction angle
ω	Circular frequency, or rotational speed
ω_d	Circular natural frequency for damped system
ω_n	Circular natural frequency for un-damped (and also damped) system

Subscripts

0	At origin
0	initial state
0, 1, 2,…	At location 0, 1, 2,…, etc., or at condition 0, 1, 2,…, etc.
a	Allowable
allow	Allowable
a, b,…	at point a, b,…, etc.
b	bending
b	branch pipe
c	circumferential direction
c	cold or ambient temperature
c	critical condition
D	discharge side
e	equivalent
f	friction
H	hoop direction
h	hub of flange
h	hot or operating temperature
hp	hoop pressure
i	in-plane
i	inside surface of the pipe
L	longitudinal direction
l	longitudinal direction
lp	longitudinal pressure
m	mean value
max	maximum value
n	natural vibration
n	nominal
n	normal
o	out-plane
o	outside surface of the pipe
p	pressure
R	radial direction
R	resultant
R	rigid body response
R	run pipe
r	ring stiffener
r	run pipe

s	suction side
s	shear
st	static
T	test condition
t	tangential
t	torsion
xy	on x plane and in y direction
x, y, z	components in x, y, and z directions, respectively
y	yield condition
y	at y distance away

Abbreviations

ABS	absolute
ANSI	American National Standards Institute
API	American Petroleum Institute
ASME	American Society of Mechanical Engineers
ASTM	American Society of Testing and Materials
AWWA	American Water Work Works Association
B&PV	Boiler and Pressure Vessel
CEN	Comité Européen de Normalisation (European Standard)
cps	cycles per second
DLF	dynamic load factor
ft	foot or feet
in.	inch or inches
Hz	Hertz = cycles per second
K	Kelvin = 273 + °C
ksi	kilo pounds per square inch = 1000 psi
lb	pound (weight or force)
lbf	pound force
m	meter
MDOF	multi degrees of freedom
MSS	Manufacturer Standardization Society of the Valve and Fitting Industry
N	Newton
NRC	Nuclear Regulatory Commission
Pa	Pascal = N/m^2
psi	pounds per squire inch
R	Rankin = 460 + °F
RPM	revolution per minute
SDOF	single degree of freedom
SIF	stress intensification factor
SRSS	square root sum of the squares
WRC	Welding Research Council

CHAPTER 1

INTRODUCTION

A piping system is the most efficient and common means of transporting fluids from one point to another. Within a petrochemical complex, acres and acres of piping can be seen running in every direction and at many different levels. Piping constitutes 25% to 35% of the material of a process plant, requires 30% to 40% of the erection labor, and consumes 40% to 48% of the engineering man-hours [1]. The actual importance of piping, however, can far exceed these percentages. An entire piping system is composed of a large number of components. The failure of just one single component has the potential to shut down the entire plant or, worse yet, cause serious public safety problems. In spite of this, piping is generally considered a low-technology subject in the academia. Very few colleges teach the subject, leaving engineers to gain this knowledge only through actual practice in the field.

To find out exactly where pipe stress fits in the piping design process, let us first find out what procedures are involved in designing a piping system. A piping system is designed in the following steps by different engineering disciplines:

(1) Process engineers, basing on process requirements and plant capacity, determine, among other things, the flow path, the flow medium and quantity, and operating conditions. They then put all this information into process flow diagrams.

(2) Material specification engineers assign suitable categories of specifications for the piping system based on the process flow and reactivity of the contained fluid. Each specification is applicable to certain combinations of fluid types, temperature ranges, and pressure ranges. Material specifications normally include pipe material, pipe wall thickness for each pipe size, the corrosion and erosion allowances, flange class, valve types, fitting and branch connection type, bolt material, gasket type, etc.

(3) System engineers combine process flow diagrams, material specifications, and equipment data sheets to create operational piping diagrams. They select the applicable material specification and determine the size for each line based on flow quantity, allowable pressure drop, and flow stability. Piping diagrams are generally combined with the necessary instrument and control circuits to become piping and instrument diagrams (P&IDs). Special items such as potential two-phase flow and slug-flow zones are also identified on these diagrams for special consideration in design and analysis. In addition to the P&IDs, a line list covering all pipe spools is also constructed. This line list contains most of the design, upset, and operating parameters to be used in the layout, analysis, and fabrication of the piping system.

(4) Piping designers, in coordination with other disciplines, conceive an overall plant layout, perform a piping routing study, determine the pipe rack locations, and place the actual piping that connects to designated points. They lay out and support the piping by following the rules and procedures set up by each individual company. In general, three sets of drawings are prepared. The first set is the schematic planning drawings, used as a communication board between different departments. Actions and comments from related disciplines are all resolved and recorded in these drawings. Pipe supports are also recorded in this set of drawings. The second set is the composite drawings, consisting of to-scale drawings of all pipes and equipment in the area. These drawings, to be used in the construction, are evolved from the planning

drawings. The third set of drawings is the isometrics of the piping, used for stress checks and shop fabrications.
(5) Piping mechanical engineers check the stresses and supports of the systems. Using the P&IDs, they develop operating modes so that all the expected operating conditions are properly analyzed. Proper supports and restraints are selected and placed to optimize the overall cost and performance of the systems. They also design or specify piping specialty items, such as expansion joints, flue heads, special connections, spring hangers, vibration supports, and so forth.

It may be a bit surprising that designing a piping system is so involved. Indeed, on a large project, not only is every discipline required, but the effort also engages quite a few people. Piping design and piping mechanical are the two disciplines that require the most number of personnel. However, for a small project handled by a small outfit, generally only one or two people are assigned to take care of all the work of various disciplines. In such cases, pipe stress activity is often neglected. Due to the resilient nature of ductile piping systems, the piping will work most of the time, even without going through proper stress checks. This may be acceptable for small non-hazardous piping, but not for most of public and industrial piping systems, which require the piping system to be safe and operational all the time.

Item (5) is the main subject of this book. The task of a piping mechanical engineer is generally called pipe stress and support. The scope of the pipe stress and support activity has increased exponentially in the past three decades. This is due to the stringent requirements of the modern plant. For instance, in the 1960s, the pipe stress and support manpower used for a petrochemical plant was about 4000 man-hours. A nuclear power plant in that era would have used about the same amount of pipe stress man-hours. Nowadays, the pipe stress and support manpower required for a petrochemical plant has increased by ten times to about 50,000 man-hours. The effort required by a nuclear power plant has grown 1000-fold to reach as high as 2 million man-hours [2]. With this exponential growth in the man-hours involved, the probability of getting sub-standard output from some of the engineers is very high. A time-saving tool, such as an efficient pipe stress computer program, not only significantly reduces the cost of designing a plant, but also greatly improves the quality of the plant. Good work starts with a good grasp of the scope of the jobs that need to be done. In this chapter, the non-mathematical concepts of pipe stress activities will be discussed.

1.1 SCOPE OF PIPE STRESS ANALYSIS

Pipe stress analysis is, of course, the analysis of the stress in the pipe. However, if we ask what is to be analyzed, many will hesitate to answer. In the 1950s and 1960s, when engineers started analyzing piping systems, they had only one thing in mind: calculating the stress due to thermal expansion. In other words, they checked the piping layout to see if the piping system was flexible enough to absorb the thermal expansion due to temperature change. The analysis was referred to as piping flexibility analysis. Books [3–6] and articles [7–9] written during this period dealt mainly with flexibility analysis. Later on, as technology progressed, pipe stress analysis encompassed much more than just checking flexibility; yet nowadays many engineers still refer to pipe stress analysis as flexibility analysis. This slight mix-up in terms is not important. However, the concept that flexibility is the only consideration in piping stress analyses can lead to an expensive, and unsafe, sub-standard design. For instance, many engineers tend to consider that providing additional flexibility in the piping is a conservative approach. In reality, additional flexibility not only increases the material cost and pressure drop, it also makes the piping prone to vibration, the biggest problem area of the piping in operation. Since the publication of the 1955 piping code [10] and Kellogg's [3] book, failures due to insufficient flexibility have become very rare. Nowadays, most failures are caused by vibration, thermal bowing, creep, thermal fatigue not related to flexibility, steam/water hammer, expansion joints, and so forth [11, 12]. These facts should serve as clues toward designing better piping systems.

Consider, for example, the piping installed to move the process fluid from the storage tank to the process unit as shown in Fig. 1.1. First, we have to deal with tank shell displacement and rotation

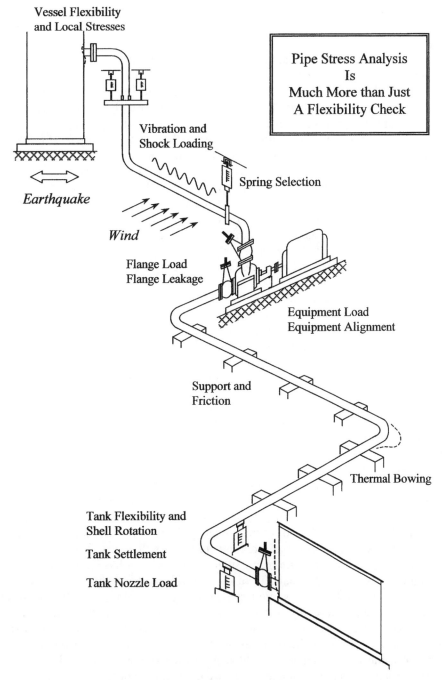

**FIG. 1.1
TASKS OF PIPE STRESS ANALYSIS**

due to the hydrostatic bulge of the shell. This temperature-independent displacement and rotation will exert a great influence on the connecting piping. Furthermore, the tank nozzle connection is far from rigid. Its flexibility has to be estimated and included in the analysis. Then, after the pipe forces and moments at the connection are calculated, they have to be evaluated for their acceptance. There

4 Chapter 1

are many items like these, which are not normally called piping flexibility, but are required to be considered in piping stress analysis. These items, using Fig. 1.1, shall be identified one by one starting from this tank connection.

The next items we come across are flanges and valves. Can they maintain the tightness under piping forces and moments? Can valves operate properly under pipe forces and moments? These need to be checked even though the pipe itself is strong enough for the same forces and moments.

We know the support friction can also have a significant effect on piping movement. The situations and methods to include the friction effect also need to be considered.

In addition to the average pipe temperature, the pipe might also have a temperature gradient across the pipe cross-section due to stratified flow or blow off of low-temperature fluid. Even radiant energy from the sun on empty un-insulated pipe can cause this type of temperature difference. This type of bowing phenomena can create a great problem in the piping and needs to be considered.

For piping connecting to the rotating equipment such as pumps, the pipe load has to be maintained to within the manufacturer's allowable range to prevent the equipment from excessive vibration, wear, and overheating. The piping connected to rotating equipment also needs the consideration of potential water hammer, pulsation, and other dynamic phenomena.

Proper spring hangers will need to be selected and placed to ensure that the piping is properly supported under all operating conditions.

At the vessel connection, again, the flexibility and displacement at the connection have to be included in the analysis. After pipe forces and moments at the connection are calculated, vessel local stresses have to be evaluated to see if they are acceptable. Then, of course, if the structure is located in an earthquake or hurricane zone, earthquake and wind loading have to be considered when designing the piping system.

In general, the purpose of pipe stress analysis can be summarized into two broad categories:

(a) *Ensure structural integrity*: This involves the calculation of stresses in the pipe due to all design loads. Necessary procedures are taken to keep the stress within the code allowable limits. This code stress check is the basic assurance that failures from breaks or cracks will not occur in the piping.

(b) *Maintain system operability*: A piping itself can be very strong, yet the system may not be operable due to problems in the connecting equipment. Flange leakage, valve sticking, high stress in the vessel nozzle, and excessive piping load on rotating equipment are some of these problems. The work required in maintaining the system operability is generally much greater than that required in ensuring the structural integrity. This is mainly attributable to the lack of coordination between engineers of different disciplines. Rotating equipment manufacturers, for instance, design non-pressure parts, such as support and base plate, based mainly on the weight and the torque of the shaft. Then they specify the allowable piping load with that design, disregarding the fact that some practical piping load always exists and needs to be accommodated. The allowable loads they provide are generally much too small to be practical. Unfortunately, these allowable values go unchallenged, mainly because the industry as a whole gives no incentives to manufacturers to produce equipment that can resist the extra piping load. If more and more engineers would request the extra strength or give preferential treatment to manufacturers that produce stronger equipment, an optimal solution might eventually be reached. Until such time comes, piping mechanical engineers should be prepared to spend three times as many man-hours in stress engineering the piping system connected to rotating equipment.

1.2 PIPING COMPONENTS AND CONNECTING EQUIPMENT

In this section, we will deal with three categories of hardware: pressure parts of the piping proper, support and restraint elements, and connecting equipment. The functions of support and restraint elements are apparent. And because a separate chapter is dedicated to pipe supports and restraints, we shall not include them in this discussion.

In the following, we use Fig. 1.2 as a guide to summarize the types of piping components and arrays of connecting equipment typically encountered in piping stress engineering.

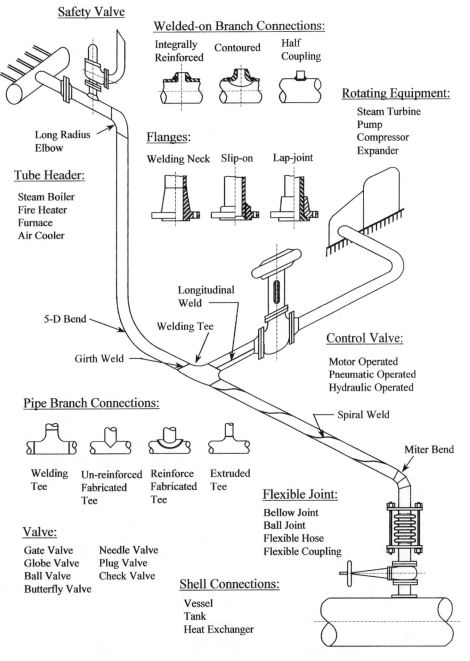

FIG. 1.2
PIPING COMPONENTS AND INTERFACE EQUIPMENT

(a) **Main pipe.** Starting with the main pipe, we have to know the pipe material, which is generally given by the American Society of Testing and Materials (ASTM) [13] specification number. The pipe is generally identified by its nominal diameter and nominal thickness. For pipes 12 in. (300 mm) or smaller, the nominal diameter is very close to the inside diameter of the pipe with standard wall thickness. However, for sizes 14 in. (350 mm) and larger, the nominal pipe diameter is exactly the same as the outside diameter of the pipe. For each pipe size, the outside diameter is fixed for a given nominal diameter. This is done so that all pipes with the same nominal size can use the same pipe support attachments such as clamps, and insulation blocks. Each pipe size also has several commercially available thicknesses called nominal thickness. The available nominal thicknesses are standardized in related standards such as ASTM A-106 [13] and American Society of Mechanical Engineers (ASME) B36.10 [14]. In the United States, these thicknesses are also represented by weight grades as standard (Std), extra strong (XS), and double extra strong (XXS), and by schedule numbers ranging from schedule 10 (Sch-10) to schedule 160 (Sch-160). Stainless steel pipes have separate schedule numbers appended with the letter "S" such as Sch-5S, Sch-10S, etc. For instance, a 6-in., Std pipe means that the pipe has a 6.625-in. outside diameter and a 0.280-in. wall thickness, whereas a 14-in., Sch-40 pipe corresponds to a pipe with a 14-in. outside diameter and 0.438-in. wall thickness. The schedule number itself dose not represent any thickness; it has thickness assigned only when tagged with a pipe diameter. See Appendix A for standard nominal pipe wall thicknesses.

The pipe material involves different manufacturing processes that might implicate some stress risers and tolerances. The pipe can be broadly classified via the method in which it was manufactured: seamless or welded. Welded pipes, which are somewhat weaker than seamless pipes due to the weld, are further classified into several categories. From the seam position, we have longitudinally welded and spirally welded (two main types). From the welding process, we have furnace butt welded, electric resistance welded, and electric fusion welded (three methods). Each type of welding has its unique joint efficiency that needs to be included in the pressure design. For tolerances, we are mostly concerned with thickness under-tolerance, which can reduce the pressure resisting capability of the pipe. Under-tolerance is the allowable amount of thickness in which a manufactured pipe can be made thinner than the nominal thickness. In general, seamless pipes and electric resistance welded pipes have a higher under-tolerance of about 12.5% of the nominal wall thickness. Electric fusion welded pipes have a somewhat different under-tolerance determined by the plate used; however, API 5L suggest 10% to 12.5% for the fusion welded pipe also. This under-tolerance needs to be included in the stress design of the pipe.

(b) **Welds.** To connect piping components together, we use either circumferential weld or flange connections. The circumferential weld, also called girth weld, is for a permanent connection, whereas the flange is used for locations requiring occasional separation. The girth weld also has a joint efficiency that works against longitudinal stress. This joint efficiency does not affect pressure design, which is controlled by circumferential hoop stress. For loadings other than pressure, this girth weld joint efficiency is implied in the stress intensification factor. Therefore, in most piping codes, circumferential weld joint efficiency is seldom mentioned. It is also often overlooked by stress engineers.

(c) **Weld strength reduction factor.** In addition to the joint efficiency that affects the general strength of the piping, the weld also hastens creep failure at creep temperature. The additional reduction of creep strength over the non-weld-affected body is called the weld strength reduction factor. This is the factor applied, over the joint efficiency, at high temperature ranges. The same factor is applied at both longitudinal welds and circumferential welds. However, longitudinal weld affects only the calculation of wall thickness, which is governed by the circumferential hoop stress. On the other hand, circumferential weld affects only the sustained longitudinal stress due to pressure, weight, and other mechanical loads. The weld strength reduction factor is not applicable to occasional stress due to the generally short duration of the stress. It also does not affect thermal expansion and displacement stress range due to the self limiting nature of the stress. Generally, the temperature that requires the application of the weld strength reduction factor starts from 950°F (510°C). However, B31.1 and

B31.3 treat it slightly differently. Details of the application are given in Chapter 4, which deals with code stress requirements.

(d) **Flanges.** Flanges are available in several different types. From the structural construction point of view, it can be classified as welding neck, slip-on, or lap joint (three types). Each type has its length, weight, and stress intensification factor, all of which have to be identified and considered in the stress analysis. Flanges are also identified by classes. Each class has its set of pressure-temperature ratings, relating allowable pressures with operating temperatures. At a given operating temperature, the selected flange class shall offer an allowable pressure that is either equal to or greater than the design pressure. Table 1.1 shows the pressure-temperature rating for forged A-105 carbon steel flanges. The table also shows the flange thickness required for each class, for 6-in. and 12-in. flanges. The data, taken from ASME B16.5 [16], is presented mainly to show the general trend of the allowable pressure versus temperature and class. It also gives some idea about the magnitude of flange thickness, excluding hub. For Class 2500 flanges, flange thickness equals roughly 2/3 of the pipe diameter.

It is also interesting to note that the flange classes were originally called pounds. Class 300 flanges were called 300-pound flanges, and so forth. This was because, originally, Class 300 flanges were rated for 300 pounds per square inch (psi) pressure at a benchmark temperature. Pound (lb) meant psi, and had nothing to do with the weight of the flange. The benchmark temperatures used were different for different materials. For A-105 carbon steel, the benchmark temperature was 500°F for Class 150, and 850°F for all other classes. However, as more accurate material data became available and more accurate stress calculations became possible, the original ratings also appeared less accurate. Currently, for A-105 flanges, Class 150 has a rating pressure of 170 psi at 500°F benchmark temperature, and Class 300 has a rating pressure of 270 psi at 850°F benchmark temperature. It is obvious that current pressure-temperature ratings no longer correspond to the original benchmark idea. Although the pound classification is not very meaningful right now, it is still used by many engineers. See Appendix E for layout dimension and weight of valves and flanges.

TABLE 1.1
PRESSURE-TEMPERATURE RATING FOR A-105 CARBON STEEL FLANGES [16]

Temp. °F	Allowable Pressure, psi Classes						
	150	300	400	600	900	1500	2500
-20 to 100	285	740	990	1480	2220	3705	6170
200	260	675	900	1350	2025	3375	5625
300	230	655	875	1315	1970	3280	5470
400	200	635	845	1270	1900	3170	5280
500	170	600	800	1200	1795	2995	4990
600	140	550	730	1095	1640	2735	4560
650	125	535	715	1075	1610	2685	4475
700	110	535	710	1065	1600	2665	4440
750	95	505	670	1010	1510	2520	4200
800	80	410	650	825	1235	2060	3430
850	65	270	355	535	805	1340	2230
900	50	170	230	345	515	860	1430
950	35	105	140	205	310	515	860
1000	20	50	70	105	155	260	430
Flange Thickness, in — 6-in Flg	1.00	1.44	1.62	1.88	2.19	3.25	4.25
Flange Thickness, in — 12-in Flg	1.25	2.00	2.25	2.62	3.12	4.88	7.25

(e) **Bends.** The turning of the pipe is accomplished by bends. Bends have several general types. The most common bend is the so-called long-radius elbow, which has a bend radius equal to 1.5 times the nominal pipe diameter. This is a fitting manufactured by the code-approved standard ASME B16.9 [15] and others. The same standard also includes a short-radius elbow, which has a bend radius equal to the nominal pipe diameter. Short-radius elbows are used in tight spots where available space is not enough for long-radius elbow. The cited factory-made forged elbows are quite expensive and also incur high flow friction loss due to the small bend radius. One alternative is to make the bend directly by bending the pipe. To avoid excessive thinning and potential wrinkling, the bend radius of this type is generally bigger than three nominal pipe diameters. The one shown in the figure is a 5-D bend. Another alternative, mainly to save cost, is to cut the pipe into angled miters and bring them together to form the bend. This type of bend is called a miter bend. All these different bends have different wall thickness requirements, flexibility factors, and stress intensification factors to be considered in the design and analysis.

(f) **Branches.** Branch connections are used to form the branches of the piping. The most common full-size branch connection is the forged welding tee, which is made according to ASME B16.9 [15] and other standards. Generally, the welding tee is quite expensive, but provides the smoothest flow passages and least stress intensification factor among all branch connections. Besides the welding tee, the most economical and readily available branch connection is the un-reinforced fabricated tee. Generally called a stub-in connection, this type of connection is made simply by cutting a hole on the run pipe and welding the branch pipe to it. A stub-in connection is cheap and easy to make, but can handle only about one-half of the pressure that the pipe can. It also has a very high stress intensification factor. To improve both the pressure resisting capability and stress intensification, proper reinforcement is required. When designed properly, the reinforced fabricated branch connection can take the same pressure as a run pipe can, and also substantially reduce stress intensification from the un-reinforced branch connection. Other branch connections include extruded tee, integrally reinforced weld-on, contoured weld-on, and half coupling. Each of these branch connections has its pressure design requirements and stress intensification that need to be considered in the design and analysis. See Table 3.1 for the flexibility factors and stress intensification factors of piping components and connections. See also Appendix B for layout dimension of butt-welding fittings.

(g) **Valves.** The piping also consists of many types of valves. For valves, we have to know the type, end-to-end length, and weight, for inclusion in the analysis. The valve itself is generally approximated to an equivalent pipe of the same length with three times the stiffness of the connecting pipe. For valves with a heavy operator, such as motor-operated ones, the operator weight and off-center location has to be included in the design analysis. This is especially important in analyzing dynamic effects such as earthquakes. For safety-relief valves, we also have to consider the dynamic effect due to the sudden discharge of the fluid when the safety valve pops open. Valves share the same classification as flanges. They have the same pressure-temperature rating as flanges for a given class and material.

(h) **Flexible joints.** To increase the flexibility of the system, the piping may also include some types of flexible connections. The one shown in the figure is a tied bellow expansion joint. Other flexible connections that might be used are hinged bellow joints, gimbaled bellow joints, ball joints, flexible couplings, and flexible hoses. Flexible connections are generally considered engineered items, which are not simply picked out of a catalog. Their selection involves some engineering and operational considerations.

(i) **Terminal connections.** In addition to piping components, we also have to consider the effect of the piping on connecting equipment. Because most units of the connecting equipment are weaker than the piping itself, the piping load satisfactory to piping may not be tolerable to the equipment at all. In fact, to engineer the piping reaction to the acceptable level of the equipment is the most difficult task for a piping stress engineer. The items listed in the following are some of the most common types of equipment connected to a piping system.

(j) **Shell connections.** A piping system is often connected to the shell of a pressure vessel, tank, drum, or heat exchanger. Although these shell connections are considered fixed interfaces, they do impose some movements to the piping at the connection due to expansion of the shell. The bulging of the shell, as in a low-type tank connection, may also impose some rotation to the piping. The acceptability of the piping load is determined by the local shell stress produced. In some cases, manufacturers of equipment, such as a heat exchanger, may specify the allowable forces and moments for the connection on its equipment. Normally, piping stress engineers will work with vessel engineers to decide if the piping load is acceptable. However, to expedite the validation process, piping stress engineers will first calculate the local vessel stress, which is integrated in the piping stress analysis, of the connection before giving the piping load to vessel engineers for approval. When calculating the piping load at a shell connection, there is often a disagreement between the vessel engineer and the piping engineer about the flexibility of the shell. Some vessel engineers insist that the flexibility of the shell should not be included in the piping analysis, so the resulting piping load is artificially increased to protect the vessel. This is actually very short sighted and may result in a very shaky design of the piping. Detailed discussions of shell connections are given in the chapter dealing with stationary interfaces.

(k) **Tube bundle header connections.** Another type of stationary interface is the header connected to tube bundles. Steam boilers, fired heaters, and air coolers are some examples. This type of interface generally has a given set of allowable forces and moments, which are provided either by the manufacturer or by an applicable industry standard. Because this type of interface generally does not have a precise boundary point, the piping may have to include all or part of the tube bundles in the analysis. Detailed discussions on this type of interface are given in the chapter dealing with stationary interfaces.

(l) **Rotating equipment connections.** The biggest challenge to pipe stress engineers is the piping connected to rotating equipment such as pumps, compressors, and turbines. Because of its extremely low permissible force and moment, a rotating equipment piping system is normally stress-engineered by an experienced piping mechanical engineer. Yet, it still requires considerably more effort to accomplish the stress engineering than to accomplish the same for a system that is not connected to rotating equipment. The problem is that rotating equipment cannot endure even a very slight deformation, or else risk the consequence of shaft misalignment. To maintain smooth operation of the rotating equipment, the shaft needs to be kept in perfect alignment without causing binding at the bearings and interference of the internal parts. Therefore, the acceptance criterion of rotating equipment piping is the strain rather than the stress. In the case of rotating equipment piping, the allowable strain is only equivalent to about one-fifth of the allowable piping stress criterion for a medium-size piping. The allowable is even smaller for larger pipes. See Chapter 9 for details.

1.3 MODES OF FAILURE

The main purpose of the piping mechanical work is to prevent piping failure. Therefore, it is important to find out how the pipe fails. The pipe can fail in many different modes with many different mechanisms. Some of the common modes of failure are discussed in the following subsections.

1.3.1 Static Stress Rupture

The pipe will fail when it is stressed to beyond its strength, as measured by testing. That is the definition of the strength of the material. Static stress means no time element is involved. The failure occurs as soon as stress reaches the limit. Because the pipe material will not take any stress higher than this limit, the limit is also called ultimate strength of the material. Other modes of failure may take considerably less stress than the ultimate strength. As the pipe material is statically stressed, it begins to deform (Fig. 1.3). The area under the curve represents the energy required for the failure. This area is also referred to as the energy absorbing capacity, or the toughness. In general, static rupture can be further classified into two categories: ductile rupture and brittle rupture.

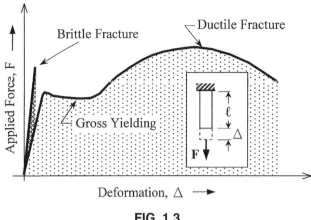

FIG. 1.3
STATIC FRACTURE

Ductile rupture. In ductile rupture, as the load increases, the material first yields producing a considerable plastic deformation and fails after going through fairly large amount of elongation or contraction. The material involved is called ductile material. A ductile piping material elongates about 25% (one-fourth of the original length) before the failure. Therefore, it has a very large energy absorbing capacity. The energy absorbing capacity has little effect on the slowly applied static loads, but is very important in resisting impact loads. Without this large energy absorbing capability, a very small impact can translate to a very high damaging stress.

Ductility is one of the most important considerations for the material used in piping. In addition to its ability to mitigate the impact load, ductility also redistributes the high concentrated stresses into a more favorable stress distribution through yielding. We often hear the saying that a piping system is very forgiving. This forgiveness of the piping system is mainly attributed to ductility.

The failure of the material mostly starts at the notches formed by defect or shape discontinuity. Under a given load, the stress produced at the notches is considerably higher than at the bulk of the material. The yielding of ductile material will smooth out the stress concentration near the notch, and thus increase the toughness of the material.

Figure 1.4 shows the stress near a circular hole in a plate. The stresses are evenly distributed at the area remote from the hole. As the area gets closer to the hole, the stress field also changes to follow the passage formed by the material. At the horizontal diametrical direction of the hole, the net area is smallest and the stress flow squeezes near the hole, producing a peak stress much higher than the average stress of the net area represented by *b-d*. As the load is increased, the peak stress eventually reaches the yield strength of the material. At this point, the yielded portion can no longer take any additional stress, prompting the nearby un-yielded portion to pick up more stress. This essentially limits the highest stress to the yield strength until all stresses at the neck reach the yield strength. The yielding prevents the stress from reaching the elastic peak stress as shown by the dotted lines. Therefore, ductile material is substantially less notch-sensitive than the non-yielding material.

Because of yielding, a ductile piping system will also shift the load to other parts of the system, if the stress at a certain portion of the system reaches the yield point. This load-shifting capability effectively uses the whole system to resist the load instead of depending on certain local locations. This increases the system reliability and the tolerance to abnormalities.

It is important to remember that the ductility of the pipe is the presumption of many design rules and specifications. Without ductility, many of the design calculations are meaningless.

Brittle rupture. If the pipe does not yield or does not produce plastic deformation, the energy absorbing capacity, represented by the area enclosed by the force-deformation curve, is very small as shown in Fig. 1.3. The material that fails without any yielding is called brittle material.

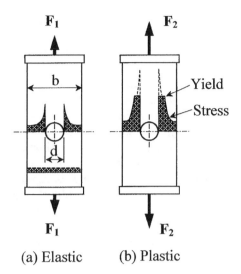

**FIG. 1.4
EFFECT OF NOTCHES**

Brittle failure often occurs unexpectedly and suddenly. This is due to two main factors. First, because of the low energy absorbing capability, a slight impact translates to a very high stress thus causing the failure. For instance, we can break a glass cup rather easily just by hitting the edge of the cup on a hard object with a small impact. Another point is that without yielding, the material is unable to relieve the high stress concentration at the crevices of a notch. Many of us have observed how a crack develops on the car windshield. A crack is produced when a tiny hard object, such as a piece of airborne pebble, hits the glass. The crack forms sharp star-like fissures, which just keep growing under normal driving conditions.

Brittleness is the inherent nature of some materials such as glass and gray cast iron. These materials, when used, require strict control of the stress and the nature of the loading. They cannot be used in the environment with either thermal or mechanical shock loading. The use of brittle materials requires extreme caution and care. A glass cup, for instance, can easily break in a minor incident, such as when it is dropped on the floor.

Some materials, on the other hand, become brittle because of temperature change. Most piping materials loss their ductility as the temperature drops below a certain limit. For instance, most carbon steels are susceptible to brittle failure at temperatures lower then −20°F (−29°C), whereas other materials (e.g., austenitic stainless steel, aluminum, copper, and brass) do not become brittle at temperatures as low as −425°F (−254°C). Some materials can also become brittle at high temperature due to metallurgical change. Mild carbon steel may loss its ductility at temperatures above 800°F (427°C) due to its susceptibility to graphite formation. For this discussion, however, we are mainly concerned with the loss of ductility due to cold temperature.

All the design philosophy and approaches we commonly use are based on the assumption that the material is ductile. Except for inherently brittle materials, we have to be very careful when using a material at a temperature range that is conducive to brittle failure. Most piping codes list −20°F (−29°C) as the minimum temperature for which the material is normally suitable without impact testing other than required by the material specification. Nevertheless, due to its wider temperature application range, the B31.3 process piping code [17] sets very detailed rules and requirements on low temperature application of the materials.

B31.3 sets the minimum temperature for each material listed in the allowable stress tables. Again, this minimum temperature for each material is the lowest temperature that the material can withstand without impact testing other than that required by the material specification. However, there are some exceptions. For instance, for austenitic stainless materials with carbon content higher than 0.1% or not solution heat treated, impact testing is required for a design temperature lower than −20°F although the listed minimum temperature is −425°F. For the most commonly used carbon steels, B31.3 uses four curves (A, B, C, and D) to determine the impact testing requirements. We will use curve B, which is applicable to mild carbon steels such as A-53, A106, and A-135, as an example to explain the rules and requirements. Curve B is re-plotted in Fig. 1.5. The region above the curve does not require impact testing other than that required by the material specification. The figure shows that the higher the wall thickness, the higher the temperature under which the impact testing is required. This thickness correlation is mainly because a thicker wall creates a higher uneven stress distribution and higher probability of containing bigger size defects. Impact testing, as stipulated by the code, indicates using the full allowable stress as given in the allowable stress table. Without impact testing, the material in question can still be used but at a reduced allowable stress. The rate of allowable stress reduction starts at 1% per each Fahrenheit degree (5/9 Celsius degree) of temperature lower than the non-impact test temperature. The reduction rate maintains the same level for the initial 40°F temperature difference, and then slows down asymptotically to a maximum total reduction of 70% at a temperature difference of 217°F (120°C). In Fig. 1.5, a pipe with a 1.5-in. (38.1 mm) wall thickness and operating at 32°F (0°F) would require impact testing in order to use the full allowable stress. At 1.5-in. wall thickness, the temperature not requiring an impact testing is 51°F. The operating temperature, in this case, is 51−32 = 19°F lower than the non-impact testing temperature. Therefore, if impact testing is not performed, the allowable stress has to be reduced by 19%.

The loss of ductility is a very serious concern for a piping system. Therefore, the minimum acceptable temperature, the impact test requirements, and the stress reduction provisions as outlined by the code should be followed closely.

1.3.2 Fatigue Failure

The pipe can fail under a stress lower than the ultimate strength of the material if the stress is cyclic. Bending a paper clip back and forth repeatedly can easily reproduce this effect. This type of failure is due to fatigue of the material. Fatigue failure is attributed to the combination of the stress amplitude

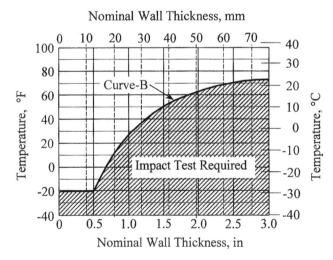

FIG. 1.5
IMPACT TEST REQUIREMENT OF MILD CARBON STEEL PIPE (A-53, A-106, A-135, ETC.) [17]

and the number of load cycles. Higher applied stress amplitude needs fewer cycles to rupture, and vice versa. The general design curve is as shown in Fig. 1.6. The allowable stress amplitude reduces asymptotically to the limit called the endurance limit. For a polished specimen, the endurance limit is reached at somewhere between 10^6 and 10^7 cycles, but a commercial pipe generally reaches the limit at much higher cycles. By looking at the curve, we might wonder why the allowable stress at 10 cycles, for instance, is almost as high as 10^6 psi. This is more than ten times higher than the ultimate strength of a common pipe material. This is because the allowable stress amplitude from fatigue tests is measured by the strain, which makes very little sense to ordinary engineers. Expressing the allowable in stress makes it easier to connect with engineers. However, this stress is just an elastic equivalent stress obtained by multiplying the allowable strain with the modulus of elasticity. When the strain exceeds the yield strain, the stress becomes a fictitious stress. In other words, the stress is not a real stress when it exceeds the yield strength of the material.

In fatigue, we are dealing with stress range. One-half of the stress range is called stress amplitude. This stress amplitude is used mainly from the typical mathematical formula, which can express the stress as a type of cyclic function such as $A(\sin \omega t)$, where A is stress amplitude and ω is circular frequency of the stress cycle.

Fatigue failure is generally classified as low-cycle fatigue and high-cycle fatigue. This classification is required due to the somewhat different nature of these two types of fatigue. In practice, fatigue curve can be idealized into two straight lines. Line L-L represents the low cycle fatigue and line H-H represents the high cycle fatigue. L-L is a straight line in log-log scale, whereas H-H has a constant value equivalent to the endurance limit. The actual curve is rounded at the transition zone, which seldom participates in actual applications. Because of the different basis used by different codes in calculating the stresses, it is important to remember that the fatigue curve used in one code may not be applicable to another code.

Low cycle fatigue normally applies to stress changes due to normal start-up and shutdown cycles, major load fluctuation cycles, and occasional load cycles. The thermal expansion stress range is the most common and well-known stress that may produce low-cycle fatigue failure. At low cycle fatigue, the stress range and the number of cycles to failure have a log-log straight line experimental relation

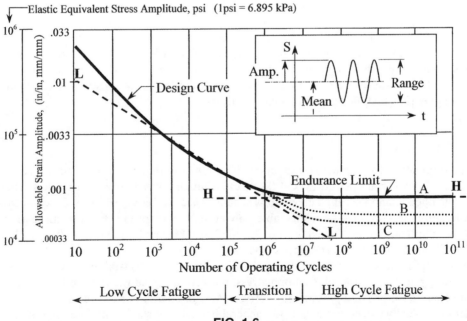

FIG. 1.6
LOW-CYCLE AND HIGH-CYCLE FATIGUE CURVES

$$SN^m = C \tag{1.1}$$

where

S = stress range
N = number of cycles to failure
m = negative slope of the log-log straight line
C = a constant, which is the elastic equivalent failure stress for $N = 1$

The allowable stress curve takes the same form with proper safety factor applied. Based on the results of his tests, Markl [7, 18] suggested that the log-log slope, m, is 0.2 for common piping components and 0.25 for stainless steel expansion joint bellows. In thermal expansion, we normally evaluate the stress range instead of the amplitude, but we also calculate the stress that is just one-half of the theoretical stress. We have to remember these double-amplitude and half-stress situations in preparation for the subjects discussed in Chapters 3 and 4.

It is important to note that the pipe has a definite life with low cycle fatigue and should be taken into account in the design of the plant. It is also important to note that low cycle fatigue has an application limit for elastic analysis. When stress is higher than this limit, which is the yield strength for the stress amplitude and twice that of the yield strength for the stress range, the commonly used elastic analysis is not applicable. Some modifications on the elastically calculated stress are required [19] for the stress exceeding this limit before it is checked with the fatigue curve.

High cycle fatigue normally results from steady-state vibration and rapidly fluctuating thermal shock. The number of cycles is generally too large to be a concern. A pipe vibrating at 1 cycle/second, for instance, will accumulate 86,400 cycles/day and 3.15×10^7 cycles/year. These values fall within the endurance limit area. Therefore, the endurance limit stress is always used in the design against high cycle fatigue.

In high cycle fatigue, we are dealing with a stress that is not only lower than the ultimate strength but also considerably lower than the yield strength of the material. As discussed in Section 1.3.1, without yielding, the peak stress will persist and hasten fatigue failure. Static stress magnitude, highly localized stress, and residual stress, which have little effect on low-cycle fatigue, become very significant factors on high cycle fatigue. A fatigue design curve may further divide into curves A, B, and C in high cycle fatigue range. Curve A may be used for materials with low mean stress, properly stress relieved welds, and little potential stress risers. On the other hand, curve C may be used for materials with maximum mean stress, non-stress-relieved welds, and high potential stress risers. Higher mean stress results in a higher magnitude of static stress with a given stress amplitude. Vibration stress is discussed in more detail in Chapter 13, which deals with dynamic analyses.

1.3.3 Creep Rupture

At high temperature environments, the pipe will continue to deform under a sustained stress. The pipe may fail after a certain period even if the stress is much lower than the ultimate strength of the material. The phenomenon is called creep and the failure is called creep rupture. Creep occurs only at high temperature, or — rather — is detectable only at high temperature. For this discussion, we will focus on creep for temperatures starting from 700°F (370°C) for carbon and low alloy steels and from 800°F (430°C) for high alloy and stainless steels.

At creep temperatures, the material has a definite life against a sustained stress. The science of creep is finding out the relationships among stress, temperature, and the time to failure of a material with various compositions and physical properties. Most of us do not need to know the details of all these relations. What we need to know is the general behavior of steel at high temperatures and how to design the piping system to avoid premature creep failure. The general behavior of creep can be explained with the typical creep curves as shown in Fig. 1.7. The figure shows three curves representing the time-deformation relations of three different constant loads applied at the same type of specimen under a given temperature. As soon as the load is applied, an initial deformation is immediately generated. This is a non-creep-related static deformation that consists of elastic or elastic plus plastic components, but is roughly proportional to the load. As time goes by, the specimen continues to deform, leading to eventual rupture. This time-dependent progressive deformation is the nature of creep.

Creep can be categorized into three stages. Stage 1, also called primary stage, involves the beginning region of creep, having a deformation rate that is very high at the beginning and decreasing gradually to the minimum level at the end. This is the only stage in which the creep rate decreases against time. The minimum creep rate continues on to stage 2 and more or less maintains the same rate until stage 3 is reached. Stage 2, also called the secondary stage, is the main platform of creep tests. Stage 3, also called the tertiary stage, is characterized by a reduction in the cross-section and increase in creep rate. Stage 3 is regarded as the failure region and should be avoided in service.

In the design against creep failure, two criteria are generally used. One is the creep rate, a/b, of stage 2 and the other is the rupture stress at the end of service life. Tests are required to establish these criteria. However, it is not practical to perform a real-time laboratory simulation for 30 or 40 years of service life. Therefore, extrapolations are performed on data acquired from tests lasting for less than, for instance, a year or two. Although the validity of an extrapolation is always cautioned, there are several general trends of creep that make extrapolation reasonably acceptable. From the vast experimental data gathered, we have observed the following general trends of creep:

(1) In log-log scale, applied stress and secondary stage deformation rate have a directly proportional straight-line relationship.
(2) In log-log scale, rupture stress and time to rupture have an inversely proportional straight-line relationship.
(3) In log-log scale, elongation at rupture and time to rupture have an inversely proportional straight-line relationship.

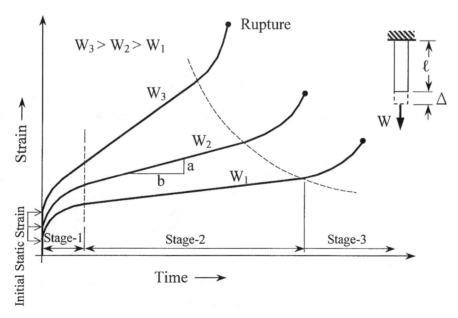

FIG. 1.7
CREEP AT A GIVEN TEMPERATURE

Based on these general trends, each material requires only three test data to reasonably draw a straight line, which can be used for the extrapolation. In piping, the design allowable stress for creep condition is based on two benchmarks. One benchmark is the rupture stress at the end of 100,000 hours, and the other is the stress producing a secondary creep rate of 0.01% per 1000 hours. The latter benchmark is roughly equivalent to 1.0% deformation at the end of 100,000 hours. By applying suitable factors over the two benchmarks, the allowable stress is determined for a given material at a given temperature. These allowable stresses are the main working tools for designing the piping system.

Figure 1.8 shows the allowable stresses taken from B31.3 for a number of materials. In the figure it is clear that the allowable stress drops sharply after the temperature reaches a certain point. This is mainly due to the transition from static failure to creep failure at high temperature. The figure also shows that the transition point and creep strength can be significantly increased by alloying the carbon steel with molybdenum, chrome, and nickel.

Because considerable thought has been put into them, the allowable stresses are pretty much all that we need to design a piping system. However, as the pipe has some definite life in the creep temperature range, we still want to know the relation between the allowable stress and the expected life at a certain temperature. To find the relationship between temperature and service life, we can use the famous Larson-Miller [20] parameter (LMP) as defined by

$$\text{LMP} = T(C + \log t) \tag{1.2}$$

where

T = absolute temperature in degree Rankin (R = 460 + °F) or Kelvin (K = 273 + °C)
t = is time to failure in hours
C = material constant (C = 20 for carbon, low, and intermediate alloy steels; C = 15 for austenitic stainless steels and high nickel alloys)

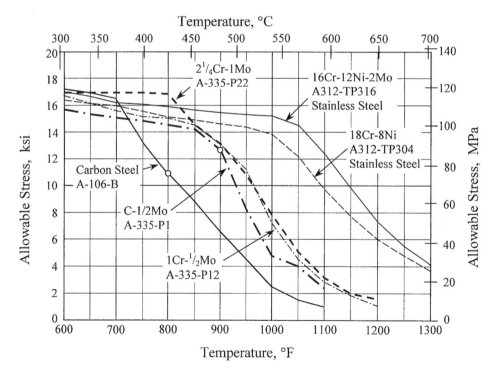

FIG. 1.8
ALLOWABLE STRESS AT HIGH TEMPERATURE [17]

The log is base 10 logarithm. From the available test data, Larson and Miller had discovered that for a given material the same LMP corresponded to the same rupture stress, and thus the same allowable stress. For instance, for a given design with a given allowable stress, we can determine any (T, t) combination that is suitable for the service. This is done by expressing the (T, t) condition as $(LMP)_2$, and setting $(LMP)_2 = (LMP)_1$, a known condition like the original design condition. From this relation, service life can be determined for any given operating temperature, and vice versa. Appendix V of B31.3 adopts this relationship in evaluating temperature and time variations for high temperature service. In this regard, it should be remembered that B31 allowable stress is based on 100,000 hours of rupture life.

1.3.4 Stability Failure

The pipe can also fail due to stability, which is caused mainly by compressive stress. The stability problem occurs mainly on large thin wall shells and pipes. However, it may also occur on thick pipes in a deepwater environment. Figure 1.9 shows several situations that may have stability problem.

Under external pressure, the first priority is to get some idea about the potential buckling of a long segment of un-stiffened pipe. A long segment of pipe produces two-lobe buckling with an allowable external pressure as

$$P_{\text{allow}} = \frac{2E}{3(1 - v^2)}(t/D)^3 \tag{1.3}$$

where

E = modulus of elasticity
t = thickness
D = outside diameter
v = Poisson's ratio

The above equation includes a safety factor of 3 applied over the theoretical formula derived by Bresse-Bryan [21]. If the allowable external pressure is smaller than the design external pressure, either the pipe thickness is increased or stiffening rings are applied to increase the allowable pressure. When the un-stiffened pipe segment is short, the buckling may occur with two or more lobes depending on the diameter, thickness, and length. The procedure for evaluating the buckling in this case is complicated, but fairly standardized. The allowable external pressure is generally checked according to the procedure outlined in UG-28 through UG-30 of ASME Boiler and Pressure Vessel (B&PV) Code, Section VIII, Division 1 [22].

Stability is very sensitive to imperfection and out-of-roundness of the shell. The ASME procedure for external pressure design is based on 1% diametrical out-of-roundness of the shell. This is the same tolerance for most of the standard specifications on shells and pipes. However, there are some specifications, such as API-5L and ASTM A-53, which have a diametrical out-of-roundness tolerance of 2% (±1%). Some adjustments are required when using the ASME procedure for external pressure design of pipe having a diametrical out-of-roundness greater than 1%.

Figure 1.9b shows some situations of instability due to axial compressive stress. The compressive load may create an overall column-buckling problem as in all structural systems. The main concern of column buckling is the sustained load such as the bellow expansion joint pressure end forces. The effect of the self-limiting compressive load, such as the thermal expansion force, is generally benign because the force is readily reduced from slackening of the buckled pipe. For local shell buckling, we have full circle wrinkling, square wave buckling, and bending wrinkling. Within the elastic range, the allowable compressive stress may be taken as [22]

$$S_{\text{Allow}} = \frac{0.125}{2}E(t/R) \tag{1.4}$$

18 Chapter 1

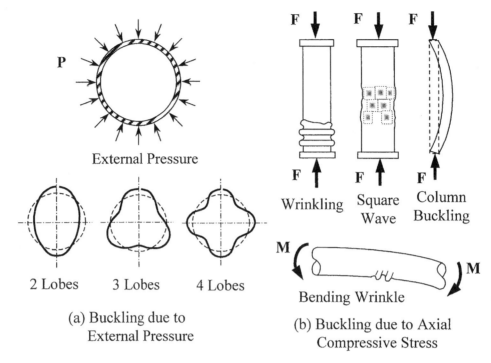

**FIG. 1.9
STRUCTURAL STABILITY PROBLEMS IN PIPING**

where R is outside radius of the pipe. It should be noted that the above allowable compressive stress represents only about one-tenth of the theoretical buckling stress, which is modified by the imperfection factor of commercial pipe, weakness due to potential square wave buckling [23], and a safety factor. Although not mentioned by the code, the allowable compressive stress is generally increased by 30% for bending stress.

1.3.5 Miscellaneous Modes of Failure

Other modes of failure include corrosion, erosion, stress corrosion, hydrogen attack. These modes of failure are mainly caused by material selection and usage. We will only discuss them briefly.

All piping is designed with a predetermined corrosion allowance. This allowance mainly compensates for the loss of material due to surface corrosion and is not related to the inter-granular corrosion, which will be discussed later in this section. Corrosion affects the piping in two areas. One is, of course, the reduction in wall thickness. The pipe has a definite life because of corrosion, which progresses continuously. Once the wall thickness is reduced to the limit that is unable to resist the design pressure, the life of the pipe to serve the given design condition is exhausted. The other effect of corrosion is the stress intensification of the small corroded pits. These pits can greatly reduce the fatigue strength of the pipe against cyclic loading.

The rubbing of the flowing fluid against the pipe wall will erode the pipe wall to some extent. Slurry flow erodes the wall the most due to the abrasiveness of the solids carried. Most two-phase flows also cause significant erosion. Cavitation at pump suction is notorious for producing corrosion erosion. Even the single-phase water flow can generate serious erosion [11] at carbon steel elbows when the flow velocity and the water chemistry are right. Turbulence, low pH value, and low oxygen content are some of the parameters that favor erosion on carbon steel pipe. Mild oxidation appears to provide a protective coating against erosion in some cases. Erosion failure is static rupture due to insufficient wall thickness at a certain portion of the system.

The pipe may also encounter some stress corrosion failures. As the term implies, it is the combination of corrosion and stress. However, the stress can be due to manufacturing process such as forming and welding or from service loads. This corrosion is an inter-granular chemical reaction at some grain boundaries. The cases that are most often mentioned are the cracks of certain austenitic stainless steel pipes in contact with fluid containing traces of chlorine. The fluid can be the internal process fluid, hydrostatic test fluid, external environment fluid, or incidental fluid such as sweat and cleaning agents. Failure usually occurs very suddenly and involves little plastic deformation.

Failure due to hydrogen brittleness occurs often enough in the power and process industries that it merits a mention here. Due to their small size, hydrogen atoms can easily diffuse from outside into or through the steel. This diffusion occurs in pipe carrying hydrogen or hydrogen-laden fluids with a diffusion rate generally higher at a higher temperature or pressure. Steel may also contain hydrogen through the decomposition of water during welding, electroplating, corrosion, or other manufacturing processes. Once inside the steel, hydrogen can weaken the steel by changing the alloying structure of the steel and increasing the internal stress. When hydrogen atoms accumulate in the voids of the pipe material, they form hydrogen molecules and become trapped. If diffusion from the outside continues, high internal pressure is developed against the wall of the voids. This localized stress intensification may not be very harmful if the steel is ductile and readily yields locally. However, this localized stress intensification can produce brittle failure in less ductile high-strength materials. Under favorable conditions, the trapped hydrogen may grab the carbon atom from iron carbide (Fe_3C) to form methane (CH_4). The de-carbonized steel generally becomes weaker than the original alloyed steel. The most damaging effect is the production of methane. Because methane cannot readily diffuse out, the increased gas volume increases the internal pressure and generates fissures along the grain boundaries. The fissures can eventually lead to cracks and failure. Ductile steels alloyed with carbide stabilizing elements such as chromium, molybdenum, vanadium, tungsten, and titanium are generally used to prevent failure from hydrogen attack.

There are many other modes of failure involving materials and operations. The book by Thielsch [24] is an excellent source of reference.

1.4 PIPING CODES

A piping system is designed and constructed based on codes and standards. Therefore, it is important that we have a good understanding of the applicable codes and standards.

Engineers have always strived to build the safest plant at the lowest cost. Because it is impossible for them to build an absolutely safe plant at any cost, they have to settle for a reasonably safe plant at a reasonable cost. The question then remained — what is reasonable? Economists and statisticians were able to come up with some formulas to achieve the optimal median cost by assessing the cost of safety problems involved throughout the life of a plant. To find the cost, all factors involved in a safety problem, such as loss of human life, intoxication of the environment, and so forth, were given a price. The problem is that the assessments of these prices were highly debatable. For instance, we just do not have any way of assessing the cost of the damage caused by discharging pollutants to the public atmosphere or water body. Furthermore, putting a price on human life is considered reprehensible in a modern society. At the very least, all these cost assessments are very controversial. Therefore, a strict cost analysis on the requirements is not practical. A sufficient level of protection for both the investor and the general public shall be achieved by a consensual opinion from all related parties. On behalf of the piping industry in the United States, the ASME took the lead in forming action committees, consisting of experts from engineering companies, academic institutions, government agencies, equipment manufacturers, plant owners, insurance companies, and independent consultants. These committee members represented different and sometimes opposing interests, and their recommendations resulted in a set of specifications now referred to as the Piping Code, or the Code.

The piping code is a set of minimum requirements used to ensure that the plants built accordingly are safe. It specifies the permissible materials, acceptable designs and fabrications, and the inspection requirements and procedures. Although the Code is a stand-alone document with no enforcing power, it becomes a part of the law when it is adopted by a regulatory agency. Moreover, when it is adopted to lay down the specifications of a contract, the Code naturally becomes part of the contract. Nowadays, many state governments in the United States have adopted the piping code as their requirements for constructing a plant. Figure 1.10 shows the role of industry codes and standards.

As stated earlier, the task of the Code is to set the proper degree of safety factor to be used. This setting of safety factor is a delicate balance of safety and cost. Originally, there was just one code used to cover all the piping systems. Later on, it was realized that using just one code seemed impractical, because the importance and cost proportions of the piping differed from industry to industry. Today, the Code has many sections, each applicable to a specific type of piping system.

In the United States, the piping code is divided into two main categories: (1) *nuclear power plant piping*, which is governed by the ASME B&PV Code, Section III [25]; (2) *non-nuclear piping*, which is governed by ASME B31 [26]. The organization header for B31 code has gone through American Standard Association (ASA), American National Standards Institute (ANSI), ANSI/ASME, ASME/ANSI, to the current ASME over the years. Many engineers still call it ANSI B31 to this day. The same situation also applies to many standards for piping components.

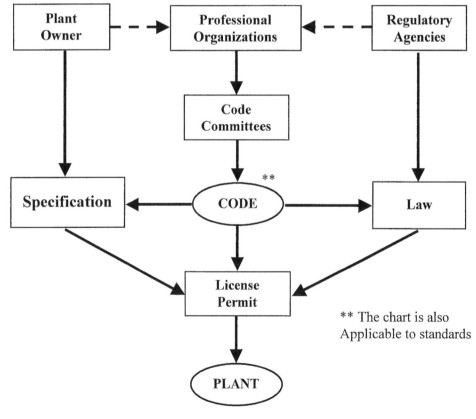

FIG. 1.10
ROLE OF CODES AND STANDARDS

Nuclear piping is further divided into three classes, and B31 is presently divided into seven active sections. Each section and class has its unique characteristics. We might wonder why the same piece of pipe can be and shall be designed in so many different ways. Naturally, there are rationales and purposes. The essences of these sections and classes of codes are summarized below.

- Power Piping (B31.1)

  The piping systems in a power plant include main steam, reheat steam, feed water, condensate water, and some utilities. Compared to the cost of heavy equipment (e.g., turbine, boiler, pumps, heat exchangers, and pollution control facility), the cost of piping is just a small part of the total cost of the plant. Because of this low cost (in proportion to the rest) and the fact that an unexpected plant shutdown can create public chaos, it is logical to make the piping system as safe and reliable as possible. The safety factor used is about 3.5 against the ultimate strength of the pipe. Due to the lack of extreme corrosive fluids involved, the corrosion allowance is considered only in the calculation of wall thickness. All other calculations are mostly based on the nominal wall thickness. The long service life of a power plant also warrants a more conservative approach in the design and construction. B31.1 also opts to use simpler and more conservative formulas in calculating pipe stresses. The resultant moments are used for all categories of stress calculations, and the stress intensification factors are applied to all components of the moment including torsion moment, which is generally not applied with a stress intensification factor.

- Process Piping (B31.3)

  A process plant, such as a petrochemical complex, normally constitutes many processing units spread out in a very large area. The interconnecting piping is also necessarily spread out all over the area. Because the cost of the piping can be as high as 35% of the cost of the entire plant, and also because the public does not pay as much attention to shutdowns at process plants, the safety factor can be reduced somewhat to pare down the overall cost of the plant. The safety factor used is about 3.0 against the ultimate strength of the pipe. Because some of the fluids in a process plant are highly corrosive, the Code requires that the corrosion allowance, as well as the manufacturing under-tolerance, needs to be included in all calculations involving sustained loadings. The calculation of pipe stress is more precise in B31.3, which has different stress intensification factors for in-plane and out-plane bending moments, and does not apply any stress intensification on torsion moment.

  Some process plants have to deal with toxic fluids. These fluids require special treatment and are classified as category-M piping. Similar to nuclear piping, which will be discussed later, double containments are sometimes used for category-M piping. Process piping may also involve very cold, very hot, and very high-pressure applications. Special stipulations are provided for these cases.

- Pipeline Transportation Systems for Liquid Hydrocarbons and Other Liquids (B31.4)

  Cross-country petroleum pipelines run through mainly unpopulated areas. Its cost is mostly in the pipe itself. Therefore, a somewhat lower safety factor is acceptable and desirable. The safety factor can be as low as 2.0 against the ultimate strength of the pipe. Because the effective use of the pipe material is essential in this type of construction, high yield strength materials are often used. The allowable stress is mostly based on the Specified Minimum Yield Strength (SMYS). In addition, the stresses are calculated based on nominal wall thickness. This extreme usage of the material is made possible because of the simplicity of the system and the effective control of corrosion. Corrosion of the pipe, mainly external, is controlled by suitable coating, cathodic protection, and constant monitoring.

- Gas Transmission and Distribution Piping Systems (B31.8)

  This piping system has many similar characteristics to B31.4 liquid transportation systems. However, in gas transmission, due to its explosive nature and the necessity of routing through highly populated areas, the piping is divided into four location classes: ranging from Location Class 1 (areas where any 1-mile section has ten or fewer buildings intended for human occupancy) to Location

Class 4 (areas where multi-story buildings are prevalent, where traffic is heavy or dense, and where there may be numerous other utilities underground). Higher allowable stress, thus lower safety factor, is used for areas with lower location classes. Location Class 1 is further divided into Division 1 and Division 2, two types of construction depending on the pressure used in the hydrostatic test. Higher allowable stress is permitted for Division 1 for testing at higher pressure. Stress calculations are mainly based on nominal wall thickness of the pipe, and the allowable stresses are mainly based on the SMYS.

- Class 1 Nuclear Piping (B&PV Code, Section III, Subsection NB)

    The Class 1 nuclear piping code adopts an entirely different philosophy from other piping codes. It emphasizes the "Design by Analysis" approach, and strict "Quality Control" and "Quality Assurance" attitude. The intent is to eliminate all uncertain guesswork. The code requires that all minor, as well as major, transients and operating cycles be considered. Primary stress, secondary stress, peak stress, local structural discontinuity, and thermal gradient stress all need to be calculated. Actual stress intensification factors are used in the calculation of both primary and secondary stresses. This is in contrast to B31, which uses only one-half of the theoretical stress intensification factor in the calculation of thermal expansion stress (see Chapters 3 and 4 for further discussions). Extensive dynamic analyses on earthquake, steam/water hammer, and vibration are required for most of the piping systems. A fatigue analysis is also required for highly stressed components. Because of the large amount of calculations involved, it is not uncommon for an engineer or a group of engineers to spend more than a year designing and analyzing a single Class 1 piping system.

    The basic safety factor used by the Class 1 piping is 3.0 against the ultimate strength of the pipe. This may appear to be less conservative than the B31.1 power piping, which uses a safety factor of 3.5. However, by combining the additional calculations performed, higher stress intensification factor used, and strict quality assurances required by Class 1 piping, the "real safety factor" for Class 1 piping is much higher than the safety factor of the B31.1 power piping.

- Class 2 and Class 3 Nuclear Piping (B&PV Code, Section III, Subsections NC and ND)

    Basically, the requirements for Classes 2/3 nuclear piping are similar to the requirements of the B31.1 power piping. However, as for all nuclear piping, the Classes 2/3 nuclear piping also requires more quality control and quality assurance than the B31.1 piping. In addition, a separate set of stress intensification factors is used for sustained and occasional stresses. This is a basic departure from B31.1, which simply uses 75% of the stress intensification factor of the displacement stress (or self-limiting stress) as the stress intensification factor for the sustained and occasional stresses. (Stress intensification factors are defined and further explained in Chapters 3 and 4.)

All the sections and classes of codes are designed and formulated independently. Each section or class starts out with its own requirements governing the following areas: methods of calculation, amount of testing and inspection, stress basis, and so forth. Afterward, the allowable stresses are set up for different categories of calculated stress. The stipulations and rules within a code are all interrelated. Therefore, it is important that each section and class of the code be used in its entirety. It is not permitted to use separate criteria from different sections of the code on any given piping system. For instance, it may be permissible to design a Class 2 nuclear piping with Class 1 requirements, but we cannot simply use the higher allowable stress given by the Class 1 piping. Moreover, the stresses calculated in each different class of piping are not always calculated with the same basis. This makes cross use of the code class and section very hazardous.

1.5 INDUSTRY PRACTICE

In the design of a piping system, although more and more emphasis is placed on "Design by Analysis", there are many items that are either impossible to analyze or simply too expensive to analyze.

A piping system is generally designed and constructed with common industry practices. This section describes some of the common practices widely used in the non-nuclear industries. For nuclear piping systems, the subject is a little different and has been well discussed [2, 27].

A pipe stress engineer often confronts questions from field engineers regarding some unusual but ingenious designs. Field engineers and project engineers often like to mention that the piping that was designed in the traditional manner has been working satisfactorily for many years. The new special designs are, at most, proven to work only on paper. The piping code recognizes these facts, and considers copies of a successfully operating piping system acceptable. For instance, ASME B31.3, paragraph 319.4.1, stipulates, "No formal analysis of adequate flexibility is required in systems which are duplicates of successfully operating installations or replacements without significant changes of systems with a satisfactory service record." A similar wording is given in B31.1, par.119.7.1 (A). The implication of all these stipulations is that "if something works, it works." In the following, the scenarios of some common industry practices are listed. Detailed discussions are given in related subjects in subsequent chapters.

1.5.1 Load Cases

Unless otherwise specified, only design pressure, operational weight, and maximum thermal expansion are included in the basic code stress compliance analysis. Although the plant might be located in a seismic zone or a hurricane prone area, the engineer may not be aware of the site characteristics due to the remoteness of the site. Even with thermal expansion, it is customary to calculate only the condition when all the piping segments involved are at their maximum or minimum temperature, which may include upset temperatures. Different operating modes with various temperature combinations are not investigated. These extra analyses are performed only when required by the design specification, which will be discussed in Section 1.6.

For earthquake analysis, normally the static equivalent approach is used. In this approach, horizontal forces proportional to the weight are applied. The weight shall include all the operation weight such as pipe, insulation, fluid content, refractory, etc. The proportional constant is referred to as the g factor, which is defined in the design specification based on seismic zone classification. Unless specified otherwise, only two directions of horizontal forces are analyzed. The vertical force, which can be significant, is not analyzed. This is somewhat justified because the piping has to be supported vertically for its weight in the first place. Past experience has also indicated that an earthquake can shake in all three directions randomly at the same time. Therefore, each direction of the force is analyzed independently, and the results are combined by the square root of the sum of squares (SRSS) method to arrive at the combined seismic effect. Different combination methods can be specified to satisfy special situations.

For wind analysis, static forces proportional to the piping projection area are applied. The projection area covers the outside envelope of the piping including the insulation. Areas shielded by building or other objects are not included. Normally, only the two horizontal directions of force are analyzed. Forces in the vertical direction are ignored, unless specified otherwise. Each direction of force is analyzed separately. The combined wind effect on each piping element is determined by selecting the greater, or the greatest, of the effects of all analyzed directions. This combination method is based on the idea that the wind does not blow in all three directions at the same time. Again, the combination method can be specified differently. For instance, there is a concern that the weakest direction is not one of the directions analyzed.

Piping systems identified as vibration lines are normally checked for their natural frequencies to prevent structural resonance by vibratory frequencies. The supports are designed with special attention to rigidity and damping effects. However, vibration analysis is not performed due to the lack of definite forcing functions. It is performed only for troubleshooting after the piping experiences noticeable vibration.

24 Chapter 1

Although analyses on some steam/water hammer effects, such as steam turbine trip and open discharge safety valve opening, have become more common, they are still not usually performed unless explicitly specified.

1.5.2 Local Support Stresses

Some local support stresses are either impossible or too time consuming to calculate. The stresses involved are typically at pipe clamp locations, support lug connections, and at pipe shells that are in direct contact with the support steel. Take the pipe resting on the steel beam as an example. The pipe shell normally has a line contact with the support steel. In many cases, the pipe is supported with a cross rod or angle steel, making the contact a point contact as shown in Fig. 1.11. This type of point contact is created to prevent crevice corrosion on the pipe surface. Although it is possible to apply a line load or a point load on the pipe shell to calculate the stresses on the shell, the calculation is time consuming. Even if the stress is calculated, there remain some unsettled issues. First, the significance of local stresses is not well defined. Second, the allowable load calculated based on the calculated shell stresses is often too low compared to the actual load experienced in the field. However, this is not to say that we can entirely ignore the local support stress. We still have to develop some type of company or, better still, an industrial standard as safe guidance for placing proper supports. See also "standard support details" to be discussed later in this chapter.

The same situation applies to clamps and support lugs. Not only are the clamping stresses not calculated, they are actually ignored in code stress compliance evaluations. The practice is to follow the manufacturer's guidance. If the manufacturer's catalog says a certain clamp can take a specified load, then any load below that given load is safe for that clamp and presumably also for the pipe shell. For the support lugs, only the shear stress at the attachment weld is calculated. The stress at pipe shell is not evaluated.

1.5.3 Local Thermal Stresses

Although they contribute more than their fair share of cracks, thermal gradient stresses are generally not calculated. Gradient stresses at pipe support lugs and shoes, for instance, are handled via design rules set up by each company. The rules are implied in the standard support details issued and used by the company.

In general, local thermal stresses generated at the welds joining dissimilar materials are also not calculated. For instance, the weld joining carbon steel and austenitic stainless steel can create large local thermal stresses due to the difference in thermal expansion rates. However, the stresses are not calculated in the routine piping stress analysis. They are taken care of by design rules set up by individual

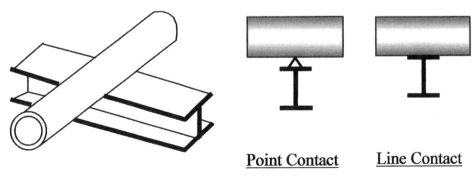

FIG. 1.11
PIPE ON SUPPORT STEEL

companies. Only when called for specifically, will a specialist be engaged in the actual calculation of these local thermal stresses.

1.5.4 Pressure Effect on Flexibility

There are two areas where pressure can affect piping flexibility: (1) pressure deformation and (2) stiffening effect on the bends.

Pressure deformation has two aspects: axial elongation and rotation at the bends [28]. Under the normal operating range, pressure elongation represents an effect equivalent to about 15°F (8°C) of additional thermal expansion. This magnitude of additional expansion is insignificant in the design of plant piping, but important in the design of cross-country pipelines. Therefore, pressure elongation is normally not included in plant piping analysis, but is always included in cross-country pipeline design.

Bend rotation due to pressure is more controversial, because it can virtually turn the whole piping system around without any valid cause. Intuitively, engineers tend to think that pressure has the tendency to open up the bend. This is actually not the case, as can be observed from a coiled garden hose that is not straightened by water pressure. Engineers also often incorrectly refer to the potential opening effect as the Bourdon tube effect. This is not true, because the Bourdon tube has an oval cross-section that promotes the opening effect. The pipe cross-section is circular in general. The actual rotation of the bend depends on the roundness of the cross-section, and also on the thickness variation around the cross-section. For a circular cross-section with uniform thickness, the bend may open somewhat. However, the bend, manufactured either by forging or by forming from straight pipe, has a thicker wall thickness at the crotch area. This thicker thickness will reduce the longitudinal elongation at the crotch causing the bend to close instead. In fact, tests have shown [29] that a miter bend, when pressurized, will close (shutting) the bend angle in the elastic region. Owing to these facts and other uncertainties, bend rotation due to pressure is normally considered insignificant and is not included in the analysis.

The Code has given specific equations to use for the pressure effect on bend flexibility and stress intensification factor. An internal pressure reduces both flexibility and stress intensification at a bend. However, the inclusion of these reductions varies from application to application. In general, the reduction in flexibility is included, but the reduction in stress intensification factor may not be allowed as in Class 1 nuclear piping. The reason behind this approach is because the pipe may be depressurized while still at a fully expanded state.

1.5.5 Stress Intensification for Sustained Loads

The code stress intensification factors, derived from low cycle fatigue tests, are intended to be used for evaluating self-limiting stresses. These stress intensification factors are used for thermal expansion and support displacement analyses. Their application to the sustained loads, such as weight, wind, earthquake, and other mechanical loads, is not straightforward [30]. B31.1 Power Piping has clearly stipulated the use of 75% of the code stress intensification factor for the sustained loads. For other codes, the application is not uniform, and can be a potential item of controversy between the parties involved in the project.

1.5.6 Support Friction

Support friction is important in many areas. This type of friction is important in estimating the potential axial movement of long pipelines. It also has significant effect on piping systems connected to large rotating equipment. Inclusion of the friction effect in the analysis significantly increases the complexity of calculation and the man-hours required. Therefore, support friction is generally not included in the analysis, unless otherwise specified. However, it should definitely be considered in the design and analysis of large rotating equipment piping and cross-country pipelines.

1.5.7 Guide and Stop Gaps

To ensure the free movement of the pipe in unrestrained directions, guides and stops are generally constructed with small gaps. These so-called construction gaps range from 1/16 in. to 1/8 in. (2 mm to 3 mm). The construction gaps, except for those located close to the connecting equipment, are generally ignored in the analysis. In general, gaps at the first support from the equipment need special attention. A detailed analysis may show that these gaps are too large, making the restraints ineffective, in which case the function of the guide or the stop should be ignored.

1.5.8 Anchor and Restraint Stiffness

The stiffness of the restraint is also called the spring constant of the restraint. Anchors and restraints are normally treated as rigid if no specific spring constant is specified. They are assumed as perfectly rigid with infinite stiffness in some computer programs. However, they are more often assigned with a huge, but finite, spring constants in most computer programs. The magnitude of this huge spring constant is set to be proportional to the moment of inertia of the pipe cross-section. It will allow a deflection of less than 0.001 in. (0.03 mm) under the normal range of the piping load.

The rigid assumption works fine for most of the piping under vertical load due to the inherent stiffness required for supporting the weight. However, for other directional restraints, the rigid assumption may result in inaccurate, and sometimes meaningless, analyses. This is more so in the case of large pipes. Because there are few structural members that are stiffer than a 12-in. pipe, the restraint stiffness of pipes larger than 12 in. in size has to be investigated on a case-by-case basis.

1.5.9 Small Piping

Piping systems with pipes 2 in. or smaller are generally field routed. They are shown only schematically on the construction drawings to give directional routing and to offer guidance for material take-off. Because the routing of these small piping systems can be changed in the field, they are normally not analyzed formally. Their integrity is ensured to some extent by the walk-through or walk-down activity to be discussed later in this chapter.

1.6 DESIGN SPECIFICATION

Design specification is a set of instructional and contractual documents to be followed in the design of the plant. In most cases, there are two sets of specifications. One is the owner's design specification outlining the requirements by which their plants are to be built. The other is the project design specification prepared by the constructors for their engineers to follow in carrying out the actual design of the plant. The project specification includes several inter-related documents as shown in Fig. 1.12. Portions of the specification pertinent to piping mechanical work are explained in the following subsections.

1.6.1 Owner's Design Specification

The owner's design specification mostly covers the general requirements for ensuring structural integrity and maintaining system operability of the plant. It also covers the regulatory requirements for obtaining construction and operation permits. Additional specifications based on the owner's

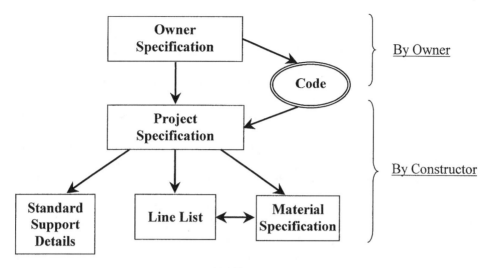

**FIG. 1.12
DESIGN SPECIFICATION**

cost/risk philosophy are also included. The items most often included in the owner's design specification are:

- *The Code and the specific edition to be used.* ASME B31.1 is normally used in power plants and B31.3 for process plants. However, some piping systems in a process plant may also be required to use the B31.1 code because they connect to a steam boiler. Steam boilers are governed by ASME B&PV Code, Section I — Power Boiler. The power boiler code requires that the piping systems under its jurisdiction be designed by B31.1 code. When specifying the code, the proper code edition is also specified. Because the code requires that it shall be applied in its entirety, mix use of code editions is generally not allowed. Therefore, the applicable edition of the code is determined at the time of the contract and is properly recorded. Any extra consideration beyond the code requirement shall also be specified.
- *Occasional loads to be applied.* This involves the magnitude of the wind and earthquake, if any, to be designed for. The code normally allows wind and earthquake effects to be dealt with separately. There are, however, some details that need to be specified: the combination method to be used for the multi-directional earthquake and wind loads; the credit, if any, of the support friction to be taken; the duration factor to be used; if any special dynamic method is to be applied.
- *Hydrodynamic loads.* Piping systems that are subject to steam/water hammer shall be identified. Systems affected by steam turbine trip, safety valve opening, emergency valve closure, and so forth, are subject to significant hydrodynamic forces. Other considerations, such as condensate lines, operating at near-saturate temperature might see large shaking forces due to slug flow. These loads, if specified, are normally analyzed with the static method. It should be specified if a special dynamic method, such as time-history method, is to be used.
- *Pneumatic test.* Piping systems should be designed for the heavier of operational weight or hydro-test weight. For large, low-pressure gas piping systems, it may be desirable to test with air to reduce the support cost. Pneumatic test may also be used in places where water is not readily available. The lines permissible to test with air are normally identified in the specification. A pneumatic test involves a huge amount of stored energy of the compressed gas, which might damage the plant if failure occurs during the test. The test is normally performed using small steps in pressure at a time.

- *Support friction.* Inclusion of support friction in the design analysis is costly. Therefore, support friction is included only in the analysis of critical piping systems specified. Piping systems connected to large rotating equipment are normally analyzed with support friction included.
- *Stress intensification factors for sustained loads.* Use of the stress intensification factor on sustained loads is not explicitly stipulated in some codes. In such cases, the owner may have to specify the policy to be used in the design analysis.
- *Special allowable stresses.* This includes allowable stresses for special conditions, such as one-time foundation settlement.
- *Standards to be used for rotating equipment and stationary equipment.* The allowable piping loads given in the equipment standard are generally applicable to normal operating conditions. Greater allowable loads, if permitted, for the standby, idle, upset, and occasional conditions need to be specified.
- *Stresses at active valves.* To ensure the proper function of the safety related valves, special requirements are imposed on the active valves such as safety relief valves and control valves. A reduced allowable stress and/or a minimum flange class can be specified.
- *Stiffness of fittings and equipment.* The stiffness of valves and flanges is normally considered as three times as stiff as the connecting pipe of equal length. The stiffness of the rotating equipment is normally considered rigid, which is set to be 1000 times as stiff as the connecting pipe of the same length. These or other desirable numbers have to be specified.
- *Special limitations and requirements.* Limitations are normally placed on the use of rod hangers, springs hangers, expansion joints, flexible hoses, snubbers, and other special supports and components.

1.6.2 Project Specification

Project specification is also called job specification, or job requirements in some cases. It combines the requirements given in the owner's specification and those of the constructor into one document for the design engineers to follow. The basic document has the same features as the owner's specification. The project specification, as a whole, also includes material specification, line list, and standard support details. Aside from covering the general requirements, it also defines the specific requirements for every piping system in the plant.

(1) Material specification

The materials required for piping systems in the project are classified into material specification groups. Each group covers a certain portion of similar piping systems, and is assigned a specification number or name. Each company uses its own set of identification symbols, usually consisting of two or three characters for easy reference. Some companies use names that are related to the class of flanges and valves used in the group, such as 150-A, 150-B, etc. Material specification generally contains the following items:

- Group Name: Name to be referenced by other documents, such as the line list and isometrics
- Service: Type(s) of fluid(s) handled
- Pipes: The wall thickness, given as schedule number or weight designation, and the ASTM or ASME specification name for the pipes. All pipe sizes are covered. Different specifications may be used for different pipe sizes.
- Corrosion Allowance: This is the additional thickness to be allowed for combined corrosion and erosion. Manufacturing under-tolerance is given in the applicable ASTM or ASME specification, and is not specified here.
- Service Limits: Each material specification has limits placed on its services by pressure and temperature combinations. This is mainly determined by the allowable stress of the pipe material. A

lower pressure is allowed at a higher temperature due to lower allowable stress. In general, this service limit is the same as the pressure-temperature rating of the flanges.
- Valves: The class, material, and construction of the valves including gate valves, globe valves, check valves, and other valves to be used in each size of the pipe.
- Flanges: The class, material, and types of flanges to be used in each size of the pipe.
- Bolting: The ASTM or ASME specification for the bolts and nuts.
- Gaskets: Type of gaskets.
- Fittings: Ratings and types of elbows, tees, couplings, unions, and other fittings.
- Branch Connections: The type of branch connection depends on the combination of the run pipe size and branch pipe size. A table is provided showing the branch connection type for each branch pipe size at a given run pipe size. This table ensures that branch connections are appropriate for the pressures given in the service limits specified.

(2) Line list

The line list contains material and operational information for every spool of pipe to be installed in the plant. This list not only contains the basic data for the analysis, it also serves as the checklist for the work completed. The following items are contained in the line list:

- Line Number: This is actually an alphanumerical identification tag consisting of service type, size, and material specification group, but it is traditionally called a number. Every company has its own unique way of naming the line numbers. For instance, 12-FW-20123-B1A may be used to identify 12 in., feed water, located in area 20, pipe spool no. 123 that uses material specification B1A.
- Fluid: The type of fluid handled.
- From Point: Either the equipment name tag or the line number of the pipe from which the spool starts.
- To Point: Either the equipment name tag or the line number of the pipe to which the spool ends.
- Operating Temperature: This is the normal operating temperature.
- Operating Pressure: This is the normal operating pressure. Normal operating pressure and normal operating temperatures exist concurrently.
- Design Temperature: The maximum sustained temperature expected during operations. This may not include the downtime service temperature experienced in dry out or steam out.
- Design pressure: This is the maximum sustained pressure expected.
- Flexibility Temperature: The temperature to be used in thermal flexibility analysis. This may include dry out or steam out temperature as well as upset temperatures.
- Material Specification: The material group to be used.
- Insulation: Insulation type and thickness. Sometimes, an insulation specification is called out instead.
- Fluid Density: Fluid density or specific gravity under normal operating conditions.
- Test Type: Pressure test to be conducted by water, air, or other means.
- Flow Diagram: The drawing number of the flow diagram (or P&ID) in which the spool is included.
- Special Remarks: Potential slug flow, vibration, and other conditions that require special attention.

(3) Standard support details

A plant requires thousands of pipe supports in many different varieties. Needless to say, it is impractical to prepare the details for each individual support. To facilitate the design process and for effective transmission of the required information between different disciplines and vendors, a set of standard pipe support details is prepared. This set of support details is a company standard that is applicable to all projects. It is not job specific, although some special standard details may be prepared for a specific project.

Current standard pipe support details have evolved from more than a century of experience. Throughout the years, owners and constructors have exchanged design ideas with many different related parties. Also due to spin-offs and mergers of companies over the years, divisions between companies have become less and less obvious each year. These are some of the reasons why nowadays most of the standard support details used by one company appear to be similar to those used by other companies. This type of standardization through evolution is good for the industry as a whole.

Standard support details range from basic components such as bolts, clamps, and rods, to whole assemblies such as spring hanger assemblies, rod hanger assemblies, clamped shoes with stops, and vessel brackets. Each standard detail is assigned a unique number for identification. All communications between different in-house disciplines are conducted using these numbers. For instance, once the piping mechanical engineer locates where a support is required and puts the standard pipe support detail number on the planning drawing, the piping designer knows what is needed to be done on the pipe, and the civil structure engineer knows what type of support structure is needed in that area.

Standard pipe support details also provide guidance to stress engineers as to what type of supports should be used with certain types of pipes. For instance, they inform the engineer what type of support shoe is required for a given pipe size and temperature combination, and if the pipe can be supported directly on the pipe shell, through an inversed T-shape shoe, an H-shape shoe, or a saddle. They also give guidance on whether to use a welded shoe or a clamped shoe. The support type given in the standard support details ensures that the local stress produced by the support load is not excessive, and no detailed local stress calculation is required. To some extent, these details also ensure that the thermal gradient stresses at support attachment points are not excessive.

1.7 PLANT WALK-DOWN

Plant walk-down or walk-through by an experienced pipe stress engineer before the startup of a plant is an important procedure required for most projects. In a typical project setup, design activities are conducted in the office away from the construction site. Design engineers normally do not have first-hand knowledge of how the plant is constructed. Therefore, it is important for design engineers to conduct on-site inspection so they can get a good first-hand look at the actual layout of their designs. The purpose of this plant walk-down is twofold: (1) to allow design engineers to catch any installation that is not in line with their design; and (2) to let the design engineers get a "feel" of the installed design. The feel from seeing the design on paper can be quite different from the feel of the actual installation. Better designs in the future can be expected from this type of experience.

This walk-down activity may be conducted at the same time the as-built drawings are being drafted in the field. As discussed earlier, small pipes (2 in. or smaller) are generally field routed. This is the time to check if any modification is required from a stress concern, and also to put the final layouts in the as-built drawings. This work is generally conducted by the engineer or engineers who are experienced stress engineers and who are also familiar with the process flow of the plant. They should have a good idea of the temperature and the amount of movement involved in each piping. Their first priority is to check if the piping and its supports are installed as designed. They will also call out the designs that appear to be improper based on their observations.

REFERENCES

[1] Rase, H. F., 1963, *Piping Design for Process Plants*, page vii, Preface, John Wiley & Sons, New York.
[2] Welding Research Council Bulletin 300, 1984, "Technical Position on Industry Practice," Welding Research Council, New York, Dec. 1984.
[3] Kellogg, The M. W., Company, 1956, *Design of Piping Systems*, revised 2nd ed., John Wiley & Sons, Inc., New York.

[4] Spielvogel, S. W., 1943, *Piping Stress Calculations Simplified*, McGraw-Hill, Inc., New York.
[5] Blaw-Knox Company, 1947, *Design of Piping for Flexibility with Flex-Anal Charts*, Blaw-Knox Company, Power Piping Division, revised 5th ed., Pittsburgh.
[6] Tube Turns, Inc., 1986, *Piping Engineering*, 6th ed., Tube Turns, Inc., Louisville, KY.
[7] Markl, A. R. C., 1955, "Piping-Flexibility Analysis," *Transactions of the ASME*, 77(2), 124–149.
[8] Brock, J. E., 1952, "A Matrix Method for Flexibility Analysis of Piping Systems," *Journal of Applied Mechanics*, 19(4), pp. 501–516.
[9] Chen, L. H., 1959, "Piping Flexibility Analysis by Stiffness Matrix," *Journal of Applied Mechanics*, 26(4), pp. 608–612.
[10] American Standard Association, 1955, *Code for Pressure Piping (B31.1-1955)*, American Society of Mechanical Engineers (ASME), New York.
[11] Bush, S. H., 1988, "Statistics of Pressure Vessel and Piping Failures," *Transactions of the ASME*, 110, pp. 225–233.
[12] Bush, S. H., 1992, "Failure Mechanisms in Nuclear Power Plant Piping Systems," *Transactions of the ASME*, 114, pp. 389–395.
[13] *Annual Book of ASTM Standards*, "Part 1, steel-piping, tubing, fittings," American Society for Testing Materials (ASTM), Philadelphia, PA.
[14] *ASME B36.10M, Welded and Seamless Wrought Steel Pipe*, ASME, New York.
[15] *ASME B16.9, Factory-Made Wrought Buttwelding Fittings*, ASME, New York.
[16] *ASME B16.5, Pipe Flanges and Flanged Fittings*, ASME, New York.
[17] *ASME B31.3, Process Piping, ASME Code for Pressure Piping*, B31, ASME, New York.
[18] Markl, A. R. C., 1964, "On the Design of Bellows Elements," Paper presented at the National District Heating Association's 55th Annual Meeting, June 15–18, 1964, Niagara Falls, Ontario.
[19] Tagart Jr., S. W., 1968, "Plastic Fatigue Analysis for Pressure Components," ASME Paper 68-PVP-3, presented at Joint Conference of ASME Petroleum Division with Pressure Vessel and Piping Division, Sep. 1968, Dallas.
[20] Larson, F. R., and Miller, J., 1952, "A Time-Temperature Relationship for Rupture and Creep Stresses," *Transactions of the ASME*, 74, pp. 765–775.
[21] Windenburg, D. F, and Trilling, C., 1934, "Collapse by Instability of Thin Cylindrical Shells Under External Pressure," *Transactions of the ASME*, 56, p. 819.
[22] *ASME Boiler and Pressure Vessel Code*, Section VIII, "Pressure Vessels", Division 1, ASME, New York.
[23] von Karman, T., and Tsien, H. S., 1941, "The Buckling of Thin Cylindrical Shells under Axial Compression," *Journal of the Aeronautical Sciences*, 8(8), pp. 303–312.
[24] Thielsch, H., 1965, *Defects and Failures in Pressure Vessels and Piping*, Chapters 16 and 17, Reprinted in 1977 with new material, Robert E. Krieger Publishing Co., Huntington, NY.
[25] *ASME Boiler and Pressure Vessel Code*, Section III, "Rules for Construction of Nuclear Facility Components", Division 1, ASME, New York.
[26] *ASME Code for Pressure Piping*, B31, An American Standard, "B31.1 Power Piping", "B31.3 Process Piping", etc. ASME, New York.
[27] Antaki, G. A., 1995, "Analytical Considerations in the Code Qualification of Piping Systems," PVP-Vol. 313-2, *International Pressure Vessels and Piping Codes and Standards: Vol. 2 — Current Perspectives*, pp. 3–17, ASME, New York.
[28] Peng, L. C., 1982, "An Interpretation on Pressure Elongation in Piping Systems," *Current Topics in Piping and Pipe Support Design*, pp. 71–78, ASME, New York.
[29] Roche, V. R., and Baylac, G., 1973, "Comparison Between Experimental and Computer Analysis of the Behavior Under Pressure of a 90-degree Bend with an Elliptical Section," *Proceedings of the 2nd International Conference on Pressure Vessel Technology*, San Antonio, TX, 1973, ASME, New York.
[30] Peng, L. C., 1979, "Toward More Consistent Pipe Stress Analysis," *Hydrocarbon Processing*, 58(5), pp. 207–211.

CHAPTER 2

STRENGTH OF MATERIALS BASICS

One of the prerequisites to becoming a pipe stress engineer is being familiar with the principles of strength of materials. In this chapter, some of the basics of this subject relevant to pipe stress engineering will be discussed. The reader should consult standard textbooks on strength of materials for general theories and principles.

Although only the basics of strength of materials are discussed here, some of the items are unique. For instance, the significance of the roundhouse stress-strain curve, the purpose of stress intensity, the meaning of equivalent stress, and so forth are seldom discussed in a standard textbook. Yet, these items are very important in analyzing the stress in piping. This chapter is presented in plain language as much as possible. Therefore, some of the explanations here may not be as rigorous as they would be in an academic setting. The sequence of the discussions will sometimes be in reverse order to what is customarily done in other, more traditional, treatments on the subjects.

2.1 TENSILE STRENGTH

The standard method of measuring the strength of a material is to test its tensile strength by stretching the specimen to failure. Because test results vary considerably with different specimens and procedures, the American Society for Testing and Materials (ASTM) [1] has published a standard for testing and interpretation of results. The test not only determines the ultimate strength of the material but, by stepping up the force gradually, also establishes the relationship between the applied force and the elongation of the specimen.

Figure 2.1 shows the relationships between applied force, in terms of stress, and the corresponding elongation produced. At a given stage of the testing, stress and strain are calculated as

$$S = \frac{F}{A} \qquad (2.1)$$

$$e = \frac{\ell}{L} \qquad (2.2)$$

where

F = applied force
A = cross-section area of the specimen
L = length of the specimen
ℓ = elongation of the specimen

34 Chapter 2

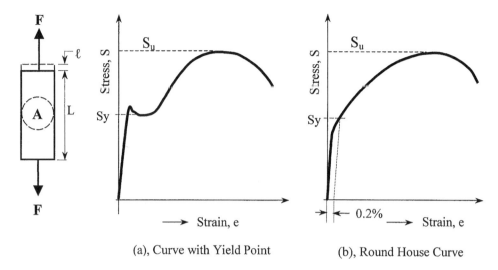

FIG. 2.1
BASIC STRESS STRAIN RELATIONS

These two equations are the direct expressions of the following definitions:

Stress (*S*) is the amount of force per unit cross-section area.
Strain (*e*) is the amount of elongation per unit length of the specimen.

2.1.1 Modulus of Elasticity

In the elastic range, the stress/strain ratio is constant. This relationship is referred to as Hook's law. The proportional constant is called the modulus of elasticity or Young's modulus, and presented as

$$E = \frac{S}{e} \tag{2.3}$$

The stresses and strains throughout the testing process are plotted in a chart called the stress-strain curve. For piping materials, stress-strain curves can be grouped into two categories. The most familiar one, shown in Fig. 2.1(a), has a pronounced yielding point when an abrupt large elongation is produced without the application of additional force. Most low carbon and low alloy steels have this characteristic. The other category, shown in Fig. 2.1(b), does not have this apparent yield point. The curve runs rather smoothly throughout the process. This type of curve is generally referred to as a roundhouse stress-strain curve. Austenitic stainless steel is one of the most important materials to exhibit this type of stress-strain curve. The stress-strain curve defines the following important design parameters.

2.1.2 Proportional Limit

The curve starts out with a section of straight line. Within this section, the material strictly follows Hook's law, and this portion of the curve is generally referred to as the perfect elastic section. The highest point of this perfect elastic section is called the proportional limit. Young's modulus is defined by this straight section of the curve.

2.1.3 Yield Strength, S_y

The point at which the specimen generates a large deformation without the addition of any load is called the yield point. The corresponding stress is called the yield stress, or yield strength, S_y. The yield point is easy to recognize for materials with a stress-strain curve similar to that shown in Fig. 2.1(a). For materials with a roundhouse stress-strain curve as shown in Fig. 2.1(b), there is no apparent yield point. For these materials, the common approach is to define yield strength by the amount of stress required to produce a fixed amount of permanent deformation. This is the so-called offset method of determining the yield point. ASTM specifies an offset strain of 0.2% to be used for common pipe materials covered by its specifications. This value (0.2%) is quite arbitrary. One rationale is to have yield strength determined in this manner so as to make these values comparable with those of pipe materials with pronounced yield points. It also ensures that the same offset method can be used for materials with pronounced yield points (Fig. 2.1a) and still result in the same yield strength.

Because of this rather arbitrary 0.2% offset definition, the yield strength so determined is not as significant as the real yield strength of materials with pronounced yield points (Fig. 2.1a). This is the reason why some common austenitic stainless steels often have two sets of allowable stresses [2] with different limitations against yield strength. For pressure containing capability, the stress is allowed to reach higher percentage of yield strength if a slightly greater deformation is acceptable. This is mainly because the yield strength of roundhouse materials does not signify an abrupt gross deformation. However, when dealing with seating or sealing, such as with flange connection applications, the 0.2% offset criterion represents sizable damaging deformation. In this case, yield strength is treated similarly to the one with the abrupt yield point, and the allowable stress is limited to the same percentage of the yield strength regardless of material type.

Some piping codes [3] also include a benchmark stress at 1.0% offset to be used as a complement to the 0.2% yield strength. For some materials whose permanent deformation point is difficult to determine, the total deformation may be used. The American Petroleum Institute (API) 5L [4] adopts a 0.5% total elongation to set its yield strength.

2.1.4 Ultimate Strength, S_u

The highest stress on the stress-strain curve is called the ultimate strength. As the material is stretched, the cross-section area will be reduced. However, the stresses given on the curve are determined by dividing the applied force with the original cross-section area of the un-stretched specimen. This explains why the curve shows a drop in stress near the break point toward the end of the curve. The use of the original cross-section area is required as all design calculations are based on original cross-section.

The above tensile test stress is taken at the cross-section plane that is perpendicular to the applied force. In general, stress is not uniform across the whole area. However, with careful arrangement of the specimen shape and test equipment, we can pretty much consider the stress uniform.

2.1.5 Stresses at Skewed Plane

As the force is applied to the specimen, stress is produced not only on the plane perpendicular to the applied force, but also on all other imaginable planes. Figure 2.2 shows the stress generated on the plane that is not perpendicular to the applied force. For simplicity, we consider the specimen to be a small rectangular prism. On an arbitrary plane *m-m* that is inclined with an angle, θ, from the normal plane *m-n*, the applied force, F, decomposes into normal force, F_n, and shear force, F_s. The normal force is perpendicular to the plane and the shear force runs parallel to the plane. The normal force creates the normal stress as given in Eq. (2.4), and the shear force produces the shear stress as given in Eq. (2.5).

36 Chapter 2

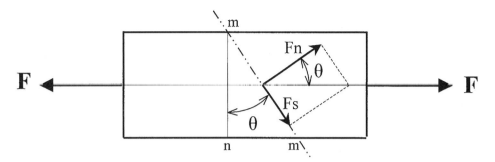

**FIG. 2.2
STRESSES AT SKEWED PLANE**

$$S_n = \frac{F_n}{A_m} = \frac{F\cos\theta}{A/\cos\theta} = S\cos^2\theta \qquad (2.4)$$

$$S_s = \frac{F_s}{A_m} = \frac{F\sin\theta}{A/\cos\theta} = S\sin\theta\cos\theta = \frac{S}{2}\sin 2\theta \qquad (2.5)$$

2.1.6 Maximum Shear Stress, $S_{s,max}$

The magnitudes of normal stress and shear stress at a given plane depend on the angle of inclination. For the shear stress, the maximum value is reached when $\sin 2\theta$ is equal to 1.0. That is, the shear stress is greatest when $2\theta = 90$ deg. or $\theta = 45$ deg. Substituting $\theta = 45$ deg., we have the maximum shear stress equal to one-half of the maximum normal stress, S, as shown in Eq. (2.6)

$$S_{s,max} = \frac{1}{2}S \qquad (2.6)$$

The fact that maximum shear stress is one-half of the tensile testing stress leads us to set a stress intensity as twice the value of the maximum shear stress. The stress intensity puts shear stress on the same footing as tensile testing stress and makes them directly comparable to each other. More about the stress intensity will be discussed later in this chapter.

2.1.7 Principal Stresses

Normal stress reaches its maximum level when $\cos^2\theta = 1.0$. This is equivalent to $\theta = 0$ deg. On the other hand, normal stress will be zero, or at its minimum value, when $\theta = 90$ deg. These maximum and minimum normal stresses, which are perpendicular to each other, are called principal stresses. Principal stresses are the basic stresses used in evaluating the damaging effect on the material. The planes on which these principle stresses act upon are called principal planes. There are no shear stresses in principal planes.

2.2 ELASTIC RELATIONSHIP OF STRESS AND STRAIN

The definition of modulus of elasticity as given in Eq. (2.3) is applicable only to the portion of the stress-strain curve that is below the proportional limit. Above the proportional limit, stress and strain do not have a simple mathematical relationship. The stress analysis that follows this constant stress-strain relationship is called elastic analysis. Elastic means that the pipe will return to its original unstressed state when the applied load is completely removed. Most piping stress analyses are either elastic or elastic equivalent analyses. The elastic equivalent analysis uses the elastic method to evaluate the inelastic behavior of the piping. For practical purposes, the stress below the yield strength is generally considered elastic stress.

In elastic analysis, the modulus of elasticity, E, is considered constant throughout the entire stress-strain range. Therefore, in addition to calculating directly from the applied force, the stress can also be calculated from the strain generated. Equation (2.3) can be rearranged for calculating the stress from the strain, and vice versa. From this equation, one might be tempted to calculate the stress by multiplying the modulus of elasticity with all the strain that can be found in the piping. This might lead to an incorrect result, because not all strains are stress-producing strains, as will be discussed in the following section.

2.2.1 Poisson's Ratio

When a specimen is stretched in the x-direction, as shown in Fig. 2.3, a stretching elongation, ℓ_x, is produced in the same direction. This is equivalent to producing a strain of $e_x = \ell_x/L_x$, where L_x is the length of the specimen in the x-direction. While this x-elongation is being created, a measure of shrinkage, ℓ_y, takes place in the y-direction at the same time. This shrinkage at the perpendicular direction to the applied force is a relief effort of the material to maintain a minimum general internal distortion. This phenomenon generates a shrinkage strain of $e_y = -\ell_y/L_y$ in the y-direction. The same thing also happens in the z-direction, which is perpendicular to the paper. The example uses the tensile force as the applied force. A compressive force will develop equivalent strains with all the signs reversed. The absolute value of the ratio of e_y and e_x, resulting from an x-direction force, is a constant called Poisson's ratio.

$$\text{Poisson's ratio}, \nu = \left|\frac{e_y}{e_x}\right| = \frac{\ell_y/L_y}{\ell_x/L_x} \qquad (2.7)$$

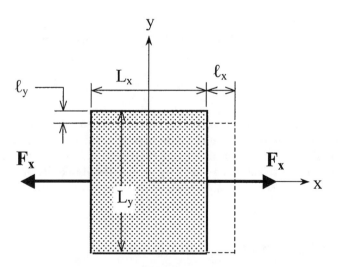

FIG. 2.3
RELIEF (POISSON'S) STRAIN

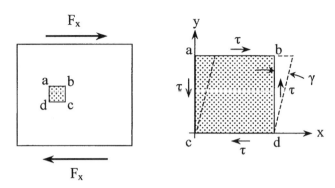

**FIG. 2.4
SHEAR DEFORMATION**

For metallic piping materials, the Poisson's ratio is roughly equal to 0.3. This is the value used by the American Society of Mechanical Engineers (ASME) B31 Piping Codes [2] when the exact value is not available.

The term e_y, the by-product of the applied strain e_x, is called Poisson's strain or the relief strain. This strain originated from the natural phenomenon of the material. It does not create any stress with its deformation. However, if this deformation is restrained, a stress will be generated. For a general three-dimensional stress situation, the strain in one direction is expressed by the stresses of all three directions as shown in Eq. (2.8).

$$e_x = \frac{S_x}{E} - v\frac{S_y}{E} - v\frac{S_z}{E}, \quad e_y = \frac{S_y}{E} - v\frac{S_z}{E} - v\frac{S_x}{E}, \ldots \text{etc.} \tag{2.8}$$

2.2.2 Shear Strain and Modulus of Rigidity

Shear strain is actually an angular deformation produced by shear stress. Figure 2.4 shows the deformation of an element *a-b-c-d* subject to shear force F_x. The shear stress τ is generated along with a rotation γ. The ratio between them is called the shear modulus of elasticity, or the modulus of rigidity. The modulus of rigidity maintains a constant value within the elastic range. This is generally referred to as an extension of Hook's law. The modulus of rigidity and the modulus of elasticity also have the relation as shown below.

$$G = \frac{\tau}{\gamma}, \quad G = \frac{E}{2(1+v)} \tag{2.9}$$

In practical manual piping stress analyses, we seldom have to deal with shear strain. However, shear strain is very important in the implementation of pipe stress analysis computer software, which can easily include the shear deformation that is otherwise too tedious for manual calculations. This distinction between computerized and manual analyses may result in quite different results if the pipe segment is very short. However, the computerized analysis is always more accurate by including the shear deformation.

2.3 STATIC EQUILIBRIUM

To calculate the stress at piping components, the first order of business is to calculate the forces and moments existing at a given cross-section. Calculating these forces and moments at an actual piping system is a very complicated process requiring specialized knowledge and experience. This task nowadays is mostly done with computer software packages developed by analytical specialists. What the practicing stress engineer requires are the basic principles for checking and confirming the results generated by the computer analysis. One of the most important basic principles is that of static equilibrium of forces and moments.

2.3.1 Free-Body Diagram

A free-body diagram is usually used to investigate the internal forces and moments at a given location of the piping component. A free-body is actually a portion of the pipe material enclosed by an arbitrary boundary. However, for the purpose of easy manipulation, the boundary is generally taken to be either a rectangle or a circle. For piping stress analyses, we usually take a section of the pipe, shown in Fig. 2.5, as a free-body. On the boundary of the free-body, we have to include all the forces and moments that exist. For a piping section, there are three directions of forces and three directions of moments at each end of the free-body. In addition, there are body forces that must also be included. The most common body forces are gravity and inertia forces. Figure 2.5 also includes the gravity force, which is equivalent to the weight, W, of the body. Inertia force will be discussed in the chapter that deals with dynamic analysis. Surface forces such as wind, friction, and pressure are skipped for the time being.

The forces and moments shown are all acting on the body. Those at the ends are internal forces and moments that are common to the bodies sharing the same boundary. For instance, the forces and moments at end-b of this a-b body are the same forces and moments acting on the end-b of the b-c body, except that the directions are reversed.

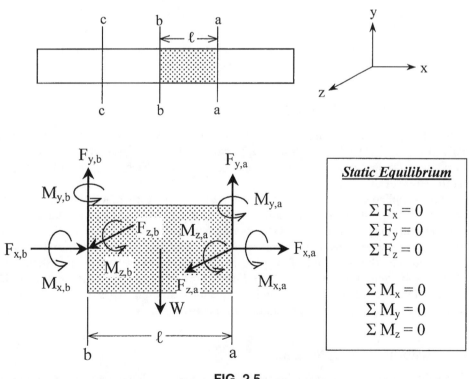

FIG. 2.5
FREE-BODY DIAGRAM

2.3.2 Static Equilibrium

To satisfy the condition that the body should not move ceaselessly, the resultant forces and moments of the entire body have to be zero (balance). This is the main principle of static equilibrium. Mathematically, we are used to calling it $\Sigma F_x = 0$, $\Sigma F_y = 0,\ldots, \Sigma M_z = 0,\ldots$, and so forth. Verbally, it states that the summation of forces in each direction is zero, and the summation of moments in each direction is also zero. The summation of forces is straightforward; however, summing the moments is quite different. In moments, we are talking about the summation of moment effects rather than the algebraic summation of the apparent moments. These equilibrium relations are often used to calculate the forces and moments at a given location from the known forces and moments at another location. For instance, if we know the forces and moments at end-a, then the forces and moments at end-b can be determined by the following relations:

(a) Forces: They are calculated using the following straightforward algebraic summations. Note that weight, W, is the gravity force acting toward the negative y direction.

$$\Sigma F_x = 0: \quad F_{x,a} + F_{x,b} = 0; \quad \text{or } F_{x,b} = -F_{x,a}$$
$$\Sigma F_y = 0: \quad F_{y,a} + F_{y,b} - W = 0; \quad \text{or } F_{y,b} = -F_{y,a} + W$$
$$\Sigma F_z = 0: \quad F_{z,a} + F_{z,b} = 0; \quad \text{or } F_{z,b} = -F_{z,a}$$

(b) Moments: Because the total moment effect about any given point has to be zero, any point in the body can be chosen to check the equilibrium relation. A point with expected unknown forces is normally chosen for this purpose to simplify the calculation. The following are calculated by taking the moment about point b.

$$\Sigma M_x = 0: \quad M_{x,a} + M_{x,b} = 0; \quad \text{or } M_{x,b} = -M_{x,a}$$
$$\Sigma M_y = 0: \quad M_{y,a} + M_{y,b} - F_{z,a}\ell = 0; \quad \text{or } M_{y,b} = -M_{y,a} + F_{z,a}\ell$$
$$\Sigma M_z = 0: \quad M_{z,a} + M_{z,b} + F_{y,a}\ell - W\ell/2 = 0 \quad \text{or } M_{z,b} = -M_{z,a} - F_{y,a}\ell + W\ell/2$$

In the descriptions of piping stress analysis topics, the generic term "force" implies both force and moment. By the same token, the generic term "displacement" implies both displacement and rotation.

2.4 STRESSES DUE TO MOMENTS

The stress at any point on a given plane of the material is calculated by dividing the force with the area as defined in Eq. (2.1). The equation is applicable only to uniform stress distributions. For a non-uniform stress distribution, a very small area is considered at each given point for the purpose of calculating the stress at that point.

When a force is applied on a solid bar of practical size, we can pretty much assume that the stress is uniform and calculate the stress by dividing the applied force with the cross-sectional area. However, for moment loads, the magnitude of the stresses changes across the entire cross-section, thus requiring different treatment. There are two types of moment loads: bending moment and torsion (or twisting) moment. Assuming that the centerline of the pipe element is lying in the x-direction, then the bending moments, M_y and M_z, bend the pipe in the lateral directions, whereas the torsion moment, M_x, twists the pipe in the axial direction. Bending and twisting produce different types of stresses and stress distributions. The stresses due to bending moments will be discussed first.

2.4.1 Stresses due to Bending Moments

Figure 2.6 shows a moment, M, acting on a section of a beam, which has a rectangular cross-section of h by b. The moment has the tendency of bending the beam into an arc shape. The basic beam

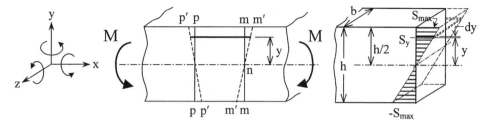

FIG. 2.6
BENDING STRESS DISTRIBUTION

formula, based on experience, assumes that a transverse plane of the beam remains plane and normal to the longitudinal fibers of the beam after bending. For instance, planes *m-m* and *p-p* will remain straight planes, *m'-m'* and *p'-p'*, after bending. The stress created is proportional to the deformation represented by the movement *m-m'*. With the moment direction as shown, tensile stress is generated at the top portion, and compressive stress is generated at the bottom portion of the cross-section. By further assuming that the material has the same tensile and compressive characteristics, it becomes apparent that the identical distribution of tensile stress and compressive stress is required to balance the axial forces. With identical tensile and compressive stress distributions, a zero stress is established at the center surface. The surface of zero stress is called the neutral surface, and the intersection of the neutral surface with any cross-section is called the neutral axis.

The bending stress at the beam is calculated by equalizing the moment generated by the stress with the moment applied. In Fig. 2.6, assuming that the maximum stress at the outer fiber is S_{max}, then the stress at y distance away from the neutral axis is $S_y = S_{max} y/(h/2) = (2S_{max} y)/h$. Taking the moment about the Z-neutral axis, the moment generated by the stress consists of two equal parts, one due to tensile stress and the other due to compressive stress. The sum of the moments is, therefore, twice the moment generated by the tensile stress. That is,

$$M = 2\int_0^{h/2} bS_y * y * dy = 2\int_0^{h/2} b(2S_{max}/h)y^2 dy$$

$$= \frac{2S_{max}}{h} * 2\int_0^{h/2} b*y^2 dy = \frac{2S_{max}}{h} * I$$

$$S_{max} = M\frac{h}{2I} = \frac{M}{Z} \tag{2.10}$$

where

$$I = 2\int_0^{h/2} b*y^2 dy = 2b\left[\frac{y^3}{3}\right]_0^{h/2} = \frac{bh^3}{12}$$

is the moment of inertia, about the *z*-axis, of the rectangular beam cross-section. And

$$Z = \frac{I}{h/2} = \frac{bh^2}{6}$$

is the corresponding section modulus.

42 Chapter 2

In piping stress analysis, most common cross-sections are circular. Equation (2.10) is also applicable to a circular cross-section, using a different moment of inertia and section modulus. For a solid circular cross-section of diameter D, the moment of inertia and section modulus are [5, 6]:

$$I = \frac{\pi D^4}{64}, \quad Z = \frac{\pi D^3}{32}$$

These formulas can also be easily derived from the polar moment of inertia to be discussed later in this section.

The maximum bending stress calculated is located at the extreme outer fiber of the cross-section. Stresses at the inner locations are smaller and are proportional to the distance from the neutral axis. Because of this localized nature of the maximum stress, the damaging effect of bending stress is considered to be not as severe as a uniform stress of the same magnitude.

2.4.2 Moment of Inertia

The moment of inertia and section modulus have been introduced in the calculation of bending stress. The moment of inertia about a given axis is formally defined as the sum of the products of each elementary area of the cross-section multiplied by the square of the distance from the area to the axis. Figure 2.7 shows the definitions and formulas for the moment of inertia, $I_{z\text{-}z}$ and $I_{y\text{-}y}$, around z-z and y-y axes, respectively. $I_{z\text{-}z}$ resists the M_z moment that is acting around the z-z axis, whereas $I_{y\text{-}y}$ resists the M_y moment acting around the y-y axis. For a symmetrical cross-section, areas with the same moment arm are lumped together for easier integration. In a rectangular cross-section (Fig. 2.6), we have lumped the area with the same moment arm, y, to form a strip of width b. The summation is then only required to integrate these strips starting from the neutral axis to the outer surface.

2.4.3 Polar Moment of Inertia

With twisting moments, we are dealing with the polar rotation arm, r, rather than the cranking arms, z and y, used in the bending moment of inertia. The moment of inertia determined by the polar arm is

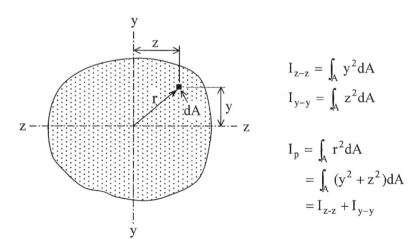

FIG. 2.7
MOMENT OF INERTIA OF CROSS SECTION

called the polar moment of inertia, I_p. Because $r^2 = z^2 + y^2$, the polar moment of inertia is the sum of the two bending moment of inertias about the two mutually perpendicular axes. This is given as

$$I_{z-z} = \int_A y^2 dA, \qquad I_{y-y} = \int_A z^2 dA$$
$$I_p = \int_A r^2 dA = \int_A (y^2 + z^2) dA = I_{z-z} + I_{y-y} \qquad (2.11)$$

Equation (2.11) is often used to determine the polar moment of inertia by calculating the bending moment of inertia. It is also a convenient tool for calculating the moment of inertia of a circular cross-section from the polar moment of inertia. For a circular cross-section, the calculation of the moment of inertia is quite cumbersome, but the calculation of the polar moment of inertia around the center point is fairly straightforward. Because I_{y-y} and I_{z-z} are equal for circular cross-section, from Eq. (2.11) we have $I_{y-y} = I_{z-z} = I_p/2$. The polar moment of inertia is calculated by integrating the concentric rings, $2\pi r\, dr$, as follows:

$$I_p = \int_0^R r^2 * 2\pi r * dr = 2\pi \int_0^R r^3 dr = 2\pi \left[\frac{r^4}{4}\right]_0^R = \pi \frac{R^4}{2} = \frac{\pi D^4}{32}$$

$$Z_p = \frac{I_p}{R} = \frac{\pi D^3}{16}$$

2.4.4 Moment of Inertia for Circular Cross-Sections

The preceding polar moment of inertia can be used to calculate the moment of inertia of circular cross-sections as

$$I = I_{z-z} = I_{y-y} = \frac{I_p}{2} = \frac{\pi R^4}{4} = \frac{\pi D^4}{64}$$

$$Z = \frac{I}{R} = \frac{\pi D^3}{32}$$

2.4.5 Stresses due to Torsion Moment

The twisting of a bar with a non-circular cross-section is a very complicated process. In this book, only the twisting of a bar with a circular cross-section will be discussed. As shown in Fig. 2.8, when a circular bar is twisted by the torsion moment, M_t, a rectangular element a-b-c-d will be deformed into a skewed parallelogram. This angular deformation produces shear stress in response to the torsion moment.

When calculating the torsion shear stress, the circular cross-section is assumed to remain circular and the plane perpendicular to the axis remains perpendicular to the axis, after the twisting. This intuitive assumption, based on the all-around symmetric nature of the circular cross-section, is also verified by experiments. Based on this assumption, the stress on the plane perpendicular to the axis is pure shear without any axial component. For the rectangular element a-b-c-d, the shear stress is proportional to the angular deformation shown in the dotted lines. This angular deformation diminishes toward the centerline of the bar and reduces to zero at the center. Assuming that the maximum stress at the surface is τ_{max}, then the stress at the location r distance from the center is $\tau = \tau_{max} r/R$. The shear stress can be calculated by the equilibrium of the moment generated by the stress and the moment applied. That is,

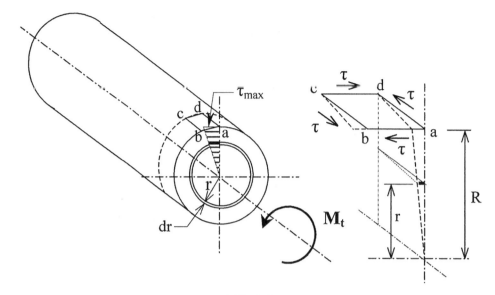

**FIG. 2.8
STRESSES DUE TO TORSION MOMENT**

$$M_t = \int_0^R \tau * 2\pi r dr * r = \int_0^R \frac{\tau_{max} r}{R} * 2\pi r dr * r = \frac{\tau_{max}}{R} \int_0^R 2\pi r dr * r^2$$
$$= \frac{\tau_{max}}{R} * I_p$$

$$\tau_{max} = M_t \frac{R}{I_p} = \frac{M_t}{Z_p} \qquad (2.12)$$

where

I_p = polar moment of inertia of the cross-section
$Z_p = I_p/R$ = torsion section modulus

Equation (2.12) is analogous to Equation (2.10), except that the value of Z_p is exactly twice that of Z for circular cross-sections such as pipe cross-sections.

Figure 2.8 also shows that with pure shear stress, τ, on face *a-b*, there will also be a shear stress of the same magnitude in the perpendicular face *b-c*. The directions of these shear stresses are so determined that the net twisting effect of the element is balanced. Because a non-circular cross-section does not have the capability to maintain this perpendicular shear stress at the corners, a warping will be initiated at the corner surfaces. Due to warping of the cross-section, a plane perpendicular to the axis does not remain perpendicular after the twisting; Eq. (2.12) is not applicable to bars with non-circular cross-sections. Outside these *a-b-c-d-a* surfaces, where only shear stress exists, are other surfaces with normal stresses. At 45 deg. from the *a-b* plane, the shear stress is zero and the normal stress is at maximum. These maximum normal stresses have the same magnitude as the maximum shear stress. They are tension at one plane and compression at the perpendicular plane.

2.5 STRESSES IN PIPES

A piping component experiences two main categories of stresses. The first category of stress comes from the pressure, either internal or external. The second category of stress comes from the forces and moments generated by weight, thermal expansion, wind, earthquake, and so forth.

2.5.1 Stresses due to Internal Pressure

The most common and important stress at a piping component is the stress due to internal pressure. When a pipe is pressurized, its inside surface is exposed to the same pressure in all directions. The pressure force is acting in the normal direction of the surface. However, we generally do not have to deal with every detail of the internal surface to determine the effect of the pressure and to calculate the stress due to pressure. Figure 2.9 shows a leg of piping spool subject to an internal pressure, P. Because of this pressure, the pipe wall is stretched in every direction. From the symmetry of the circular cross-section, we intuitively assume that there are two principal stresses, axial and circumferential, developed uniformly along the circumference of the pipe wall. These two stresses acting on a typical pipe wall element are designated as S_{lp} and S_{hp}, respectively, as shown in Fig. 2.9(a). The stress in the axial direction is called the longitudinal pressure stress, and the one in circumferential direction is called the hoop pressure stress. To calculate the magnitude of these stresses, a ring *m-m-n-n* containing the element is taken as the free-body. The stresses are then calculated by the equilibrium of the pressure force and the stress force acting at the boundaries.

Longitudinal pressure stress, S_{lp}. The longitudinal stress on the pipe is generally considered uniform in both circumferential and diametrical locations. Because it is required to have uniform longitudinal strain across the cross-section and across the thickness to prevent the pipe from being flared into a funnel shape, elastically, the longitudinal stress varies slightly with the circumferential stress due to Poisson's effect. However, a uniform longitudinal stress distribution is considered for both straight and bend sections in practical applications. From Fig. 2.9(b), we know that the longitudinal forces

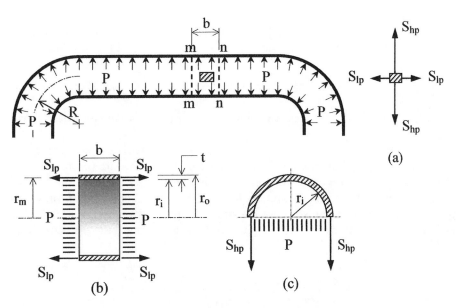

FIG. 2.9
STRESSES DUE TO INTERNAL PRESSURE

at both sides of the free-body are identical. Therefore, the forces balance out on each side of the free-body. By taking the equilibrium of the forces in the axial direction over the entire circular cross-section, we have

$$\text{Pressure force} = \pi r_i^2 P$$
$$\text{Stress force} = \pi(r_o^2 - r_i^2) S_{lp} = \pi(r_o - r_i)(r_o + r_i) S_{lp} = \pi t(2r_m) S_{lp}$$

Because the stress force balances the pressure force, we have

$$r_i^2 P = t(2r_m) S_{lp}$$

$$S_{lp} = \frac{r_i^2 P}{2r_m t} < \frac{r_m^2 P}{2r_m t} = \frac{r_m P}{2t} < \frac{r_o P}{2t} \tag{2.13}$$

Equation (2.13) represents several variations of the formulas that might appear in different articles on this subject. The first expression is the most accurate one, whereas the last expression, $r_o P/2t$, is the most conservative one yielding the highest stress.

Hoop pressure stress, S_{hp}. The distribution or variation of the hoop pressure stress is not as uniform as the longitudinal stress. In the diametrical direction, the stress is higher at the inner surface and lower at the outside surface. The piping code [2] has given a design formula that properly takes into account this non-uniform stress distribution. This section will only discuss the general simplified formula that assumes that stress is uniform across the thickness. The code requirements will be discussed later in a separate chapter.

By balancing the forces acting on the semi-circular band as given in Fig. 2.9(c) for a band width, b, we have

$$\text{Pressure force} = 2r_i b P, \text{ which should be equal to}$$
$$\text{Stress force} = 2bt * S_{hp}$$
$$\text{or } S_{hp} = \frac{r_i}{t} P$$

The above relation assumes a uniform hoop stress across the thickness. In reality, the stress is not uniform and is greater near the inside surface. To compensate for this non-uniform stress distribution, the design equation normally uses the outside radius instead of the inside radius as

$$S_{hp} = \frac{r_o}{t} P \tag{2.14}$$

From Eqs. (2.13) and (2.14), it is clear that hoop pressure stress is roughly twice the value of longitudinal pressure stress.

Hoop pressure stress at a bend of uniform thickness has some variation across the circumference. If a radial strip is taken as a free-body, it is clear that the inner curvature (crotch) area has less area to resist the pressure force compared to the outer curvature area. Therefore, a higher hoop stress is expected at the crotch area. The theoretical hoop stresses at a bend of uniform thickness can be calculated by multiplying the hoop stress of the straight pipe as given by Eq. (2.14) with the following factors (Derivations of these factors are given in Chapter 4 where the pressure design of piping components is discussed.):

$$\frac{(2R - r_m)}{(2R - 2r_m)} \cong \frac{4(R/D) - 1}{4(R/D) - 2} \quad \text{at intrados (crotch)}$$
$$\frac{(2R + r_m)}{(2R + 2r_m)} \cong \frac{4(R/D) + 1}{4(R/D) + 2} \quad \text{at extrados (crown)}$$

These factors, occasionally called the Lorenz factors, are sometimes used as stress intensification factors due to pressure at the bends. They are also used to gauge the actual wall thickness required at different locations of a bend. However, in most practical applications, they tend to be ignored. This is partially because a forged or hot rolled bend generally has a thicker wall at the crotch area, thus neutralizing the effect of these factors. The safety factor normally included in the design code and specification also covers this type of minor deviations.

2.5.2 Stresses due to Forces and Moments

Besides the pressure that generates the pressure stress discussed in the preceding subsection, other loadings can also produce internal forces and moments that generate significant stress in the pipe. These internal forces and moments acting at a given pipe cross-section (Fig. 2.10) are the result of thermal expansion, weight, wind, earthquake, and other internal and external loads applied to the piping system.

Stresses due to forces. In a piping component, forces can be divided into two categories: shear force, F_s, which acts in a direction perpendicular to the pipe axis; and axial force, F_a, which acts in the axial direction of the pipe. With coordinate axes selected as in Fig. 2.10, the shear force comprises two forces, F_y and F_z, each of which produces a shear stress at the pipe cross-section. The stress is not uniform, and is greatest at the diametrical centerline perpendicular to the force. The ratio of the maximum value and the average value is called the shear distribution factor. For most pipe cross-sections, the shear distribution factor is very close to 2.0. Therefore, we can write

$$\tau_{xy,\text{max}} = 2\frac{F_y}{A}, \quad \tau_{xz,\text{max}} = 2\frac{F_z}{A}, \quad A = \pi(r_o^2 - r_i^2) = 2\pi r_m t$$

τ_{xy} implies that the shear stress is parallel to the y-axis and acting on the plane perpendicular to the x-axis. Normally, we are only interested in the combined maximum shear stress at the cross-section.

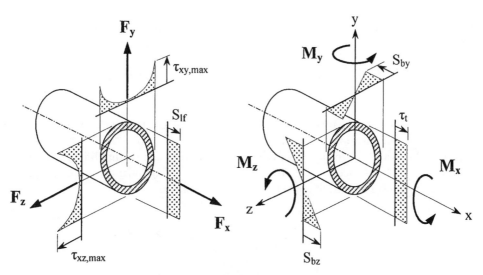

FIG. 2.10
STRESSES DUE TO FORCES AND MOMENTS

In this case, the two shear forces are first combined into a resultant shear force as $F_s = \sqrt{F_y^2 + F_z^2}$. The stress is then calculated using this resultant force.

For the axial force, $F_a = F_x$, the stress is either pure tension or compression, depending on the sign of the force. The axial stress generated is uniform across the cross-section. Therefore, at a given cross-section of the pipe we have the stresses due to forces as

$$S_{lf} = \frac{F_a}{A}, \quad \tau_f = 2\frac{F_s}{A}, \quad A = 2\pi r_m t \tag{2.15}$$

where

S_{lf} = longitudinal stress due to the force
τ_f = shear stress due to the force

Both stresses are acting on the same cross-section plane. Stresses due to direct forces as given in Eq. (2.15) are generally insignificant when compared with stresses due to moments. Therefore, stresses due to direct forces are often ignored in the evaluation of the piping system.

Stresses due to moments. Moment loads are also divided into two categories: bending moment and torsion moment. The bending moment is further divided into two components, M_y and M_z, around the conveniently selected y-axis and z-axis, respectively. As discussed in Section 2.4, each bending moment creates a linear distribution of stress with the highest stress occurring at the outer surface farthest from the bending axis and is equal to

$$S_{by} = \frac{M_y}{Z}, \quad S_{bz} = \frac{M_z}{Z}, \quad Z = \frac{\pi}{4r_o}(r_o^4 - r_i^4)$$

S_{by} and S_{bz} are located at the outside surface of the pipe, but are 90 deg. apart from each other. They are normally combined together to become the total bending stress. That is,

$$S_b = \sqrt{S_{by}^2 + S_{bz}^2} = \frac{1}{Z}\sqrt{M_y^2 + M_z^2}$$

The stress created by the torsion moment, $M_t = M_x$, is shear stress that is linearly distributed in the diametrical direction with the maximum at the outer surface. The shear stress, however, is uniformly distributed along the circumferential direction. The magnitude of the highest shear stress is calculated by

$$\tau_t = \frac{M_t}{Z_p} = \frac{M_t}{2Z}$$

where τ_t is shear stress due to the torsion moment. This stress needs to be combined with bending and other stresses to evaluate the piping system.

For pipes with thin walls, the formulas for the section modulus can be simplified. In a thin wall pipe, we can assume that the cross-section area is concentrated to a ring having a radius of r_m. Then, from the definition of the polar moment of inertia, we have the following

$$I_p = 2\pi r_m t * r_m^2 = 2\pi r_m^3 t, \quad Z_p = \frac{I_p}{r_m} = 2\pi r_m^2 t, \quad Z = \frac{Z_p}{2} = \pi r_m^2 t$$

These approximate formulas appear occasionally in the code book [2]. The preceding lays the background.

2.6 EVALUATION OF MULTI-DIMENSIONAL STRESSES

A piping component is generally exposed to multi-axial stresses. The evaluation of these stresses is not as simple as our intuition may lead us to think. For instance, Fig. 2.11 shows a few combinations of three directional principal stresses. The absolute magnitudes of the stresses are the same. The body in case (a) has all three directions subject to tension, in case (b) all in compression, in case (c) with one direction in compression, and in case (d) with one direction at zero stress. Intuitively, case (b) appears safest to us because all the stresses are trying to keep the body together. This, in fact, proves to be the case as proved by experiments on a body subjected to a large external hydraulic pressure. However, our assessments of the other cases are not as clear. For instance, we may think case (a) is bad, because all the stresses have the tendency of tearing the body apart. But in reality, it is as good as case (b), except that it is hard to test in real life. How about case (c), where one direction of stress is in compression? This is where our intuition fails to reach the right conclusion. This may surprise us, but case (c) is actually considerably worse than case (a) in terms of safety. In case (d), one direction of tensile stress is removed. This case (d) fares just as bad as, but slightly better than, case (c). As will be derived later in this section, damage to the body is determined by the difference of the maximum and minimum principal stresses.

Presentation of a general three-dimensional stress field is very tedious and complicated. In the following, we will discuss the general two-dimensional stress field to get acquainted with the general philosophy of the code stress evaluation. A three-dimensional stress field can generally be evaluated using the two-dimensional approach, investigating one plane at a time.

2.6.1 General Two-Dimensional Stress Field

Figure 2.12 shows a general two-dimensional stress field on the piping component. The dimension in the direction perpendicular to the paper is immaterial and can be considered the thickness of the pipe in most cases. In this general stress field, there are normal stress and shear stress, as denoted, at each side of the rectangular free-body. For normal stress, tension is considered positive and compression negative. The sign convention for shear stress is not as well defined as for normal stress. Fortunately, this ambiguity of sign in shear stress does not materially affect the outcome of our discussion. The following discussion and derivation are based on the assumption that signs are positive as shown. The body is assumed to align with the x and y coordinates as shown.

The first step of the evaluation requires the calculation of maximum normal stress and shear stress that might exist at any plane of the body. This shall start with the determination of the normal and shear stress at any plane m-m located with a θ angle from the x-axis. From the stresses on this general plane, the maximum stresses from all conceivable planes can be determined. To find the stresses at

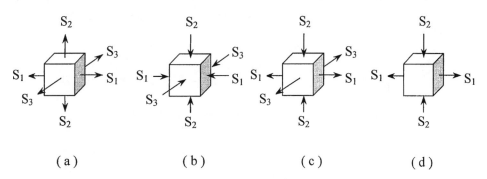

FIG. 2.11
ELEMENTS SUBJECT TO THREE-DIMENSIONAL STRESSES

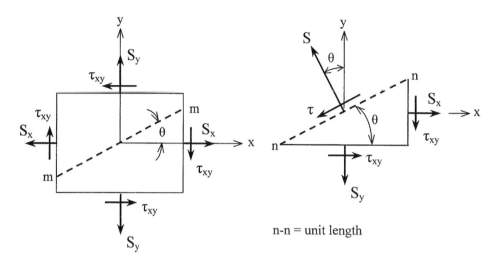

**FIG. 2.12
TWO-DIMENSIONAL STRESS RELATIONS**

plane *m-m*, a right triangle free-body with *m-m* as the cord is constructed. Taking a unit length *n-n* along the *m-m* plane, we have the base equal to $\cos\theta$ and the height equal to $\sin\theta$. By balancing the forces along the normal and parallel directions to the plane, we have the following relations:

$$S = S_x \sin\theta \sin\theta + S_y \cos\theta \cos\theta + \tau_{xy} \sin\theta \cos\theta + \tau_{xy} \cos\theta \sin\theta$$
$$= S_x(1-\cos2\theta)/2 + S_y(1+\cos2\theta)/2 + \tau_{xy}\sin2\theta$$
$$= \frac{1}{2}(S_x + S_y) - \frac{1}{2}(S_x - S_y)\cos2\theta + \tau_{xy}\sin2\theta \tag{2.16a}$$

$$\tau = S_x \sin\theta \cos\theta - S_y \cos\theta \sin\theta - \tau_{xy}\sin\theta\sin\theta + \tau_{xy}\cos\theta\cos\theta$$
$$= \frac{1}{2}(S_x - S_y)\sin2\theta + \tau_{xy}\cos2\theta \tag{2.16b}$$

The θ of the plane where the maximum and minimum normal stresses occur can be found by differentiating Eq. (2.16a) with respect to θ, and setting the derivative to zero. Maximum stress is then calculated by back-substituting this θ into the equation. Maximum shear stress can also be determined in the same manner. However, an elegant pictorial representation can be developed by first moving $(S_x + S_y)/2$ term in Eq. (21.6a) to the left-hand side, and then squaring both (2.16a) and (2.16b) and summing them together as follows:

$$\left[S - \frac{1}{2}(S_x + S_y)\right]^2 + \tau^2 = \left[\frac{1}{2}(S_x - S_y)\cos2\theta - \tau_{xy}\sin2\theta\right]^2$$
$$+ \left[\frac{1}{2}(S_x - S_y)\sin2\theta + \tau_{xy}\cos2\theta\right]^2$$

or

$$\left[S - \frac{1}{2}(S_x + S_y)\right]^2 + \tau^2 = \left(\frac{S_x - S_y}{2}\right)^2 + \tau_{xy}^2 \tag{2.16c}$$

2.6.2 Mohr's Circle for Combined Stresses

Equation (2.16c) is a general equation of a circle in S-τ coordinates. It shows that the relationship between normal stress, S, and shear stress, τ, at any given plane can be expressed with a circle. The center of the circle is located at $(S_x + S_y)/2$ away from the origin on the S-axis, and the radius of the circle is the square root of the right-hand side of Eq. (2.16c). This relationship can be expressed with Mohr's stress circle as shown in Fig. 2.13. The procedure for drawing the circle is as follows: (1) Draw the horizontal axis as the S coordinate and the vertical axis as the τ coordinate. (2) Locate S_x and S_y on the S coordinate. Here, S_x is assumed to be greater than S_y. (3) At S_x or S_y points, draw a vertical line that is equal to τ_{xy}. (4) Locate the mid-point between S_x and S_y to serve as the center point, and draw a circle with the distance between the center point and the τ_{xy} point defined in (3) as the radius.

From the figure and the equation, the maximum and minimum normal stresses are found by setting $\tau = 0$, and taking the square roots of both sides of the equation. The results are as follows:

$$S_1 = \frac{S_x + S_y}{2} + \sqrt{\left(\frac{S_x - S_y}{2}\right)^2 + \tau_{xy}^2} \tag{2.17a}$$

$$S_2 = \frac{S_x + S_y}{2} - \sqrt{\left(\frac{S_x - S_y}{2}\right)^2 + \tau_{xy}^2} \tag{2.17b}$$

The maximum shear stress is equal to the radius of the circle. That is,

$$\tau_{max} = \sqrt{\left(\frac{S_x - S_y}{2}\right)^2 + \tau_{xy}^2} = \frac{S_1 - S_2}{2} \tag{2.18}$$

The significance of the combined stress is determined by its potential for damaging the structure. Its assessment would require the understanding of the theory of failure. For this discussion, the general three-dimensional body with three principal stresses S_1, S_2, and S_3 related by $S_1 > S_2 > S_3$ is used as an example. Compressive stresses are regarded as negative stresses.

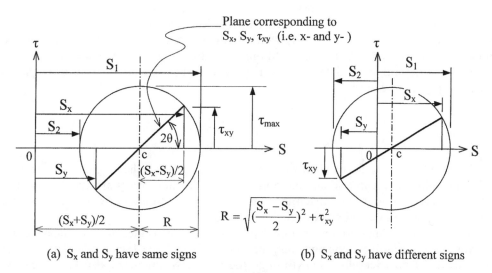

FIG. 2.13
MOHR'S STRESS CIRCLE FOR TWO-DIMENSIONAL STRESSES

2.6.3 Theories of Failure

There are several different theories of failure that have been proposed and used [7]. These theories include: (1) *maximum stress theory*, which predicts that the material will yield when the absolute magnitude of any of the principal stresses reaches the yield strength of the material; (2) *maximum strain theory*, which predicts that the material will yield when the maximum strain reaches the yield point strain; (3) *maximum shear theory*, which predicts that the material will yield when the maximum shear stress in the material reaches the maximum shear stress at the yield point in the tension test; (4) *maximum energy theory*, which predicts that the material will yield when the strain energy per unit volume in the material reaches the strain energy per unit volume at yielding in the simple tension test; and (5) *maximum distortion energy theory*, which predicts that the material will yield when the distortion energy per unit volume in the material reaches the distortion energy per unit volume at yielding in simple tension test.

The maximum stress theory fits very well with brittle materials such as concrete and non-ductile cast iron. For the ductile materials that are prevalent in piping, both the maximum shear theory and the maximum distortion energy theory agree very well with the experiments. The maximum distortion energy theory is slightly more accurate, but the maximum shear theory is simpler and easier to apply. ASME has adopted the maximum shear failure theory in its piping and pressure vessel codes.

2.6.4 Stress Intensity (Tresca Stress)

In the maximum shear failure theory, the condition for yielding occurs when the maximum shear stress in the material equals the maximum shear stress at the yield point in the tension test. From Eq. (2.6), the maximum shear stress at the yield point in the tension test is equal to one-half of the tensile yield strength. In other words, twice the maximum shear stress at yield in the tension test is the same as the tensile yield strength. Therefore, by *defining stress intensity as twice the maximum shear stress*, it can then be directly compared to the tensile yield strength and other data obtained in the tension test. In Eq. (2.18), for a two-dimensional stress field, we have

$$\text{Stress intensity, } S_i = 2\tau_{max} = S_1 - S_2$$

The above expression can be extended to a three-dimensional stress field as $S_i = S_1 - S_3$, where $S_1 > S_2 > S_3$. Compressive stresses are considered negative stresses in the above comparative quantity. This provides the basis for the comparative damaging effect of the stress systems shown in Fig. 2.11.

In piping stress analysis, we deal mostly with two-dimensional stress fields with the stress at the third dimension either zero or insignificant. In this case, it is simpler to calculate the stress intensity directly from the general stress field given in Fig. 2.12, without calculating the principal stresses. Again, from Eq. (2.18) we have

$$S_i = 2\tau_{max} = 2\sqrt{\left(\frac{S_x - S_y}{2}\right)^2 + \tau_{xy}^2} = \sqrt{(S_x - S_y)^2 + 4\tau_{xy}^2} \qquad (2.19)$$

Assuming that x-axis is in the axial direction of the pipe, then S_x is the longitudinal stress resulting mainly from bending moments and internal pressure, S_y is the hoop stress mainly from internal pressure, and τ_{xy} is mainly due to the torsion moment. Stresses due to direct axial and shear forces are generally insignificant and are often neglected.

There are a couple of things to note in applying Eq. (2.19). One is the nature of the bending stress, which always presents both tension and compression at a given cross-section. The other is the plane that includes the diametrical direction (through the thickness direction) also has to be included. The normal stress in this diametrical direction is either zero or equal to the acting pressure, which is compressive.

2.6.5 Effective Stress (von Mises Stress)

The maximum distortion energy failure theory agrees with the nature of ductile materials the most. This theory is very popular in the European piping community. Based on distortion energy theory, the condition for yielding is [7]

$$(S_1 - S_2)^2 + (S_2 - S_3)^2 + (S_1 - S_3)^2 = 2S_{y_p}^2$$

where S_{y_p} is the yield point tensile stress in the tension test. To make a stress directly comparable to the yield point stress and other data obtained in the tension test, the effective stress is coined and defined as

$$S_e = \sqrt{\frac{1}{2}[(S_1 - S_2)^2 + (S_2 - S_3)^2 + (S_1 - S_3)^2]} \qquad (2.20)$$

For a two-dimensional stress field, as is commonly the case in practical piping system, S_3 can be set to zero, and we have

$$S_e = \sqrt{S_1^2 - S_1 S_2 + S_2^2}$$

The principal stresses S_1 and S_2 can be found using Eqs. (2.17a) and (2.17b), respectively, for two-dimensional stress systems. Substituting Eqs. (2.17a) and (2.17b) to the above, the effective stress for two-dimensional stress field becomes

$$S_e = \sqrt{(S_x - S_y)^2 + S_x S_y + 3\tau_{xy}^2} \qquad (2.21)$$

By comparing Eq. (2.21) with Eq. (2.19), it is difficult to assess whether the effective stress is larger or smaller than the stress intensity. In other words, it is not possible to say which theory is more conservative. However, one thing is clear: when either one of the normal stresses is zero, the stress intensity is somewhat larger than the effective stress. The degree of difference depends on the ratio of the normal stress and the shear stress. At the extreme condition when both normal stresses are zero, the stress intensity is bigger than the effective stress by a factor of $\sqrt{4/3} = 1.155$. However, in most practical piping applications, both stress intensity and effective stress can be considered equivalent to each other.

2.7 BASIC BEAM FORMULAS

A piping system is essentially a group of beams connected together to form the shape required for transporting fluids from one point to another. Therefore, the behavior of the beam is the basic component of the pipe stress analysis. Table 2.1 shows some basic beam formulas that stress engineers are all familiar with. The following are a few important notes derived from these basic formulas:

(1) With a given loading, the bending moment is proportional to the length of the beam, but the displacement is proportional to the cube of the length. A slight increase in length creates a large increase in displacement, which translates to a large increase in flexibility.
(2) For a given configuration, the displacement is inversely proportional to EI. This EI is generally referred to as the *stiffness coefficient* of the cross-section. If a simulation is required for a non-standard cross-section such as refractory lined or concrete lined pipe, the EI of the simulating pipe has to match the combined EI of the pipes being simulated.

TABLE 2.1
BASIC BEAM FORMULAS

(a) Simple Beam – Uniform Load

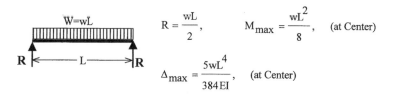

$$R = \frac{wL}{2}, \qquad M_{max} = \frac{wL^2}{8}, \quad \text{(at Center)}$$

$$\Delta_{max} = \frac{5wL^4}{384EI}, \quad \text{(at Center)}$$

(b) Simple Beam – Concentrated Load

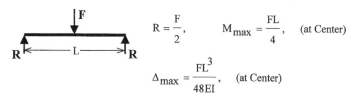

$$R = \frac{F}{2}, \qquad M_{max} = \frac{FL}{4}, \quad \text{(at Center)}$$

$$\Delta_{max} = \frac{FL^3}{48EI}, \quad \text{(at Center)}$$

(c) Fixed Beam – Uniform Load

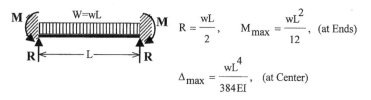

$$R = \frac{wL}{2}, \qquad M_{max} = \frac{wL^2}{12}, \quad \text{(at Ends)}$$

$$\Delta_{max} = \frac{wL^4}{384EI}, \quad \text{(at Center)}$$

(d) Fixed Beam – Concentrated Load

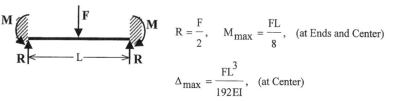

$$R = \frac{F}{2}, \qquad M_{max} = \frac{FL}{8}, \quad \text{(at Ends and Center)}$$

$$\Delta_{max} = \frac{FL^3}{192EI}, \quad \text{(at Center)}$$

(3) The displacement formulas include only the term EI. The terms involving shear modulus G and cross-section area A, associated with shear deformation, are not included. The formulas are good for practical lengths of beams. However, they are not accurate for short beams whose length is shorter than ten times the cross-sectional dimension of the beam. The formulas used by most computer programs include the shear deformation, thus making a short beam more flexible than that calculated by the formulas in Table 2.1. This is one of the potential discrepancies between a computer result and a hand calculation result. In such cases, the computer result is more accurate.

(4) Items (a) and (c) can be used to determine the stress due to weight under normal supporting spans. Because the actual piping in the plant is supported somewhere between simple support and fixed support [8], the following average formulas are generally used, instead, for evaluating weight supports.

$$S = \frac{M}{Z} = \frac{wL^2}{10Z}, \qquad \Delta = \frac{3wL^4}{384EI}$$

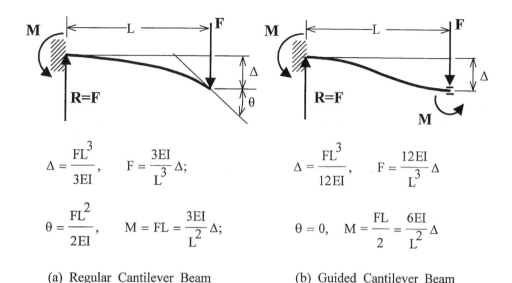

(a) Regular Cantilever Beam   (b) Guided Cantilever Beam

FIG. 2.14
DIFFERENT TYPES OF CANTILEVER BEAMS

2.7.1 Guided Cantilever

In piping flexibility analysis, the guided cantilever method is one of the often-mentioned approximate approaches [8]. Figure 2.14 shows two different types of cantilever beams that may be used in piping stress analysis. The regular cantilever beam shown in Figure 2.14(a) is actually half of the simple beam subject to a concentrated load as given in Table 2.1(b). The fixed end is equivalent to the mid-point of the simple beam. The guided cantilever beam shown in Figure 2.14(b) is actually half of the fixed beam subject to a concentrated load as given in Table 2.1(d). These beam models are often used to calculate the forces and moments, approximately, in a given length of pipe subject to a displacement resulting from thermal expansion.

The regular cantilever model requires an accompanying rotation, θ, while absorbing the displacement, Δ. This rotational relief reduces the forces and moments generated by a given displacement. The problem is that an actual piping system does not freely offer this rotational relief. It does offer some relief, but much less than the amount shown in Fig. 2.14(a). A more realistic and conservative approach is to assume that no rotation is taking place, as shown in Fig. 2.14(b). This is the so-called guided cantilever approach. For a given expansion or displacement, the force created by the guided cantilever is four times as great as that by a regular cantilever, and the moment, thus the stress is twice as large as that by a regular cantilever.

2.8 ANALYSIS OF PIPING ASSEMBLY

The analysis of a piping assembly is very complicated and is accomplished mostly by using computer programs. In general, an engineer is not required to have knowledge of the computing methods implemented in the computer programs in order to use the program. However, some common sense regarding general analytical approaches can help analysts to better understand the procedure and better interpret the results.

A piping system consists of many components laid out in all directions. Analysis of the piping system is normally idealized into a combination of straight pipe segments and pipe bend elements. In other words, the analysis is performed using a finite element method consisting of two types of elements: the straight pipe and the pipe bend.

2.8.1 Finite Element

The body of a structure generally has stresses and strains that vary continuously throughout the body. It is very difficult, if not impossible, to calculate exactly these stresses and strains. However, to obtain a practical result, the body can be divided into many sub-bodies, each with a finite size. Each small body is considered to have a predictable stress and strain distribution over it. These small bodies are called finite elements. In piping analysis, these bodies are actually fairly large compared to the general sense of a finite element. Here, we use straight pipe and curved pipe, two types of beam elements. Each element has two nodes, N1 and N2, as shown in Fig. 2.15(a). N1 is the beginning node, and N2 is the ending node.

The characteristics of the element are expressed in the local coordinates aligned with the element geometry. For a straight pipe element, the local x-axis is always in the axial direction pointing from the beginning node toward the ending node. The local y-axis and z-axis are perpendicular to each other in the lateral directions. For a curved pipe element, the local axis convention differs slightly among different investigators. One popular convention, as shown in Fig. 2.15(a), assigns the local x-axis as connecting the two nodes and pointing from the beginning node to the ending node. The local y-axis is perpendicular to the x-axis and pointing from the mid-cord point toward the bend tangent intersection point.

In a general three-dimensional environment, each node has six degrees of freedom, three in translation and three in rotation. At each element, the forces and displacements have a fixed relationship in local coordinates and are designated with a prime (′) notation. Each node of an element is associated with three displacements, D_x', D_y', D_z', and three rotations, R_x', R_y', R_z'. Correspondingly, each node is also associated with three forces, F_x', F_y', F_z', and three moments, M_x', M_y', M_z'. In general, the term displacement is used to cover both displacement and rotation. By the same token, the term force is

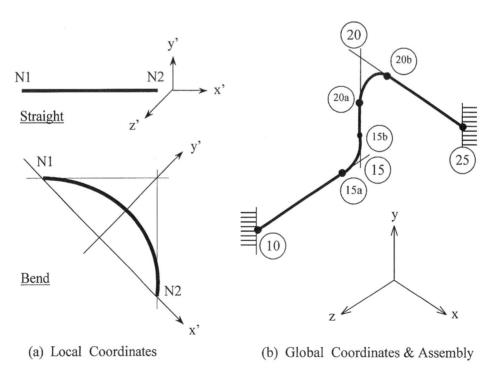

(a) Local Coordinates   (b) Global Coordinates & Assembly

FIG. 2.15
COORDINATE SYSTEMS

used to cover both forces and moments. The finite element method is built under the premise that on each element there is a relationship between these forces and displacements. That is, for each element we have the relation in local coordinates as

$$\{\mathbf{F}'\} = [K']\{\mathbf{D}'\} \tag{2.22}$$

where $\{\mathbf{F}'\}$ is the force vector representing 12 forces and moments at both nodes. That is, $\{\mathbf{F}'\} = \{F'_{x_1}, F'_{y_1}, F'_{z_1}, M'_{x_1}, M'_{y_1}, M'_{z_1}, F'_{x_2}, F'_{y_2}, F'_{z_2}, M'_{x_2}, M'_{y_2}, M'_{z_2}\}^T$. Subscript 1 represents node N1 and subscript 2 represents node N2. Superscript T denotes transpose, meaning a column vector. $\{\mathbf{D}'\}$ is the displacement vector representing 12 displacements and rotations at both nodes. That is $\{\mathbf{D}'\} = \{D'_{x_1}, D'_{y_1}, D'_{z_1}, R'_{x_1}, R'_{y_1}, R'_{z_1}, D'_{x_2}, D'_{y_2}, D'_{z_2}, R'_{x_2}, R'_{y_2}, R'_{z_2}\}^T$. $[K']$ is a 12×12 symmetric stiffness matrix. The exact terms in $[K']$ are too complex to be included here. Interested readers may find them in related publications [9–12]. There are several different forms of these terms, using somewhat different local coordinate conventions.

2.8.2 Data Points and Node Points

Before a piping system is analyzed, every element in the system has to be identified. Customarily, these elements are identified with point numbers. These numbers serve as the communication addresses just like real-life house numbers. Up to three different point numbers may be associated with an analysis. Take, for example, the simple assembly shown in Fig. 2.15(b). The first set of numbers needed is the set of data point numbers used to describe the geometry of the system. The following are the locations that need to be assigned with data point numbers:

(1) Terminal points such as anchors, free ends, vessel, and tank connections, etc. Points 10 and 25 belong to this category.
(2) Bend tangent intersection points (working points). Points 15 and 20 belong to this category.
(3) Branch intersection points.
(4) Key flange face points.
(5) Restraint and loading points.
(6) Other points where the response of the system is of interest.

The above data points are required for a precise description of the system. From these essential data points, the computer program may also generate some other data points required for the analysis. At bend 15, for instance, the required input data point is the tangent intersection point 15, but the points required for the analysis are the end points 15a and 15b of the bend. Point 15 is not located at any physical part of the piping system; therefore, it is not used in the analysis.

From the information provided by the data points, the computer program will re-number the entire piping system to assign analysis node numbers. The analysis node numbers have to be consecutive starting from 1. They are generated following the input sequence of the data. In our example system, we have nodes 1 = 10, 2 = 15a, 3 = 15b, and so forth. In an actual analysis, these sequentially generated node numbers may be once again re-numbered to achieve the so-called bandwidth optimization. The bandwidth is essentially the difference between the node number of a given point and the node number of its adjacent connecting point or points. For a multi-branched system, the bandwidth greatly depends on the re-numbering. The result of this re-numbering is another set of node numbers, which are called re-numbered node numbers. The last re-numbering is intended to achieve the minimum bandwidth possible. Smaller bandwidths require less data storage and also less computing time. The whole numbering effort may appear very confusing, but it should not be a concern of the analyst. All analysis results are given in reference to the original data point numbers assigned.

2.8.3 Piping Assembly

To mathematically assemble the piping system, all individual elements have to use a common coordinate system. This common coordinate system is called the global coordinate system. The global y-axis is generally fixed in the vertical direction pointing upward. This practice is partly due to piping tradition [8] and partly for the convenience of specifying the gravitational load. This y coordinate assignment also makes it easier to correlate it with the elevation that is often used in a piping system. The global x- and z-axes are normally aligned with plant's major orientations, which are often called plant-North and plant-East, etc.

The local coordinate stiffness matrix of each element has to be first converted to global coordinates before a mathematical assembly can be performed. Depending on the orientation of the element, a rotational matrix consisting of directional parameters is used to convert the local force and local displacement vectors into global force and global displacement vectors. That is,

$$\{F'\} = [L]\{F\}, \quad \{D'\} = [L]\{D\}$$

where

- $[L]$ = 12 × 12 transformation matrix consisting of four sets of 3 × 3 rotational parameter matrices placed at the diagonal
- $\{F\}$ = force vector in global coordinates
- $\{D\}$ = displacement vector in global coordinates

Substituting the above relations to Eq. (2.22), we have

$$[L]\{F\} = [K'][L]\{D\} \quad \text{or} \quad \{F\} = [L]^{-1}[K'][L]\{D\}$$

The global stiffness matrix, $[K]$, of each element is then created by applying the rotational transformation on the local stiffness matrix $[K']$ as

$$\{F\} = [K]\{D\}, \quad \text{where } [K] = [L]^{-1}[K'][L] = [L]^{T}[K'][L] \tag{2.23}$$

$[L]^{-1}$ is the inverse of the transformation matrix $[L]$, and $[L]^{T}$ is the transposition of $[L]$. For rotational transformation matrices, $[L]^{-1}$ and $[L]^{T}$ are equal. The transposition is obtained by interchanging the rows and columns of the original matrix, $[L]$. In the actual assembling process, the force/displacement relation given in Eq. (2.23) is partitioned to

$$\begin{vmatrix} F_1 \\ F_2 \end{vmatrix} = \begin{vmatrix} K_{11} & K_{12} \\ K_{21} & K_{22} \end{vmatrix} \begin{vmatrix} D_1 \\ D_2 \end{vmatrix} \tag{2.24}$$

where subscripts 1 and 2 denote node N1 and node N2, respectively. Each F_* and D_* sub-vector represents six components corresponding to six degrees of freedom.

For a system with n-node points, the total number of degrees of freedom is $N = 6n$. These N degrees of freedom are all potentially related. Therefore, the system has ($N \times N$) overall stiffness matrix relating the forces at all degrees of freedom with the displacements at all degrees of freedom. The assembly of the overall stiffness matrix uses the force vector and displacement vector, both representing all degrees of freedom, as the starting template. Each element matrix as given by Eq. (2.24) is placed inside the overall stiffness matrix at the location representing the proper degree of freedom in the force and displacement template. All element stiffness values placed at the same location inside the overall stiffness are simply added together algebraically. The assembly is completed when all element matrices are included in the overall matrix. The overall matrix of a piping system generally forms a narrow band around the diagonal. The width of this band is called the bandwidth of the matrix. Again, because a smaller bandwidth requires less storage and expedites calculation, it is generally desirable to reduce the bandwidth by re-numbering the node numbers. The assembly of the overall matrix is based on the renumbered nodes. Restraints and anchors are treated as additional stiffness added to the diagonal of the corresponding node location. The final overall equation looks like

$$\begin{vmatrix} F_1 \\ F_2 \\ \cdot \\ \cdot \\ \cdot \\ \cdot \\ F_N \end{vmatrix} = \begin{vmatrix} K_{11} & K_{12} & \cdot & \cdot & \cdot & \cdot & K_{1N} \\ K_{21} & K_{22} & K_{23} & \cdot & \cdot & \cdot & K_{2N} \\ \cdot & \cdot & K_{33} & \cdot & \cdot & \cdot & \cdot \\ \cdot & \cdot & \cdot & \cdot & \cdot & \cdot & \cdot \\ \cdot & \cdot & \cdot & \cdot & \cdot & \cdot & \cdot \\ \cdot & \cdot & \cdot & \cdot & \cdot & \cdot & \cdot \\ K_{N1} & \cdot & \cdot & \cdot & \cdot & \cdot & K_{NN} \end{vmatrix} \begin{vmatrix} D_1 \\ D_2 \\ \cdot \\ \cdot \\ \cdot \\ \cdot \\ D_N \end{vmatrix} \qquad (2.25)$$

where N is the total number of degrees of freedom, which is equal to six times the number of node points. This is also often referred to as the number of equations.

For the example system shown in Fig. 2.15(b), there are six node points. Therefore, the force vector and displacement vector each has 36 components. The size of the overall stiffness matrix is 36 × 36. The displacement vector is the unknown quantity, but the force vector is associated with the load and is considered as the known quantity. Therefore, the analysis is reduced to the problem of solving for the displacements at all node points based on the known forces at all node points. This is the equivalent of solving $6n$ simultaneous linear equations defined by Eq. (2.25). The discussion of the solution technique is beyond the scope of this book. Interested readers can read, for instance, the book by Bathe and Wilson [12].

After the nodal displacements are solved for each load case, the forces and moments (in global coordinates) at each element can be found by using Eq. (2.24). These global forces and moments have to be transformed back to the local coordinate before the stresses can be calculated.

REFERENCES

[1] American Society for Testing and Materials (ASTM), "Standard Methods and Definitions for Mechanical Testing of Steel Products," ASTM A-370.
[2] American Society of Mechanical Engineers (ASME), *ASME Code for Pressure Piping, B31, An American National Standard*, ASME, New York.
[3] European Standard EN 13480-3, 2002, *Metallic Industrial Piping — Part 3: Design and calculation*, CEN European Committee for Standardization.
[4] API Spec 5L, *API Specification for Line Pipe*, American Petroleum Institute, Washington, D.C.
[5] Young, W. C., 1989, *Roark's Formulas for Stress & Strain*, 6th ed., McGraw-Hill Book Company, New York.
[6] American Institute of Steel Construction, Inc., *Manual of Steel Construction*, AISC, Chicago.
[7] Timoshenko, S., 1956, *Strength of Materials, Part-II*, 3rd ed., Van Nostrand Co., New York.
[8] M. W. Kellogg Company, 1956, *Design of Piping Systems*, revised 2nd ed., John Wiley & Sons, New York.
[9] Przemieniecki, J. S., 1968, *Theory of Matrix Structural Analysis*, McGraw-Hill Book Company, New York.
[10] Brock, J. E., 1952, "A Matrix Analysis for Flexibility Analysis of Piping Systems," *Journal of Applied Mechanics*, *19*(4), pp. 501–516.
[11] Brock, J. E., 1955, "Matrix Analysis of Piping Flexibility," *Journal of Applied Mechanics*, *22*(3), pp. 361–362.
[12] Bathe, K.-J., and Wilson, E. L., 1976, *Numerical Methods in Finite Element Analysis*, Prentice-Hall, Englewood Cliffs, NJ.

CHAPTER

3

THERMAL EXPANSION AND PIPING FLEXIBILITY

The most basic requirement of piping stress analysis is to ensure adequate flexibility in the piping system for absorbing the thermal expansion of the pipe. Prior to the nuclear power era, almost all discussions and treatments of pipe stress were concentrated in piping flexibility analysis. The works by Spielvogel [1], Kellogg [2], Olson and Cramer [3], and Brock [4] are some of these earlier examples. Even nowadays, despite the fact that stress analysis covers much more than flexibility analysis, engineers still tend to regard pipe stress analysis as just a fancy term for pipe flexibility analysis. In any case, piping flexibility remains as one of the most important tasks in piping engineering and stress analysis.

3.1 THERMAL EXPANSION FORCE AND STRESS

The pipe expands or contracts due to temperature changes in the pipe. In this discussion, the term expansion implies either expansion or contraction. When a pipe expands it causes the entire piping system to make room for its movement. This creates forces and stresses in the pipe and on its connecting equipment. If the piping system does not have enough flexibility to absorb this expansion, the force and stress generated can be large enough to damage the piping and the connecting equipment. An idealized straight pipe as shown in Fig. 3.1(a) will be used to investigate the potential magnitude of the force and stress that can be generated [5].

3.1.1 Ideal Anchor Evaluation

Figure 3.1(a)(1) shows a straight pipe connected to two ideal anchors. An ideal anchor has infinite stiffness such that no anchor movement is generated regardless of the magnitude of the force applied. The anchors implemented in computer programs are mostly ideal anchors. In our example, when the temperature of this two-anchored straight pipe changes, it causes the pipe to expand. However, the anchors at the ends prevent it from expanding. The resistance of the anchor generates force on the anchors from which the same force is reflected back to the pipe. Figure 3.1(a)(2) shows that when one end of the pipe is loose, the pipe has a free expansion equal to $\Delta = \alpha L(T_2 - T_1)$, where α is the thermal expansion rate of the pipe and $(T_2 - T_1)$ is temperature change. There is no force or stress generated in the free expansion state. However, because neither end is loose in this two-anchored case, the force generated on the anchor is equivalent to the force required to push the free expanded end back to its original position. The squeezing movement is equal to the strain times the length. That is,

$$\Delta = \alpha(T_2 - T_1)L = eL = \frac{S}{E}L = \frac{F}{EA}L$$

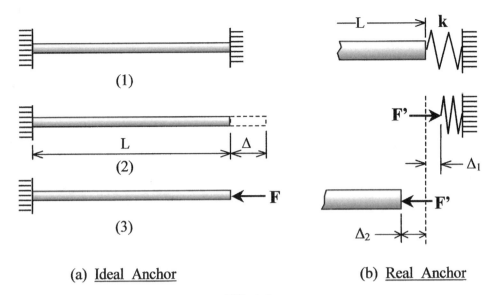

(a) Ideal Anchor (b) Real Anchor

**FIG. 3.1
THERMAL EXPANSION FORCE**

or

$$F = EA\alpha(T_2 - T_1), \quad S = E\alpha(T_2 - T_1) \tag{3.1}$$

where E is the modulus of elasticity of the pipe material, A is the cross-section area of the pipe, F is the anchor force, and S is the axial stress. With ideal anchors, the force and stress generated are independent of the length of the pipe. The force and stress generated are huge even for a small pipe section with only a moderate temperature change. For example, increasing the temperature of a 6-in. standard wall carbon-steel pipe from 70°F ambient condition to 300°F operating condition creates an axial stress of 45,000 pounds per square inch (psi) and an axial force of 250,000 pounds (lbs) in the pipe and on the anchors.

The force and the stress are excessive even though the temperature is only 300°F. This shows that a straight-line direct piping layout is generally not acceptable. The ideal anchor simulation also shows that the magnitudes of the forces and the stress are the same regardless of pipe length. This idealized conclusion sometimes becomes a handicap when designing a closely spaced and tight piping system.

3.1.2 The Real Anchor

Equation (3.1) shows the potential magnitudes of the force and the corresponding stress that can be generated by a temperature change in the pipe. These quantities are independent of the length of the pipe. This is the conclusion that engineers normally obtain from a theoretical treatment on the subject. In the real world, however, the stiffness of the anchor is limited. A stiffness of 10^6 lb/in. is already very difficult to achieve due to the flexibility of the structure, foundation, and attachments themselves. Depending on pipe size, the practical anchor stiffness more likely ranges from 10^4 to 10^7 lb/in.

Figure 3.1(b) shows an interaction of a real anchor with an expanding straight pipe. Here, only one end of the pipe is connected to a real anchor with a stiffness k. The other end is still connected to an ideal anchor of infinite stiffness. Because of its flexibility, the real anchor will absorb part of the pipe expansion, Δ_1, and the pipe itself absorbs the rest of the expansion, Δ_2. The combination of Δ_1 and Δ_2 is the total expansion, Δ. Because both anchor and pipe receive the same force, F', we have

$$\Delta_1 = \frac{F'}{k}, \quad \Delta_2 = \frac{F'L}{EA}$$

$$\Delta = \alpha L(T_2 - T_1) = \Delta_1 + \Delta_2 = \frac{F'}{k} + \frac{F'L}{EA} = F'\left(\frac{1}{k} + \frac{L}{EA}\right)$$

or

$$F' = \frac{\alpha L(T_2 - T_1)}{L\left(\dfrac{1}{kL} + \dfrac{1}{EA}\right)} = \frac{\alpha(T_2 - T_1)EA}{\left(\dfrac{EA}{kL} + 1\right)} \tag{3.2}$$

Equation (3.2) shows that with a real anchor, the force generated is dependent on the length and anchor stiffness. For short pipes, the force is roughly proportional to the length of the pipe. The equation includes an additional term, *EA/kL*. Another equivalent term would have been likewise included if both ends of the pipe were connected to real anchors. This term is important when dealing with short pipes. Take the 6-in. pipe discussed previously, for example; a 5-ft pipe with one end connected to an anchor having a stiffness of 10^5 lb/in. will result in a force of 8800 lb and a stress of 1600 psi. When both ends are connected to real anchors, the force is further reduced to 4500 lb and the stress to 800 psi. These numbers are roughly just 1/30 and 1/60, respectively, of the numbers obtained by ideal anchors.

From the above, it is clear that the stiffness of the anchor has a very great effect on the forces and stresses generated. This is especially true for close layouts at moderate temperature ranges. In an actual analysis, it is important to obtain as accurate as possible the stiffness of the anchors and restraints. However, obtaining accurate anchor and restraint stiffness is a very complicated process involving the structural analysis of the support structure, foundation, and attachment. It is not feasible for a routine piping stress analysis to include the accurate anchor and restraint stiffness. Generally, the ideal anchor and restraint are used in conjunction with engineering judgment.

3.2 METHODS OF PROVIDING FLEXIBILITY

There are two main categories of methods for providing piping flexibility: the flexible joint method and the pipe loop method. Flexible joints, including expansion joints, ball joints, and others, are discussed in Chapter 7. This chapter discusses the method of using pipe loops and offsets to provide flexibility.

From Fig. 3.1 we know that the huge thermal expansion force and stress on an anchored straight section of pipe are the result of squeezing the free expansion axially back to the pipe. This is very difficult, as we can experience by squeezing the ends of a wooden stick. Instead of this direct squeezing, we can absorb the same amount of movement much easier by bending the stick sideways. This is the principle of providing piping flexibility. The flexibility is provided by adding a portion of the piping that runs in the direction perpendicular to the straight line connecting two terminal fixation points. Figure 3.2 shows an expansion loop used in a long straight pipe run. With the loop, the pipe expands into the loop by bending the legs of the loop instead of squeezing the pipe axially. The longer the loop leg, the lesser the force generated in absorbing a given expansion. From the basic beam formula given in Table 2.1, we know that the required force is inversely proportional to the cube of the leg length, and the generated stress is inversely proportional to the square of the leg length. A small increase in loop leg length has a considerable reduction effect on force and stress.

3.2.1 Estimating Leg Length Required

The required leg length can be estimated via the guided cantilever approach. The method is explained by using the L-bend given in Fig. 3.3 as an example.

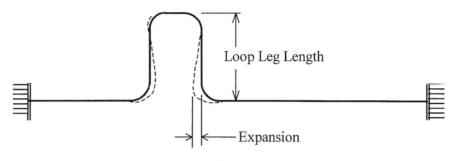

FIG. 3.2
PIPE EXPANSION LOOP

When the piping system is not constrained and is free to expand as in Fig. 3.3(a), points B and C will move to B' and C', respectively, due to thermal expansion. The end point C moves Δ_x and Δ_y amounts in x and y directions, respectively, but no internal force or stress is generated in the absence of a constraint. However, in the actual case, the ends of the piping are always constrained as in Fig. 3.3(b). This is equivalent to moving the free expanded end C' back to the original point C, and forcing point B to move to point B''.

The deformation of each leg can be assumed to follow the guided cantilever shape shown in Fig. 2.14(b). From a flexibility point of view, this is conservative because the end rotation is ignored. The force and stress of each leg can now be estimated by the guided cantilever formula. For this simple L shape, leg AB is a guided cantilever subject to Δ_y displacement, and leg CB is a guided cantilever subject to Δ_x displacement. The stress at each leg is mainly the beam bending stress caused by the expansion displacement. From the cantilever beam formula, we can estimate the stress at each leg as follows

$$S = \frac{M}{Z} = \frac{1}{Z}\frac{6EI}{L^2}\Delta = \frac{1}{\pi r^2 t}\frac{6E\pi r^3 t}{L^2}\Delta = \frac{6Er}{L^2}\Delta = \frac{3ED}{L^2}\Delta \tag{3.3}$$

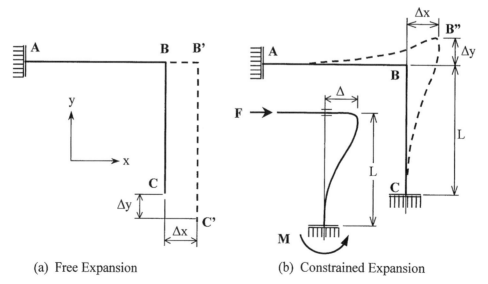

(a) Free Expansion   (b) Constrained Expansion

FIG. 3.3
EXPANSION STRESS BY GUIDED CANTILEVER APPROACH

The approximate formulas $Z = \pi r^2 t$ and $I = \pi r^3 t$ are used, respectively, for the section modulus and moment of inertia of the pipe cross-section. Equation (3.3) is a convenient formula for quick estimation of the expansion stress. By substituting $E = 29.0 \times 10^6$ psi and $S = 20,000$ psi, Eq. (3.3) reduces to Eq. (3.4a) for finding the leg length required for steel pipes

$$L = \sqrt{\frac{3ED}{S}\Delta} = 66\sqrt{D\Delta} \tag{3.4a}$$

Equation (3.4a), although derived from psi units of E and S, is applicable to all consistent units. L, D, and Δ can be expressed in inches or in millimeters.

3.2.2 Inherent Flexibility

Piping in a plant generally runs through a few turns before connecting to a fixation point such as a vessel or rotating equipment. These turns and offsets generally provide enough flexibility to absorb the expansion displacement without causing excessive stress in the pipe. The American Society of Mechanical Engineers (ASME) B31 piping code [6] has provided a criterion as a measure of adequate flexibility, subject to other requirements of the code. The code states that no formal thermal expansion flexibility analysis is required when "The piping system is of uniform size, has not more than two anchors and no intermediate restraints, is designed for essentially non-cyclic service (less than 7000 total cycles), and satisfy the following approximate criterion:"

(1) English units

$$\frac{DY}{(L-U)^2} \leq 0.03 \tag{3.4b}$$

(2) SI units

$$\frac{DY}{(L-U)^2} \leq 208.3 \tag{3.4c}$$

Equations (3.4b) and (3.4c) are in conventional units

where

D = nominal pipe size, in (mm)
Y = resultant of movement to be absorbed by piping system, in (mm)
L = developed length of piping system between two anchors, ft (m)
U = anchor distance (length of straight line joining anchors), ft (m)

Equations (3.4a) and (3.4b) are practically equivalent when consistent units are used. If the surplus length $(L - U)$ is considered as the leg length perpendicular to the line of expansion, Eq. (3.4b) can be converted to Eq. (3.4a) with a constant of 69 instead of 66.

3.2.3 Caution Regarding Quick Check Formulas

During the era when flexibility analysis of a rather simple system could take a couple of weeks of hard work by a specialist engineer to accomplish, quick formulas such as Eq. (3.4a) would mean the difference between whether a plant could be constructed on schedule. These formulas have been extensively taught in various training classes. However, in this age of high-speed computers, these formulas have only very limited use. They might be used by field specialists when surveying a plant for problem installations, or occasionally by design engineers at a remote site. However, at an operating plant or an engineering office, the analysis is better performed by using a quick and accurate computer program. An accurate analysis of a fairly complicated piping system takes only an hour or so to prepare the data and to run with a computer program.

By limiting the pipe stress to about 20,000 psi, Eqs. (3.4a), (3.4b) and (3.4c) are adequate for protecting the piping itself. However, because piping is always connected to certain equipment, this 20,000-psi stress will most likely generate too much load for the equipment to take. Moreover, as will be discussed later in this chapter, some piping components are associated with stress intensifications that are not included in the equations — although a component, such as a bend, that has significant stress intensification may also possess significantly added flexibility. This mutual compensation of stress intensification and the added flexibility validates the usefulness of the equations. However, it is worth noting that not all stress intensifications come with added flexibility.

3.2.4 Wall Thickness and Thermal Expansion Stress

Because expansion stress is calculated by dividing the moment, M, with the section modulus, Z, engineers might be wrongly tempted to increase the wall thickness to reduce the expansion stress. An increase in wall thickness increases the section modulus, but also proportionally increases the moment of inertia. The section modulus is defined as $Z = I/r_o$, which is directly proportional to the moment of inertia. Therefore, the first consequence of increasing the wall thickness is an increase in bending moment under a given thermal expansion. This increased moment divided by the proportionally increased section modulus ends up with the same stress as before, prior to the increase of the thickness. The thicker wall thickness does not reduce the thermal expansion stress. It only unfavorably increases the forces and moments in the pipe and at the connecting equipment. Therefore, as far as thermal expansion is concerned, the thinner the wall thickness the better it will be for the system.

3.3 SELF-LIMITING STRESS

The stress generated by thermal expansion is self-limiting in nature. This self-limiting stress behaves quite differently from the sustained stress caused by weight and pressure. Figure 3.4 shows the differences between these two types of stresses.

Figure 3.4(a) is a pipe subject to weight load, thus generating sustained stress in the pipe. By increasing the weight gradually, the stress and the accompanying displacement also increase accordingly. When the stress reaches the yield point of the material, the stress maintains the same magnitude, but the displacement increases abruptly by a large amount. (This is an oversimplification, because the bending stress does not exactly behave this way.) In sustained loading, the stress generated has to be in static balance with the applied load — that is, stress is always proportional to the load, but displacement depends on the characteristics of the material. In this case, if the stress, S_w, exceeds the yield strength slightly, a large strain, e_w, will be created. This strain is so large that it results in a gross deformation in the system. Therefore, sustained stress is generally limited to values lower than the yield strength of the material.

Figure 3.4(b) shows the case for self-limiting stress. When the pipe is subject to thermal expansion or other displacement load, the mechanism of balance shifts to the strain — that is, the strain always corresponds with the amount of expansion or displacement. Once the displacement reaches its potential magnitude, the whole thing stops because the displacement has only that much. The generated stress depends on the characteristics of the material. For instance, when the strain corresponding to the displacement exceeds the yield strain, it stays there without any further abrupt movement. This position fixing nature is called self-limiting. The stress due to thermal expansion is self-limiting stress.

Self-limiting stress has the following characteristics: (1) Generally, it does not break ductile pipe in one application of the load. (2) Its mode of failure is fatigue requiring many cycles of applications. (3) Fatigue failure depends on stress range, measured from the lowest stress to the highest stress in the

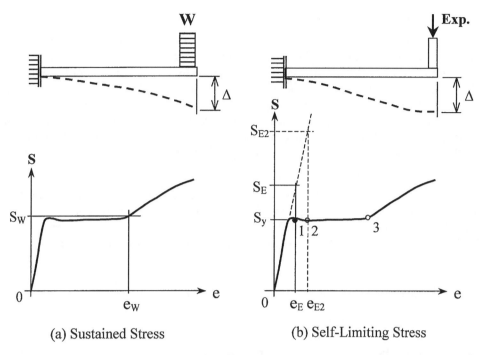

**FIG. 3.4
SUSTAINED VS. SELF-LIMITING STRESSES**

operating cycles. (4) Stress can reach yield strength without causing failure or gross system deformation. (5) When dealing with stresses at or beyond yield strength, the failure is evaluated by the strain range rather than the stress range.

3.3.1 Elastic Equivalent Stress

The actual stress of a self-limiting stress rarely exceeds the yield strength. As shown in Fig. 3.4(b), even if the self-limiting strain exceeds the yield strain, the actual stress is still equal to the yield strength. This nature makes it difficult to evaluate the effect of self-limiting stress using the actual stress. For instance, points 1, 2, and 3 all correspond to the same stress, yet we know point 3 is much more critical than the others. Therefore, it is natural that self-limiting stress should be evaluated by the strain rather than by the stress. However, because the analyses are traditionally set up to calculate stress and they are also done by elastic methods, the determination of the actual strain is very complicated. To overcome this shortcoming, the elastic equivalent stress is used to measure the strain. The elastic equivalent stress is simply the product of the strain and modulus of elasticity. This is equivalent to replacing the actual stress-strain curve with a straight line extending from the initial straight portion of the curve — that is, the elastic equivalent stress for strain e_E is S_E, and for e_{E2} it is S_{E2}, etc. This elastic equivalent stress can be several times higher than the yield stress, yet the actual stress is still the same as the yield stress.

3.4 STRESS INTENSIFICATION AND FLEXIBILITY FACTORS

A piping system consists of many different components such as bends, elbows, reducers, tees, valves, and flanges. However, in the analysis we normally idealize these various components into two types of elements: the straight pipe beam element and the curved pipe beam element.

3.4.1 Ovalization of Curved Pipes

A piping system depends mainly on its bending flexure to absorb thermal expansion and other displacement loads. When a straight pipe is subject to bending, it behaves like any straight beam: its cross-section remains circular and the maximum stress occurs at the extreme outer fiber. However, under a bending moment, a curved pipe element behaves differently from that of a solid curved beam. When subject to a bending moment, the circular cross-section of the bend becomes oval. This is the famous ovalization we are all aware of. Figure 3.5 shows the ovalization associated with in-plane bending. An out-of-plane bending, on the other hand, produces an oblique ovalization inclining at an angle with the major axes. The ovalization tendency of the curved pipe has resulted in the following peculiar phenomena:

(1) Increase of flexibility. Ovalization is caused by the relaxation of the extreme outer fiber of the bend. Without the proper participation of the extreme outer fiber, the effective moment of inertia of the cross-section is reduced. This reduction in effective moment of inertia increases the flexibility of the bend over the non-ovalized theoretical bend by a factor [6] of

$$k = 1.65/h \qquad (3.5)$$

where k is the flexibility factor and $h = tR/r_m^2$ is the bend flexibility characteristic.

(2) Increase of longitudinal bending stress. The relaxation of the extreme outer fiber has shifted the maximum longitudinal stress due to bending to a location away from the extreme outer fiber location. This reduction in the moment resisting arm of the high stress portion is equivalent to reducing the effective section modulus of the cross-section. The maximum longitudinal stress is, therefore, greater than the maximum stress obtained by the elementary bending theory. The ratio of the two stresses is the stress intensification factor (SIF). That is,

$$i = \frac{S}{M/Z} \qquad (3.6)$$

The theoretical longitudinal SIFs are related to the bend flexibility characteristic as [2]

$$i_{\text{Li}} = \frac{0.84}{h^{2/3}} \quad \text{for in-plane bending} \qquad (3.7)$$

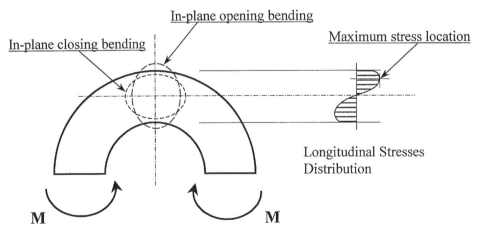

**FIG. 3.5
OVALIZATION OF BEND UNDER EXTERNAL BENDING**

$$i_{Lo} = \frac{1.08}{h^{2/3}} \quad \text{for out-of-plane bending} \tag{3.8}$$

(3) Creation of circumferential shell bending stress. Squeezing the circular cross-section into an oval shape generates bending on the pipe wall. This, in turn, creates a high circumferential bending stress on the pipe wall. Because this shell bending stress is non-existent on a circular cross-section, there is no direct comparison with the non-ovalized bend. For the sake of convenience, the stress is compared with the flexure bending stress of a circular cross-section as shown in Eq. (3.6).

The theoretical SIFs for the circumferential stresses are [2]

$$i_{Ci} = \frac{1.80}{h^{2/3}} \quad \text{for in-plane bending} \tag{3.9}$$

$$i_{Co} = \frac{1.50}{h^{2/3}} \quad \text{for out-of-plane bending} \tag{3.10}$$

The flexibility factor given in Eq. (3.5) is used by ASME codes for both in-plane and out-of-plane bending. The theoretical SIFs given by Eqs. (3.7), (3.8), (3.9), and (3.10) are used only in Class 1 nuclear piping [7]. For other ASME piping codes, only one-half of the theoretical values are used instead.

3.4.2 Code SIFs

Earlier piping stress analyses were mainly concerned with the flexibility of piping subject to thermal expansion. As previously discussed, the failure mode of self-limiting expansion stress is fatigue due to repeated operations. Therefore, to validate these SIFs, the most direct and logical approach is the fatigue test. After many tests and researches, Markl and others [8–10] have found that theoretical SIFs are consistent with the test data. However, tests performed on commercial pipe also revealed an SIF of almost 2.0 against a polished homogeneous tube with regard to fatigue failure. This factor is mainly due to the unpolished weld effect, or clamping effect at fixing points, combined with the less than perfectly homogeneous commercial pipe. To simplify the analysis procedure, the applicable SIFs are taken based on the commercial girth welded pipe as unity. This, in effect, reduces the applicable SIFs to just one-half of the theoretical SIFs given by Eqs. (3.9) an (3.10). That is, for bends we have:

- In-plane bending SIF

$$i_i = \frac{0.90}{h^{2/3}} \tag{3.11}$$

- Out-plane bending SIF

$$i_o = \frac{0.75}{h^{2/3}} \tag{3.12}$$

The preceding equations are for bends. For other components, Markl [10] has succeeded in using equivalent bends as shown in Fig. 3.6 [12] to arrive at SIFs that are comparable with the test results. Using the equivalent elbows and making adjustments for actual crotch radius and thickness, a set of SIFs [11] for various components was constructed with a single flexibility characteristic parameter, h. The SIF for a welding tee, for example, can also be expressed by Eq. (3.11) by setting the flexibility characteristic, $h = 4.4t/r$ (recently revised to $h = 3.1t/r$), from the equivalent elbow characteristics. However, in contrast to smooth bends, the SIF for out-of-plane bending is generally greater than that for in-plane bending in miter bends, welding tees, and other branch connections.

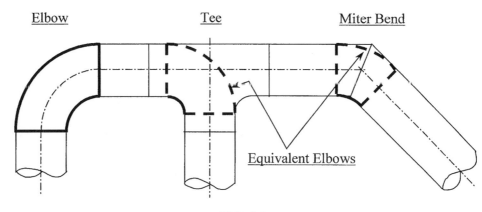

**FIG. 3.6
EQUIVALENT ELBOWS**

Although they remain mostly unchanged, these SIFs have nevertheless been continuously revised through the years. The values given in the current edition of the applicable code should always be used.

Stress intensification factors. SIFs are only one-half of the theoretical factors, and are intended for use only on self-limiting stresses. By using the commercial pipe with an unpolished girth weld as the basis, the code SIFs as given by Eqs. (3.11) and (3.12) are only one-half of the theoretical SIFs. The adoption of this basis is mainly attributed to practicality. If the theoretical SIF were used, then an analysis would have to identify all girth weld locations for applying the SIF. This is not very practical when large amounts of piping components are involved. Currently, only Class 1 nuclear piping uses the theoretical SIF.

When an SIF is involved, the stress calculated using the ASME B31 code formula is only one-half of the theoretical stress. This does not cause problems if everything is done within the range specified by the code, because the allowable stress has also been adjusted accordingly. However, there are occasions when something outside the code has to be referenced. For instance, when dealing with steady-state vibrations, pipe stress has to be evaluated with a fatigue curve that is generally constructed with theoretical stresses. In this case, the stress calculated by the B31 code has to be doubled before being applied to the fatigue curve.

By comparing Eqs. (3.11) and (3.12) with Eqs. (3.9) and (3.10), it is clear that the code SIF is the measure of the circumferential stress. Failure locations on specimens used in fatigue tests also showed that the SIF is due to the circumferential stress. Because circumferential stress is a shell bending stress that does not provide any static equilibrium to the load applied, it has little significance in the sustained load. Therefore, the code SIFs derived from fatigue tests and theoretical circumferential stresses are only applicable to self-limiting loads that produce fatigue in the pipe.

Sustained loads. For sustained loads, a separate set of SIFs is required. This separate set of SIFs for the sustained load has been used in Class 2 and Class 3 nuclear piping [7]. However, for non-nuclear piping systems, a separate set of SIFs is not provided for sustained loads. To this end, there are several practices used in various industries to deal with this matter. One of these practices also uses the same code SIF for sustained loads. This is a conservative approach used by non-discriminating engineers. Another practice completely ignores the SIF for sustained loads. This is suggested mainly by engineers who have been involved in the earlier development of SIFs. The rationale is that code SIFs are for fatigue only. However, it is recognized that some type of SIF is needed for sustained loading. One approach is to use the same set of SIFs intended for self-limiting loads (code SIFs) applied with a constant modification factor, which is somewhat less than 1.0. This approach, although not accurate, is in the proper practical range.

The SIF for sustained load is more closely related to the load-resisting longitudinal stresses given by Eqs. (3.7) and (3.8). These longitudinal stresses have the same mathematical format as the circumferential stresses given in Eqs. (3.9) and (3.10), except that their relative strengths are switched. On circumferential stresses, in-plane bending produces greater stress, whereas with longitudinal stresses, out-of-plane bending produces a larger stress. Therefore, theoretically, it is not possible to use a constant modification factor to relate these two sets of stresses. However, if only the greater stress intensifications on each set are used, then it is possible to conservatively use a constant modification factor. This is the approach used by ASME B31.1 [6], which uses a modification factor of 0.75. For other ASME B31 codes that use different in-plane and out-of-plane SIFs, the constant modification factor approach may not be suitable.

Tables for flexibility factors and SIFs. Each code has a table or tables that list and explain the flexibility factors and SIFs for most of the common components used in a piping system. Although flexibility factors, used to calculate the piping forces and moments, are all similar, SIFs differ slightly between the codes. This is attributed to the unique characteristics of the jurisdictional industry served by each code.

To show the function of these tables, we use ASME B31.3 tables as shown in Tables 3.1 and 3.2 as an example. The B31.3 table is somewhat more complicated than the others. It shows in-plane and out-plane categories of SIFs for each type of component. The purpose is to apply different stress intensification on different orientation of moment. These tables are used as follows. For each component, we first have to calculate the flexibility characteristic, h. From this flexibility characteristic, the flexibility factor and SIFs are calculated. It is clear from the table that for all the different components only the flexibility characteristics are calculated differently. The flexibility factor and SIFs are calculated more or less the same way for all components. This is the nice thing about the equivalent bend approach mentioned previously. The flexibility factor is included in structural analysis to obtain piping forces and moments. SIFs are then applied to piping moments to calculate pipe stresses.

ASME B31.1, on the other hand, provides only one stress intensification for each component. The value used is equivalent to the greater of the in-plane and out-plane values. Because of this one stress intensification approach, it is not necessary to distinguish in-plane moment or out-plane moment. The moments are considered the same regardless of their orientation. In fact, B31.1 stress intensification is also applicable to torsion moment, which is not applied with any stress intensification by B31.3 and some other codes. More on stress calculation is given in the next chapter, which deals with code stress requirements.

The stress intensification and flexibility factors at elbows and bends are sensitive to flanged ends and internal pressure. The effect of flanged end is discussed below and the effect of internal pressure will be discussed later in Section 3.7.

Effect of flanges on bend flexibility and SIFs. The flexibility factor and SIF at bends are mainly due to ovalization of the cross-section. Therefore, it is natural to expect these factors to be reduced by the stiffening effect of the flange connections. ASME code stipulates that when one or both ends of a bend are attached with flanges, the bend flexibility factor and SIF shall be multiplied by the factor C, where $C = h^{1/6}$ is for one end flanged and $C = h^{1/3}$ is for both ends flanged. For simplicity, it does not matter which end is flanged when only one end is flanged. The reduction factor is uniformly applied to the whole bend regardless of which end is flanged.

3.5 ALLOWABLE THERMAL EXPANSION STRESS RANGE

As discussed in the previous section, the failure mode of thermal expansion is fatigue failure requiring many cycles of repeated operations. In evaluating fatigue failure, the stress range encompassing the minimum and the maximum stresses in each cycle has to be considered. This is the reason why

TABLE 3.1
B31.3 FLEXIBILITY FACTORS AND STRESS INTENSIFICATION FACTORS [6]

Description	Flexibility factor, k	Stress intensification factor Out-plane, i_o	Stress intensification factor In-plane, i_i	Flexibility characteristic, h	Sketch
Welding elbow or pipe bend	$\dfrac{1.65}{h}$	$\dfrac{0.75}{h^{2/3}}$	$\dfrac{0.90}{h^{2/3}}$	$\dfrac{t_n R}{r^2}$	
Closely spaced miter bend $s < r(1 + \tan\theta)$	$\dfrac{1.52}{h^{5/6}}$	$\dfrac{0.90}{h^{2/3}}$	$\dfrac{0.90}{h^{2/3}}$	$\dfrac{\cot\theta}{2} * \dfrac{t_n s}{r^2}$	
Single miter bend or widely spaced miter bend $s \geq r(1 + \tan\theta)$	$\dfrac{1.52}{h^{5/6}}$	$\dfrac{0.90}{h^{2/3}}$	$\dfrac{0.90}{h^{2/3}}$	$\dfrac{1 + \cot\theta}{2} * \dfrac{t_n}{r}$	
Welding tee per ASME B16.9	1	$\dfrac{0.90}{h^{2/3}}$	$3/4\, i_o + 1/4$	$3.1 \dfrac{t_n}{r}$	
Reinforced fabricated tee with pad or saddle	1	$\dfrac{0.90}{h^{2/3}}$	$3/4\, i_o + 1/4$	$\dfrac{\left(t_n + \dfrac{t_r}{2}\right)^{5/2}}{r\, t_n^{3/2}}$	
Un-reinforced fabricated tee	1	$\dfrac{0.90}{h^{2/3}}$	$3/4\, i_o + 1/4$	$\dfrac{t_n}{r}$	
Extruded welding tee	1	$\dfrac{0.90}{h^{2/3}}$	$3/4\, i_o + 1/4$	$\left(1 + \dfrac{r_x}{r}\right)\dfrac{t_n}{r}$	
Welded-in contour insert	1	$\dfrac{0.90}{h^{2/3}}$	$3/4\, i_o + 1/4$	$3.1 \dfrac{t_n}{r}$	
Branch welded-on fitting (integrally reinforced)	1	$\dfrac{0.90}{h^{2/3}}$	$\dfrac{0.90}{h^{2/3}}$	$3.3 \dfrac{t_n}{r}$	

stress range — rather than static stress — is considered in evaluating thermal expansion and other displacement stresses. As for the allowable stress, suggested values and their rationales have been thoroughly discussed by Rossheim and Markl [11, 13]. Their suggested values have been used by the codes [6] since 1955 with only minor intermittent modifications.

As discussed in Section 3.3, self-limiting stress dose not cause the abrupt gross structural deformation when it reaches the yield strength of the pipe. Therefore, if allowed by fatigue, the stress range

TABLE 3.2
FLEXIBILITY FACTOR AND STRESS INTENSIFICATION FACTOR FOR JOINTS [6]

Description	Flexibility factor, k	Stress intensification factor, i
Butt welded joint, or weld neck flange	1	1.0
Double welded slip-on flange	1	1.2
Fillet welded joint, or socket weld fitting	1	1.3–2.1
Lap joint flange (with B16.9 lap joint stub)	1	1.6
Threaded pipe joint, or threaded flange	1	2.3
Corrugated straight pipe, or corrugated or creased bend	5	2.5

can exceed the yield strength. The starting point of investigation is an elastic equivalent stress range at twice of the yield strength. (The sum of the hot yield strength plus the cold yield strength, instead of twice of the yield strength, would have been a more accurate statement here.) This is also equivalent to a strain range, e_E, of twice the yield strain. The background for setting the allowable stress range can be explained by the help of three possible operating conditions (Fig. 3.7), described as follows:

(1) *No cold spring and no stress relaxation.* As shown in Fig. 3.7(a), this piping system heats up from the zero stress, zero strain point, 0, expanding gradually to reach the yield point at hot condition, a, and continues on to the final point, b. This final point corresponds to e_E with an elastic equivalent stress of the sum of the cold and hot yield strengths. At moderate operating temperature with no, or very little, stress relaxation expected, the piping stays at point b for the whole operating period.

When the system cools down, the pipe contracts, elastically reducing the stress from point b to point c where stress is zero, but the cooling process is only halfway complete. As the cool-down continues, the sign of the stress reverses. An initially tensile stress becomes compressive from this point on. The operation cycle ends at point d when the temperature returns to the pre-operation temperature. The total contraction strain is the same as the total expansion strain. The final pipe stress is equal to the cold yield strength, assuming the compressive stress-strain curve is the same as the tensile one. The next operating cycle starts from point d, and goes elastically to point b without producing any yielding. The subsequent operating cycles all follow the elastic line from d-b and back with b-d. Except for the initial yielding created by the first operating cycle, no yielding is produced in any subsequent operating cycle. This could signify that an expansion stress range below the sum of hot yield strength and cold yield strength probably would not produce a fatigue failure for an ideal material.

When the pipe cools down to ambient condition at point d, the system retains a reversed stress equivalent to the cold yield stress. This stressing at cold condition is equivalent to the effect of a cold sprung system. This situation is called *self-spring*. Self-spring causes flanges to suddenly spring apart as the last bolts are removed, on a line being disconnected after a period of service. A gap is created between the flanges when the line is disconnected. The gap is equivalent to the cold spring gap that will be discussed in Section 3.6, although the line was not initially cold sprung.

(2) *Fifty percent cold spring with no stress relaxation.* Because it is generally desired that no yielding in the piping is produced during the cold spring process, a 50% or less cold spring is done for an expected expansion stress range that is twice the yield strength. Figure 3.7(a) shows the case with 50% cold spring. The piping is first cut short to form a gap that is equal to 50% of the

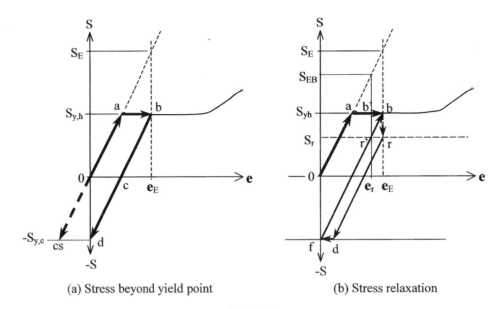

FIG. 3.7
STRESS BEYOND YIELD POINT, COLD SPRING, AND STRESS RELAXATION

expected expansion, and then the ends of the gap are pulled together to pre-stress the system. With the expansion stress range of twice the yield strength as the benchmark, a 50% cold spring will produce a reverse stress equal to the cold yield strength at point cs. The system starts operating from this point, cs, elastically in a straight line all the way to the hot yield point, *a*, without creating any yielding on the pipe. The stress at this final point is one-half of the expansion stress range. The system cools down elastically from point *a*, again back to original point cs. The same cycle repeats through the life of operation without creating any yielding on the pipe.

By comparing line cs-*a* and line *d-b*, a cold sprung system operates essentially the same as a non-cold-sprung system, except without the initial yielding. Because the initial yielding of the non-cold-sprung system has very little effect on fatigue, cold spring has no effect on fatigue. Therefore, no credit on the expansion stress reduction is allowed for cold spring. The expansion stress range is the one that is counted.

(3) *With stress relaxation.* In high temperature environments, the stress relaxes with time when yielding or creep occurs in the pipe. Figure 3.7(b) shows a non-cold-sprung system with a stress relaxation at operating condition. As given in case (1), the operation reaches point *b*, at operating temperature. Owing to temperature effect, the stress gradually relaxes to point *r* while operating. When the pipe cools down, both stress and strain reduce elastically from point *r* to point *d* to reach the cold yield strength before the cool-down is completed. As the cool-down continues, it produces plastic strain *d-f* to reach the ambient temperature. In the subsequent operating cycle, the process follows *f-b-r-d-f*, again producing a *d-f* plastic strain. Because our goal is to avoid producing plastic strain in every cycle, the cold yield strength plus hot yield strength benchmark stress range is not suitable for situations with stress relaxation.

To prevent the generation of plastic strain in every operating cycle under stress relaxation condition, the benchmark stress range has to be reduced to cold yield strength plus the stabilized relaxation residual stress, S_r. This benchmark stress range is shown as S_{EB} in Figure 3.7(b). With this modified benchmark stress range, the pipe will operate along *f-r′* pass indefinitely. However, the problem is that the stabilized relaxation residual stress is not readily available for most materials.

From the preceding deductions, the benchmark allowable expansion stress range can be set as the sum of the cold yield strength and hot yield strength at below creep range. At creep range, it will be the sum of the cold yield strength and the creep strength at expected plant life. Markl [11] suggested that it appears to be conservative to make it the sum of the cold yield strength and 160% of the stress producing 0.01% creep in 1000 hours at the operating temperature. The benchmark stress range can be written as

$$S_{EB} = (S_{yc} + S_{yhx}) \qquad (3.13a)$$

where

S_{yc} = yield strength at cold condition
S_{yhx} = is the lesser of the yield strength at hot condition and 160% of the stress producing 0.01% creep in 1000 hours at the operating temperature

The stress producing 0.01% creep in 1000 hours is one of the criteria of setting the hot allowable stress. In terms of ASME B31.1 allowable stresses, which was originally set at no greater than 5/8 of the yield strength at corresponding temperature, the above equation can be written as

$$S_{EB} = 1.6(S_c + S_h) \qquad (3.13b)$$

where S_c is cold allowable stress and S_h is hot allowable stress. The benchmark stress range, S_{EB}, is considered the maximum stress range to which a system could be subjected without producing plastic flow at either cold or hot limit. In ASME B31.3 code and the recent B31.1 code, because of their use of higher allowable stress based on 2/3 of the yield strength, the benchmark stress range has become

$$S_{EB} = 1.5(S_c + S_h) \qquad (3.13c)$$

The above derivation of the non-yielding benchmark stress range requires further justification before being used. The first thing that needs to be clarified is the fact that the code stress is only one-half of the theoretical stress when an SIF is involved, as discussed in Section 3.4. Therefore, theoretically the constants in Eqs. (3.13b) and (3.13c) have to be halved to avoid any yielding. However, because only the local peak stress and shell bending stress are halved, the difference between the code stress and the theoretical stress affects only the final fatigue evaluation. Equations (3.13b) and (3.13c) ensures that the main load resisting membrane stress would not exceed the yield. This is important in avoiding gross structural deformation, and in ensuring the validity of elastic analysis commonly used.

The other thing that needs to be worked out is the actual application of the benchmark stress range. This mainly involves the setting of the safety factor. If the ASME B31 code had used the theoretical SIF, the benchmark stress range would have represented the stress range limit for unlimited operating cycles. In actual applications involving limited number of operating cycles, Markl suggested a total allowable stress range of

$$S_A + S_{PW} = 1.25(S_c + S_h) \qquad (3.14)$$

where S_A is the basic allowable stress range for calculated thermal expansion stress and S_{PW} is the sustained stress due to pressure and weight. By comparing with Eqs. (3.13b) and (3.13c), this represents 78% and 83% of the benchmark stress range. This may not be very significant as we are comparing the calculated stress with a theoretical stress. However, the important thing is whether this allowable stress provides enough safety factor for systems operating at limited number of cycles. Because the longitudinal stress due to pressure and weight is generally allowed to reach hot allowable stress, that is, $S_{PW} = S_h$, the allowable stress range for thermal expansion only, S_A, becomes

$$S_A = f(1.25S_c + 0.25S_h) \tag{3.15}$$

where f is a stress-range reduction factor varying from $f = 1.0$ for $N < 7000$ cycles, to $f = 0.5$ for $N > 250,000$ cycles. However, the newer fatigue data has shown that the f value may be considerably smaller than 0.5 when $N > 250,000$ cycles. The exact value used by the code will be discussed in another chapter dealing with code design requirements. See also Chapter 13 on vibration analysis for high cycle fatigue. The stress-range reduction factor, f, has been recently renamed stress range factor because it reduces the allowable stress, and not the stress range.

Based on the experimental data, the allowable expansion stress range given in Eq. (3.15) represents an average safety factor of 2 in terms of stress, and a safety factor of 30 in terms of cyclic life. However, because of the spread in individual test data, the potential minimum safety factor can be as low as 1.25 in terms of stress and 3 in terms of life. This emphasizes the need for making a conservative estimate of the stress and the number of operating cycles. It should be noted, however, that these safety factors were based on the original (1955) allowable stress, which was taken as $5/8$ of the yield strength instead of the current $2/3$ of the yield strength.

The stress-range reduction factor cuts off at 7000 cycles with no increase in allowable stress-range permitted when the operating cycles are less than 7000. With $f = 1.0$ at 7000 cycles, the allowable stress has already reached the benchmark stress limit. Any stress beyond that might produce gross yielding in the system, thus invalidating the elastic analysis. Therefore, the actual safety factor increases as the number of cycles reduces. The 7000 operating cycles represents 1 cycle/day for 20 years. This number is more than most piping systems experience nowadays when batch-operating processes requiring daily turnaround are rare.

Some engineers may be puzzled that the allowable stress range as given in Eq. (3.15) can reach as high as the ultimate strength of the pipe material. We want to re-emphasize here that the stress calculated and referred to is the elastic equivalent stress as discussed in Section 3.3. The actual stress is always less than the yield strength of the material. It is also be emphasized that corrosion in the pipe can substantially reduce fatigue life at an unpredictable rate. Therefore, proper control of corrosion with some adjustment of the allowable stress may be required. Furthermore, when a piping system is connected to delicate equipment, such as a pump, turbine, or compressor, the allowable stress is governed by the allowable reaction forces and moments of the equipment. The allowable equipment reaction corresponds to the static stress, which may be reduced by cold spring. The allowable stress for the piping connected to a piece of delicate equipment can be as small as only a fraction of the allowable stress range given in Eq. (3.15).

3.6 COLD SPRING

Cold spring, pre-spring, and cold pull all refer to the process that pre-stresses the piping at installation or cold condition in order to reduce the force and stress under the operating or hot condition. Cold spring is often applied to a piping system to: (1) reduce the hot stress to mitigate the creep damage [14]; (2) reduce the hot reaction load on connecting equipment; and (3) control the movement space. However, at the creep range the stress will be eventually relaxed to the relaxation limit even if the pipe is not cold sprung. The general belief is that the additional creep damage caused by the initial thermal expansion stress is insignificant [11] if the total expansion stress range is controlled to within the code allowable limit. The real advantage of cold spring has become the reduction of the hot reaction on the connecting equipment. For instance, a 50% cold spring would split the reaction force into halves. One-half of it is realized at cold condition as soon as the pipe is cold sprung. The more critical hot reaction is reduced to only one-half of the total reaction that otherwise would have been imposed by a non-cold-sprung system.

Although cold spring offers unquestionable benefits, its adoption varies among industries. In power plant piping, cold spring is done normally for most of the major piping systems such as main steam,

Thermal Expansion and Piping Flexibility   77

hot reheat, and cold reheat lines. On the other hand, cold spring is seldom used in petrochemical process piping. The petrochemical industry shies away from cold spring mainly due to the belief that the intended benefits of cold spring may not be realized due to inadequate specifications and poor supervision of field erection procedures. This has become the culture of the two industries partly due to the much larger amount of piping involved in a petrochemical plant than in a power plant. Without proper specifications and installation procedures, no cold spring should be attempted. In some cases, even the self-sprung gap is closed by cutting-and-pasting the piping rather than cold springing it back for connection. Nevertheless, even in the petrochemical industry, a one-dimensional cut-short on yard piping to control the corner movement is occasionally performed. This cut-short on yard piping requires no special procedure due to the small force involved in a very flexible system.

The cold spring process involves laying out the piping somewhat shorter than the installing space to create a gap at the final weld or final bolting location, as the system is erected without straining. The system is then pulled or pushed according to a predetermined procedure to close the gap and to finish the final joint.

3.6.1 Cold Spring Gap

The size of the cold spring gap depends on the cold spring factor desired. The size of the gap for a 100% cold sprung system is the total amount of system expansion modified with the differential anchor movement. For example, take the simple system shown in Fig. 3.8, where a three-dimensional gap is generally required to achieve a uniform cold spring. Gap size can be calculated form the overall dimensions, L_x, L_y, and L_z, plus the differential effect of the anchor displacements. The procedure can be quite confusing in a general thee-dimensional system. Therefore, it is recommended to use a systematic procedure that does not require visual interpretation. Before calculating the gap, we need to define the gap orientation first. This gap orientation is required for entering the cold spring gap data into a computer program. Generally, the gap is considered as running in the same direction as the sequence of node points. In this sample piping system, we are describing the system from point A to point B; therefore, the gap is running from point G_A to point G_B. A minus gap means running from G_A to G_B in the negative coordinate direction. For the x direction, the gap is calculated by

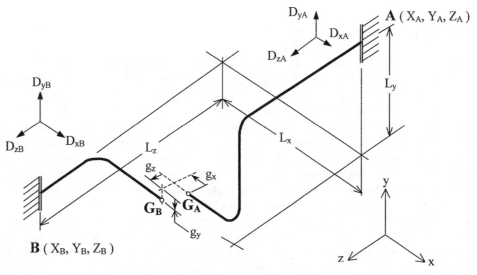

FIG. 3.8
COLD SPRING GAP

$$g_x = C\{\alpha(T_2 - T_1)(X_B - X_A) - (D_{xB} - D_{xA})\}$$

where C is the cold spring factor ranging from 0.0 for no cold spring to 1.0 for 100% cold spring; X_A, X_B are x coordinates for anchor A and anchor B, respectively; and D_{xA} and D_{xB} are x direction anchor displacements for anchor A and anchor B, respectively. The gaps in y and z directions are calculated with similar procedures.

The case assumes that the piping is operating at a temperature higher than the construction temperature. A 100% cold sprung system, if installed properly, will have the expansion stress reduced to zero when the system reaches the operating temperature. It will be free of any thermal expansion stress under the hot operating condition. This assumes that the stress range is less than the yield strength. For cases where stress range is greater than yield strength, the reduction of reaction and stress at hot condition is not that straightforward.

3.6.2 Location of Cold Spring Gap

The gap can be located anywhere in the system with the same gap size. However, it is important to locate the gap at the place where it is easy to work and achieve the intended result. Unless other factors have demonstrated otherwise, the gap is normally placed at the location near one of the terminal anchors. This allows the entire line, except a small part at the anchor side, of flexibility to be available for closing the gap. No pulling is generally required on the short side of the line. The free end can be simply clamped down into position. Locating the gap close to the terminal end also ensures a minimum rotation of the pipe. Besides the cited benefits of locating the gap at points near one of the terminal ends, the accessibility and the amount of force required are also consideration factors. From the standpoint of the forces and moments required, the most desirable location is the one that produces the minimum internal pipe forces and moments, especially moments under operating temperature without cold spring. Because the cold spring process essentially has to create the reverse mirror image of the operating condition, a gap at a high internal force and moment location requires a large pulling effort to create the proper cold spring. The most difficult location is the one that requires a large moment, especially the torsion moment twisting around the pipe axis. An analysis of the operating condition needs to be performed so the proper location can be selected. Another factor that needs to be considered is whether there is enough working space for pulling and welding.

3.6.3 Cold Spring Procedure

The success of cold spring depends on the specification and execution of the procedure. Preparation of cold spring starts with the assignment of the cold spring gap. Next, a set of drawings is prepared showing the fabrication dimensions versus the routing dimensions of the piping. The hanger and support attachment locations are laid out in reference to the final cold sprung locations. During the erection of the cut-short piping, a spacer equivalent to the cold spring gap is generally clamped to the ends of the gap to ensure the proper gap size. A marking line in the axial direction is drawn across the gap on both ends of the pipe. This line is used to ensure the proper alignment against the twisting between the ends. Based on the magnitude of the forces and moments expected at the gap location, some strategic locations are selected for applying the pulling forces. In general, more than one pulling locations are needed for each side of the piping.

Before starting the cold spring, an analysis with the gap applied and piping at ambient temperature has to be performed. This analysis includes the piping on both sides of the gap. Displacements from the analysis are the movements expected from the cold spring. Displacements near each end of the gap show the amount of cold spring required from each side of the piping. The expected cold spring movements are then noted on the drawing. These movements are subsequently marked in the field as guides for the pulling. An analysis for each side of the piping, with the cut-off at the gap, can also be performed to estimate the pulling forces required. This analysis is accomplished by applying a unit force on each pulling location in the selected direction to see the amount of the gap movement due to each force. A proper combination of forces can then be determined by matching the total movements against the cold spring required on each side of the piping. As a rule of thumb, a three-dimensional cold spring gap generally requires three pulling forces located at strategic locations. The directions of these pulling forces are more or less perpendicular to each other. Additional effort may be required to prevent twisting between two ends. During the pulling, in addition to the permanent hangers and locked spring hangers, some temporary supports may also be required. As the pipe moves up and down, the hanger rods need to be adjusted accordingly to match the expected cold spring movements. The final welding is performed with both sides of the piping clamped together at the gap.

All spring hangers and rod hangers are adjusted after the cold spring is completed. They are adjusted in the same way as would be done had the piping not been cold sprung. The spring hangers are locked at their regular cold positions until the piping is hydro-tested and ready to operate.

3.6.4 Multi-Branched System

Cold spring for a multi-branched system is considerably more complicated than that for a two-anchor system. In general, more than one gap is required to achieve a uniform cold spring [15] of a multi-branched system. This would require a close integration of design, analysis, fabrication, and erection activities. The procedure can overwhelm even a well-organized company. Therefore, one alternative is to cold spring just the main portion of the piping instead of uniformly cold springing the entire system. From the basic beam formula, we know that the force generated by a displacement is inversely proportional to the cube of the beam length that is perpendicular to the direction of the displacement. The force generated by the expansion of L_x is inversely proportional to roughly the cube of the resultant length of L_y and L_z. Similarly, the forces generated by the expansion of L_y and L_z are inversely proportional to the cube of the resultant lengths of (L_x, L_z) and (L_x, L_y), respectively. This means that the longest leg in a system dominates the response of the entire system. The long leg produces the large expansion, which has to be absorbed by the short perpendicular legs in the system. Because these short perpendicular legs provide very limited flexibility, very high forces and stresses are generated in the system mainly due to just one long pipe leg. Thus, a cold spring only in the direction of the longest leg serves approximately the same purpose as a uniform cold spring in all three directions.

Figure 3.9 shows a simple three-branch system that needs to be cold sprung to reduce the anchor reaction. If a uniform cold spring were required, three cold spring gaps would have to be provided and complicated procedures would have to be developed and executed, which would be a real challenge to accomplish. Instead, because the major leg length is in the x direction, we can simply cold spring the x direction with the gap as shown. This will achieve the main purpose of the cold spring with a much simpler procedure.

3.6.5 Analysis of Cold Sprung Piping System

The piping code has specifically stipulated that the cold spring cannot be credited for reducing the expansion stress, because the main failure mode of the displacement stress is the fatigue failure, which is dictated by the stress range. A cold spring will shift some stress from the hot operating condition to the cold condition, but the stress range remains unchanged. The code further stipulates that although cold spring can reduce the hot reaction to the connecting anchor or equipment, only $^2/_3$ of the theoretical reduction can be taken. This is because cold spring is a very complicated procedure. A perfect cold spring is seldom achieved.

For a uniformly cold sprung system, an analysis under operating conditions is all that is required to obtain all the reaction information. However, because a uniform cold spring is seldom performed

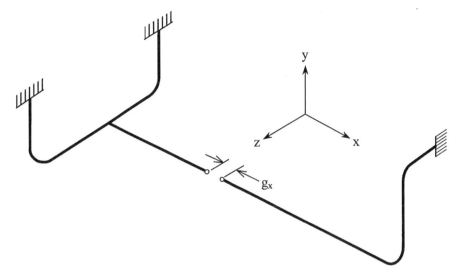

FIG. 3.9
LOCAL COLD SPRING

in actual installations, more analyses are required. In general, the following three analyses regarding thermal expansion are required:

(1) Under operating temperature, but no cold spring gap. This is used to check the thermal expansion stress range to compare with the code allowable stress range.
(2) Under operating temperature with $^2/_3$ of the cold spring gap. This is used to find the credible anchor reaction at operating condition. The $^2/_3$ factor is the fraction of the cold spring effect allowed by the code.
(3) Under ambient temperature with full cold spring gap. This is used to find the anchor reaction at the cold condition after the cold spring. This analysis is also used to provide displacement guides for executing the cold spring process.

These three analyses can be performed one by one, or all together, in an integrated computer analysis. Although the piping code does not stipulate the uncertain pulling effect on the cold reaction, it should be noted that due to less-than-perfect pulling, a much higher cold reaction than the one given by analysis (3) above may be obtained.

3.7 PRESSURE EFFECTS ON PIPING FLEXIBILITY

Because of its nature as a sustained load, internal pressure is often ignored in the evaluation of piping flexibility. This tendency has been rectified recently, but is still not widely appreciated. For instance, pressure is also a cyclic load that needs to be included in the fatigue evaluation. This cyclic pressure, which is implied in Eq. (3.14), is seldom appreciated. Furthermore, pressure also has very significant effects on piping flexibility itself. Flexibility effects occur in two areas: (1) pressure elongation of the piping element and (2) pressure effect on the bend flexibility factor and bend SIFs.

3.7.1 Pressure Elongation

In a pressurized pipe, the entire inside surface of the pipe sustains a uniform pressure. This pressure loading develops a tri-axial stress in the pipe wall. The stress component normal to the pipe wall is generally small and will be ignored in this discussion. The pipe wall is subjected to stresses S_{hp} and

S_{lp} in circumferential and longitudinal directions, respectively, as shown in Fig. 3.10. Stress in the circumferential direction is generally referred to as hoop stress. For practical purposes, S_{lp} is equal to one-half of S_{hp}, as shown in Eqs. (2.13) and (2.14). The hoop stress and the longitudinal stress generate strains in all major directions. The relation between stress and strain is given by Eq. (2.8). By substituting $S_{lp} = 1/2 S_{hp}$, the strains in hoop and longitudinal directions become

$$e_H = \frac{S_{hp}}{E} - v\frac{S_{lp}}{E} = \frac{S_{hp}}{E}(1 - 0.5v) \tag{3.16}$$

$$e_L = \frac{S_{lp}}{E} - v\frac{S_{hp}}{E} = \frac{S_{hp}}{E}(0.5 - v) \tag{3.17}$$

Elongation in the hoop direction increases the pipe diameter. This diametrical change progresses freely without creating any additional resistance to the piping system. Elongation in the longitudinal direction, on the other hand, behaves like thermal expansion. This elongation, given by Eq. (3.17), has a squeezing effect on the piping system just like thermal expansion.

The significance of the pressure elongation can be better visualized by converting it to an equivalent temperature rise [16]. Figure 3.10 shows the relationship between pressure elongation in terms of pressure hoop stress and its equivalent temperature rise, for low carbon steel pipe. For plant piping, hoop stress is normally maintained at less than 15,000 psi (103.4 MPa). In this case, pressure elongation is equivalent to a temperature increase of 17.5°F (9.72°C). In view of the high temperature range normally experienced in plant piping, the effect of pressure elongation in plant piping is insignificant. On the other hand, hoop stress in a cross-country transportation pipeline can reach 30,000 psi (206.8 MPa) or higher for high-strength pipe. The equivalent temperature of this hoop stress range can easily exceed 35°F (19.44°C), which represents a high percentage of the design temperature rise. Therefore, pressure elongation is an important factor to be considered in transportation pipeline design.

3.7.2 Potential Twisting at Bends

The longitudinal elongation discussed in the preceding section is also applicable to piping bends, but the curvature at the bends might also produce different type of deformations. One of the impressions that engineers would intuitively form is that the bend will tend to open when pressurized. Equations have been developed for this opening rotation and are implemented in some computer programs. This opening effect is often misquoted as the Bourdon tube effect. It has artificially generated, on

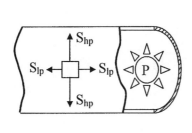
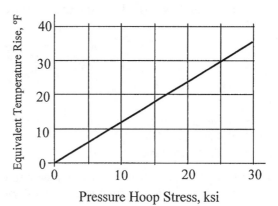

FIG. 3.10
PRESSURE ELONGATION

paper, much bigger twisting effects on the piping than was experienced in the field. Actually, a Bourdon tube is an arc shape tube with an oval cross-section. This oval cross-section is the key for the opening rotation. The general belief in the piping community on piping bends is that if the cross-section out-of-roundness is insignificant, the rotation of the bend is not expected [17].

Detailed analyses and tests on special miter bends [18] have shown that in the elastic stress range, the bends actually close rather than open after pressurization. It starts to open only after the pressure stress exceeds the yield strength. It appears that until more rigorous theories and tests prove otherwise, the opening of the pipe bend by the pressure shall be ignored.

3.7.3 Pressure Elongation Is Self-Limiting Load

Pressure elongation is often mistreated as a sustained load because of its association with pressure, which is a sustained load. Just like thermal expansion, pressure elongation generates a displacement that is a self-limiting load. Its effect on the piping system is determined by the potential axial displacement of each leg of the piping. Once the displacement reaches the potential elongation amount, it stops regardless of whether or not the yielding occurs in the piping. This pressure elongation is generally included in the flexibility analysis the same way as thermal expansion is. In general, pressure elongation is added to thermal expansion to become the total displacement load in the analysis.

3.7.4. Pressure Effect on Bend Flexibility and SIFs

The flexibility factor and SIF at a bend are mainly caused by the ovalization of the bend cross-section. Internal pressure tends to reduce ovalization, thus reducing flexibility and stress intensification. The pressure effect on bend flexibility and SIFs has been well investigated. One of the most recognized treatments is the use of modification factors established by Rodabaugh and George [17]. These factors have been adopted by the ASME piping codes, and are summarized bellow.

In large-diameter, thin-wall elbows and bends, pressure can significantly affect the magnitudes of the flexibility factor, k, and the SIF, i. Under pressurized conditions, the value of k calculated from Eq. (3.5) should be adjusted by dividing it with

$$\left[1 + 6 \left(\frac{P}{E} \right) \left(\frac{r_m}{t_n} \right)^{7/3} \left(\frac{R}{r_m} \right)^{1/3} \right] \qquad (3.18)$$

And the SIFs calculated by Eqs. (3.11) and (3.12) may be reduced by dividing them with

$$\left[1 + 3.25 \left(\frac{P}{E} \right) \left(\frac{r_m}{t_n} \right)^{5/2} \left(\frac{R}{r_m} \right)^{2/3} \right] \qquad (3.19)$$

where

r_m = mean radius of the pipe
t_n = nominal thickness of the pipe
R = radius of the bend curvature

The use of Eqs. (3.18) and (3.19) varies from case to case. Some specifications require the application of the flexibility reduction factor given by (3.18), but not the stress intensification reduction factor given by (3.19). The rationale is that piping tends to maintain its temperature longer than it can maintain its pressure during shutdown. In other words, the stress intensification reduction factor may disappear while the piping is still under full expansion state.

3.8 GENERAL PROCEDURE OF PIPING FLEXIBILITY ANALYSIS

In previous sections, we have discussed the fundamentals of piping flexibility. In this section, we will focus on the general procedure for performing the piping flexibility analysis. A computer program is generally used in the analysis. The simple system shown in Fig. 3.11 will be used to explain the general procedure.

Although most computer program will check the adequacy of the pipe wall thickness against design pressure, wall thickness is actually determined before the analysis is performed. Before the analysis, isometric drawings, together with all the data, (e.g., pipe material, diameter, thickness, design pressure, design temperature, and upset temperature), have to be collected. This data is generally contained in the line list and material specification. The analysis covers the following steps.

3.8.1 Operating Modes

The system may have several different operating modes. Although it is possible to select one mode that represents the most critical situation, normally more than one operating mode has to be considered. In this sample system, there are three possible operating modes: (1) both heat exchanger loops are operating; (2) only HX-1 heat exchanger is operating; and (3) only HX-2 heat exchanger is operating. The analyst decides which of these three modes needs to be analyzed. At first sight, it may appear

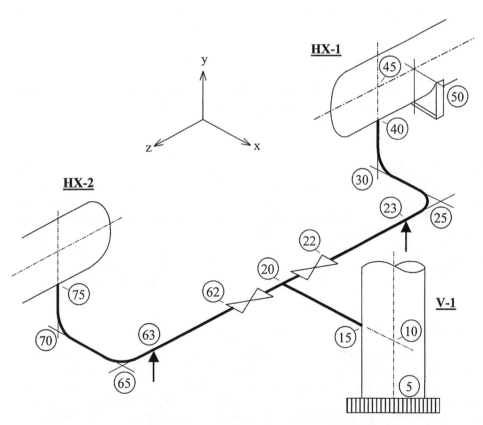

FIG. 3.11
SAMPLE ANALYSIS PROCEDURE

that only the first mode needs to be analyzed. However, a detailed investigation will show that all three modes are required, and they have to be analyzed in one integrated computer run.

It is expected that the first operating mode with both loops operating will result in the highest stress and reaction load at heat exchanger connections 40 and 75. This first operating mode, however, essentially generates no stress and load at vessel connection 15. In the second operating mode with only HX-1 operating, the tee point 20 will be pushed toward HX-2 producing significant stress and load at point 15. Symmetrically, the third mode, with HX-2 only operating, will produce the same amount of stress and load as mode 2 at point 15, but in a reverse direction. Because the stress range at point 15 is the difference of the stresses generated from operating modes 2 and 3, analyses on both modes are required. Differential moments from both operating modes are taken to calculate the stress range required for expansion stress evaluation. The stress range cannot be calculated by subtracting the calculated stresses. This is due to the normal practice of calculating the effective stresses and stress intensities, which have lost the sign of the stress.

3.8.2 Anchor Movements

Points 15, 40, and 75 are terminal points that can be considered anchors or vessel connections. Either way, there are movements at these points attributed mainly to the expansion of the vessels. These movements are calculated by multiplying the corresponding active length with the expansion rate per unit length. Take point 15, for instance, where movement is produced by the expansion of the vessel shell from vessel anchor point 5 to point 15. In the y (vertical) direction, it has a movement equivalent to the expansion from point 5 to point 10. In the x direction, it has a movement equivalent to the expansion of 10-15. These movements are called free movements. They are the same as the final displacements at point 15 when an anchor is used. However, they can be slightly different from the final displacements at point 15 if a vessel connection is used. Vessel connections are discussed more fully in a separate chapter. The x movement in this case is negative, because it is moving in the negative x direction.

Anchor movements at points 40 and 75 are determined the same way as for point 15. However, for heat exchangers and horizontal vessels that are supported with two saddles, the support arrangement plays an important role in piping flexibility. The thermal expansion stress in piping differs considerably, depending on the selection of the fixed support point. To ensure positive control of the movement and to offer positive resistance to horizontal loads such as wind and earthquake, one of the saddle supports is generally fixed and the other is allowed to slide axially with slotted bolt holes. The vessel will then expand axially from the fixed saddle support. In this sample system, it is advantageous to fix the support at point 50, so that it is closer to connection point 40 to reduce the z direction movement. The z direction movement is in the positive direction, and is moving against the piping expansion. For the y movement, the actual zero-displacement location has to be estimated. In this case, it can be assumed to be at the bottom of the vessel shell. Because this elevation is roughly the same as the elevation of point 40, the y movement can be considered as zero. In this example, points 5, 10, 45, and 50 are not required for the analysis. They are used just for calculating the anchor displacements.

3.8.3 Assignments of Operating Values

The system has three major operating modes to be analyzed. This requires a proper assignment of pipe data groups and operating values. It is not difficult to realize that three groups of pipe and operating data are required. The first one covers the main line from point 15 to 20 continues on to point 22 and 62, the second one covers the first branch from point 22 to 40, and the third one covers the rest of the piping. The three modes of operation are defined by the assignment of the temperatures for each of the modes, on each set of pipe data. Each anchor or vessel connection point should also be assigned with three sets of anchor movements, each corresponding to one of the operating modes. For instance, if a given branch is not operating, then the corresponding movements can either be zero or something corresponding to the idle temperature of the heat exchanger shell.

The multi-operating mode analysis is required for the calculation of stress range when a stress reversal is likely to occur through the change of operating modes. It is preferable to analyze all modes in one computer run, so the signs of the moments between the modes can be automatically checked.

3.8.4 Handling of Piping Components

As discussed in Chapter 2, a piping system is generally treated as a series of straight pipe elements and curved pipe elements in the analysis. Therefore, all components in the system have to be simulated with these two types of elements. Furthermore, analyses are based on the centerline of the pipe, with all the lengths defined up to the centerline intersections. The manner in which the components are treated will be discussed in the text that follows, starting sequentially from point 15.

Point 15 is considered an anchor or a vessel connection. If it is assumed as an anchor, then the point is considered perfectly rigid, unless the stiffness or spring rate in any of the six degrees of freedom, three in translation and three in rotation, is specified. If it is treated as a vessel connection, then the magnitudes of the stiffness in all six directions are calculated based on the vessel data. In this case, the vessel diameter, thickness, and reinforcement dimensions have to be specified. The use of a vessel connection represents a more accurate analysis, and generally results in a smaller thermal expansion stress. More detailed discussions on vessel connections are given in a separate chapter. Regardless of the type of end simulation used, the same anchor movements have to be specified.

Point 20 is a forged welding tee, or a fabricated branch connection, as the case may be. The connecting pipe is considered as extending to the centerline intersection point, without considering the existence of the branch connection. The stresses are calculated at this centerline location using the data of the connecting pipe. The only distinction between the branch connection and the regular pipe is that in a branch connection a keyword or flag has to be placed to inform the computer to include the proper SIF. A more sophisticated analysis will consider the branch connection the same as a vessel connection for large thin pipes when the run and branch diameter ratio is greater than 3.

Point 22 is the end of a valve. The valves, flanges, and other fittings are considered regular straight pipe with an additional stiffness added. It is more or less a standard industry practice to consider the valves to be three times as stiff as the connecting pipe of the same length. The valve is therefore treated as a pipe with three times the stiffness. To increase the stiffness, EI, either wall thickness or modulus of elasticity can be increased. The increase in thickness will proportionally increase the section modulus of the pipe, thus artificially reducing the stress calculated. An increase in modulus of elasticity does not have this shortcoming and is the preferred approach.

Point 23 is a support point. A rigid support does not allow the pipe to move in the supporting direction, and thus may significantly reduce the system flexibility. Normally, it is preferred to support the piping system with rigid supports as much as possible. This is mainly due to economy as well as reliability of the system. The rigid support is inexpensive, easy to install, and offers positive resistance to occasional loads such as wind or earthquake. However, there are locations where rigid supports are not suitable due to potentially large loads and stresses that might be generated. A large calculated force at the support normally communicates to the engineer that the location is not suitable for a rigid support. In such cases, a spring support may be required. Occasionally, it may also mean that support at this location is not required at all. A more detailed discussion of supports is given in a separate chapter.

Point 25 is a bend. As discussed earlier in this chapter, the bend has added flexibility and SIFs as compared to a theoretical circular beam with a non-ovalizing circular cross-section. These factors are all automatically included in a computer program if the geometrical dimensions of the band are specified properly. Most computer programs require the dimensions of the tangent intersection point. Therefore, the data point is assigned at the tangent intersection point. From the geometrical information of the tangent intersection point and the bend radius specified, the program locates the beginning and ending points of the bend. These end points, which may be called 25a and 25b in some cases, are the points used in the analysis.

Point 40 is another terminal point that can either be assumed as an anchor or treated as a vessel connection. For heat exchangers, due to its smaller shell diameter, the connections are most often considered anchors. The discussion for point 15 is generally also applicable to points 40 and 75.

3.8.5 The Analysis

The information gathered (discussed above) has to be organized, in a format required by the computer program of choice, to become the input data. The first thing that needs to be done is to assign a global coordinate system that will allow all pipe elements and loads to refer to a common coordinate system. Most programs require that the y coordinate should be in the vertical direction pointing upward. This convention follows the tradition of piping practices. With the y-axis in the vertical direction pointing up, it has a direct relation with the elevation scale normally used in the plant piping drawings. It also simplifies the weight load calculation, as the weight direction is automatically tied to the negative y direction. With a modern desktop or laptop computer, an analysis should not take more than a few minutes to complete once the data is prepared. In general, it requires more than one analysis to reach a satisfactory design of the system.

3.9 PROBLEMS WITH EXCESSIVE FLEXIBILITY

Because everyone is aware of the importance of piping flexibility, engineers have a tendency to provide more flexibility than required in a piping system. Indeed, it is widely believed that the more flexibility that is provided, the more conservative is the design. This is actually a very serious misconception. Today, it is pretty accurate to say that more piping problems have been created by excessive flexibility than by insufficient flexibility.

One obvious consequence of excessive piping flexibility is the added cost due to additional piping provided and the plant space required to accommodate it. In addition to the obvious one, there are many less obvious, but serious, consequences. For example, a flexible system is weaker in resisting wind, earthquake, and other occasional loads. It is also prone to vibration. Although some restraints and snubbers can be used to increase resistance to occasional loads, the added cost can be very substantial. The nature of piping flexibility can be explained with a railroad-construction analogy. There are four situations encountered in piping flexibility that resemble the four types of railroad construction methods as shown in Fig. 3.12.

Figure 3.12(a) shows the case with the rails laid end to end without any clearance. When the weather gets hot and the rail wants to expand but cannot expand, it generates a large amount of force between the ends. This can eventually buckle the rails making it inoperable. This is similar to a situation where the pipe is lying straight between two anchors. This illustrates the case of no flexibility, which can cause problems in piping and connecting equipment.

Figure 3.12(b) shows the rails laid with a small gap provided between two joining ends. This gap allows the rails to expand into it without generating any load. With this small, but sufficient gap provided, the buckling due to thermal expansion is eliminated. However, with a gap, no matter how small, a slight bouncing occurs every time the wheel passes over the gap. The rail is safe, but the ride is somewhat bumpy. This is a compromised arrangement nobody should complain about. The same thing happens to a piping system with just enough loops and offsets for the required flexibility. The flexibility in the piping eliminates the potential damage due to thermal expansion. Meanwhile, the piping system also becomes a little shaky, but it is tolerable. This is the design we need, and can live with.

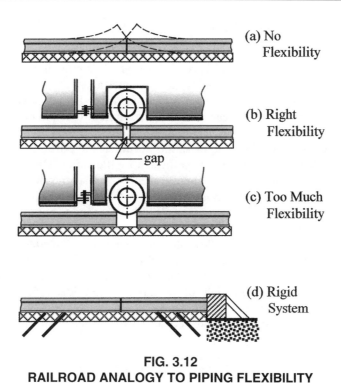

FIG. 3.12
RAILROAD ANALOGY TO PIPING FLEXIBILITY

Figure 3.12(c) shows the consequence of too much flexibility. Realizing that a proper gap between the ends of adjacent rails is necessary for the safety of the railroad, some engineers might get the idea of substantially increasing the gap size to enhance safety. When the gap becomes unnecessarily large, the potential damage due to thermal expansion is eliminated but the wheels of the railcar may not pass the gap safely. This creates a hazardous situation to the operation. In the design of piping systems, many design engineers believe that it is conservative to provide more flexibility in the piping. This indeed has become a guideline in some design specifications. However, just like the railroad with large gaps, this so-called conservative approach can also create operational hazardous in the piping.

Figure 3.12(d) shows a fully restrained continuously welded railroad construction. In a fully restrained situation, the rail generates about 200 psi/°F (2.48 MPa/°C) of axial stress due to temperature change. The stress is compressive if the operating temperature is higher than the construction temperature. With a maximum expected axial stress below about $^2/_3$ of the yield strength range, the railroad can be constructed continuously without gaps. This continuous construction allows for the smooth operation of the railcar when operating at high speeds. However, the thermal expansion problem is still there no matter what type of construction is adopted. In this type of construction, the thermal expansion problem is treated differently. It requires that heavy anchors be placed to prevent the ends of the rail from moving. Sufficient lateral restraints are also placed along the line to prevent the rail from lateral buckling. With all associated problems taken care of, the rigid railroad system offers a very smooth, safe operation for the railcars even at very high speeds.

There are similar situations in the construction of some piping systems. In cross-country pipelines, the fully restrained construction is generally adopted to reduce the cost and increase the reliability. No expansion loops or offsets are used. Similar to the continuous railroad construction, the pipeline requires heavy anchors or equivalent anchors placed at ends to control the end movements, and sufficient lateral restraining efforts are provided to prevent lateral buckling in pipelines. Details on cross-country pipelines are discussed in a separate chapter.

3.9.1 Problems Associated With Excessive Flexibility

Excessive piping flexibility comes from excessive loops and offsets in the piping system. Externally, the piping with excessive loops is flimsy and prone to vibrate. It is weak in resisting occasional loads, such as wind and earthquake. Internally, these excessive pipes and bends generate excessive pressure drop and excessive bumps to the fluid flowing inside. Aside from the increased operating cost due to the extra power required, pressure drop can cause vaporization of the fluid operating at near-saturate temperature. The near-saturate condition occurs at all condensate lines and bottom fluid lines from distillation towers, and so forth. When vaporization occurs at the portion of piping away from the pumps, it generates shaking forces due to slugs formed by the vapor. Once slug formation occurs, the whole piping system bangs and shakes violently, even if the system is not unsafe to operate. More discussions on slug flow are given in the chapter dealing with dynamic analysis.

The biggest problem associated with excessive flexibility occurs at pumping circuits. The excessive piping loops in this case are add-ons in an attempt to reduce piping loads on the pump. These loops are not in the scope of original fluid flow planning. Therefore, the added pressure drop is often not accounted for in the original system design. To ensure a smooth and efficient pump operation, a certain minimum amount of net positive suction head (NPSH) must be maintained at the pump inlet. Otherwise, vaporization and cavitation might occur at the suction side. This can substantially reduce the pump capacity and pump efficiency. The vapor eventually collapses when pressure inside the pump reaches above the saturation point. This collapse of vapor generates a huge impact force that can damage the internals of the pump.

Many pumps, such as condensate and boiler feed water pumps, are operating at near-saturate temperature. It is already very difficult to provide enough NPSH for them without the unexpected addition of flow resistance. Very often, storage tanks on the suction side have to be elevated to the top portion of the building to provide this required NPSH. In the layout of a petrochemical plant, the engineers generally do not have the luxury of elevating a storage tank as high as needed. Because the storage tank is often the bottom of a distillation or some other tower or vessel, to raise it means elevating the whole tower together with all other piping connected to it. It might also involve raising other towers as they are normally related by a fixed elevation difference. It is indeed very complicated even to raise the elevation by just a foot or two. Because of the lack of awareness regarding the need for loop, or due to the difficulty in foreseeing the size of the loop required, the extra pressure drop from the added loops is seldom accounted for in the original flow design. This naturally generates all types of operational problems for the pump. More discussions are given in Chapter 9.

3.10 FIELD PROVEN SYSTEMS

There are many installations that cannot be accurately analyzed even with current technology. It is either due to the shortfall of the technology or due to economic impracticality. The authors of the piping code had the vision to stipulate that "No formal analysis of adequate flexibility is required for a piping system which duplicates, or replaces without significant change, a system operating with a successful record;..." (ASME B31.3, par. 319.4.1; also B31.1, par. 119.7.1).

However, many modern-day engineers do not appreciate this foresight of the code. Not so many years ago, when it was customary to construct a plastic model of the plant for review by all associated parties, the engineers from an operating plant could easily spot a poorly designed system. They would say that they had never seen one like that, or that some other systems should be used because they have already proved themselves in the field. However, these comments were often not appreciated by computer-oriented engineers who would have insisted that the computer analyses had concluded otherwise.

The computer is mathematically correct in interpreting the data supplied. In piping flexibility, some systems in the field may not be readily analyzed by the computer due to a lack of accurate information. These systems include some low-temperature pump piping systems, short equipment connecting pipes, and others that involve small amounts of expansions or displacements. As discussed in Section 3.1, the analysis results of these systems depend entirely on the accurate stiffness of the end connections. If these systems were analyzed through a computer using the idealized data, they would have evolved to a design with excessive loops that may hamper the proper operation of the system.

This is not to say that the computer is not useful, but rather to emphasize that only accurate data will produce meaningful results. For systems involving small expansion or displacement, the exact stiffness of the boundary condition is very important. An analysis assuming an anchor here and another anchor there will not do the job correctly. In these cases, past experience is of the utmost importance.

Some words of caution are in order, however. The duplication of successfully operating system is only valid for systems do not involve time-dependent failure modes, such as creep or medium cycle fatigue. For a system with a time-dependent life cycle, it takes years to judge whether it is successful or not.

REFERENCES

[1] Spielvogel, S. W., 1951, *Piping Stress Calculations Simplified*, published by author, Lake Success, NY (previous editions published by McGraw-Hill Book Company).
[2] M. W. Kellogg Company, 1956, *Design of Piping Systems*, revised 2nd ed., John Wiley & Sons, New York.
[3] Olson, J., and Cramer, R., 1965, *Pipe Flexibility Analysis Program MEC 21S*, Report No.35-65, San Francisco Bay Naval Shipyard, Mare Island Division (other versions include MEC-05, 1957; MEC-21, 1960; MEL-21, 1966; MEL-40, 1968).
[4] Brock, J. E., 1967, "Expansion and Flexibility," Chapter 4, *Piping Handbook*, 5th ed., R. C. King, ed., McGraw-Hill Book Company, New York.
[5] Peng, L. C., and Peng, T. L., 1998, "Piping Flexibility Basics," *Chemical Processing*, 61(5), pp.63–69.
[6] ASME Code for Pressure Piping, B31, An American National Standard, *B31.1 Power Piping*; *B31.3 Process Piping*, The American Society of Mechanical Engineers, New York.
[7] ASME Boiler and Pressure Vessel Code, Section-III, "Nuclear Power Plant Components", Division 1, Subsection NB, Class 1 Components; Subsection NC, Class 2 Components; Subsection ND, Class 3 Components.
[8] Markl, A. R. C., 1947, "Fatigue Tests of Welding Elbows and Comparable Double-Mitre Bends," *Transactions of the ASME*, 69(8), pp. 869–879.
[9] Markl, A. R. C., and George, H. H., 1950, "Fatigue Tests on Flanged Assemblies," *Transactions of the ASME*, 72(1), pp.77–87.
[10] Markl, A. R. C., 1952, "Fatigue Tests of Piping Components," *Transactions of the ASME*, 74(3), pp. 287–299 (discussion pp. 299–303).
[11] Markl, A. R. C., 1955, "Piping-Flexibility Analysis," *Transactions of the ASME*, 77(2), pp. 124–143 (discussion pp. 143–149).
[12] Peng, L. C., 1979, "Toward More Consistent Pipe Stress Analysis," *Hydrocarbon Processing*, 58(5), pp. 207–211.
[13] Rossheim, D. B., and Markl, A. R. C., 1940, "The Significance of, and Suggested Limits for, the Stress in Pipe Lines Due to the Combined Effects of Pressure and Expansion," *Transactions of the ASME*, 62(5), pp. 443–464, July.
[14] Robinson, E. L., 1955, "Steam-Piping Design to Minimize Creep Concentrations," *Transactions of the ASME*, 77(7), pp. 1147–1158 (discussion, pp. 1158–1162).
[15] Peng, L. C., 1988, "Cold Spring of Restrained Piping Systems," presented at ASME PV&P Conference, Pittsburgh, PA, June, 1988, published in *PVP-Vol. 139, Design and Analysis of Piping, Pressure Vessels, and Components-1988*.

[16] Peng, L. C., 1981, "An Interpretation on Pressure Elongation in Piping Systems," *PVP-Vol. 53, Current Topics in Piping and Pipe Support Design*, ASME Publications.
[17] Rodabaugh, E. C., and George, H. H., 1957, "Effect of Internal Pressure on the Flexibility and Stress Intensification Factors of Curved Pipe or Welding Elbows," *Transactions of the ASME, 79*(4), pp. 939–948.
[18] Roche, V. R., and Baylac, G., 1973, "Comparison Between Experimental and Computer Analysis of the Behavior Under Pressure of a 90° Bend with an Elliptical Section." *Proceeding of the 2nd International Conference on Pressure Vessel Technology*, San Antonio, TX, 1973, ASME.

CHAPTER

4

CODE STRESS REQUIREMENTS

The main purposes of piping stress analysis are: (1) to ensure structural integrity and (2) to maintain system operability. The first part is the main objective of the industrial piping codes. To ensure the structural integrity of the piping systems, the piping codes have assembled a set of procedures and specifications covering the minimum requirements for material, design, fabrication, erection, inspection, and testing. The piping is ensured of proper safety factor on structural integrity when all code requirements are followed and satisfied. This chapter covers the stress requirements of the codes.

This chapter will try to look at the code from a slightly different angle to help the reader better understand its requirements. Some rationales of the requirements may be presented to help the reader understand and comply with the code. However, these rationales should not be interpreted as the intent of the code. They are just the personal interpretations of the authors. The exact intent of the code can be obtained through the code committee. The code book has detailed information of the official channels for submitting questions, comments, and requests for interpretation. Furthermore, this chapter contains only the general requirements related to piping stress. The reader should consult the most recent edition of the applicable code for the complete and updated requirements of the code.

4.1 "DESIGN" CHAPTER OF THE PIPING CODES

The code stress requirements are given in the chapter or part dealing with designs. Because there are many different sections of the codes, it is not possible to discuss the requirements of all the codes in detail. We will only discuss some of the key requirements. In this section, we will list some of the applicable codes and briefly explain their background. Currently, there are two main systems of standards covering the design of piping systems: the American National Standards published by the American Society of Mechanical Engineers (ASME) and European Standards published by Comité Européen de Normalisation (CEN). The following is the list of piping codes:

(a) ASME Code for Pressure Piping, B31 [1]

   B31.1   Power Piping
   B31.3   Process Piping
   B31.4   Pipeline Transportation System for Liquid Hydrocarbons and Other Liquids
   B31.5   Refrigeration Piping and Heat Exchanger Components
   B31.8   Gas Transmission and Distribution Piping Systems
   B31.9   Building Services Piping
   B31.11   Slurry Transportation Piping Systems

(b) ASME Boiler and Pressure Vessel Codes, Section-III Rules for Construction of Nuclear Facility Components (B&PV Section-III) [2]

   Subsection NB, Class 1 Components, NB-3600 Piping Design
   Subsection NC, Class 2 Components, NC-3600 Piping Design
   Subsection ND, Class 3 Components, ND-3600 Piping Design

(c) CEN Metallic Industrial Piping [3]

EN 13480-3, Metallic industrial piping — Part 3: Design and calculation

Among the codes listed above, there are several general characteristics and similarities:

- NB-3600 for Class 1 nuclear piping stands out as a unique code. It has very little inter-relationship with other codes. The design is based mostly on the "Design by Analysis" philosophy. Extensive calculations on local and peak stresses are required. All potential dynamic phenomena are included. This is the only code that uses theoretical stress intensifications. These theoretical elastic stress intensifications, in contrast to the adjusted ones used in other codes, are called stress indices in NB-3600.
- B31.1 power piping is the father of all piping codes. It was initially issued as the "American Standard Code for Pressure Piping" in 1942. Most of the current stress requirements were in existence in the 1955 edition.
- NC-3600 and ND-3600 nuclear piping codes, for Class 2 and Class 3 piping, respectively, are similar to ASME B31.1 power piping. However, since the 1983 edition of the codes, evaluations of sustained stress and occasional stress have diverged from that of B31.1. Currently, these stresses are evaluated in accordance to the procedure similar to the one used in Class1 nuclear piping.
- The European code is similar to B31.1. However, there are differences in some areas. For instance, the criteria for setting the allowable stress and the formula for calculating the wall thickness are different.
- B31.3 evolved from B31.1 and was first issued in 1959 as "Petroleum Refinery Piping." Currently, it has become a unique code that caters to the unique industry it serves. The allowable stress level, the stress evaluation methods, and the application of stress intensification are all somewhat different from that of the B31.1 code, from which it evolved.
- B31.4 and B31.8 are transmission pipeline codes. Their characteristics are derived mainly from the fact that they use numerous pipes with simple piping configurations. They tend to use higher allowable stress compared to other codes. In the calculations of basic stresses, they mostly adopt the approaches used in B31.3.
- All other codes follow either B31.1 or B31.3 approach. B31.9 is a simplified version of B31.1. Both B31.5 and B31.11 are simplified versions of B31.3.

From the brief summary above, some guidelines are established for preparing this chapter to present most of the relevant information. The following are the outlines of the subjects to be discussed.

- The discussions will be based mainly on B31.1 and B31.3 with cursory mentions of the differences in other codes.
- B31.4 and B31.8 cross-country pipeline codes are unique by themselves. They are only briefly mentioned in this chapter, but will be treated in detail in a separate chapter.
- Requirements for NB-3600 Class 1 nuclear piping will be discussed briefly and separately at the end of this chapter.

4.2 LOADINGS TO BE CONSIDERED

The code generally tabulates a list of loads that need to be considered in the design of the piping system. Some of the loads, such as internal pressure, weight, and temperature related to longitudinal gross expansion, are the common ones that are automatically considered in all routine analyses. Other loads, of which the analysts may or may not be aware, are generally included in the analysis only when they are specified in the design specification. The potential loads are discussed in the following subsections.

4.2.1 Pressure

This type of load includes internal and external pressures. The design pressure is generally set at the most severe condition of concurrent internal (or external) pressure and temperature expected during service. For ductile materials, the codes do allow occasional cases of the pressure exceeding the design value. Depending on the codes, frequency of occurrence, and duration of the occurrence, the allowance varies from 15% to 20% in B31.1 and from 20% to 33% in B31.3.

Test pressure. Test pressure generally is not a design parameter. It is a quality control measure used to weed out poor-quality components, especially poor-quality welds, in a properly designed system. The test pressure is generally set as 150% of the design pressure at equivalent operating condition. The test pressure at testing temperature is calculated as

$$P_T = \frac{1.5 P S_T}{S} \tag{4.1}$$

where S_T is the allowable stress of the pipe material at test temperature, S is the allowable stress at design temperature, and P is the design pressure.

However, the test stress shall not exceed 90% of the yield strength (100% yield strength for B31.3) of the pipe at test temperature. When the test hoop stress at test temperature exceeds the above limit, the test pressure shall be reduced to the maximum pressure that meets this test stress limit. If the longitudinal stress, resulting from the test pressure and the test weight, exceeds the above limit, then temporary supports are required to reduce the stress to within the limit.

4.2.2 Temperature

Temperature in a piping component consists of many different categories as shown in Fig. 4.1. These temperature distributions create different temperature effects on the piping. The temperature effect most often referred to is the average temperature across the whole cross-section, as shown in Fig. 4.1(a). This temperature generates a longitudinal growth and is used in common thermal flexibility analysis. The temperature at the most severe condition of concurrent temperature and pressure expected during service is generally selected as the design temperature. Other service temperatures, such as for dry out or steam out, may be higher than the design temperature, but at a considerably lower pressure. These service temperatures are also considered in the flexibility design and are used to determine the flexibility temperature. Another temperature that might cause gross displacement

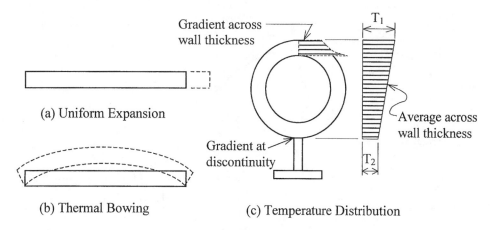

FIG. 4.1
TEMPERATURE LOADING

in the system is bowing temperature. Analysis of the piping system subjected to these gross thermal displacements is discussed in Chapter 3.

Bowing temperature. A stratified flow and other operating conditions may cause the average wall temperature to vary across the pipe cross-section creating the thermal bowing as shown in Fig. 4.1(b). The stratified flow in a horizontal line, for instance, can create a higher temperature at the top wall of the cross-section. This temperature difference, $T_1 - T_2$, between the top and bottom areas of the cross-section makes the mid-span of the pipe move up in response to the larger amount of expansion at the top. This bowing effect can generate a very large lateral displacement and rotation and create significant stress in the pipe. More significantly, it can often tear up supports and small connections. Rotation is one of the prime causes of leakage at flange connections.

The diametrical temperature difference across the cross-section is called the bowing temperature. The bowing temperature occurs mostly in top and bottom directions, and is mainly generated by stratified flow created by rapid quenching of high-temperature gas engaged in petrochemical production, emergency cooling injection, cold re-circulation, start-up of a liquefied natural gas transfer line, rapid deployment of cryogenic liquid fuel, and so forth. It can also be generated by the startup of a large steam pipe. The share of operational difficulties due to thermal bowing is very high. Therefore, it should be properly treated in the design. The effect of the bowing temperature is analyzed by applying the angular rotation at each affected member of the piping. This angular rotation is calculated from the arching radius created for each member, as shown in Fig. 4.2. The radius of the curvature, the end rotation, and the potential lateral movement of a simple supported pipe with span L are given in the following equations:

$$R = \frac{D}{\alpha(T_1 - T_2)} \tag{4.2}$$

$$\theta = \sin^{-1}\left(\frac{L/2}{R}\right) \tag{4.3}$$

$$y = R - \sqrt{R^2 - (L/2)^2} \tag{4.4}$$

The lateral displacement, y, is often large enough to lift the pipe off some of the supports. These equations assume that temperature varies linearly from top to bottom of the cross-section. The actual temperature variation is generally non-linear and often resembles a stepwise change. A non-linear

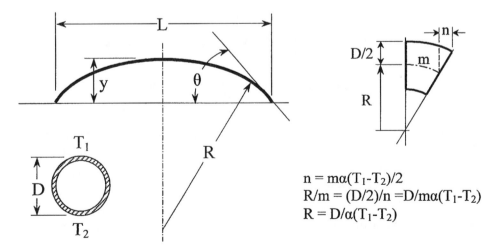

FIG. 4.2
ANGULAR ROTATION DUE TO THERMAL BOWING

temperature variation has to be separated into the linear distribution superimposed with an off-linear part [4]. These equations deal with the linear part of the cross-sectional temperature variation that generates gross displacements in the piping system. The resulting movement is combined with the normal longitudinal thermal expansion to obtain the expansion or displacement stress. The off-linear part generates only internal thermal stress similar to that across the wall temperature gradient discussed below. (See Chapter 11 for more details on thermal bowing).

Across the wall temperature gradient. Because of the heat transfer resistance, the temperature at the inside pipe wall is generally different from that of the outside wall. The amount of difference depends on the heat loss or gain through the wall. This difference can be large if the heat input or output rate is high. One major cause of large temperature gradient is the rapid startup of the piping. The through-wall temperature gradient will tend to flare up to a cone shape at the free end of the pipe. However, in a regular pipeline this tendency to flare up is suppressed, creating a thermal gradient stress. This stress is local and does not cause any movement in the piping system. It is generally referred to as thermal stress, in contrast to the expansion stress created by uniform thermal expansion and bowing. Besides Class 1 nuclear piping, which will be discussed later in this chapter, this thermal gradient stress is generally not included in the design calculation of non-nuclear piping. In non-nuclear piping, its significance is considered in the operation procedure and the standard design details implemented.

Thermal discontinuity. The temperature gradient at structural discontinuity also generates local thermal stress due to the structural restraining effect. This stress and also the stress due to dissimilar materials are again generally not included in the calculation of non-nuclear piping. They are more or less dealt with by the standard rules and procedures adopted by each company. More details about the stress due to dissimilar materials will be discussed in Chapter 5.

Weather effect. A low ambient temperature can affect the selection of the pipe material. It also needs to be considered when calculating the expansion stress range. A rain shower normally has a significant effect on the thermal gradient stress of the un-insulated portion of the piping. It can also lead to thermal bowing on bare piping.

4.2.3 Weight Effects

The weight of the piping, including pipe material, insulation, refractory, fluid, snow, and ice, should also be considered in the design. In general, computer software packages do not automatically include the weight of support attachments, such as clamps and hanger rods. The weight of support attachments can be significant and needs to be manually included in the analysis. Potential variations, such as inherent overweight of elbows and the pipe itself, also need to be considered. Test weight of gas piping likewise requires special consideration. Temporary supports may be required for larger gas or vapor pipes under hydro test condition. Because no pipe movement is involved, temporary supports for hydrostatic test are all fixed types such as rigid hangers and cradles. However, these temporary supports have to be removed after testing to prevent them from interfering with proper pipe movement.

4.2.4 Wind Load

Wind load can be specified either with basic wind speed or the actual wind force. If the basic wind speed is specified, American Society of Civil Engineers (ASCE) 7 [5] or the Uniform Building Code [6] can be used to derive the wind force. Wind force is applied at the outer surface of the pipe including insulation. The area to be applied is the projection area perpendicular to the wind direction in consideration. Normally, only two mutually perpendicular horizontal directions are considered. The vertical wind component is considered only if it is stipulated in the design specification. The wind effect on the part of piping that is either inside an enclosed building or shielded by a proper object can be ignored. The stress produced from wind load is generally classified as an occasional stress.

4.2.5 Earthquake

Earthquake load is analyzed via either the dynamic approach or the static approach. The dynamic approach generally uses the response spectra method, which will be discussed in Chapter 13. The static approach applies an acceleration factor in each of the three major geometrical directions to simulate the effect. Therefore, the earthquake load is proportional to the weight of the piping, including all the components discussed for the weight load above. The magnitude of the acceleration, commonly called g factor, can be determined according to ASCE 7 or the Uniform Building Code. Normally, the earthquake load applies to all three directions with somewhat smaller magnitude in the vertical direction.

The earthquake load and wind load are generally not required to be considered as acting concurrently. However, in the design of offshore piping, acceleration normally comes from the water wave caused by the wind. Therefore, wind and acceleration loads are considered as acting concurrently. The stress generated by the earthquake load is generally classified as an occasional stress.

4.2.6 Dynamic Fluid Loads

Piping may be subjected to various types of dynamic loads. Figure 4.3 shows some of the well-known dynamic effects. Analysis of dynamic loads will be discussed in a separate chapter. Here, only the phenomena are briefly discussed, along with a description of the nature of their stresses.

Vortex shedding. When the fluid flows past an object or a cavity, it generates vortices at the tail end of the object. These vortices shed away alternatively from each side of the object, as shown in Fig. 4.3(a). Each shedding of the vortex imposes a force on the object. The force has a large component in the direction perpendicular to the flow direction. The alternative shedding of the vortices from the alternative side of the object generates an alternative force perpendicular to the flow direction. This alternative force has a characteristic frequency determined by the flow velocity, fluid density, and size of the object. If this frequency coincides with one of the lower natural frequencies of the piping,

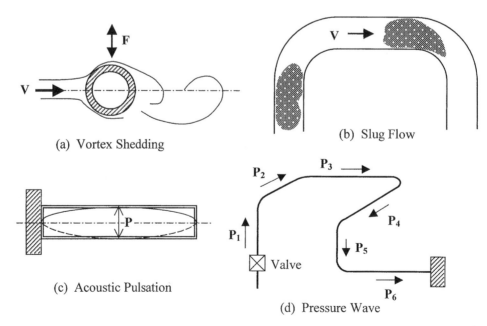

**FIG. 4.3
DYNAMIC EFFECTS ON PIPING**

a severe vibration may occur. The acceptance of the vibration stress is evaluated with the criteria of high-frequency fatigue whose details are discussed in the chapter dealing with dynamic analyses.

Slug flow. A slug of liquid formed in a vapor line or a vapor pocket formed in a saturated liquid line creates a density discontinuity in the line as shown in Fig. 4.3(b). This density discontinuity produces an unbalanced shaking force in an otherwise steady flow with a fairly uniform velocity along the line. When the flow passes a turning bend, it imposes an impact force to the bend due to a change in the flow direction, and thus a change in momentum. It also exerts a reaction force to the bend upon leaving it. In a steady flow with same fluid density, the forces at all bends are the same, equalizing each other at every leg of the piping system. Therefore, no net shaking force is present in this situation. However, when the flow contains slugs, the force changes whenever the fluid density changes at the bend. This generates a force differential between two bends located at both ends of a pipe leg. These unbalanced forces can severely shake the piping. Slug flow most often exists at saturated liquid lines. A condensate line, for instance, can easily produce a local vapor pocket due to an excessive pressure drop in the line. Slugs are also assumed to exist in the flare lines used for both liquid and vapor discharges. The shaking forces generated by slug flow are largely random in nature. They shake the piping, causing the piping to vibrate more or less with the natural frequencies of the piping system. The stress generated has to be evaluated with the criterion of high frequency fatigue.

Acoustic pulsation. In a pulsation flow, such as the one that comes from reciprocating compressor or reciprocating pump, there are two phenomena that need to be investigated. One is the traveling pressure wave created by the volumetric change of flow. The other is the acoustic effect of the piping in response to this volumetric change. The traveling pressure wave can be significant for liquid lines, but is generally not significant in gas lines. In gas lines, the main concern is the piping acoustic response to the volumetric change. In a reciprocating machine, the inlet/outlet flow rate changes periodically with dominant frequencies proportional to the rotational speed of the machine. To push this variable volume into the piping, a periodic pressure pulsation is created at the inlet/outlet nozzle. This initial pressure pulsation forms a pressure wave traveling down the pipeline as the fluid is transported. It is also the source of acoustic pulsations that can have a more severe effect on the piping than the initial pressure excitation.

As shown in Fig. 4.3(c), acoustic pulsation is viewed considering the whole piping system as an organ pipe with a pressure excitation applied at the end. Depending on the length of the piping and the dominant frequencies of the pressure excitation, an acoustic resonance may be established, generating a high-pressure pulsation. This pressure pulsation along the line is either at the same phase or at an opposite phase, but the magnitude is different along the line. This means that the pressures at the two bends forming a pipe leg are generally different. This differential pressure at two adjacent bends creates shaking forces to the piping. These shaking forces have a frequency similar to the pulsation frequency. A severe vibration in the piping can occur when one of the lower natural frequencies of the piping system coincides or is near the shaking frequency. The stress generated is a steady-state vibration stress that needs to be evaluated with the criterion of high frequency fatigue.

Pressure wave. A shock wave is created when the flow in a piping is abruptly altered. This shock wave, when traveling down the line, can create all types of disturbances, such as impacting the bends, separating the liquid column, rejoining the separated liquid columns. These actions come with noisy banging sounds that mimic hammering. The phenomena is generally called water or steam hammer depending on the type of the fluid involved. Under a normal operating environment, the flow rate is changed gradually by the proper stroking of the valve. However, there are occasions that require an abrupt change in the flow rate. One of those occasions is the loss of electric power to the motor of a pump and another is the loss of electric generator load at a steam power plant. In a power plant, for instance, when the electric generator load is lost due to some reason, the steam turbine inlet steam has to be cut off immediately to prevent the turbine from over-speeding. This abrupt shutoff of inlet steam creates a large pressure wave at both the up-stream and down-stream sides of the valve. As shown in Fig. 4.3(d), at any given instant this pressure wave has different magnitudes along the line. This differential pressure generates shaking forces, which shake the piping system to potentially create high stresses in the piping and high loads at supports and connected equipment. The stress generated by water/steam hammer is generally classified as an occasional stress.

4.2.7 Harmonic Anchor Displacement Loads

When a process tower is subjected to an internal turbulent flow, or an offshore platform is subjected to a random wave force, it will vibrate at one or two of their fundamental natural frequencies. The amplitudes of these vibrations are not large enough to be a concern to the equipment or structure itself, but can cause severe vibration to the connecting piping if the piping resonates with the anchor excitation.

4.2.8 Passive Loads

There are loads that exist only when the piping moves. These loads, including support friction force, spring hanger resistance, soil resistance and others, are called passive loads. Passive loads can add significant stress to the piping and also significant loads to the support and equipment. Details of these forces will be discussed in Chapter 6. Passive loads are generated because of the movements of the piping. They themselves prevent the pipe from moving, and do not have the capacity to move the pipe to generate stress.

4.3 BASIC ALLOWABLE STRESSES

The basic allowable stresses are also called code table stresses because they are tabulated in the code book. They are used directly in the calculation of pipe wall thickness subject to design pressure. We call them basic allowable stresses because they are the design stresses for the most important but basic sustained loads. For other loadings, the design stresses are modified from these basic allowable stresses by applying factors and/or combinations. These allowable stress tables are the backbones of the code. They are established based on the following criteria.

4.3.1 Bases for Establishing Allowable Stresses

Allowable stresses for piping materials are established differently for different groups of materials, which are generally categorized as: (1) bolting materials, (2) cast iron, (3) malleable iron, and (4) other materials. Group 4, which covers most of the common piping materials, will be discussed as a comparison of different codes. The basic allowable stress values at a given temperature for materials other than bolting materials, cast iron, and malleable iron are taken as the lowest of the following:

(1) The lower of one-third of ultimate strength at room temperature and one-third of ultimate strength at temperature. B31.1 and Class 2 nuclear piping use a 1/3.5 factor instead of 1/3. EN-13480 uses a 1/2.4 factor instead of 1/3.
(2) Except as provided in item (3) below, the lower of 2/3 of yield strength at room temperature and 2/3 of yield strength at temperature.
(3) For austenitic stainless steels and nickel alloys having similar roundhouse stress-strain behavior, the lower of 2/3 of yield strength at room temperature and 90% of yield strength at temperature. This 90% yield allowable is not recommended for flanged joints and other components in which slight deformation can cause leakage or malfunction.
(4) 100% of the average stress for a creep rate of 0.01% per 1000 hours.
(5) 67% (2/3) of the average stress for rupture at the end of 100,000 hours.
(6) 80% of the minimum stress for rupture at the end of 100,000 hours.

(7) For structural grade materials, the basic allowable stress shall be 0.92 times the lowest value determined from items (1) through (6).

The allowable stresses for bolting materials are generally smaller. This is partially because bolt stresses are somewhat more unpredictable. It is also a well-known fact that bolts are often stressed above the design value in the field to ensure a proper seating of the gasket.

The above bases for allowable stress provide safeguards against (1) gross deformation and excessive strain follow-up, (2) rupture, and (3) creep. To safeguard against gross deformation, the stress is limited to 2/3 of the yield strength, and to guard against rupture, the stress is limited to 1/3.5 to 1/3 of the ultimate strength. The margin against rupture is generally referred to as the rupture safety factor. In other words, the allowable stress is set with a rupture safety factor of 3 or 3.5 depending on the code used.

The safety factor against creep for systems operating in the high temperature domain is not easy to put a number on. However, because a system normally experiences the full design temperature in only a small portion of its operating life, the 100,000-hour benchmark is generally considered conservative, although it is much lower than the expected plant life. The time to rupture at a given stress level is very sensitive to the actual operating temperature. A 20°F temperature decrease at a design temperature of 1000°F will more than double the 100,000 hours of time to rupture. Creep rupture is time-dependent. By combining the low creep allowable stress together with the expected corrosion allowance throughout the operating life, an unusually thick pipe wall may be initially required if the pipe is designed for full operating life. Some high temperature systems, therefore, are purposely designed to have them replaced, according to schedule, once or twice throughout the life of the plant.

Allowable stresses for B31.3 are generally higher than those for other codes. This is due to the unique industries it serves as discussed in Chapter 1. However, in high temperature ranges where creep damage governs, allowable stress becomes the same for all codes. It should also be noted that the allowable stress for Class 1 nuclear piping is set higher than the allowable stress for B31.1 power piping and Class 2 and Class 3 nuclear piping. This is mainly due to the additional design calculations, quality controls, and quality assurances required by Class 1 piping. The combined effort makes the real safety factor in Class 1 piping higher than that of Class 2 and Class 3 piping, even though the basic allowable stress value is higher in Class 1 piping. The extra calculations and quality assurances of Class 1 piping eliminate many of the uncertainties that might compromise the quality of the piping system.

4.3.2 Code Allowable Stress Tables

The basic allowable stresses for the approved material are tabulated in the code book as appendices. These tables list the approved materials together with the established allowable stresses at selected temperature marks. The materials not listed are unapproved, and shall be used only through special qualifications. The materials are applicable only in the temperature ranges that have allowable stresses listed subject to limitations given by the accompanied notes. Besides the allowable stresses, the table also contains related information as follows:

(1) Spec. No.: These are the American Society for Testing and Materials (ASTM) [7] specification numbers. The ferrous materials are prefixed with the letter "A," and the non-ferrous materials are prefixed with the letter "B." The ASTM specification comprises a set of rules for manufacturing and testing a group of pipe materials of similar characteristics. Each specification number covers numerous different materials. For instance, the most common specification, A-106, configured for "seamless carbon steel pipe for high-temperature service" has several grades. Therefore, in addition to Spec. No., we also need other classifications to identify a given material.

(2) Grade: Most specification numbers have more than one grade. Generally, the grade is associated with material compositions and strength. The plate material used in welded pipe may also be classified as grade in some codes.

(3) Type or Class: Types and classes are associated with different manufacturing methods and the scope of inspections required. For instance, Type S means seamless, Type E means electric resistance welded, and Type F means furnace butt-welded. In B31.3, the Type or Class is not listed in the main table. The type and class together with joint efficiency are listed in a separate table in B31.3.

(4) Material Composition: This shows the alloy type and compositions of relevant materials.

(5) P-Number: This the welding qualification number to which the material belongs. This number corresponds to the one given in ASME B&PV Code Section IX [8].

(6) Notes: The table contains many special notes. The applicable notes are listed in this column. These notes generally are related to the limitations and extra attentions required for the given material.

(7) Specified Minimum Tensile Strength

(8) Specified Minimum Yield Strength

(9) Joint Efficiency or Quality Factor: For welded pipe, this is the longitudinal or spiral weld joint efficiency. For castings, this is the casting quality factor. These factors are to be applied to the allowable stress for calculating the wall thickness. They are not used in the evaluation of the piping flexibility. In B31.3, this column, together with type and class, is listed in a separate table.

(10) Minimum Temperature: This the design minimum temperature for which the material is normally suitable without impact testing beyond that required by the material specification. This column is not available in B31.1, which generally considers −20°F (−29°C) the minimum applicable temperature. (See also 1.3.1 (b) Brittle Rupture.)

(11) Maximum Allowable Stress in Tension: The stress values are given for the benchmark temperature points generally spaced by every 100°F at low temperature ranges and spaced by every 50°F at high temperature ranges. The stress at a temperature in between the benchmark points can be linearly interpolated. The material should not be used for temperatures outside the two extreme temperatures within which the allowable stresses are given.

In most code tables, the given allowable stresses include also the corresponding longitudinal joint efficiency, E, and casting quality factor, F. The values are generally referred to as *SE* values. However, the values given in B31.3 allowable stress tables do not include the joint efficiency and quality factor. For the pressure design of B31.3 components, the allowable stress value given in the code table has to be multiplied with the applicable joint efficiency or quality factor. B31.3 gives the joint efficiencies and quality factors in a separate table.

To identify a material for design and analysis, we only need the specification number, grade, and type or class, if applicable.

4.3.3 Weld Strength Reduction Factor

The material at weld-affected zone is weaker than the unaffected zone. This weakness at weld is already taken care of by applying a joint efficient on the allowable stress, and also by applying a stress intensification factor or stress index on the calculated stress.

At high temperature ranges, the weld weakens further, thus accelerating creep failure. This deterioration against creep is attributed to weld residual stress, weld material discontinuity stress, and weld shape discontinuity stress. The effect of these minor stresses is not significant when the pipe fails at a stress higher than the yield strength. However, when the pipe fails at a stress much lower than the yield strength, these minor stresses have a very significant effect such as in the case of high cycle fatigue discussed in Section 1.3.2. At creep range, the pipe also fails at a stress much lower than the yield strength. Therefore, the contribution of these minor weld stresses is expected to be significant too. The fraction of the weld strength reduction, in comparison with base pipe material, at creep range is called weld strength reduction factor, or simply, W factor.

The weld strength reduction factor is a very complex function of temperature, pipe base material, weld material, welding process, smoothness of the weld, and heat treatment of the weld. The code provides certain W factors for some materials. The listed factors should be used if available. As a general idea, the W factor for austenitic stainless steels and creep strength enhanced ferritic (CSEF) steels with normalizing plus tempering ($N + T$) post-weld heat treatment (PWHT) is 1.0 for 950°F (510°C) and below, and is 0.5 at 1500°F (815°C). The factor can be linearly interpolated for temperatures between 950°F and 1500°F.

The application of W factor depends on weld location and load category. The longitudinal weld and spiral weld, which are used in the production of the pipe, work only against the hoop stress in pressure design of the pipe and its components. Circumferential welds, which are used to assemble the piping, work only against longitudinal stress. Therefore, the W factor for circumferential weld is used in reducing the allowable stress for the sustained longitudinal stress. However, the W factor does not apply to occasional loads due to the load's short duration, which has little effect on creep failure. The W factor is also not applicable to expansion and displacement stresses due to their non-sustained natures, which have little effect on creep failure.

Application of the circumferential W factor would require the identification of all weld locations in the design analysis. This is simply impractical for non-nuclear industries with short project schedules. Therefore, the same approach used in setting the evaluation procedure for thermal expansion stress might be used here. In the development of thermal expansion stress evaluation procedure [9], we have assumed that the piping has a circumferential weld everywhere. With this assumption, we simply ignored the existence of circumferential weld and adjust the stress intensification factor and allowable stress accordingly. For the W factor, we can also assume that the circumferential weld exists at every point. In the case of W factor, we do not even have to adjust anything. The factor is used "as is" for all points.

4.4 PRESSURE DESIGN

The most basic and important requirement of pipe stress analysis is to ensure that each piping component is strong enough to contain the service pressure. However, most engineering design procedures delegate this important task to the systems engineering group rather than to the piping mechanical group. This is a necessity so that the systems engineering group can properly determine the pipe size for the flow rate needed for the process. Wall thickness, together with the required flow area, determines the pipe size. However, regardless of who determines the wall thickness, it is the pipe stress engineer's responsibility to ensure that the thickness is sufficient for the design pressure and other loads to be considered. Pressure design can be accomplished by several different ways.

- Use of standard components having established ratings

  Pressure-temperature ratings for certain piping components have been established in some of the standards approved by the applicable code. Typical examples are flanges (ASME B16.5, ASME B16.47, etc.) and valves (ASME B16.34). In using these components, the designer only needs to select the proper material type and the class of which the allowable pressure at the design temperature is equal to, or higher than, the design pressure. This is known as design by selection. No calculation is involved.

- Use of standard components not having specific ratings

  Some of the code approved standards for components, such as butt-welding fittings (ASME B16.9, ASME B16.11, etc.), do not provide specific pressure-temperature ratings, but state that the pressure-temperature ratings are based on the straight seamless pipe. Because the standards specify that the components shall be furnished in nominal thickness, the manufacturing under-tolerance, generally 12.5% for seamless pipe, has to be considered. When using these components, the designer only has to select the components that have the material and nominal wall thickness same as those of the connecting pipe. To ensure that the components have the same rating as the connecting pipe, local reinforcements are applied in the manufacturing of the components. For instance, an elbow or a tee has a considerably thicker wall at crotch areas than at the connecting ends.

- By design calculations

Piping components (e.g., straight pipe) with no established pressure ratings can be rated by design calculations. The code provides pressure design calculations for basic components such as straight pipe, smooth bend, miter bend, branch connection, reducers, blank, etc. Components whose design calculations are not given by the code are designed based on calculations consistent with the design criteria of the code.

4.4.1 Straight Pipe

The pressure resisting capability of the pipe is determined by its thickness. Calculation of the thickness required for a straight pipe is the first step toward the pressure design of all the components. This straight pipe thickness not only serves the important function of selecting the main portion of the piping, it also serves as the reference thickness for other components. Pipe thickness is selected in three steps: (1) From the design pressure, the minimum required net thickness is calculated. (2) The net thickness is then added with allowances, such as corrosion and erosion allowance, thread allowance, and manufacturing under-tolerance, to become the minimum nominal thickness. (3) The final step is to select a commercially available nominal thickness that provides this minimum required nominal thickness.

Minimum required net thickness. The minimum required net thickness of the straight portion of the piping is calculated via Eqs. (4.5), (4.6), and (4.7)

$$t = \frac{PD}{2(SEW + Py)} \tag{4.5}$$

$$t = \frac{P(d + 2c)}{2[SEW - P(1 - y)]} \tag{4.6}$$

$$t = \frac{PD}{2SE} \tag{4.7}$$

where

- t = minimum required net thickness
- P = design pressure
- D = outside diameter of the pipe
- d = inside diameter
- S = allowable stress of the pipe material at design temperature
- E = longitudinal joint efficiency or quality factor
- W = weld strength reduction factor
- c = allowance for corrosion, erosion, and others
- y = coefficient value as given in Table 4.1.

Equation (4.5) is the modified Lame equation with the y values taken based on empirical data allowing for some initial yielding at higher temperature ranges. Equation (4.6) is the inside diameter formula used by ID pipes with a fixed inside diameter. This inside diameter formula is obtained by substituting $D = (d + 2c + 2t)$ to Eq. (4.5). The allowance, c, is supposed to include manufacturing

TABLE 4.1
VALUES OF COEFFICIENT y (FOR t < D/6)

Temperature, °F	900 and below	950	1000	1050	1100	1150	1200	1250 and above
Temperature, °C	482 and below	510	538	566	593	621	649	677 and above
Ferritic steels	0.4	0.5	0.7	0.7	0.7	0.7	0.7	0.7
Austenitic steels	0.4	0.4	0.4	0.4	0.5	0.7	0.7	0.7
Other ductile materials	0.4	0.4	0.4	0.4	0.4	0.4	0.5	0.7
Cast iron	0.0							

The value of y may be interpolated for intermediate temperature.
For $t \geq D/6$, $y = (d + 2c)/(D + d + 2c)$.
The table taken from B31.3. B31.1 uses $y = 0.4$ for cast iron.
CEN uses an equivalent y value of $(1 - 0.5E)$, where E is joint efficiency.

under-tolerance that is generally a function of the nominal wall thickness. Therefore, a re-check may be required after the nominal wall thickness is determined. The inside diameter formula in B31.1 is expressed differently in terms of the thickness that includes the allowance.

Equation (4.7) is the simplified conservative formula generally referred to as Barlow's formula. This equation is the same as Eq. (2.14). Equation (4.7) can also be considered a special form of Eq. (4.5) by considering the y coefficient as zero. Equation (4.7) is very conservative and is generally not used in creep range application. Due to its simplicity, this equation is used extensively in piping literatures.

The above wall thickness calculation formulas have the allowable stress and joint efficiency grouped together. The (SE) have become a combined quantity. This is the value given in the code allowable stress tables in most codes. However, in B31.3 the code allowable stress table lists only the stress (S). Joint efficiency (E) is listed in a separate table. The joint efficiency (E) includes the efficiency of all welded seams that are not perpendicular to the axis of the pipe. This includes spiral welds as well as longitudinal welds. It also includes the quality factors of forgings and castings.

For $t \geq D/6$ or for $P/SE > 0.385$ high-pressure pipe. For high-pressure pipe, the calculation of thickness for pressure design generally involves the non-linear stress distribution through yielding of the pipe material. Although it can be conservatively used for any pressure, Eq. (4.5) becomes economically impractical and sometimes mathematically impossible for very high pressure pipes. Because the pipe has to have a passage to channel the fluid, the maximum thickness allowed is $\frac{1}{2}D$, which corresponds to zero inside diameter. By substituting $t = \frac{1}{2}D$ into Eq. (4.5), we have the maximum pressure corresponding to zero inside diameter as

$$P = \frac{(SE)}{1 - y} \quad \text{for zero inside diameter with Eq. (4.5)}$$

From this equation, it is clear that a different equation is needed for designing high pressure piping. The current approach is to look into stress distribution under the condition preceding rupture. At the instant before rupture, the yielding of pipe material redistributes the stress to the shape that greatly enhances the pressure resisting capability of the pipe. Because the design involves plastic flow of material, it requires special consideration of certain factors, such as theory of failure, effects of fatigue due to loading and unloading, and thermal stress due to temperature gradient and others. These considerations are too involved to be included in this book. Readers should consult Chapter IX [10] of B31.3 or other applicable codes for detailed requirements.

Straight pipe under external pressure. Straight pipe under external pressure is primarily designed with consideration to the stability against pipe shell buckling, in addition to the above limitation on hoop stress, which is in compression under external pressure. For long un-stiffened pipe, the allowable external pressure can be calculated with Eq. (1.3). The general procedure for determining the

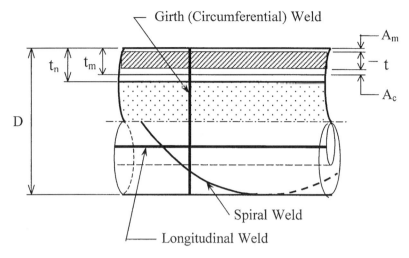

FIG. 4.4
STRAIGHT PIPE WALL THICKNESS

wall thickness and stiffening requirements for straight pipe under external pressure involves charts and figures too complex to be included in this book. Designers should consult ASME Boiler and Pressure Vessel Code, Section VIII, Division 1 [11], UG-28 through UG-30 and Appendix 5 for details.

Selection of commercially available nominal wall thickness. Selection of commercially available nominal wall thickness is the final step of the wall thickness calculation. The minimum required nominal thickness is calculated as follows

$$t_m = \frac{t + c}{(1 - u/100)} + v \quad (4.8)$$

where u is the manufacturing under-tolerance given by the percentage of the nominal thickness and v is the manufacturing under-tolerance specified by absolute thickness. For seamless steel pipes, u is generally specified as 12.5%. The value of v is applicable to some welded pipes rolled from plates with under-tolerance expressed in absolute thickness. Once the required minimum nominal thickness, t_m, is determined, a commercially available nominal thickness is selected to cover it. This means that the pipe will have some extra thickness not required by the pressure design. The involved quantities are graphically expressed in Fig. 4.4. In this graph, A_c represents c allowance, which includes corrosion, erosion, threading, and other mechanical allowances. A_m represents manufacturing allowance that includes u or v given in Eq. (4.8). t_n is the commercially available nominal thickness to be ordered. ($t_n - t_m$) is the extra thickness resulting from the selection of the commercially available thickness. For all calculations relating to nominal wall thickness, it refers to the selected and purchased commercially available nominal thickness t_n.

4.4.2 Curved Segment of Pipe

As discussed in Chapter 2, hoop stress in a curved segment of pipe varies around the circumference of the cross-section. With a uniform thickness around the cross-section, hoop stress is higher at the intrados (inside bend radius or crotch area) and is lower at the extrados (outside bend radius or crown area). This means that, for the same allowable stress, a thicker wall is required at the intrados and a thinner wall can be used at the extrados. Because the wall thickness required is proportional to the stress generated from the pressure, the stress intensification factor due to pressure can be directly used

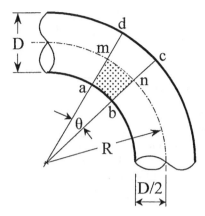

FIG. 4.5
PRESSURE STRESS AT BEND

to obtain the thickness required. The stress intensification factor can be derived from Fig. 4.5. Take a radial pie segment as a free body and divide it into two zones by the centerline *m-n*. This division by centerline is just an intuitive approach, but works real well as proved by other, more sophisticated methods. At the inside zone near the crotch, we have the pressure force acting in *a-b-n-m-a* area to be resisted by the pipe wall *a-b*. By equating the pressure force and stress force, we have

$$\left(\frac{1}{2}\theta R^2 - \frac{1}{2}\theta(R - D/2)^2\right) P = \theta(R - D/2)t\, S$$

or

$$t = \frac{4(R/D) - 1}{4(R/D) - 2} \frac{PD}{2S}$$

The last term on the right-hand side of the equation is equivalent to the thickness required for straight pipe as given by Eq. (4.7). A similar relationship can be derived for the outside zone of the pipe segment. Therefore, we have bend stress intensification factors due to internal pressure for both the stress and the required thickness as

$$I = \frac{4(R/D) - 1}{4(R/D) - 2} \quad \text{at intrados (crotch)} \quad (4.9a)$$

$$I = \frac{4(R/D) + 1}{4(R/D) + 2} \quad \text{at extrados (crown)} \quad (4.9b)$$

These stress intensification factors are used differently in different codes. In B31.3, these factors are used to reduce the allowable stress value in Eq. (4.5) by dividing the *SEW* value by the *I* factor. In CEN code, these factors are multiplied directly on the thickness calculated by Eq. (4.5). These codes allow a thinner wall thickness than that of the straight pipe at extrados. B31.1, on the other hand, requires that minimum thickness after the bending shall not be smaller than the thickness required for the straight pipe. Because thinning due to bending occurs at the extrados, the low stress area, this approach requires thicker wall at the extrados than the one theoretically required by B31.3 and CEN codes. B31.1 does not require the checking of the extra thickness required at the intrados area. It is assumed that once the thickness at the extrados area is maintained at straight pipe level, the extra thickness required at the intrados area is automatically satisfied due to bending thickening. Table 4.2 is the guideline, given by B31.1 and other codes, for pipe thickness to be ordered for satisfying the

TABLE 4.2
PIPE THICKNESS FOR BENDING

Radius of bends	Minimum thickness recommended before bending
Six pipe diameters or greater	$1.06 t_m$
Five pipe diameters	$1.08 t_m$
Four pipe diameters	$1.14 t_m$
Three pipe diameters	$1.25 t_m$

Interpolation is permissible for intermediate radii.
t_m is the minimum required nominal thickness for straight pipe.
Pipe diameter is the nominal diameter.

minimum wall thickness requirement after the bending. Allowances and manufacturing tolerances should be properly included when ordering.

Forged elbows. Forged elbows are the most common curved segments used in providing turns in a piping system. These elbows, manufactured by the approved standards listed in the code, are suitable for the design pressure given by the pressure-temperature rating of such standards. They have either satisfied the minimum thickness requirements given above or passed the proof test requirements of the applicable standard.

4.4.3 Miter Bends

The direction changes of a piping are sometimes accomplished by connecting mitered pipe ends together to form an angular offset. These mitered connections are called miter bends. B31.3 states that an offset of 3 deg. or less does not require the design consideration as a miter bend, whereas B31.1 considers a joint with θ angle (one-half of the offset) less than that given by Eq. (4.10) to be equivalent to a girth butt-welded joint, and the rules of the miter joints do not apply.

$$\theta < 9\sqrt{\frac{t_n}{r}} \quad (\theta \text{ in deg.}) \tag{4.10}$$

The above offset allowance is applicable only to welds that are located sufficiently apart, thereby satisfying the criterion of widely spaced miter joints defined later in this section.

A miter bend usually consists of many mitered joints generally referred to as miter cuts. A mitered joint can be considered an equivalent bend in the design calculations. Figure 4.6 shows the parameters used in miter bend design calculations. First, each miter welded offset is considered as an equivalent bend having a bend radius of $R_e = r(1 + \cot \theta)/2$ [9]. This also represents an equivalent bend tangent width of $w = R_e \tan \theta = r(\tan \theta + 1)/2$, where r is the mean radius of the pipe cross-section.

Widely spaced miter bend. Widely spaced miter bend is a miter bend having a miter distance, s, greater than $2w$. In this case, the ends of the equivalent bends do not touch each other. The whole miter bend is considered a series of equivalent bends interconnected by short straight pipes. The analysis is also performed in such a way that considers it as an assembly of bend-straight-bend-straight, etc.

Closely spaced miter bend. Closely spaced miter bend is a miter bend having a miter distance, s, shorter than $2w$. In this case, the ends of the equivalent bends either touch or overlap each other. The whole miter bend is considered one single equivalent bend. Because of the overlapping of the equivalent tangent width, the equivalent bend radius is reduced. The actual equivalent radius is determined by the miter distance and miter angle as $R_e = (s/2)/\tan \theta = (s \cot \theta)/2$. For closely spaced miter bends, the equivalent bend radius equals the construction radius, R. That is, $R_e = R$.

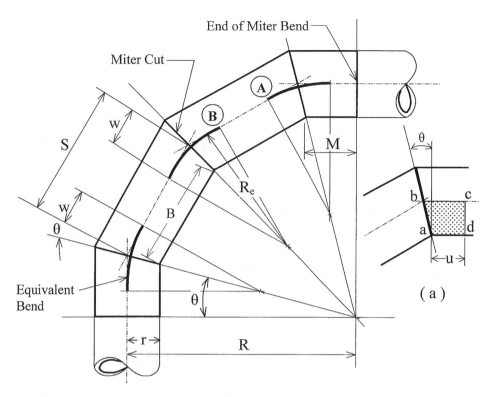

**FIG. 4.6
MITER BEND CHARACTERISTICS**

Miter bends have several general limitations. Miter bends have several general limitations in their applications. First, to avoid excessive degradation of the heat affected zone, there is a minimum requirement of the crotch distance, B. B31.1 states that for an internal pressure higher than 10 psi (70 kPa), distance B shall be greater than $6t_n$. That is, $B > 6t_n$ for B31.1. B31.3 and CEN, on the other hand, state that the construction bend radius, R, shall not be less than that given by Eq. (4.11)

$$R = \frac{A}{\tan \theta} + r \tag{4.11}$$

where A has empirical values given in Table 4.3. Because $A = (R - r)\tan \theta$, the value of A is equivalent to one-half of B used by B31.1.

Another limitation is that a miter bend with θ greater than 22.5 deg. can be used only when the pressure is less than 10 psi (70 kPa). Although B31.3 allows the use of a miter bend with θ greater than 22.5 deg. at higher design pressure, an increased safety factor is applied in the thickness calculation of those miter bends.

Miter bend thickness is calculated by assuming an effective bridging width, u, to resist the pressure at the zone near the miter weld. As in the case of a smooth bend, the critical stress of a miter bend is also located at the crotch area. At the crotch area, the effective pipe wall width u (length a-d), as shown in Fig. 4.6(a), is assumed to resist the pressure force acting on the trapezoid area a-b-c-d. Based on the theory of beams on elastic foundation (see the chapter dealing with discontinuity stresses for details of β and others), this effective width is set by using $u\beta = 1$. That is,

$$u = \frac{1}{\beta} = \sqrt[4]{\frac{t^2 r^2}{3(1 - v^2)}} = 0.781 \sqrt{rt}, \quad \text{for } v = 0.3$$

TABLE 4.3
VALUES OF A FOR MITER BEND

(1) For U.S. customary units		(2) For SI metric units	
$(t_n - c)$, in.	A, in.	$(t_n - c)$, mm	A, mm
≤ 0.5	1.0	≤ 13	25
$0.5 < (t_n - c) < 0.88$	$2(t_n - c)$	$13 < (t_n - c) < 22$	$2(t_n - c)$
≥ 0.88	$[2(t_n - c)/3] + 1.17$	≥ 22	$[2(t_n - c)/3] + 30$

$(t_n - c)$ is the net thickness excluding corrosion and other allowances.

where β is the characteristic parameter of the pipe treated as a beam on elastic foundation. The pressure force is equal to pressure times the trapezoid area.

$$\text{Pressure force} = Pr\left(0.781\sqrt{rt} + 0.5r\tan\theta\right)$$

The stress force equals the stress times the effective pipe wall area.

$$\text{Stress Force} = St\left(0.781\sqrt{rt}\right)$$

By equating pressure force and stress force, and substituting the stress $S = SEW$ as the allowable stress, we have

$$Pr\left(0.781\sqrt{rt} + 0.5r\tan\theta\right) = SEW(t)\left[0.781\sqrt{rt}\right]$$

From this relationship, we have the following two working equations for determining the allowable pressure for a given nominal wall thickness and for determining the minimum net thickness required for a given design pressure

$$P = \frac{SEW(t_n - c)}{r}\left(\frac{t_n - c}{(t_n - c) + 0.64\tan\theta\sqrt{r(t_n - c)}}\right) \quad (4.12)$$

$$t = \frac{Pr}{SEW}\left(1 + 0.64\sqrt{r/t}\tan\theta\right) \quad (4.13)$$

Equation (4.12) uses $t_n - c$ instead of t. Equation (4.13) requires iterations because thickness t exists at both sides of the equation.

Equations (4.12) and (4.13) may not be applicable to closely spaced miter bends. This is mainly because B width can be less than twice of the effective u width in closely spaced miter bends. However, since B and u do not have any definite relations, it is not possible to set an exact rule as to when these two equations are not applicable. The codes adopt the following approaches:

- For B31.1, the division is whether the miter bend is closely spaced. If it is, thickness is calculated based on the smooth bend formula. It is the thickness required for the straight pipe multiplied with the stress intensification factor at intrados as given by Eq. (4.9a). If it is a widely spaced one, then the above equations apply.
- For B31.3 and CEN, they only distinguish whether it is a multiple miter bend or a single miter bend. If it is a single miter bend, then the above equations apply. If it is a multiple miter bend, then both the above equation and the smooth bend approach should be investigated. The thicker wall resulting from the two approaches is used.

Although B31.3 permits the use of miter bends with θ greater than 22.5 deg. for pressures higher than 10 psi (70 kPa), the 0.64 constant in Eqs. (4.12) and (4.13) has to be doubled to 1.25 when θ is greater than 22.5 deg.

Extension of the miter bend thickness. The miter bend thickness should extend at least a width M from the intrados weld as shown in Fig. 4.6. M should be greater than the value given by the following Eq. (4.14).

$$M = \text{The greater value of } \left[2.5\sqrt{rt_n} \text{ and } \tan\theta(R-r)\right] \tag{4.14}$$

Curved or mitered segment of pipe under external pressure. The curvature at a bend and the miter joints at a miter bend serve as buckling arrestors against external pressure. They have a similar effect as stiffening rings. Therefore, with the same cross-section and material, a curved or miter segment of pipe is more stable than the straight segment of pipe under external pressure. B31.3 states that the wall thickness of a curved or mitered segment of pipe under external pressure may be determined as specified for straight pipe.

4.4.4 Branch Connections

Branching of a piping system begins from a branch connection. Figure 4.7 shows some of the common types of branch connections used in piping systems. Type (a) is the un-reinforced fabricated tee, commonly referred to as a stub-in connection. This type of connection is feasible at low-pressure applications when the pipe is generally thicker than required by the design pressure. The allowable working pressure of this type of connection depends on the branch to run diameter ratio. For a large diameter ratio, the allowable pressure is roughly only about 50% of the pressure allowed by the pipe. Type (b) is the reinforced fabricated tee, which has a reinforcement pad welded at the junction to strengthen the pressure resisting capability. This type of joint can be designed to resist the design pressure up to the same full pressure resisting capability of the pipe. The reinforcement pad thickness is generally the same as the pipe thickness so it can be cut from the same batch of pipe material. A pad thickness of 1.5 times the pipe thickness is occasionally used if required either by pressure or by moment stress. Type (c) is the forged welding tee manufactured under the code approved list of standards such as ASME B16.9. A welding tee can take the same pressure as the matching pipe. Type (d) is the extruded header outlet or tee. This type of outlet is formed by extrusion of a segment of pipe, which is generally thicker than the main connecting pipe. Multiple outlets can be formed in the same segment of pipe to become a header. Type (e) is the integrally reinforced branch connection. It includes specially contoured forgings as well as heavy wall half-couplings. These are commercial items that are designed to provide 100% of the pressure resisting capability as the connecting pipe.

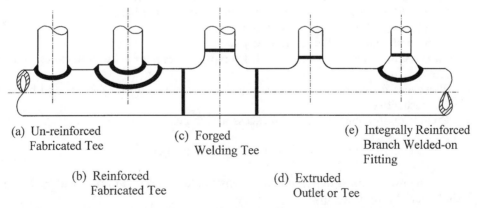

FIG. 4.7
DIFFERENT TYPES OF BRANCH CONNECTIONS

Welding tees and integrally reinforced fittings are normally used without special calculations. They are implemented in the design by selecting the ones with pressure rating at the corresponding design temperature, equal to or higher than the design pressure. Fabricated tees and extruded outlets, on the other hand, require calculations to ensure that they are suitable for the design pressure. Calculations are based on the age-old rule-of-thumb approach called the "area replacement" method. In this method, the area of the pipe wall removed by the opening has to be compensated with an equal amount of surplus area located near the opening. This is a very low-technology approach that probably would not pass a strict scrutiny based on modern technology. However, this approach, which has been used very successfully for close to a century, can be considered one of the most important rules invented by the piping and pressure vessel community.

This section mainly discusses the pressure design of the components. However, it should be noted that even though a connection is designed to withstand the same full pressure as the connecting pipe, there are still stress intensifications associated with the connection when stresses due to bending and twisting moments are being calculated.

(a) Reinforcement of Fabricated Branch Connections

Fabricated branch connections are not very easy to make for full-size outlets, but are very popular for reduced outlet connections. They are less expensive than welding tees and are also readily available in the field and at fabrication shops. This availability is very important especially for a remotely located project. The main task in designing the fabricated branch connection is ensuring that the connection is good enough to resist the design pressure. This is achieved by providing proper reinforcement.

Based on the area replacement rule, the reinforcement requirement of the fabricated branch connection is checked by calculating the area removed by the opening and the surplus areas available. The surplus areas include the inherent surplus pipe wall and the added reinforcement from the pads and the welds. When calculating the areas, all the corrosion, erosion, and manufacturing under-tolerance have to be considered. Figure 4.8 shows a general outline of a fabricated branch connection. The following are the definitions of the key areas:

A_1 = $t_h d_1 (2 - \sin \beta)$, is the reinforcement area required. It is the opening multiplied with the $(2 - \sin \beta)$ factor to compensate for the higher stress intensification due to oblique junction of a non-perpendicular connection. This factor is 1.0 for perpendicular connections

d_1 = $[D_b - 2(T_b - c_b)]/\sin \beta$, is the longitudinal length of the opening

d_2 = half-width of the effective reinforcement zone
= d_1 or $(T_b - c_b) + (T_h - c_h) + d_1/2$, whichever is greater, but no more than D_h, the nominal diameter of the header

t_h, t_b = required net thickness as calculated by Eq. (4.5) for header and branch, respectively

L_4 = altitude of the effective reinforcement zone outside of the header
= $2.5(T_b - c_b) + T_r$ or $2.5(T_h - c_h)$, whichever is smaller

T_r = minimum thickness of the reinforcing pad
Subtracts the under-tolerance if made from pipe; uses nominal wall thickness if made from plate

c_h, c_b = Allowances for corrosion, erosion, and under-tolerance of the header and branch thickness, respectively

A_2 = available reinforcement area from excess wall in header
= $(2d_2 - d_1)(T_h - t_h - c_h)$

A_3 = available reinforcement area from excess wall in branch
= $2L_4(T_b - t_b - c_b)/\sin \beta$
If the allowable stress of the branch pipe is smaller than that of the header, then A_3 has to be reduced proportionally

A_4 = area of the reinforcement pad, if any, plus the welds inside the effective reinforcing zone defined by d_2 and L_4.

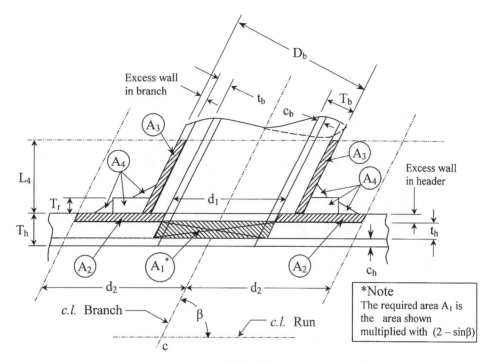

FIG. 4.8
REINFORCEMENT OF BRANCH CONNECTION

When the total of the available reinforcement areas ($A_2 + A_3 + A_4$) is greater than the required reinforcement area, A_1, the connection is considered suitable for the design pressure.

(b) Allowable Pressure for the Stub-in Branch Connection

The stub-in branch connections are popular in low-pressure applications, such as building service piping and plant utility piping. Due to the availability limitation of the commercial pipes and the requirement of supportability, the wall thickness of the low-pressure pipe is generally much thicker than required for the design pressure. Therefore, a stub-in connection without any reinforcing pad is often sufficient for the design pressure. As a rule of thumb, if the design pressure is less than 50% of the allowable pressure of the pipe, then the stub-in connection is satisfactory for the design pressure. Other than the above 50% reduction criterion, a calculation is generally required for higher pressure to validate the adequacy of the connection.

Figure 4.9 shows a 90-deg. pipe-to-pipe stub-in branch connection. The procedure given in Section 4.4.4, (a) can be used directly to check if the connection is strong enough for the design pressure. In the following, a procedure will be developed to find the maximum allowable pressure of the connection. Based on $\sin \beta = 1.0$ and $T_r = 0$, we have

$$t_h = \frac{PD_h}{2(SE + Py)}, \quad t_b = \frac{PD_b}{2(SE + Py)}$$

$$A_1 = d_1 t_h, \quad A_2 = (2d_2 - d_1)(T_h - c_h - t_h)$$

$$A_3 = 2L_4(T_b - c_b - t_b), \quad A_4 = w^2$$

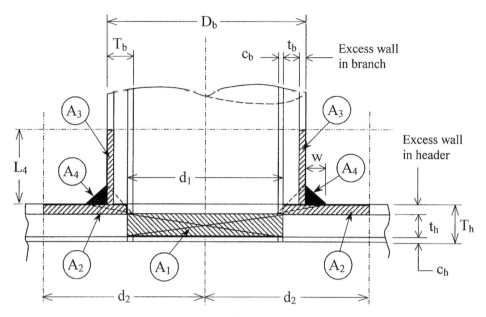

FIG. 4.9
90-DEGREE PIPE TO PIPE (STUB-IN) BRANCH CONNECTION

Equating the required reinforcement area and the available reinforcing areas, $A_1 = A_2 + A_3 + A_4$, we have

$$d_1 t_h = (2d_2 - d_1)(T_h - c_h - t_h) + 2L_4(T_b - c_b - t_b) + w^2$$

or

$$d_1 t_h = (2d_2 - d_1)(T_h - c_h) - (2d_2 - d_1)t_h + 2L_4(T_b - c_b) - 2L_4 t_b + w^2$$

or

$$2d_2 t_h + 2L_4 t_b = (2d_2 - d_1)(T_h - c_h) + 2L_4(T_b - c_b) + w^2$$

Assigning the right-hand side to $K = (2d_2 - d_1)(T_h - c_h) + 2L_4(T_b - c_b) + w^2$, and substituting the wall thickness for header and branch pipes with the corresponding formulas, the above becomes

$$2d_2 \frac{PD_h}{2(SE + Py)} + 2L_4 \frac{PD_b}{2(SE + Py)} = K$$

or

$$P = \frac{K(SE)}{d_2 D_h + L_4 D_b - Ky} \quad (4.15)$$

Equation (4.15) is just a working formula. The solution requires a couple of iterations, because d_1 and d_2 are loosely related to P. The equation does not show any simple relationship between the allowable pressure for the main pipe and the un-reinforced branch connection. One thing for sure is that the un-reinforced branch connection can take at least 50% of the pressure allowed by the un-perforated pipe. Although the stub-in connection may be strong enough for the allowable pressure calculated, it has considerably greater stress intensification factors against moment loading than other types of branch connections.

4.4.5 Pressure Design for Other Components

Besides the straight pipe, bend, and branch connections discussed above, the code also has formulas and procedures for cone, reducer, flange, blank, and others. For components whose design formulas are not given, the code provides the following general guidelines.

Listed components. The pressure containing components manufactured in accordance with the code-approved list of standards can be used according to the pressure-temperature rating given by the applicable standard.

Unlisted components. The components that do not belong to the code-approved list of standards shall be based on theoretical calculation, finite element analysis, experimental stress analysis, proof test, or their combinations. Extensive, successful service experience under comparable condition can be considered a type of proof test.

4.5 STRESSES OF PIPING COMPONENTS

Because not all stresses are created equal, we have to know what stress to calculate first before we proceed to calculate. The ASME piping and pressure vessel codes generally divide the stress into three categories: primary, secondary, and peak. However, in non-nuclear piping, we seldom calculate peak stress, which is implied in the stress intensification factors derived mainly from fatigue tests. Instead, in non-nuclear piping the stresses are divided into three main categories: sustained, displacement, and occasional. Each category of stress has its own method of calculation and evaluation. Occasional stress is generally evaluated in conjunction with sustained stress. To comply with the code stress requirements, each category of stress at every component is calculated and compared with the allowable stress. Allowable stress is generally expressed in terms of cold and hot basic allowable stresses denoted by S_c and S_h, respectively. Other than for the pressure design discussed in the previous section, the longitudinal joint efficiencies are not applicable in the evaluation of all three categories of stresses. If the allowable stress given by the code table includes a longitudinal joint efficiency, the basic allowable stress can be obtained by dividing the code table stress by the longitudinal joint efficiency.

4.5.1 Calculations of Component Stresses

The code stress category is based on the load category. A sustained load creates sustained stress, and so forth. Stress calculation procedures are all the same regardless of the load or stress category.

The first step of stress calculation is to apply the corresponding category of loads to the piping system to find out the responding internal forces and moments at each component. This is the so-called structural analysis involving finite element or other appropriate methods. The routine piping stress analysis involves only straight beam and curved beam elements. The structural analysis deals only with the centerline of the piping system. It calculates the total forces and moments acting at the entire piping cross-section of each centerline point. These forces and moments are then used to calculate the stresses.

Calculations of the forces and moments at each component follow the basic principle of structural mechanics, which is unvarying and straightforward. However, the calculation of code stresses differs fairly appreciably among the codes. Lately, there have been significant attempts [12] in ASME to keep the requirements similar among all codes, especially between B31.1 and B31.3. However, some differences will no doubt remain as they are. These differences are the driving force separating the code into different sections in the first place.

The standard stress formulas given by the codes, such as B31.1, include only the moments. In other words, the forces, except pressure force, are ignored in stress calculations. This is because the stresses due to forces are generally very small. In cases where stresses due to forces are significant, they will be reflected in the unusually huge loads generated at the anchors and restraints. Once these huge loads

114 Chapter 4

at the anchors and restraints have been reduced to an acceptable level, pipe stresses due to forces will be reduced to an insignificant level.

Figure 4.10 shows the moments acting at a bend and at a branch connection, which consists of three straight segments of pipe. Because of the different stress intensification factors involved at different component types and orientations, the moments calculated from the structural analysis have to be re oriented in accordance with the component plane before being used to calculate the stresses. The legs of a bend or a branch connection form the plane of the component. The moment causing the leg to bend in the plane is called the in-plane bending moment, M_i, and the moment causing the leg to bend in the direction perpendicular to the plane is called the out-plane bending moment, M_o. The moment twisting the leg is called the torsion moment, M_t.

The stresses are calculated at the nodes located at the ends of each element. Each node at the element is independently calculated using the forces and moments occurring at that node point. This approach sounds reasonable for non-branched elements. For a branch connection, however, the forces and moments from all connecting legs are expected to act together at the common branch point, which is the junction of the legs. This naturally raises the question of the validity of calculating the stress on each leg independently. However, this appears to be not much of a concern in non-nuclear piping. Class 1 nuclear piping, to be discussed later in this chapter, considers the forces and moments from all three legs together at a branch connection.

The stresses due to these moments are calculated as follows (see Section 2.5.2 for stress formulas and Table 3.1 for stress intensification factors):

- *In-plane bending stress* varies around the cross-section of the pipe with the maximum stress at the extreme fiber equal to

$$S_{bi} = \frac{M_i}{Z} i_i \qquad (4.16a)$$

- *Out-plane bending stress* also varies around the cross-section of the pipe with the maximum stress at the extreme fiber equal to

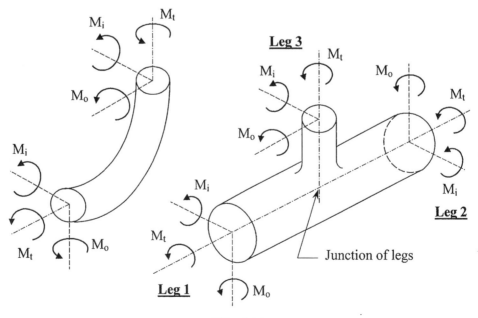

FIG. 4.10
TORSION MOMENT, AND IN-PLANE AND OUT-PLANE BENDING MOMENTS

$$S_{bo} = \frac{M_o}{Z} i_o \qquad (4.16b)$$

- *Combined bending stress* is the combination of in-plane and out-plane bending stresses. Because the maximum fiber stresses of the in-plane and out-plane bending moments are located 90 deg. apart, they can be combined by the square root of sum of squares (SRSS) method. That is,

$$S_b = \sqrt{(S_{bi})^2 + (S_{bo})^2} = \frac{\sqrt{(i_i M_i)^2 + (i_o M_o)^2}}{Z} \qquad (4.16c)$$

The combined bending stress is in the longitudinal direction of the pipe. It has the maximum tensile stress at one point and varies gradually through the cross-section. Eventually, it changes to compression and reaches the maximum compression at the diametrically opposite location. When combining with other stresses, we have to keep in mind that the bending stress always has equal magnitude of tension and compression in a cross-section.

- *Torsion stress* is a shear stress uniformly distributed around the circumference of the cross-section, but varies linearly in the diametrical direction. The maximum shear stress at the outer surface is

$$\tau = S_t = \frac{M_t}{Z_p} i_t = \frac{M_t}{2Z} i_t \qquad (4.16d)$$

The polar (torsion) section modulus, Z_p, is twice the bending section modulus, Z. The torsion stress intensification factor, i_t, is currently considered as unity for most of B31 codes.

Stresses given by Eqs. (4.16c) and (4.16d) can be evaluated separately, or in combination, depending on the allowable stress criterion given by the design specification or code.

- *Stress intensity and expansion stress.* Thermal flexibility and most modern design analyses are based on the combined effect of all multi-dimensional stresses involved. ASME B&PV and B31 codes adopt the maximum shear failure theory in evaluating the combined effect (see Section 2.6.3) The combined quantity used is called the stress intensity, which is twice the maximum shear stress. However, this combined quantity is not called the stress intensity in B31 because it uses only one-half of the theoretical stress intensification factors. Rather, it is often referred to as the expansion stress. This is because in B31 codes, this combined quantity was used first in the evaluation of thermal expansion stress range, S_E, but not in anything else.

Stress intensity can be calculated based on the general two-dimensional formula given by Eq. (2.19). In the B31 evaluation of thermal expansion stress, the cyclic effect of pressure is not included. The pressure effect is implied in the establishment of the allowable stress. Therefore, when evaluating expansion stress we have only the longitudinal bending stress and the torsion shear stress to consider. By setting S_y in Eq. (2.19) as zero, we have

$$S_E = S_i = \sqrt{S_b^2 + 4\tau^2} = \sqrt{\frac{(i_i M_i)^2 + (i_o M_o)^2}{Z^2} + 4\left(\frac{i_t M_t}{2Z}\right)^2}$$

or

$$S_E = S_i = \frac{1}{Z}\sqrt{(i_i M_i)^2 + (i_o M_o)^2 + (i_t M_t)^2} \qquad (4.17)$$

Equation (4.17) is used in B31.3 with i_t set to unity. The above equation can be simplified by conservatively using a uniform stress intensification factor, i, which is the greatest of (i_i, i_o, and i_t). Taking this conservative approach, we have a new working formula as follows:

$$S_E = S_i = \frac{i}{Z} M_R, \quad \text{where,} \quad M_R = \sqrt{M_i^2 + M_o^2 + M_t^2} \qquad (4.18)$$

Equation (4.18) is used by B31.1, CEN, and nuclear piping codes.

- *Stress calculation for branch leg.* For branch connections, pipe stresses are calculated for each leg of the pipe using the moments, acting at the centerline intersection point, and the section modulus of the leg being evaluated. However, stresses at the branch leg (leg-3) are calculated differently. For the branch leg, the moments used are still the ones that are acting on the leg, but the section modulus used is an effective section modulus. This can be explained with the help of Fig. 4.11.

Figure 4.11 shows the potential locations of failure due to branch moments [13, 14]. When the stress intensification factor is significant, the failure will most likely occur at the header wall right next to the wall-to-wall junction. This is shown as "Failure Location A" in the figure. If the stress intensification is not significant, the pipe will crack at the branch pipe wall next to the junction marked as "Failure Location B." B31 codes use the equivalent section modulus to cover both potential failure locations.

At location A, in addition to stress intensification, the strength and stress of the pipe depend on the wall thickness of the header, not the wall thickness of the branch. Therefore, it is logical to use a section modulus based on the header thickness. This is the first consideration of the effective section modulus, which is defined as

$$Z_e = \pi r_b^2 t_e \tag{4.19}$$

where

- Z_e = effective section modulus to be used, instead of the branch section modulus, Z_b, in calculating stress at branch leg
- r_b = mean radius of the branch pipe. For integrally reinforced connections with sufficient length of uniform cross-section, the reinforcement section can be treated as the branch pipe for the connection
- t_e = effective thickness. It is the smaller of T_h or (iT_b). This T_h covers location A and (iT_b) covers location B of Fig. 4.11
- T_h = header thickness, excluding reinforcement
- T_b = branch thickness, excluding reinforcement
- i = applicable stress intensification factor; can be conservatively taken as the smallest of all the factors involved

The inclusion of iT_b in the definition of t_e is supposed to take care of location B. The application of Eq. (4.19) can be very confusing in actual cases. Because each component of the moment has its own stress intensification factor, this leads to the necessity of using a different equivalent section modulus for each moment component. Theoretically, it requires a different equivalent section

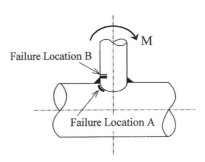

FIG. 4.11
STRESSES AT BRANCH PIPE

modulus for torsion moment, in-plane bending moment, and out-plane bending moment. Even with the approach of conservatively using a uniform stress intensification factor for all moments, as in B31.1, we still have to consider the different intensification factors for different categories of stresses. For instance, B31.1 uses $0.75i$ as the stress intensification factor for the sustained and occasional stresses. In this case, $0.75iT_b$ has to be used instead of iT_b.

The other likely confusion comes from the corrosion allowance and manufacturing under-tolerance of the wall thickness. Nominal wall thickness is used in expansion stress calculations, but the net wall thickness subtracting the allowance is generally used in sustained stress calculations. This will result in a different equivalent section modulus than that given by Eq. (4.19). This confusion can be alleviated by the use of equivalent stress intensification factors instead.

- *Equivalent stress intensification factors* can be used to simplify the process of evaluating the stresses at the branch leg. In this method, the stress at the branch leg is calculated with the unmodified original branch leg section modulus, but with an equivalent stress intensification factor, which is defined as

$$i_{e,x} = i_x \frac{Z_b}{\pi r_b^2 t_h} = i_x(Z_R) \qquad (4.20)$$

where

$i_{e,x}$ = equivalent stress intensification factor for i_x; $i_{e,x} \geq 1.0$
i_x = stress intensification factor of the applicable moment component ($=i_i$, i_o, i_t, i, $0.75i$, etc.)
Z_b = section modulus of the branch pipe
t_h = applicable header thickness
Z_R = ratio of branch pipe section modulus versus the section modulus of the equivalent header ring, as shown in the formula

This equivalent stress intensification factor takes care of location A, and the limitation of the $i_{e,x} \geq 1.0$ will automatically take care of location B.

4.5.2 Sustained Stresses

Sustained stress is the most important stress in the piping system. In the past, most piping systems were designed by considering sustained stress only. Sustained stress comes mainly from pressure, weight, and occasional loads, such as wind, earthquake, and water/steam hammer. Wind, earthquake, and water hammer loads occur only occasionally with short duration. These loads are classified as occasional loads and are treated separately.

Sustained stress due to pressure and weight is evaluated differently for different codes. Figure 4.12 shows the major components of the stresses acting on the pipe wall. The pressure produces S_{HP} and S_{LP} in the circumferential (hoop) and longitudinal directions, respectively. There is also a direct compressive stress acting at the inside surface (or the outside surface, if external pressure) due to pressure. However, this direct compressive stress, which has the same magnitude as the pressure, is ignored in most applications. For discussion purpose, S_{LP} can be regarded as one-half of S_{HP}. S_{HP} is kept below the allowable stress by the pressure design discussed previously.

The weight creates bending stress, S_{LW}, and torsion shear stress, τ. Stresses due to direct forces are again ignored. The pressure stress and shear stress are considered uniform around the pipe circumference, whereas the bending stress varies around the circumference. The point of interest is where the maximum or minimum bending stress is located.

For easy visualization, we will use the two-dimensional stress intensity formula given by Eq. (2.19) as guide for the discussion. The formula shows that the signs are very important for two mutually perpendicular normal stresses. If these two stresses have the same sign (say both of them are tensile), then

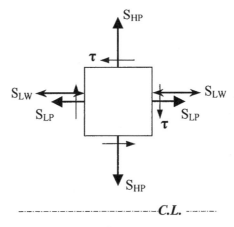

**FIG. 4.12
SUSTAINED STRESSES**

their difference will be counted as effective stress. In cases when both directions of the stresses on the plane shown have the same sign, the combination with the other dimension that is perpendicular to the pipe wall surface will govern. Because the stress at that direction is either zero (outside surface) or comparably small (inside surface), the combination with the third direction can be considered as due to either the longitudinal stress or the hoop stress alone. If the two perpendicular stresses have different signs, say one is compressive and the other tensile, then the combined effect is their absolute sum. Because bending stress can be either tensile or compressive, it is possible that the total longitudinal stress, combining S_{LW} and S_{LP}, is compressive. In such cases, the longitudinal stress will have to be combined with the hoop pressure stress, S_{HP}. However, with the common sustained stresses shown in Fig. 4.12 and the fact that $S_{HP} = 2S_{LP}$, the combination of hoop and longitudinal stresses is not required as explained below.

For simplicity, the following discussion assumes that the pipe is subjected to internal pressure. Longitudinal stress is calculated as $S_L = S_{LP} + S_{LW}$. Because S_{LW} can either be tensile or compressive, the maximum potential compressive longitudinal stress is $S_{L,min} = S_{LP} - S_{LW}$. Compressive longitudinal stress is additive to the tensile hoop stress in the combined effect. When S_{LW} is greater than S_{LP}, $S_{L,min}$ becomes compressive and is required to be combined with S_{HP}. The combined stress intensity can be taken as the difference of the two stresses. That is, $S = S_{HP} - S_{L,min} = S_{HP} - (S_{LP} - S_{LW}) = 2S_{LP} - S_{LP} + S_{LW} = S_{LP} + S_{LW}$. This is the sum of the longitudinal pressure stress and the weight bending stress, all in the longitudinal direction alone. This shows that the combination of the longitudinal stress and hoop stress is not required.

Evaluation of sustained stresses varies significantly among the codes. The main difference lies in the use of stress intensification factors [15]. The evaluation approaches can be classified into the following three groups.

(1) *B31.1 Power Piping and CEN codes.* These codes have a definitive equation setup for evaluating sustained stress.

$$S_L = \frac{PD_o}{4t_n} + \frac{(0.75i)M_A}{Z} \leq 1.0S_h \tag{4.21}$$

M_A is the resultant moment due to weight and other sustained loads. It also includes the torsion moment. Although the use of the S_L symbol has the appearance of longitudinal stress, it actually is combined stress intensity. There are two important items to note. One is the use of the nominal wall thickness; the other is the use of $0.75i \geq 1.0$ as the stress intensification factor. At creep range, the allowable stress of $1.0S_h$ has to be modified with weld strength reduction factor, W, to $1.0S_h W$ for circumferential welds.

(2) *Class 2 and Class 3 Nuclear Piping.* Although Class 2 and Class 3 nuclear piping codes follow pretty much the same philosophy and approach as B31.1 power piping, their evaluations of sustained stresses are fundamentally different. They actually adopt the Class 1 nuclear piping approach in the evaluation of sustained stress. The evaluation is given by

$$S_{SL} = B_1 \frac{PD_o}{2t_n} + B_2 \frac{M_A}{Z} \leq 1.5 S_h \qquad (4.22)$$

S_{SL} can be considered as special (or sustained) longitudinal stress. The above formula is similar to (4.21), but uses a different set of stress intensification values. These stress intensification values are called *stress indices* in Class 1 nuclear piping. The pressure term is actually based on hoop stress, adjusted by the stress index B_1 (normally, 0.5). The moment term has its own stress index, B_2. These stress indices are generally different from the stress intensification factors used in B31 codes. Because the basis for calculation is different, the allowable stress is also different.

(3) *B31.3 Process Piping and Others.* Process piping code and some of the other codes specify only the allowable for the longitudinal stress, without giving any specific formula for the calculation. B31.3, for instance, stipulates that the sum of longitudinal stresses, S_L, in any component in a piping system, due to pressure, weight, and other sustained loadings shall not exceed the basic hot allowable stress, S_h, with weld strength reduction factor, W, applied for circumferential welds. It does not explicitly give the formula for the calculation. Normally, only the bending stress as given in Eq. (4.16c) and the longitudinal pressure stress are included. The stress due to the torsion moment is not included as it is not a longitudinal stress. The stress due to direct axial force may also be included, but is generally insignificant for sustained loads. In the calculation of the sustained stress, the net wall thickness after subtracting the mechanical, corrosion, and erosion allowance is used. The load due to weight is based on nominal thickness of the pipe. Because no specific formula is given, there is a controversy in the use of the stress intensification factor. Currently, different companies tend to use different stress intensification values. Some use full code stress intensification factors, whereas other companies use no stress intensification at all. The main argument by the latter is that the code stress intensification factors are developed specifically for fatigue evaluation of expansion and other self-limiting displacement stresses.

4.5.3 Occasional Stresses

Occasional stresses are sustained stresses that occur only occasionally. They are generated by such loads as earthquake, wind, steam/water hammer, fluid transient forces and others. The code occasional stress category deals only with sustained portion of the stress. The self-limiting portion, such as the stress produced by support displacements caused by an earthquake, is combined with the expansion stress. Because occasional stress is sustained stress, its treatment has the same differences between different codes as in the sustained stress.

The occasional stresses of a piping system can involve a few potential loads. The combinations and evaluations of these loads are not standardized. Figure 4.13 shows the flow chart of the procedure normally used in the evaluation of all postulated occasional loads. The most frequently encountered loads are earthquake and wind. These loads, if present, apply to all piping systems in the plant. Other loads, such as fluid transient force and harmonic force, are applicable only to certain specific piping systems.

The evaluation of occasional stress involves many load cases. Earthquake and wind have to be investigated from three orthogonal directions. Each direction of load is one load case. Fluid transient forces can come from different events. Loads from each event are grouped into one load case. Similarly, harmonic forces with the same frequency are grouped together to become one load case. These

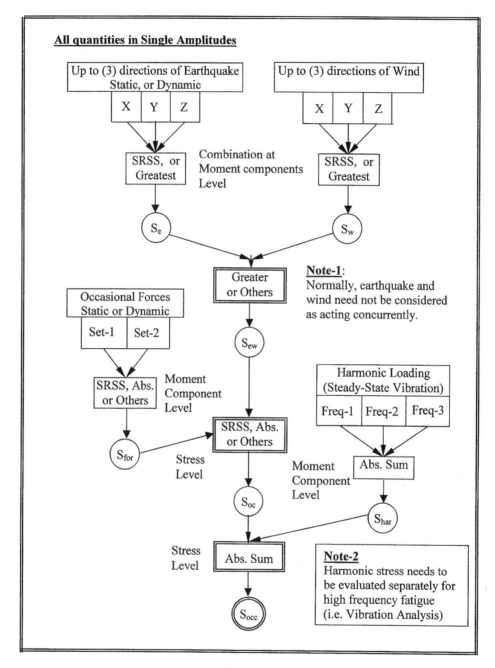

**FIG. 4.13
TYPICAL PROCEDURE FOR EVALUATING OCCASIONAL STRESS**

load cases are analyzed separately with their results combined according to the flow chart. The combination within each type of load is done at the moment component level. For instance, M_x combines with M_x, M_y combines with M_y, and so forth. The combinations of different types of loads are generally done at the stress level.

Because it is not very probable that the postulated design earthquake would occur at the same time as the postulated design wind load, the code generally does not require the consideration of earthquake and wind as occurring at the same time. In this case, the combination of earthquake and wind is done by picking the greater of the earthquake stress and wind stress for every point of the system. However, when dealing with offshore piping systems, very often the acceleration, representing earthquake, is actually the rocking motion from the wave motion. Because the greatest wind load creates the greatest wave, the simulated earthquake and wind stresses have to be combined with the SRSS or other appropriate methods.

Again, the evaluation of the occasional stress varies greatly among the codes. It can be categorized into the following three groups:

(1) *B31.1 Power Piping and CEN codes*. These codes, again, have a definite equation setup, just like in the case of sustained stress, for evaluating the occasional stress. The equation is given as follows:

$$\frac{PD_o}{4t_n} + \frac{(0.75i)M_A}{Z} + \frac{(0.75i)M_B}{Z} \leq kS_h \qquad (4.23)$$

M_B is the resultant moment due to occasional loads. Because several different types of potential occasional loads may be involved in a system, the stress from M_B is the combined occasional stress, S_{occ}, obtained from the procedure given in Fig. 4.13. Also because it is basically a sustained stress, the same $0.75i$ (≥ 1.0) sustained stress intensification is used also for the occasional stress. The allowable for the combined sustained plus occasional load is expressed with an allowable factor times the hot basic allowable stress. The allowable factor, k, depends on the duration of the loads. It varies from 1.15 to 1.2 for B31.1. The k value in CEN varies from 1.0 to 1.8, depending not only on the duration, but also the level of service to be tolerated.

(2) *Class 2 and Class 3 Nuclear Piping*. Just as in the case of sustained stress, Class 2 and Class 3 nuclear piping adopt the same approach as Class 1 nuclear piping in evaluating occasional stress. The equation has the same format as (4-23) but uses a separate set of stress intensification values. The allowable value varies with duration of the event and also the service level to be tolerated.

(3) *B31.3 Process Piping and Others*. B31.3 only stipulates that the sum of longitudinal stresses due to pressure, weight, and other sustained loadings, S_L, and of the stresses produced by occasional loads, such as wind or earthquake, may reach as much as 1.33 times the basic allowable stress at design temperature. In other words, $S_L + S_{occ} \leq 1.33 S_h$. The stress calculation method is generally considered the same as the method used in calculating sustained stress. Some of the other codes set the allowable limit of the occasional stress in combination with other categories of stresses.

4.5.4 Thermal Expansion and Displacement Stress Range

Thermal expansion and piping flexibility are discussed extensively in Chapter 3. This section discusses only the stress requirements of the codes.

Thermal expansion and other displacement stresses are self-limiting. A large thermal expansion stress can cause the piping material to yield, but once the displacement reaches the expansion value, the yielding stops. Failure due to a single application is rarely expected in ductile piping material. The main mode of failure for thermal expansion stress is low cycle fatigue. The integrity of the piping depends on the stress range and the number of operating cycles. Therefore, in expansion and displacement stress evaluation, the stress calculated is the stress range.

When a piping system has more than one operating mode, the calculation of stress range can be quite complex. The situation most often encountered is the piping system operating at high temperature, but can cool down to a much lower temperature than the installation temperature due to cold ambient conditions. In this simple case, an equivalent jack-up temperature to compensate for the low ambient temperature can be used in the analysis to obtain the proper stress range. However, this approach may result in higher anchor reactions that may be more critical than the stress itself. In a more general situation, the procedure shown in Fig. 4.14 can be used to calculate the maximum stress range regardless of the operating cycles.

The different operation modes are analyzed individually from the construction temperature to the operating temperature. The construction temperature is considered as base temperature. These individual analyses cover the individual stress ranges from the base temperature to the postulated operat-

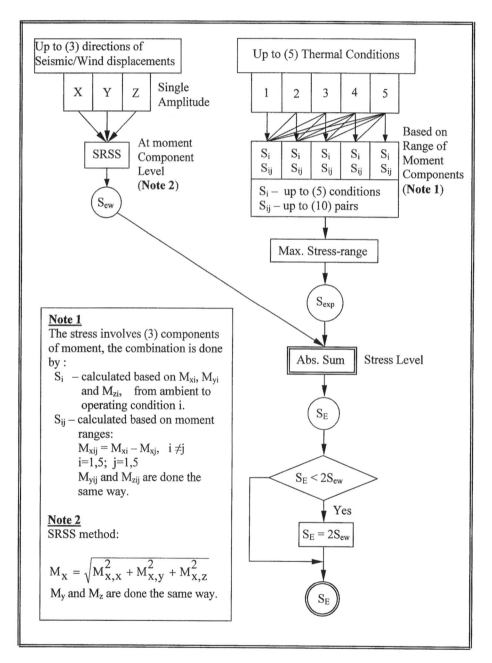

FIG. 4.14
TYPICAL PROCEDURE FOR EVALUATING DISPLACEMENT STRESS

ing temperatures. Each individual operating mode is then paired with every other operating mode to obtain the difference of the directional moment components at every piping component (element). These differential moment components are used to calculate the stress ranges between the operating modes. The overall expansion stress range is selected as the greatest stress range among the stress ranges from the base individual modes and the differential stresses between all pairs of modes.

When a system also experiences anchor and support displacement loads caused by earthquake or wind, these displacements are analyzed independently for up to three directions of earthquake or wind. It is to be noted that each direction of earthquake or wind is capable of creating three-dimensional movements at the structure. In other words, the anchor displacements caused by x-earthquake, for instance, can be in the x, y, and z directions. The forces and moments generated by each direction of earthquake or wind are then combined to calculate the occasional displacement stress. To combine with thermal expansion stress, the occasional displacement stress is calculated with the single amplitude (one-half stress range) value. It is then superimposed on the expansion stress range, as shown in Fig. 4.15 for the combination. The total displacement stress range is the absolute sum of the overall expansion stress range and the single amplitude occasional displacement stress. If the total displacement stress range so calculated is less than the occasional displacement stress range (double amplitude), then the total displacement stress range is set to the same as the occasional displacement stress range. This is reflected at the lower part of the flow chart given in Fig. 4.14.

The methods of calculating the thermal expansion stresses are fairly uniform among all the codes. The calculation is based on nominal wall thickness and cold modulus of elasticity. The evaluation is based on elastic structural analysis. If the structural analysis is performed with hot modulus of elasticity, a code preference, then the moment so calculated has to be modified with the ratio of the cold and hot modulus of elasticity, E_c/E_h, before being used to calculate the stress. For systems with non-uniform E_c/E_h, the maximum E_c/E_h covering the entire system has to be used. Code stress intensification factors are used by all codes. Currently, however, the B31 code does not specify any intensification factor for torsion moment; hence, the following two approaches emerge from the applications.

(1) *Use one stress intensification factor for all moments.* The same code stress intensification factor used for the bending moment is also applied to the torsion moment. This is the approach adopted by B31.1, CEN, and Class 2 and Class 3 nuclear piping. The stress is calculated using Eq. (4.18).

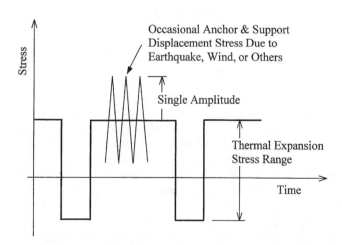

FIG. 4.15
COMBINATION OF THERMAL EXPANSION AND OCCASIONAL DISPLACEMENT STRESSES

(2) *Use different stress intensification factors for different directions of moments.* No stress intensification factor is applied to torsion moment. The stress is calculated with Eq. (4.17) using $i_t = 1.0$. This approach is adopted by B31.3 and some other codes.

Besides B31.4 and B31.8 transmission pipeline codes, which will be discussed in a separate chapter, all the other codes have the same format of allowable for the expansion and displacement stress range. The allowable for the combined total expansion and displacement stress range is expressed as

$$S_E \leq S_A + f(S_h - S_L) = f[1.25(S_c + S_h) - S_L] \tag{4.24}$$

where $S_A = f(1.25S_c + 0.25S_h)$ is the base allowable for the expansion stress as given by Eq. (3.15). The second term, $(S_h - S_L)$, is the leftover allowable from longitudinal sustained stress. Factor f is the stress range factor used to reduce the allowable for systems operating at more than 7000 cycles, the base number of cycles used in deriving S_A. Because fatigue strength can be expressed as [9] $iSN^{0.2} = C$, we can express fatigue strength as $(iS) = CN^{-0.2}$. The stress range factor becomes $f = (iS)_N/(iS)_{7000} = (7000)^{0.2}N^{-0.2} = 5.875N^{-0.2}$, where $(iS)_N = CN^{-0.2}$ is the fatigue strength for N operating cycles, etc. The codes use $f = 6.0N^{-0.2} \leq 1.0$. Generally, no higher allowable stress is permitted for systems operating at less than 7000 cycles. This is partially due to the question regarding the validity of elastic structural analyses performed on systems so highly stressed. However, for a single non-repeated application of anchor movement, such as foundation settlement, an allowable stress of $3S_c$ is given by some codes.

Starting from the 2004 edition, B31.3 allows a maximum $f = 1.2$ for ferrous materials with specified minimum tensile strength equal or greater than 75 ksi (517 MPa) and at material temperature less or equal to 700°F (371°C). The rationale is that when yield strength governs, $1.2 \times 1.25(S_c + S_h)$ equals to $(S_{y,c} + S_{y,h})$ when S_c and S_h are each allowed to no more than 2/3 of the yield strength at cold temperature, $S_{y,c}$, and the yield strength at hot temperature, $S_{y,h}$, respectively. With this stress range of twice the yield strength, the stress will shakedown to pure elastic cycling after a few cycles. The problem is that our B31-calculated thermal expansion stress is only one-half of the theoretical stress [9]; twice of the yield can translate to four times of the yield strength when theoretical stress intensification factor is included. Therefore, the use of $f = 1.2$ is discouraged. The increase of current thermal expansion allowable stress has very little consequence in the design of piping systems. The relief we need most is the allowable loading on the connecting equipment.

The allowable stress given by Eq. (4.24) can exceed the yield strength of the pipe material. This may appear unreasonable to some engineers. However, it is important to remember that the stress calculated is an elastic equivalent stress corresponding to the strain range expected. It is not the actual stress value. See also Section 3.3.

4.5.5 Code Stress Compliance Report

To validate the compliance of the code stress requirements, it is generally required to show the calculated stress versus the allowable for each category of stress at every piping component in the system. The table showing the calculated stress versus the allowable stress is called the code stress compliance report. The compliance can also be presented in a graphical form called the code stress compliance chart. The code stress compliance chart normally shows the compliance status of the whole system in a single page of graphs. Figure 4.16 shows a typical code stress compliance chart. The chart has three zones representing sustained, occasional, and expansion (or displacement) stresses. Each zone has the calculated stresses charted against the allowable. For sustained and occasional stresses, the allowable stresses are generally constant throughout the system. However, the allowable stresses for the expansion stresses vary from point to point. They depend on the magnitude of the sustained stresses as given by Eq. (4.24). A lower sustained stress results in a higher expansion stress allowable.

A code stress compliance chart contains the results of the entire system presented in one page. This often means that a few hundred points are displayed on the same page or screen. Because the page

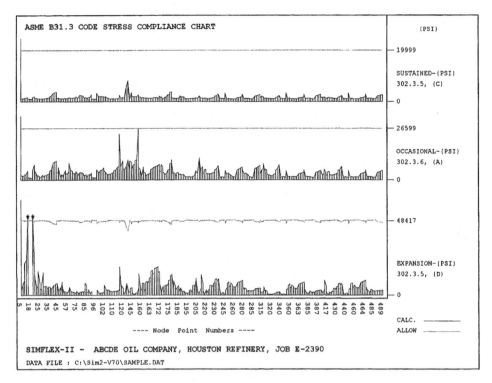

**FIG. 4.16
TYPICAL CODE STRESS COMPLIANCE CHART**

has only a limited number of columns for node numbers, the node numbers displayed are taken from equally spaced input numbers.

4.6　CLASS 1 NUCLEAR PIPING

There has been very little activity in the nuclear power industry over the past 30 years. However, due to the severity of global warming and the shortage of fossil fuels, it is likely that nuclear power may become the preferred energy source in the near future. This will be a huge challenge to pipe stress engineers, as many will have forgotten the unique requirements of Class 1 nuclear piping, referred to simply as nuclear piping.

Because space constraints prohibit a comprehensive treatment of the subject, this section will briefly summarize the stress requirements of nuclear piping, and discuss some of the main differences between these requirements and those of non-nuclear piping.

The design of nuclear piping follows not only the piping code but also the regulatory requirements. In the United States, the regulatory agency for nuclear facilities is the U.S. Nuclear Regulatory Commission (NRC). All NRC directives have to be observed in the design of nuclear plant piping, including not only Class 1 piping but also Class 2 and Class 3 piping. The NRC Regulatory Guide [16] and the NRC Standard Review Plan [17] are two of the important documents that have to be followed. The following are some of the characteristics of nuclear piping:

(1) Design philosophy

Nuclear piping adopts the "Design by Analysis" philosophy, in contrast to the "Design by Rules" approach favored by B31.1. Nuclear piping is more exact and depends mostly on theoretical analysis, whereas non-nuclear piping mostly depends on past experience and rules of thumb.

This means that many more calculations and documentations are required in the design of nuclear piping.

(2) Stress intensity

The stress quantity used by nuclear piping is called stress intensity, which is twice the maximum shear stress. The maximum shear failure theory is adopted. The stress intensification quotient is called the stress index to distinguish it from the stress intensification factor used by B31.

(3) Stress index

In contrast to B31, which uses the stress intensification factor to adjust the stress that is over the basic nominal stress calculated by dividing the moment with the section modulus, nuclear piping uses the stress index to modify the reference nominal stress to reach the actual stress intensity. B31 has only one general stress intensification factor, used mainly for thermal expansion stress calculations. Nuclear piping, on the other hand, has stress indices for pressure, moment, and thermal loads. For each type of load, it also has different stress indices for primary, secondary, and peak stress categories. Furthermore, the stress intensification factor in B31 is always greater than 1.0. However, the stress indices in nuclear piping can be less than 1.0 due to a particular reference nominal stress used. For instance, the use of hoop stress as the reference nominal stress for the pressure load results in a stress index of 0.5 for straight pipe in the primary stress calculation.

(4) Pressure design

The pressure design of nuclear components follows pretty much the same approach as in B31.1. However, the nuclear piping has a higher allowable stress due to the additional requirements on quality control and quality assurance. The origin of the material must be traceable and well documented. The lower allowable stress in B31.1 is also required to cover the imprecise calculation of the sustained stress and the loose treatment of the occasional loads. In nuclear piping, in addition to the pressure, stress limitations of the sustained and occasional loads also often determine the wall thickness of the pipe.

(5) Levels of services

The plant is designed for four levels of services. Each level of service has its permissible pressure and primary stress. Level A represents normal operation conditions, and level B is the upsetting condition without operational impairment. Level C is the condition requiring the shutdown of the plant, but under which the plant is expected to resume operation after the event. Level D is a rare occurrence in which the plant can be safely shut down, but some gross yielding in the piping component might occur and require repair or replacement. Design loads for each service level have to be clearly specified in the design specification.

(6) Occasional loads

Because all nuclear piping systems are either housed inside the main containment or shielded from open-air exposure, no wind load is generally considered in the design. However, earthquake loads have to be specified for all levels of service at all plants. The code has very specific treatments for earthquake loads.

(7) Consideration of design condition

The primary stress intensities for all service levels are calculated with the same form of formula but with the loads and allowable values changed accordingly. For design condition or level A service, the following criterion has to be met

$$B_1 \frac{PD_o}{2t} + B_2 \frac{D_o}{2I} M_i \leq 1.5 S_m \qquad (4.25)$$

B_1 and B_2 are the primary stress indices given by the code for the specific component under investigation. This is similar to Eq. (4.23) used by B31.1. However, there are several distinctive differences: (1) Nuclear piping uses a separate set of indices for the primary stress, whereas B31.1 uses the same displacement (secondary) stress intensification factor multiplied with a constant 0.75 for the primary stress. (2) The pressure term uses hoop stress as reference rather than the longitudinal stress used in B31.1. Therefore, B_1 is expected to be 0.5 for most of the components. (3) M_i is the resultant of the moment components combining the deadweight, earthquake, and other mechanical loads. The combination is done on the moment component level, whereas in B31.1 the combination is done by adding the resultant moments of the weight and the resultant moment of the earthquake as given in Eq. (4.23).

(8) Satisfaction of primary plus secondary stress intensity range

The failure mode of the self-limiting secondary stress or displacement stress is the fatigue damage due to cyclic operations. The important quantity is the stress range. However, to evaluate fatigue life, all self-limiting and non-self-limiting cyclic loads have to be considered. This is the primary plus secondary stress range.

The fatigue damage of nuclear piping is evaluated for every pair of operating conditions. These operating conditions are derived from the operating modes analysis. Each operating condition is associated with a load set of pressure, temperature, earthquake, and other factors. For convenience, an ambient condition with zero pressure is often included as one of the operating conditions.

In *B31*, the displacement stress range is calculated independently with its own allowable stress. Its combination with the primary stress, which may or may not be cyclic, is handled by adjusting the allowable stress. The allowable stress left over from the sustained stress is added to the allowable of the displacement stress.

In *nuclear piping*, the secondary stress is combined with the primary stress to become the primary plus secondary stress intensity range for each pair of operating conditions. This primary plus secondary stress is used to check if the system will cycle under elastic condition. Later, a peak stress range is calculated to determine the allowable number of operating cycles for each pair of operating conditions. Fatigue curves are used in determining the allowable number of operating cycles. Deadweight effects are not considered in determining the stress range, because they are non-cyclic.

The primary plus secondary stress intensity range is calculated without including stress concentration effects. This stress, sometimes referred to as nominal stress, S_n, is used to ensure that the system will go through elastic cycling after the shakedown. The primary plus secondary stress intensity range is calculated by the following equation.

$$S_n = C_1 \frac{P_o D_o}{2t} + C_2 \frac{D_o}{2I} M_i + C_3 E_{ab} |\alpha_a T_a - \alpha_b T_b| \leq 3 S_m \qquad (4.26)$$

C_1, C_2, and C_3 are secondary stress indices given by the code for the specific component under investigation. For common components such as butt-welding elbows and butt-welding tees, C_2 is about twice the magnitude of the stress intensification factors used by B31. (See also item (9) below.) $C_1 = 0.5$ and $C_3 = 0$ for B31 codes, which routinely do not evaluate the thermal stress covered by C_3. M_i is the range of moment loading resulting from live weight, earthquake, thermal expansion, and support displacement. This M_i is the resultant of the moment component ranges combining all the loads when the system goes from one service load set to another. In other words, each moment component is combined for all loads first, and then the resultant is taken from these individually combined components. The last term of the equation is the gross thermal discontinuity stress. It deals with differential temperature and/or dissimilar metal properties between adjacent parts. T_a and T_b are the average temperatures on side a and side b, respectively, of the gross thermal discontinuity, structural discontinuity, or material discontinuity. For generally cylindrical shapes, the average of the temperature on each side shall cover a distance of $\sqrt{Dt}$ from the interface of the discontinuity. E_{ab} is the average modulus of elasticity of the two sides at room temperature.

The limitation of primary plus secondary stress range to $3S_m$ is to ensure that the system will cycle in the elastic range after a few cycles of the shakedown process. The $3S_m$ value is less than, or equal to, twice of the yield strength of the pipe material at temperature. If Eq. (4.26) is satisfied, then fatigue evaluation can proceed with the established elastic approach. However, if Eq. (4.26) is not satisfied, the system may still prove satisfactory by using the simplified elastic-plastic discontinuity analysis given in item (10) below.

(9) Satisfaction of peak stress intensity range

The peak stress range determines the fatigue life of a component. In nuclear piping, fatigue is evaluated by using the fatigue curves based on strain-controlled tests. Stresses are the fictitious values obtained by multiplying the strain with the modulus of elasticity. This procedure is acceptable only if Eq. (4.26) is satisfied. If Eq. (4.26) is satisfied, the peak stress intensity range is calculated as follows:

$$S_p = K_1 C_1 \frac{P_o D_o}{2t} + K_2 C_2 \frac{D_o}{2I} M_i + K_3 \frac{E\alpha|\Delta T_1|}{2(1-\nu)} \\ + K_3 C_3 E_{ab}|\alpha_a T_a - \alpha_b T_b| + \frac{E\alpha|\Delta T_2|}{1-\nu} \quad (4.27)$$

K_1, K_2, and K_3 are the local stress indices given by the code for the component under consideration. These indices are normally called elastic stress concentration factors, produced mainly by the notch effects. The local stress indices are significant at all welds. Therefore, all weld locations and types have to be identified in a nuclear piping analysis. This is very different from the B31 piping, which simply considers that girth welds exist at all locations. However, $K_2 C_2$ can be considered as twice the B31 stress intensification factor for most components. In addition to the gross thermal discontinuity included in Eq. (4.26), peak stress also includes the stresses from linear and non-linear thermal gradients, ΔT_1 and ΔT_2.

Figure 4.17 shows a temperature gradient across the pipe wall thickness. The distribution is assumed to be uniform around the circumference, but varies in the radial direction across the thickness. This is mainly due to thermal resistance of the pipe material against the heat input from the fluid and heat loss to the atmosphere. The temperature distribution can be decomposed into average, linear variation, and non-linear variation, three portions. The average

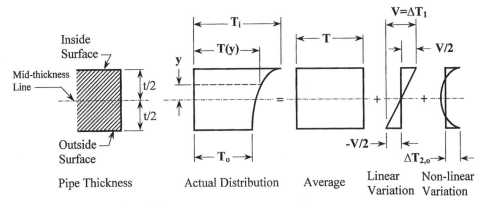

FIG. 4.17
DECOMPOSITION OF TEMPERATURE DISTRIBUTION RANGE

temperature, T, is the mean temperature taken across the thickness. This average temperature produces gross deformation in the system and is used to calculate thermal expansion movements. The linear variation ΔT_1 considers linearly equal but opposite variations from the mid-thickness line. This generates shell-bending stresses across the thickness. The non-linear variation ΔT_2 is the portion leftover from the first two portions. ΔT_2 is generally different at inside and at outside surfaces. It shall take the greater value of the two. Furthermore, if this non-linear portion is negative it shall be considered zero, because the linear portion already covers it. The non-linear portion of the temperature gradient generates localized skin stresses.

From Eqs. (4.26) and (4.27), it is apparent that considerable thermal and heat transfer analyses are required for the design of nuclear piping [18]. These thermal and heat transfer analyses may include both steady-state and transient conditions [19].

Alternating stress intensity S_{alt} is calculated as one-half of the peak stress when Eq. (4.26) is satisfied. That is, $S_{alt} = S_p/2$ when primary plus secondary stress range is within twice of the yield strength. This alternating stress intensity is used to find the allowable number of operating cycles from the applicable fatigue curve.

Cumulative damage is the fatigue damage accumulated from all pairs of operating conditions. At each pair of operating conditions, the peak stress intensity range and the alternating stress intensity is calculated to determine the allowable number of operating cycles using an applicable fatigue curve. The ratio of the numbers of the design operating cycles to the allowable cycles is called the usage of fatigue life for the given pair of operating conditions under consideration. The cumulative usage or damage is then calculated by adding the usages of all operating pairs expected. The system is satisfactory when the cumulative usage is 1.0 or less.

In calculating the cumulative damage, the application of occasional loads, such as earthquake, is not straightforward. Generally, it is desirable to combine occasional loads with the most severe operating moment range. However, the most severe operating moment range of the piping component generally occurs at different operating pairs for different components. This makes the tracking of the number of operating cycles used very difficult, because occasional loads have only a limited number of cycles that need to be considered. One simplified approach that may be considered is treating the occasional load as occurring during the normal operating condition. In this case, two load conditions — one normal operating plus occasional load and the other normal operating minus occasional load — are created. This is required because the responses from occasional load mostly do not have the proper sign. These two created load conditions are then treated just as any other load condition with the number of operating cycles the same as the number of the occasional load cycles.

(10) Simplified elastic-plastic discontinuity analysis

When the primary plus secondary stress intensity range of any of the operating pairs exceeds the $3S_m$ limit, the system may still prove satisfactory by using the simplified elastic-plastic discontinuity analysis [20]. Because the component will be subjected to a nominal stress range above twice the yield strength, a further check on the protection against plastic hinge and thermal ratchet is required. This is done by limiting the thermal expansion stress and the temperature gradient stress to certain acceptable values — that is, before the simplified elastic-plastic analysis can be performed for the pair of operating conditions with $S_n > 3S_m$, the following three conditions have to be satisfied:

(a) Thermal expansion stress range given in Eq. (4.28) shall be met:

$$S_e = C_2 \frac{D_o}{2I} M_i^* \leq 3S_m \tag{4.28}$$

where S_e is the nominal value of expansion stress and M_i^* is the resultant of the moment ranges including only moments due to thermal expansion and thermal anchor and support movements.

(b) Temperature gradient stress expressed as ΔT_1 should satisfy the relation below:

$$\Delta T_1 \text{ range} \leq \frac{y' S_y}{0.7 E \alpha} C_4 \tag{4.29}$$

where

y' = 3.33, 2.00, 1.20, and 0.8 for $[(PD_o/2t)/S_y]$ = 0.3, 0.5, 0.7, and 0.8, respectively
P = maximum pressure for the pair of conditions under consideration
C_4 = 1.1 for ferritic material and = 1.3 for austenitic material
S_y = yield strength value taken at average fluid temperature of the transient under consideration.

(c) The primary plus secondary membrane plus bending stress intensity, excluding thermal expansion stress, shall be less than $3S_m$. That is,

$$C_1 \frac{P_o D_o}{2t} + C_2 \frac{D_o}{2I} M_i + C_3' E_{ab} |\alpha_a T_a - \alpha_b T_b| \leq 3S_m \tag{4.30}$$

Equation (4.30) is the same as Eq. (4.26) except that, here M_i does not include the moments due to thermal expansion and anchor and support movement, and C_3' is an adjusted stress index given in the code for thermal loading.

When (a), (b), and (c) conditions are met, the effective alternating stress intensity can be found as

$$S_{alt} = K_e \frac{S_p}{2} \tag{4.31}$$

where K_e is the correction factor for cycling beyond the elastic limit. The K_e factor is defined as $K_e = 1.0$ for $S_n \leq 3S_m$; $K_e = 1.0 + [(1-n)/n(m-1)](S_n/3S_m - 1)$ for $3S_m < S_n < 3mS_m$; and $K_e = 1/n$ for $S_n \geq 3mS_m$. The values of the material parameters m and n are given in Table 4.4. After the alternating stress intensity is determined, the allowable number of operating cycles of the load pair can be found from the applicable fatigue curve as usual.

(11) Stresses at branch connections

Stresses at branch connections are calculated differently from those of other components. In nuclear piping, a branch connection or tee is evaluated by combining the moment effects from both branch and run together. In all stress calculations, whenever a moment is involved, as given in an equation, a combined moment effect, instead of the simple moment, is used as follows:

$$X_n \frac{D_o}{2I} M_i \left(= X_n \frac{M_i}{Z} \right) \text{ replaced with } X_{nb} \frac{M_b}{Z_b} + X_{nr} \frac{M_r}{Z_r} \tag{4.32}$$

**TABLE 4.4
VALUES OF m AND n**

Materials	m	n
Carbon steel	3.0	0.2
Low alloy steel	2.0	0.2
Martensite stainless steel	2.0	0.2
Austenitic stainless steel	1.7	0.3
Nickel-chromium-iron	1.7	0.3
Nickel-copper	1.7	0.3

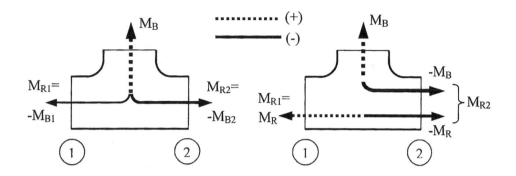

(a) No through run moment (b) With through run moment

FIG. 4.18
BRANCH AND THROUGH RUN MOMENT

where X_n is the stress index of the stress category under evaluation; X_{nb} is the stress index for branch moment and X_{nr} for the run moment; and Z_r and Z_b are the section moduli of the run and branch pipes, respectively. M_b is the resultant branch moment and M_r is the resultant of the through-run moment. As shown in Fig. 4.18, the through-run moment is not the same as the run moment. The run moment includes the through-run moment and a portion of the branch moment that comes through each side of the run.

The *through-run moment components*, M_{xr}, M_{yr}, and M_{zr}, are defined by the code as follows: (a) If M_{i1} and M_{i2} (with $i = x, y, z$, etc.) have the same algebraic sign, then M_{ir} equals zero. (b) If M_{i1} and M_{i2} have opposite algebraic signs, then M_{ir} is equal to the smaller value of M_{i1} or M_{i2}. M_{ir} has to be determined for each load pair before being combined with other load pairs. This method works properly if the signs of the moment components are maintained throughout the analysis. However, it does not work when the moment components of the load set are all positive, such as in the case of earthquake analysis by the response spectra method. (See the chapter on dynamic analysis for details on response spectra analysis.) Indeed, some of software packages have erroneously set all M_{ir} to zero in those all-positive cases. It may also cause confusion when a computer program, such as MEL-40 [21], uses a special sign convention that is not consistent with the intent of the code. A more foolproof method is needed.

Figure 4.18(a) shows the case when all run moments at both sides are needed to balance the branch moment. The signs of moments at both sides of the run are the same, but are opposite to the sign of the branch moment. In this case, the through-run moment is zero. Figure 4.18(b) shows the situation in which the moment at one end of the run and the moment at the other end of the run have different signs. The situation is created by the through-run moment. The through-run moment equals the smaller of M_{R1} ($=M_R$) or M_{R2} ($=M_B + M_R$). In this case, the through-run moment is M_{R1}. Without looking at the signs, the through-run moment can also be found from the equilibrium of the moments from all three legs as shown in Fig. 4.18(b). By taking the absolute values of all the moments, we have

$$|M_{R1}| + |M_{R2}| - |M_B| = M_R + M_R + M_B - M_B = 2M_R$$

or

$$M_R = \frac{1}{2}\{|M_{R1}| + |M_{R2}| - |M_B|\} \tag{4.33}$$

The through-run moment can be taken as one-half of the difference of the absolute sum of the moments at both sides of the run minus the absolute value of the branch moment. This is done for each direction of the moment. Because this method involves only the absolute values, it does not matter if the sign is lost or if the sign convention is not consistent with the one implied by the code.

REFERENCES

[1] ASME Code for Pressure Piping, B31, An American National Standard, "B31.1 Power Piping," "B31.3 Process Piping," The American Society of Mechanical Engineers, New York.
[2] ASME Boiler and Pressure Vessel Code, Section-III, "Rules for Construction of Nuclear Facility Components," Division 1, The American Society of Mechanical Engineers, New York.
[3] European Standard EN-13480-4, 2002, "Metallic industrial piping – Part 3: Design and calculation," European Committee for Standardization (Comité Européen de Normalisation).
[4] Flieder, W. G., Loria, J. C., and Smith, W. J., 1961, "Bowing of Cryogenic Pipelines," *Transactions of the ASME, Journal of Applied Mechanics*, pp. 409–416.
[5] ASCE 7-02, "Minimum Design Loads for Building and Other Structures," American Society of Civil Engineers, Washington, D. C., 2002.
[6] ICBO, 1997, "Uniform Building Code," International Conference of Building Officials, Whittier, CA.
[7] Annual Book of ASTM Standards, "part 1 steel-piping, tubing, fittings," etc., American Society for Testing Materials, Philadelphia.
[8] ASME Boiler and Pressure Vessel Code, Section IX, "Welding and Brazing Qualifications," American Society of Mechanical Engineers, New York.
[9] Markl, A. R. C., 1955, "Piping-Flexibility Analysis," *Transactions of the ASME*, 77(2), pp. 124–143 (discussion pp. 143–149).
[10] ASME Code for Pressure Piping, B31.3, Process Piping, Chapter IX, "High Pressure Piping"
[11] ASME Boiler and Pressure Vessel Code, Section VIII, Division 1, "Pressure Vessel," The American Society of Mechanical Engineers, New York.
[12] Koves, W. J., 2000, "Process Piping Design: A Century of Progress," *Journal of Pressure Vessel Technology*, 122, pp. 325–328.
[13] Markl, A. R. C., 1952, "Fatigue Tests of Piping Components," *Transactions of the ASME*, 74(3), pp. 287–303.
[14] Rodabaugh, E. C., 1987, "Accuracy of Stress Intensification Factors for Branch Connections," WRC Bulletin 329, Welding Research Council, New York.
[15] Peng, L.C., 1979, "Toward more consistent pipe stress analysis," *Hydrocarbon Processing*, 58(5), pp. 207–211.
[16] U.S. NRC, "Regulatory Guide," U.S. Nuclear Regulatory Commission, Office of Standard Development.
[17] U.S. NRC, "Standard Review Plan for the Review of Safety Analysis Reports for Nuclear Power Plants," U.S. Nuclear Regulatory Commission, Office of Nuclear Reactor Regulation.
[18] Munson, D. P., Keever, R. E., Peng, L. C., and Broman, R., 1974, "Computer Application to the Piping Analysis Requirements of ASME Section III, Sub-article NB-3600," *Pressure Vessels and Piping, Analysis and Computers*, ASME.
[19] Brock, J. E., and McNeill, D. R., 1971, "Speed Charts for Calculation of Transient Temperatures in Pipes," *Heating, Piping, and Air Conditioning*, 43(11).
[20] Tagart, S. W., 1968, "Plastic Fatigue Analysis of Pressure Components," presented at ASME Joint Conference of the Petroleum and the Pressure Vessels and Piping Divisions, Dallas, TX, Paper No. 68-PVP-3.
[21] Laldor, L.M., H.C. Neilson, Jr., and R.G. Howard, 1969, "User's Guide to MEL-40, A Piping Flexibility Analysis Program," U.S. Naval Ship Research & Development Center.

CHAPTER 5

DISCONTINUITY STRESSES

Pressure vessels usually contain regions where abrupt changes in geometry, material, or loading occur. These regions are known as discontinuity areas, and the stresses associated with them are called discontinuity stresses by the code [1, 2]. These codes have outlined a general procedure for analyzing the discontinuity stresses and discussed examples of some common occurrences. This chapter will discuss some of the applications relevant to the design of piping systems.

Because of dissimilar characteristics, each of the adjacent parts joining at a discontinuity area behaves differently to an applied load, such as internal pressure or temperature. The deformations of the disconnected free bodies are different from each other. Because these parts are joined together, they share a common displacement that is different from their free displacements. The difference between the free displacement and the actual joint displacement is a forced displacement, which produces forces and stresses. These additional stresses are referred to as discontinuity stresses.

Calculation of discontinuity stresses is generally based on the behavior of the longitudinal strip of the cylindrical shell. Because a longitudinal stripe of a vessel behaves like a railroad sitting on an elastic foundation, the discontinuity stresses at the vessel are generally calculated based on the theory of "Beams on Elastic Foundation" [3, 4]. A beam on an elastic foundation receives a lateral reaction force that is proportional to the displacement. The rail track is a typical example of a beam on elastic foundation.

5.1 DIFFERENTIAL EQUATION OF THE BEAM DEFLECTION CURVE

Some basic principles of strength of materials have been discussed in Chapter 2. However, beam deflection equations were skipped for simplicity. As these equations are essential for discussing beams on elastic foundation, they are summarized in this section.

The sign conventions are as given in Fig. 5.1. The positive y-axis is assumed to be pointing downward. This is different from the convention used by piping stress analyses, but follows the convention used in traditional treatment of beams on elastic foundation. All positive values of forces and moments are as shown.

Figure 5.1 shows the relationship between the changes in forces and moments at an infinitely small beam element defined by two adjacent transverse parallel planes a-b and c-d. The positive shear forces and moments are as shown. The uniform force, q, per unit length, acts upward representing the reaction to the downward movement of the beam. Because the beam is sitting on an elastic foundation, this reaction force is proportional to the downward displacement. That is, $q = ky$, where k is the spring constant of the foundation per unit length of the beam. The changes in forces and moments across these two planes can be found by the equilibrium of them acting on the entire element. First, by taking the equilibrium of the y forces, we have

$$-V - q\,dx + V + dV = 0$$

or

$$\frac{dV}{dx} = q \tag{5.1}$$

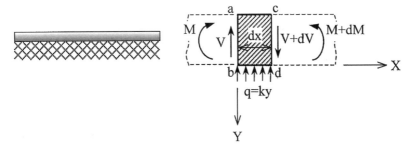

**FIG. 5.1
BEAM ON ELASTIC FOUNDATION**

The relation of the moments and forces are determined by the equilibrium of the moments. Taking the moment around the base plane *a-b* and setting the sum to zero, we have the equilibrium equation as follows

$$M - (M + dM) + (V + dV)dx - q dx \frac{dx}{2} = 0$$

By ignoring the higher-order terms of very small quantities, the above becomes

$$\frac{dM}{dx} = V \tag{5.2}$$

Combining Eqs. (5.1) and (5.2), we have

$$\frac{d^2M}{dx^2} = q = ky \tag{5.3}$$

The deformation of a slender beam can be entirely attributed to the bending moment. The shear force effect on the displacement is generally small in slender beams and can be ignored. The basic elastic curve of a beam in pure bending, as can be found in a textbook on strength of materials, is given as follows:

$$EI \frac{d^2y}{dx^2} = -M \tag{5.4}$$

By differentiating the above equation with respect to *x* twice and substituting Eq. (5.3), we have the differential equation for the beam on elastic foundation

$$EI \frac{d^4y}{dx^4} = -ky \quad \text{or} \quad \frac{d^4y}{dx^4} + \frac{k}{EI} y = 0 \tag{5.5}$$

This is a linear differential equation of fourth order with constant coefficients and right member zero. It can be solved by letting $y = e^{\alpha x}$, where α is an exponent constant yet to be determined. Differentiating and substituting $y = e^{\alpha x}$ to Eq. (5.5), and dividing both sides with *y*, we have

$$\alpha^4 + \frac{k}{EI} = 0 \quad \text{or} \quad \alpha = \sqrt[4]{-\frac{k}{EI}} = \sqrt[4]{\frac{k}{EI}} \sqrt[4]{-1}$$

The α value, involving the fourth roots of negative one, has the following four values:

$$\sqrt[4]{-1} = \frac{1}{\sqrt{2}}(\pm 1 \pm i); \quad i = \sqrt{-1}$$

The square root of 2 factor can be combined with *k*/EI to become the characteristic factor

$$\beta = \sqrt[4]{\frac{k}{4\text{EI}}} \tag{5.6}$$

The four solutions of Eq. (5.5) become

$$y_1 = e^{\beta x + i\beta x}; \quad y_2 = e^{\beta x - i\beta x}; \quad y_3 = e^{-\beta x + i\beta x}; \quad y_4 = e^{-\beta x - i\beta x}$$

The general solution of the equation is

$$y = Ay_1 + By_2 + Cy_3 + Dy_4$$

where A, B, C, and D are integration constants. After some mathematical manipulations, we have the working solution as

$$y = e^{\beta x}(C_1 \cos\beta x + C_2 \sin\beta x) + e^{-\beta x}(C_3 \cos\beta x + C_4 \sin\beta x) \tag{5.7}$$

The integration constants C_1, C_2, C_3, and C_4 are to be determined from the known boundary and physical conditions of the beam. This chapter will use Eq. (5.7) to solve some of the discontinuity stress problems related to piping design.

5.2 INFINITE BEAM ON ELASTIC FOUNDATION WITH CONCENTRATED LOAD

The infinite beam with a concentrated load is the first situation to be investigated. Equation (5.7) applies to the entire beam except at the loading point, which is considered the origin with $x = 0$. Because of this discontinuity point, the constants C_1, C_2, C_3, and C_4 will be different at the left-hand and right-hand sides of the beam. However, due to the condition of symmetry, it is only necessary to investigate one side of the beam. By investigating the right-hand ($+x$) side, the integration constants are determined by the following conditions.

Because either C_1 or C_2 would have made the deflection infinite at an infinite distance away ($x \to \infty$), both C_1 and C_2 have to be zero. That is, $C_1 = C_2 = 0$. The deflection curve for the right-hand portion of the beam becomes

$$y = e^{-\beta x}(C_3 \cos\beta x + C_4 \sin\beta x) \tag{5.8}$$

The first two constants have been determined by the boundary condition at a point infinitely away from the loading point. The other two constants can be determined by the boundary condition at the loading point. Due to the condition of symmetry, at the loading point the slope must be zero and the shear force equals one-half of the applied force. To satisfy these conditions, we have

$$y' = \frac{dy}{dx} = \beta e^{-\beta x}[(-C_3 + C_4)\cos\beta x + (-C_3 - C_4)\sin\beta x]$$

with $x = 0$, $y' = \beta(-C_3 + C_4) = 0$; hence, $C_3 = C_4$

and

$$V = \frac{dM}{dx} = -\text{EI}\frac{d^3y}{dx^3} = -\text{EI}\beta^3 e^{-\beta x}[2(C_3 + C_4)\cos\beta x + 2(-C_3 + C_4)\sin\beta x]$$

with $x = 0$, $V = -\text{EI}\beta^3[2(C_3 + C_4)] = -\frac{P}{2}$

From the above two relations we have

$$C_3 = C_4 = \frac{P}{8\text{EI}\beta^3} = \frac{P\beta}{2k}$$

From Eq. (5.8), the deflection curve for the right-hand portion of the beam becomes

$$y = \frac{P\beta}{2k}e^{-\beta x}(\cos \beta x + \sin \beta x) \tag{5.9}$$

From which we have the following relations

$$\theta = \frac{dy}{dx} = -\frac{P\beta^2}{k}e^{-\beta x}\sin \beta x \tag{5.10}$$

$$M = -EI\frac{d^2y}{dx^2} = -\frac{P}{4\beta}e^{-\beta x}(\sin \beta x - \cos \beta x) = \frac{P}{4\beta}e^{-\beta x}(\cos \beta x - \sin \beta x) \tag{5.11}$$

$$V = -EI\frac{d^3y}{dx^3} = -\frac{P}{2}e^{-\beta x}\cos \beta x \tag{5.12}$$

The above equations can be written with American Society of Mechanical Engineers (ASME) notations [1, 2] as follows:

$$y = \frac{P\beta}{2k}f_3(\beta x) \tag{5.13}$$

$$\theta = -\frac{P\beta^2}{k}f_4(\beta x) \tag{5.14}$$

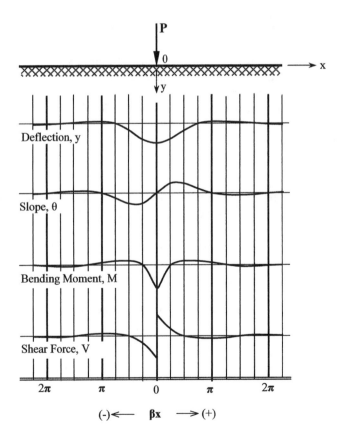

FIG. 5.2
INFINITE BEAM WITH CONCENTRATED FORCE, P CURVES REFLECT ONLY THE GENERAL SHAPES

$$M = \frac{P}{4\beta} f_2(\beta x) \tag{5.15}$$

$$V = -\frac{P}{2} f_1(\beta x) \tag{5.16}$$

where

$$f_1(\beta x) = e^{-\beta x} \cos \beta x \tag{5.17a}$$

$$f_2(\beta x) = e^{-\beta x}(\cos \beta x - \sin \beta x) \tag{5.17b}$$

$$f_3(\beta x) = e^{-\beta x}(\cos \beta x + \sin \beta x) \tag{5.17c}$$

$$f_4(\beta x) = e^{-\beta x} \sin \beta x \tag{5.17d}$$

Figure 5.2 shows the deflections, slopes, bending moments, and shear forces along the beam in terms of the dimensionless parameter βx. The bending moment curve represents the attenuation of the stress from the loading point, and the shear force curve represents the load carrying capacity of the beam element at the points away from the loading point. The attenuation behaviors of these two curves are enlarged in Fig. 5.3. This figure serves as the guideline for setting the effective zones for reinforcement, nozzle separation, load carrying area, and other factors. The effect of the concentrated load diminishes to less than 20% of that at the loading point when βx is greater than 1.0. This $\beta x = 1.0$ appears to be a good cutoff point and has been used in the pressure design of the miter bends discussed in Chapter 4.

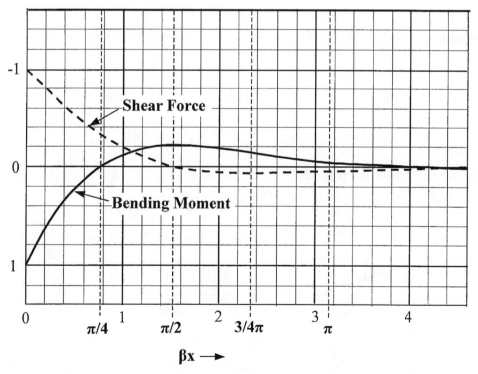

FIG. 5.3
ATTENUATION OF SHEAR FORCE AND BENDING MOMENT

5.3 SEMI-INFINITE BEAM ON ELASTIC FOUNDATION

A very long uniform beam with the loads applied at the edge is called a semi-infinite beam. Most of the discontinuity problems involve two pieces of dissimilar parts joining together. It can be the joining of two pieces of dissimilar semi-infinite pipe segments, or a semi-infinite pipe segment with a non-pipe dissimilar body. Therefore, the semi-infinite beam model is the main work force for calculating the discontinuity stresses in the piping systems.

As shown in Fig. 5.4, the semi-infinite beam generally involves a shear force and a bending moment. The general solution for the beam on elastic foundation as given in Eq. (5.7) is applicable to the entire beam. Again, because either C_1 or C_2 would have made the displacement infinite as the x value approaches infinity, both C_1 and C_2 have to be zero. Hence, the general solution is the same as the one previously given for the infinite beam (Eq. 5.8). For convenience, the general solution, as given by Eq. (5.8), is duplicated as follows:

$$y = e^{-\beta x}(C_3 \cos \beta x + C_4 \sin \beta x)$$

The constants C_3 and C_4 can be determined by the boundary conditions at the loading point where $x = 0$. From Eqs. (5.4) and (5.2), we have

$$EI\left(\frac{d^2y}{dx^2}\right)_{x=0} = -M_0$$

$$EI\left(\frac{d^3y}{dx^3}\right)_{x=0} = -V = P$$

Differentiating the general solution (5.8) and substituting the differentials into the above relations, the integration constants are found as

$$C_3 = \frac{1}{2\beta^3 EI}(P - \beta M_0); \quad \text{and} \quad C_4 = \frac{M_0}{2\beta^2 EI}$$

Substituting C_3 and C_4, the solution becomes

$$y = \frac{e^{-\beta x}}{2\beta^3 EI}[P\cos \beta x - \beta M_0 (\cos \beta x - \sin \beta x)] \tag{5.18}$$

or, using ASME notations given in Eqs. (5.17a) to (5.17d) and substituting $4\beta^4 EI = k$, the above becomes

$$y = \frac{2\beta}{k}[Pf_1(\beta x) - \beta M_0 f_2(\beta x)] = \frac{2P\beta}{k}f_1(\beta x) - \frac{2M_0\beta^2}{k}f_2(\beta x) \tag{5.19}$$

Similarly, by successive differentiation of (5.18), we have

$$\theta = \frac{dy}{dx} = -\frac{2P\beta^2}{k}f_3(\beta x) + \frac{4M_0\beta^3}{k}f_1(\beta x) \tag{5.20}$$

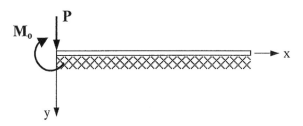

**FIG. 5.4
SEMI-INFINITE BEAM**

$$M = -EI\frac{d^2y}{dx^2} = -\frac{P}{\beta}f_4(\beta x) + M_0 f_3(\beta x) \quad (5.21)$$

$$V = -EI\frac{d^3y}{dx^3} = -Pf_2(\beta x) - 2M_0\beta f_4(\beta x) \quad (5.22)$$

Two working formulas for the condition at the edge of the beam can be found by setting $x = 0$ for Eqs. (5.19) and (5.20). Noting that $f_1 = f_2 = f_3 = 1.0$ at $x = 0$, we have

$$y_0 = \frac{2P\beta}{k} - \frac{2M_0\beta^2}{k} \quad (5.23)$$

$$\theta_0 = -\frac{2P\beta^2}{k} + \frac{4M_0\beta^3}{k} \quad (5.24)$$

By using Eqs. (5.23) and (5.24) together with the principle of superposition, many discontinuity problems regarding pressure vessels and piping segments can be solved.

5.4 APPLICATION OF BEAM ON ELASTIC FOUNDATION TO CYLINDRICAL SHELLS

One of the most important applications of the theory of beams on elastic foundation is the calculation of stresses produced at discontinuity junctions of thin-wall cylindrical shells. Before any shell application can be performed, we have to first establish that the shell behaves the same way as beam on elastic foundation and the foundation spring constant can be readily calculated.

Figure 5.5 shows a cylindrical shell subjected to a radial load uniformly distributed along any circle perpendicular to the axis of the shell. Because of the symmetrical loading, the section normal to the axis remains circular. The load causes the shell to move y-distance in the radial direction toward the center. This y displacement changes along the axis, and thus creates bending stresses on the shell. Due to the symmetry of the cross-section and loading, the situation can be investigated with a longitudinal strip of unit width, $b = 1$, as shown in the figure.

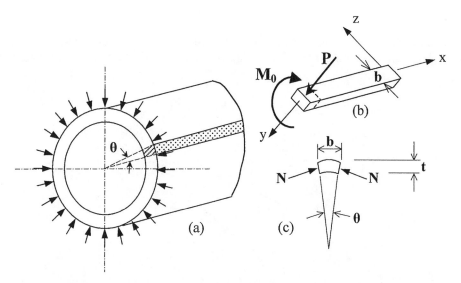

FIG. 5.5
CYLINDRICAL SHELL WITH AXIALLY SYMMETRICAL LOADING

Radial displacement y must be accompanied with a compression displacement in the circumferential direction. The situation is similar to a shell subjected to an external pressure. In conjunction with the reduction in radius, the circumference is reduced. A reduction in circumference means compression in the circumferential direction. The amount of compression strain is the same as the rate of change in radius, y/r. This gives a circumferential compressive force per unit strip length as follows

$$N = \frac{Et}{r}y$$

The strip having a width $b = r\theta$ is considered a beam supported by N furnished by the rest of the shell, $(2\pi - \theta)$. The resultant of these N forces has a radial direction component as

$$P = 2N\sin\frac{\theta}{2} \cong 2N\frac{b/2}{r} = \frac{N}{r}b = \frac{1}{r}\left(\frac{Et}{r}y\right)b = \frac{Et}{r^2}y \quad \text{for } b = 1$$

Because the support force, P, is proportional to the displacement, y, the strip is considered a beam on elastic foundation. With the unit width of the beam, the spring constant of the foundation per unit length of beam is determined by

$$k = \frac{P}{y} = \frac{Et}{r^2} \tag{5.25}$$

When the strip of the beam is subjected to a bending moment, it produces linearly varying longitudinal stresses as shown in Figure 5.6. The maximum stress equals the moment divided by the section modulus. These bending stresses produce linearly varying strains in the longitudinal direction. If the side surfaces were free to move, these stresses would have also produced circumferential varying strains as shown. However, under the symmetrical condition, the side surfaces have to remain in the radial direction, preventing the varying deformation from taking place. In the actual condition, the circumferential strain due to longitudinal bending moment is zero. Putting $e_z = 0$ in Eq. (2.8) and ignoring S_y, we have

$$e_x = \frac{S_x}{E} - v\frac{S_z}{E}; \quad e_z = \frac{S_z}{E} - v\frac{S_x}{E} = 0 \quad \text{or} \quad S_z = vS_x$$

Combining the above two equations, the stress strain relation for the bending stress becomes

$$e_x = \frac{1}{E}\left(S_x - v^2 S_x\right) = \frac{1-v^2}{E}S_x$$

This means that in a laterally constricted beam, the strain or displacement produced by a bending moment is smaller than that predicted by Hook's law by a factor of $(1 - v^2)$. The beam becomes stiffer

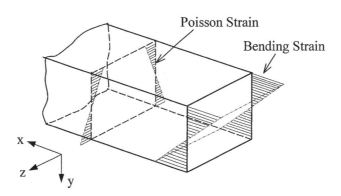

FIG. 5.6
SIDEWISE CONSTRICTION

than usual. In practical applications, either the modulus of elasticity or the moment of inertia can be increased by a factor of $1/(1 - v^2)$ to match the reduction of the strain and displacement. That is, we can choose either

$$E' = \frac{E}{1 - v^2} \quad \text{or} \quad I' = \frac{I}{1 - v^2}, \quad \text{but not both}$$

Generally, it is less confusing to modify the modulus of elasticity, because a modification of the moment of inertia may propagate to the section modulus, which remains the same. However, to avoid confusion between the modulus of elasticity used in the foundation spring constant and the one used for the bending moment, the traditional treatment of beam on elastic foundation chooses to modify the moment of inertia as

$$I' = \frac{I}{1 - v^2} = \frac{t^3 b}{12(1 - v^2)} = \frac{t^3}{12(1 - v^2)}$$

for a rectangular cross-section with width b = unity. Substituting the above I' for I and k from Eq. (5.25) in Eq. (5.6), we have

$$\beta = \sqrt[4]{\frac{k}{4EI'}} = \sqrt[4]{\frac{3(1 - v^2)}{r^2 t^2}} = \frac{1.285}{\sqrt{rt}}, \quad \text{for } v = 0.3 \tag{5.26}$$

Equation (5.26) shows the characteristic factor of pipe considered as beam on elastic foundation. This beam on elastic foundation analogy is strictly applicable only to axially symmetric and uniformly distributed circumferential loading and deformation. However, it has been extended to some applications with localized loading, presumably as a conservative approach.

5.5 EFFECTIVE WIDTHS

In dealing with discontinuity, it is important to know the extent of its effect. This includes the estimate of the width of the reinforcement required to spread out the load, the effective force-carrying zone to share the load, and the effective reinforcement zone. The effect of the discontinuity attenuates inverse exponentially with distance. It reduces very quickly extending outward, but never exactly reaches the zero point. Because the definition of the effective zone is not very clear-cut, some engineering judgment is exercised in defining it. Depending on the purpose it serves, the effective zone is defined somewhat differently for each type of application. However, they all follow the same pattern based on beams on elastic foundation. The responses of beams on elastic foundation are generally expressed in terms of the combined dimensionless location parameter βx. Considering the discontinuity as an infinite beam subjected to a concentrated load, the response curves given in Figs. 5.2 and 5.3 can be used as guidelines for determining the effective width. Figure 5.3 shows that at $\beta x = 1$, the shear force reduces to about 20% of the force at the discontinuity point. Because shear force represents the load carrying capacity, it appears reasonable to choose $\beta x = 1$ for the load carrying zone and effective reinforcement zone. Assuming $v = 0.3$, we have

$$\beta x = \left(\frac{1.285}{\sqrt{rt}}\right) x = 1.0 \quad \text{i.e., effective width } x = 0.778 \sqrt{rt} \tag{5.27}$$

This number has been adopted in the pressure design of miter bends discussed previously. As for the effective reinforcement zone, some codes use a smaller constant of 0.5 instead of 0.778.

The effective width for providing enough stress attenuation is defined based on the bending moment curve. Figure 5.3 shows that the bending moment reaches zero at $\beta x = \pi/4$, and reverses its sign after that. It reaches negative maximum at $\beta x = \pi/2$ with the peak magnitude reducing to 21% of the original moment. The moment reduces to negative 10% at $\beta x = 2.6$. With this 10% residual stress criterion, the separation or reinforcement width required is

$$\beta x = \left(\frac{1.285}{\sqrt{rt}}\right) x = 2.6, \quad \text{i.e., required width } x = 2.0\sqrt{rt} \tag{5.28}$$

In calculating the reinforcement width, the thickness also includes the pad thickness.

5.6 CHOKING MODEL

One basic application of shell as beam on elastic foundation is the choking model. In this model, a uniform radial line load is applied around the circumference of the shell as shown in Fig. 5.7. This is, in essence, the same model given in Section 5.4 for deriving the characteristic parameter β. In this section, we will use the model to find the stress produced. From this model, the load per unit length around the circumference is f, and the corresponding displacement due to this load is Δy. This is equivalent to an infinite beam on elastic foundation subjected to concentrated force as discussed in Section 5.2. By taking a unit circumferential width as the beam and assuming $\nu = 0.3$, then from Eqs. (5.15) and (5.26) we have the maximum bending moment occurring at forcing point $x = 0$ as

$$M = \frac{f}{4\beta} = \frac{f}{4*1.285}\sqrt{rt} = 0.1946 f\sqrt{rt}$$

Dividing the bending moment with the section modulus $Z = t^2/6$, we have the maximum bending stress at forcing point as

$$S_b = \frac{M}{Z} = \frac{1.167 f \sqrt{rt}}{t^2} = 1.167 \frac{\sqrt{r}}{t^{1.5}} f \tag{5.29}$$

This formula has been extensively used for calculating local attachment stresses in the design of piping systems [5].

The bending stress shown in Eq. (5.29) is in the longitudinal direction. It is also referred to as the longitudinal shell bending stress. Because its average across the shell thickness is zero, it is not included in the membrane stress category. As discussed previously, due to the nature of symmetry, the circumferential bending strain due to Poisson's effect is suppressed thus producing a circumferential shell bending stress equal to $\nu S_b = 0.3 S_b$ for $\nu = 0.3$. Because this Poisson's stress is always in the same sign as the longitudinal shell bending stress, it is subtractive to the longitudinal stress when combining

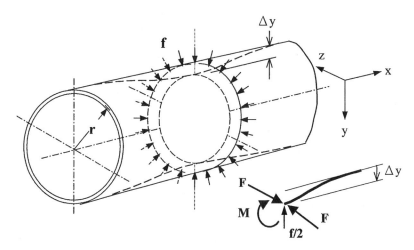

FIG. 5.7
CHOKING MODEL ON CYLINDRICAL SHELL

with other circumferential stresses. Therefore, it is conservatively ignored in most practical applications. See also Fig. 5.9.

In addition, there is also a circumferential membrane stress due to choking displacement, Δ_y. From Eqs. (5.13) and (5.25), we have the displacement at forcing point equal to

$$\Delta_y = \frac{f\beta}{2k} = \frac{f}{2}\left(\frac{1.285}{\sqrt{rt}}\right)\left(\frac{r^2}{Et}\right) = 0.643\left(\frac{r}{t}\right)^{1.5}\frac{f}{E}$$

This displacement produces a circumferential strain of Δ_y/r, thus producing a circumferential membrane stress of

$$S_{mc} = E\frac{\Delta_y}{r} = 0.643\frac{\sqrt{r}}{t^{1.5}}f \qquad (5.30)$$

This circumferential membrane stress may be additive to the pressure hoop stress depending on the direction of the force.

5.7 STRESSES AT JUNCTIONS BETWEEN DISSIMILAR MATERIALS

A piping system may involve more than one material, or may connect to an equipment of different elastic or thermal properties. The most common junctions between dissimilar materials are those involving an austenitic stainless steel section connected to a ferritic steel section. Austenitic steel is needed at high temperature zones directly in contact with the radiant flame inside a furnace or a boiler, although ferritic steel is more economical for outside piping. The same thing applies in a piping system with a combination of internally insulated and externally insulated sections. The internally insulated portion uses carbon steel, whereas the externally insulated portion requires the use of stainless or other high alloy steel pipe.

When two pipes of different materials are joined together, the joint produces additional discontinuity stresses due to differences in expansion rate, thickness, and modulus of elasticity. Figure 5.8 shows a simple bimetallic welded joint of a pipe with a uniform thickness. When the pipe temperature changes from the construction temperature, referred to as ambient, the pipe expands. Due to differences in expansion rate, the radius of Mat-a would have expanded at an amount of Δa, whereas Mat-b would have expanded by Δb, if the two sides were not joined together. Because the two pieces are joined together, the actual displacement at the junction is somewhere in between these two values.

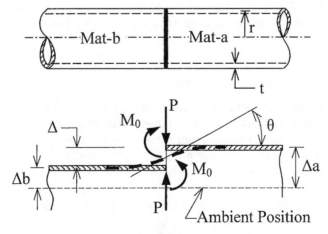

FIG. 5.8
JUNCTION BETWEEN DISSIMILAR MATERIALS

144 Chapter 5

This requires an internal squeezing force at the piece with the higher expansion rate to reduce its displacement, and an internal flaring force at the piece with the lower expansion rate to increase its displacement. The squeezing and flaring internal forces, acting in opposite directions, have the same magnitude and are denoted as P. In general, an internal shell bending moment is also produced along the pipe. The shell bending moments for both sides of the piping are the same M_0, but in opposite directions, at the junction.

Each piece of the pipe segment can be considered a semi-infinite beam on elastic foundation, subjected to the concentrated end force P and end bending moment M_0. By comparing Fig. 5.8 with Fig. 5.4, it is clear that Mat-a can use the standard formulas for a semi-infinite beam as presented, whereas Mat-b needs an adjustment in the formulas. Mat-b can be viewed as Fig. 5.4 turned upside down with the y-axis pointing upward, but with the sign of the bending moment reversed — that is, when using the standard semi-infinite beam formulas, the sign of M_0 is reversed for Mat-b. The working formulas for the dissimilar joints are then derived using two boundary conditions: (1) the slopes are the same for both joining pieces at junction; and (2) the sum of the displacements from both pieces is equal to the differential expansion $\Delta = \Delta a - \Delta b$. From Eq. (5.24) with $\theta_a = \theta_b = \theta$, we have

$$-\frac{2P\beta_a^2}{k_a} + \frac{4M_0\beta_a^3}{k_a} = -\frac{2P\beta_b^2}{k_b} - \frac{4M_0\beta_b^3}{k_b}$$

or

$$2P\left(\frac{\beta_a^2}{k_a} - \frac{\beta_b^2}{k_b}\right) = 4M_0\left(\frac{\beta_a^3}{k_a} + \frac{\beta_b^3}{k_b}\right) \qquad (5.31)$$

The sum of the displacements equals the differential thermal expansion — that is, $\Delta = \Delta a - \Delta b = r(\Delta T)(\alpha_a - \alpha_b)$. ΔT is the temperature difference between construction and operation, and α_a and α_b are thermal expansion coefficients for Mat-a and Mat-b, respectively. From Eq. (5.23), we have

$$r(\Delta T)(\alpha_a - \alpha_b) = \frac{2P\beta_a}{k_a} - \frac{2M_0\beta_a^2}{k_a} + \frac{2P\beta_b}{k_b} + \frac{2M_0\beta_b^2}{k_b}$$

or

$$r(\Delta T)(\alpha_a - \alpha_b) = 2P\left(\frac{\beta_a}{k_a} + \frac{\beta_b}{k_b}\right) + 2M_0\left(\frac{\beta_b^2}{k_b} - \frac{\beta_a^2}{k_a}\right) \qquad (5.32)$$

Equations (5.31) and (5.32) can be easily applied to junctions with two segments of pipes having identical cross-sections. They can also be applied to a pipe connecting to a relatively rigid section. Two special cases will be discussed in the following to investigate the general behaviors of the junctions between dissimilar materials. In Sections 5.7.1 and 5.7.2, we use $v = 0.3$ to obtain some comparative stress values.

5.7.1 Uniform Pipes With Similar Modulus of Elasticity

The modulus of elasticity maintains relative uniformity for all ferrous and some non-ferrous materials at temperatures within the allowable working range. Therefore, a simplified model assuming both Mat-a and Mat-b having the same EI, and hence the same k and β values will be used to quick check the general behaviors of the junction. By assuming $k_a = k_b = k$ and $\beta_a = \beta_b = \beta$, Eq. (5.31) gives $M_0 = 0$ — that is, the bending moment at the junction is zero. With $M_0 = 0$, Eq. (5.32) becomes

$$r(\Delta T)(\alpha_a - \alpha_b) = 4P\frac{\beta}{k} \quad \text{or} \quad P = \frac{1}{4}\frac{k}{\beta}r(\Delta T)(\alpha_a - \alpha_b)$$

Substituting the above M_0 and P to Eq. (5.21), we have the bending moment at a point x-distance away from the junction as

$$M = -\frac{P}{\beta}f_4(\beta x) = -\frac{k}{4\beta^2}r(\Delta T)(\alpha_a - \alpha_b)f_4(\beta x)$$

The maximum bending moment M_{max} occurs at $\beta x = \pi/4$ (i.e., $x = \pi/(4\beta) = 0.611(rt)^{1/2}$), and $f_4(\beta x) = 0.3224$. Using this βx, Eq. (5.25) for k, and Eq. (5.26) for β, the maximum shell bending stress can be calculated as

$$S_{b,max} = M_{max}\frac{6}{t^2} = \frac{0.3224 \times 6}{4t^2}\left(\frac{Et}{r^2}\frac{rt}{1.285^2}\right)r(\Delta T)(\alpha_a - \alpha_b)$$

or

$$S_{b,max} = 0.29E(\Delta T)(\alpha_a - \alpha_b) \qquad (5.33)$$

In addition to the shell bending stress, there is the hoop stress caused by the radial displacement at the junction. With the same k and β, intuitively we can conclude that the displacement contributed by each side of the pipe is one-half of the differential expansion Δ. That is, $y_{0,a} = y_{0,b} = 0.5\Delta = 0.5r(\Delta T)(\alpha_a - \alpha_b)$. This can also be obtained by substituting the above P and $M_0 = 0$ to Eq. (5.23). The choking or flaring of the radius generates a circumferential strain of $0.5\Delta/r$, and thus a circumferential membrane stress of

$$S_c = E\frac{0.5\Delta}{r} = 0.5E(\Delta T)(\alpha_a - \alpha_b) \qquad (5.34)$$

This thermal circumferential stress is tensile in the piece with the lower expansion rate, and is compressive in the piece with the higher expansion rate — that is, both tension and compression natures have to be considered, the same as in the case of bending. Because of this dual tensile and compressive nature, it is additive to the pressure hoop stress.

Because both shell bending stress and circumferential hoop stress have tensile and compressive characteristics, they will have to be added together directly to calculate the stress intensity. However, as discussed previously in this chapter, the longitudinal bending stress comes together with a circumferential bending stress due to Poisson's effect. Because Poisson's stress is always in the same sign as the longitudinal stress, it is subtractive to the longitudinal stress when we are dealing with the combined effect.

Figure 5.9 shows the stress situation at the pipe shell. Based on maximum shear failure theory, stress intensity is taken as the difference of the two perpendicular principal stresses. For the two directions of stresses to be additive, they have to be in the opposite directions — that is, one in tensile and the other in compressive. Assuming that at a given point of the pipe wall the longitudinal bending stress

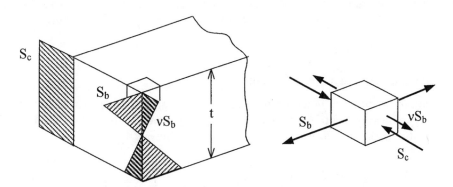

FIG. 5.9
SUBTRACTIVE NATURE OF POISSON'S STRESS

is tensile, then it will be additive to the circumferential stress only when the circumferential stress is compressive. However, a tensile longitudinal bending stress generates a tensile circumferential Poisson's bending that reduces the net compressive circumferential stress. Therefore, the combined stress intensity becomes $S_E = (1 - \nu)S_b + S_c$.

For the dissimilar material connection, the maximum circumferential stress occurs at the junction, whereas the maximum longitudinal bending stress occurs at $0.25\pi/\beta$ away. As a quick estimate, we will assume that both maximum circumferential stress and maximum longitudinal bending stress occur at the same point, in which case the potential maximum combined stress intensity is $S_E = [0.29(1 - 0.3) + 0.5]E(\Delta T)(\alpha_a - \alpha_b) = 0.703E(\Delta T)(\alpha_a - \alpha_b)$. Because their maximum values do not occur at the same place, the actual maximum stress intensity is somewhat smaller, but never less than $0.5E(\Delta T)(\alpha_a - \alpha_b)$, which occurs at the junction with zero longitudinal bending stress. We can use an average value of $0.6E(\Delta T)(\alpha_a - \alpha_b)$ as the actual maximum combined stress intensity [6]. Therefore, the thermal discontinuity stress to be added to the general thermal expansion stress is

$$S_{E,t} = 0.60E(\Delta T)(\alpha_a - \alpha_b) = 0.60E(T_2 - T_1)(\alpha_a - \alpha_b) \tag{5.35}$$

Because expansion rate changes considerably with temperature, the average value between T_1 and T_2 should be used. The piping codes generally provide the average expansion rates data, from ambient temperature to operating temperature, for most piping materials.

As an example, for austenitic stainless steel joined with carbon steel at 800°F (427°C). $\alpha_a = 10.1 \times 10^{-6}$ in./in./°F (18.18×10^{-6} mm/mm/°C); $\alpha_b = 7.8 \times 10^{-6}$ in./in./°F (14.04×10^{-6} mm/mm/°C), $E = 10^6(26.0 + 24.1)/2 = 25.05 \times 10^6$ psi (172.72×10^3 MPa), $(T_2 - T_1) = 800 - 70 = 740$°F (411.11°C), then $S_{E,t} = 0.6 \times 25.05 \times 740 \times (10.1 - 7.8) = 25{,}581$ psi (176.376 MPa). It should be noted that for uniform thin pipe connections, the stress is independent of size and thickness.

Although the B31 codes adopt a base of using one-half of the theoretical secondary stresses, the above stress occurs at the weld joint, which is the basis of the adjusted stress. Since the above stress is superimposed on a weld joint, its full value shall be used in evaluating the self-limiting stress. The thermal discontinuity stress should be either added to the general thermal expansion stress range or subtracted from the allowable value.

From the above example, it is clear that at 800°F (427°C) the thermal discontinuity stress at the austenitic-carbon steel junction is very close to the nominal allowable displacement stress range of 30,000 psi (206.8 MPa). This leaves very little margin for the expansion stress range from thermal expansion of the piping system. To reduce the thermal discontinuity stress, some critical piping may use a two-step connection by inserting a short piece of nickel-chrome-iron steel pipe (e.g., ASTM B-407) in between the austenitic-carbon steel junction. Because the expansion rate of nickel-chrome-iron is roughly the average of the expansion rates of austenitic steel and carbon steel, it can effectively reduce the junction thermal discontinuity stress by 50%. The length of insert shall at least be $5/\beta$ or 4 in. (100 mm), whichever is greater. This is to ensure the separation of the discontinuity stress fields and the weld affected zones.

5.7.2 A Pipe Connected to a Rigid Section

In comparison to the uniform junction discussed above, a pipe connected to a rigid section is another extreme of application. In this discussion, the pipe segment Mat-b is considered as rigid — that is, $k_b = \infty$. Substituting this k_b into Eqs. (5.31) and (5.32), we have

$$P = 2M_0\beta; \quad \text{and} \quad r(\Delta T)(\alpha_a - \alpha_b) = 2P\frac{\beta_a}{k_a} - 2M_0\frac{\beta_a^2}{k_a}$$

Solving the above two equations and dropping subscript a on β and k, we have

$$M_0 = \frac{1}{2}\frac{k}{\beta^2}r(\Delta T)(\alpha_a - \alpha_b) \quad \text{and} \quad P = \frac{k}{\beta}r(\Delta T)(\alpha_a - \alpha_b)$$

The maximum shell bending stress, which occurs at the junction, is

$$S_b = M_0\frac{6}{t^2} = \frac{3}{t^2}\left(\frac{Et}{r^2}\frac{rt}{1.285^2}\right)r(\Delta T)(\alpha_a - \alpha_b) = 1.817E(\Delta T)(\alpha_a - \alpha_b) \tag{5.36}$$

The pipe segment Mat-a in this case has to deflect in the radial direction the full differential expansion Δ. Therefore, the circumferential membrane stress is

$$S_h = E\frac{\Delta}{r} = 1.0E(\Delta T)(\alpha_a - \alpha_b) \tag{5.37}$$

Because shell bending stress and circumferential hoop stress occur at the same point, they need to be added directly together to become part of the self-limiting stress. After subtracting circumferential Poisson's bending, we have the stress due to discontinuity as

$$S_{E,t} = [1.817(1 - 0.3) + 1]E(\Delta T)(\alpha_a - \alpha_b) = 2.272E(T_2 - T_1)(\alpha_a - \alpha_b) \tag{5.38}$$

Equation (5.38) is applicable to nozzle inserts and socket welds. Since the stress from (5.38) is 3.7 times as large as the one given by (5.35), a dissimilar material junction at a socket weld or other insert should be avoided.

5.8 VESSEL SHELL ROTATION

Movements at boundary points are very important when calculating the stress of a piping system. Normally, it is easy to visualize and to apply the translation movements at a vessel connection. Translation movements are calculated based on direct thermal expansion of the vessel. However, due to the choking effect at the discontinuity area, a pressurized vessel shell may also have significant rotation at the vicinity of the discontinuity. The rotation can create a much more severe stress in the piping system compared to the translation movement. Under normal piping and vessel configurations, the discontinuity zone is too small to accommodate any nozzle or branch connection. However, the discontinuity zone at the bottom of a large storage tank can reach as much as 5 ft (1.524 m) in the vertical direction. In fact, most tank connections are located at the bottom discontinuity zone. Therefore, significant shell rotations may occur at the pipe connections located near the bottom portion of the tank.

Figure 5.10 shows the bottom portion of the shell of a large storage tank, which has a diameter of 180 ft (54.86 m) and a liquid height of 70 ft (21.34 m). The stored liquid produces a hydrostatic pressure that varies linearly along the height. The maximum pressure occurs at the bottom course of the shell. If the tank shell were not attached to the bottom plate, the pressure would have produced proportionally a radial displacement with the maximum occurring at the bottom. However, because the shell is attached to the bottom plate, the shell radial displacement is choked to zero at the bottom. As the choking only affects a small portion of the shell, the shell can be considered having a uniform pressure and thickness in the area being investigated. The free displacement of the shell constitutes pressure and thermal two parts. That is,

$$\Delta = \frac{S_h}{E}r + \alpha(T_2 - T_1)r = \frac{p}{Et}r^2 + \alpha(T_2 - T_1)r \tag{5.39}$$

where p is the hydrostatic pressure. For water at 70 ft deep, $p = 70 \times 62.4/144 = 30.33$ psi (0.209 MPa). The amount of displacement choked by the bottom plate depends on the mobility of the bottom plate.

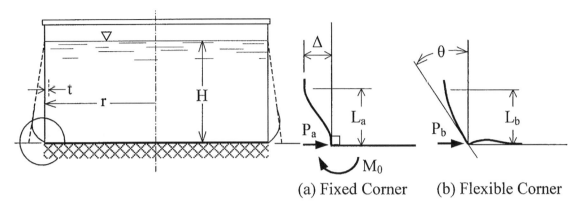

**FIG. 5.10
DEFORMATION OF TANK SHELL**

For the pressure part, the radial displacement of the bottom plate is always considered zero. For the thermal expansion part, the bottom plate has the potential to move as much as the shell does. However, because the fluid is cooler at the bottom and the friction force from the foundation can greatly restrict the expansion, no thermal expansion from the bottom plate is generally assumed. Therefore, the bottom plate will exert a full choking displacement of Δ. This is a conservative assumption for the piping stress analysis, because more choking creates more rotation, and thus more stress.

Deformation of the shell depends on the flexibility of the bottom connection. Two extreme cases (Fig. 5.10(a) and (b)) were investigated. The actual condition shall be somewhere in between these two conditions.

Case (a) shows that the bottom plate is fixed allowing no rotation at the shell. This will exert a bending moment M_0 as well as the radial force P_a at the shell junction in order to choke the expanding shell to the fixed bottom plate. This is the same as one side of the infinite beam subjected to a concentrated force. The deflection, slope (rotation), and moment curves given in Fig. 5.2 are all applicable.

Case (b) shows that the bottom plate does not offer any rotational stiffness. This is the case of the semi-infinite beam with zero bending moment at the junction. Substituting $M_0 = 0$ and $y_0 = \Delta$ to Eqs. (5.23) and (5.24), we have

$$P = \frac{k\Delta}{2\beta} \quad \text{and} \quad \theta_0 = -\frac{2P\beta^2}{k} = -\Delta\beta \tag{5.40}$$

Substituting P to Eqs. (5.19) and (5.20), we have the choked displacement and the slope at the point located x-distance away from the junction as

$$y = \frac{2P\beta}{k} f_1(\beta x) = \Delta f_1(\beta x) \quad \text{and} \quad \theta = \frac{2P\beta^2}{k} f_3(\beta x) = \Delta\beta f_3(\beta x) \tag{5.41}$$

Because the tank bottom is generally not anchored, the actual case is somewhat closer to case (b). Due to the existence of the inflection in the deflection curve, case (a) is more sensitive to the shell parameters that can only be estimated approximately. In addition, a pipe connection constitutes a rigid area on the shell, thus making a sharp curvature as in case (a) not very likely. Therefore, unless a more accurate assessment is available, a smoother curve as given by the case (b) is generally used in the design analysis.

Because of the choking at the bottom, the circumferential stress at the shell changes along the height of the shell. Figure 5.11 shows the relation between deflection and stress. At point a, the un-affected area, the shell deflects Δ, which constitutes the hydrostatic pressure portion Δ_P and thermal expan-

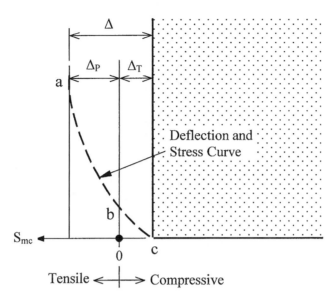

FIG. 5.11
TANK DEFLECTION AND STRESS

sion portion Δ_T. Since free thermal expansion does not produce any stress, the circumferential hoop stress is entirely due to the hydrostatic pressure. It is in tension and is proportional to Δ_P. As the shell is choked gradually, the deflection of the shell reduces gradually, thus also reducing the tensile hoop stress. Stress is reduced to zero at point *b* when the deflection equals the thermal expansion Δ_T. Further reduction of the shell deflection produces a compressive hoop stress. At point *c* the compressive hoop stress reaches maximum level. The maximum compressive hoop stress corresponds to squeezing the thermal expansion Δ_T to zero. Compressive hoop stress is often ignored in the design analysis. A design curve often assumes that the tensile hydrostatic hoop stress reduces gradually to zero at the bottom junction. This circumferential membrane stress at the choked area has to be taken into account when evaluating the interface with the piping. Further discussion on tank connections is given in Chapter 8.

REFERENCES

[1] ASME Boiler and Pressure Vessel Code, Section VIII, "Pressure Vessels," Division 2, "Alternative Rules," Article 4-7, "Discontinuity Stresses," American Society of Mechanical Engineers, New York.
[2] ASME Boiler and Pressure Vessel Code, Section III, "Rules for Construction of Nuclear Facility Components," Division 1, Article A-6000, "Discontinuity Stresses," American Society of Mechanical Engineers, New York.
[3] Hetenyi, M., 1946, *Beams on Elastic Foundation*, Chapter II, University of Michigan Press, Ann Arbor, MI.
[4] Timoshenko, S., 1956, *Strength of Materials*, Part II, 3rd edn., Chapter I, McGraw-Hill, New York.
[5] M. W. Kellogg Company, 1956, *Design of Piping Systems*, revised 2nd ed., Chapter 3, John Wiley & Sons, Inc., New York.
[6] ASME Boiler and Pressure Vessel Code, Section III, "Rules for Construction of Nuclear Facility Components," Division 1, Sub-article NB-3600, "Piping Design," American Society of Mechanical Engineers, New York.

CHAPTER 6

PIPE SUPPORTS AND RESTRAINTS

Providing sufficient pipe wall thickness and installing proper supports are two of the most important elements in ensuring structural integrity of the piping system. An adequate wall thickness is needed to contain the process fluid, and a proper support system is required for holding the pipe in place. Pipe supports are generally referred to as devices used in supporting the weight of the piping. The weight includes that of the pipe proper, the content the pipe carries, and the pipe covering, such as insulation, refractory, lining, and snow.

In addition to supports, a piping system may also need restraints to control movement, resist occasional loads such as wind and earthquake, protect sensitive equipment, increase stiffness, reduce vibration, and so forth. The purpose of using restraints is to restrict the movement of the piping in certain directions. Both restraints and supports for weight are collectively referred to as pipe supports.

6.1 DEVICE TERMINOLOGY AND BASIC FUNCTIONS

A pipe support and restraint system involves many different types of hardware components and arrangements to serve the different functions needed. In a computerized stress analysis environment, it is necessary to have a common terminology so that design engineers, computer specialists, and fabricators can all communicate with a common language. The following are some of the pipe support terminologies used by the piping community. Most of these terms originated from Kellogg's book [1], although some of the original meanings have been slightly refined. These terminologies are explained with their associated basic functions. The methods of describing these functions to the computer are also discussed.

Before discussing each device, it is important to remember that every point of the piping system is associated with six degrees of freedom: three directions of translation and three directions of rotation. Without restriction, the pipe can move in the x, y, and z directions, and can also rotate about the x, y, and z axes. The supports and restraints normally restrict the motion at one or more of the degrees of freedom. The effectiveness of the restriction to each direction depends on the stiffness of the support structure in that particular direction. This stiffness is commonly called the support spring constant or spring rate. Theoretically, a rigid support or restraint has an infinitely large spring rate. In actual applications, a sufficiently high spring rate may be used instead to avoid mathematical difficulties.

- Anchor

  An anchor fixes all six degrees of freedom. This is the most fundamental support in piping stress analysis. Most of the basic flexibility analyses are done by assuming that both ends of the piping system are anchored. Figure 6.1 shows some of the common arrangements of piping anchors. Equipment connections are normally considered anchors. However, other than equipment connections, there are very few real anchors that exist in a piping system. The places most likely to have an anchor placed are at the ends of an expansion joint and at the ends of a long pipeline.

  A theoretical anchor does not allow any pipe displacement or rotation at the anchor point. This is because the anchor is assumed to have an infinite stiffness in all six degrees of freedom.

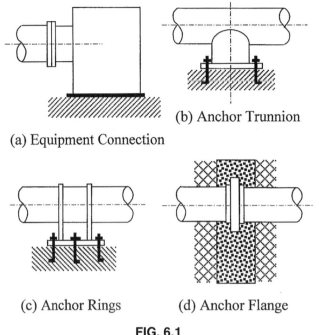

**FIG. 6.1
PIPING ANCHORS**

However, some of the equipment connections, such as vessels and tanks, do have significant flexibility in certain degrees of freedom. These connection flexibilities are handled either with a flexible connection attached to the anchor or by considering the anchor as six directional restraints, each with its own stiffness. In addition to the fixation, the equipment connection also normally exerts some displacements to the piping, due to the thermal expansion of the equipment itself. These displacements also need to be applied in the analysis. The anchor shown in Fig. 6.1(d) is normally used to limit the large end displacement of a long underground pipeline. Because it is very difficult to achieve a complete fixation under some poor soil conditions, the anchor block is occasionally designed to allow it to plow through the soil somewhat. This type of anchor offers no complete fixation. It only provides drag and prevents the pipe from moving too much, and is thus called a drag anchor.

- Supports

A support device is used to sustain a portion of the weight of the piping system and other superimposed vertical loads. However, a support is specifically referred to as the device acting from underneath the pipe, in contrast to the hanger that is working from above the pipe. As noted before, "pipe support" is also used as the generic term for all supports and restraints combined.

Figure 6.2 shows some general types of rigid hangers and supports. The hanger shown in Figure 6.2(a) will be defined and discussed separately. Support types are distinguished mainly by the pipe shoes they use. The most direct and economic supporting scheme is to rest the pipe directly on the support structure as in Figure 6.2(b). When the pipe needs to be insulated, a shoe is generally required. Figure 6.2(c) shows a shoe made from a piece of inverted-T steel. This T shoe has a single loading line placed at the weakest mid-shell location of the pipe. Detail (c) is used only for smaller pipes, up to 10 in. (250 mm) in size. For larger pipes, say between 12 in. (300 mm) through 24 in. (600 mm), the H shoe as shown in (d) can be used. The H shoe, taken from a wide flange I-shape steel, divides the load into two lines placed near the sides of the pipe shell. The

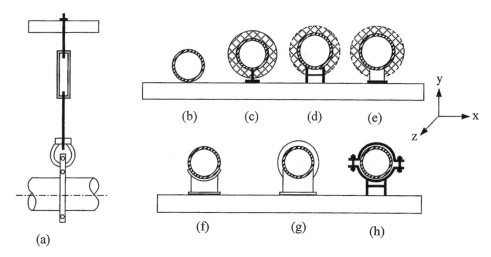

**FIG. 6.2
RIGID HANGER AND SUPPORTS**

side of the pipe shell can take a much larger load than the center portion of the shell. When the pipe temperature exceeds 750°F (400°C), the temperature gradient stress on the T and H shoes can reach excessive levels, thereby requiring the use of a trunnion shoe (e) or a clamped shoe (h). The saddle (f) and ring girder (g) are mainly for pipe sizes larger than 24 in. (600 mm).

In piping stress analysis, the hanger and supports given in Fig. 6.2 are all classified as stops in the y direction. Because these supports only stop the piping from moving downward, a more elaborate analysis will classify these as stops in the minus (−) y direction. The y direction is used by many computer software packages to represent the vertical direction pointing upward — that is, the direction directly opposing the direction of gravity.

- Hanger

Similar to the support, the hanger is used specifically to sustain a portion of weight of the piping system and superimposed vertical loads. However, a hanger sustains the piping weight from above, and its load is always in tension. Due to this tensile nature, a slender rod can be used without the concern of buckling. A hanger normally refers to a rigid hanger. Although a hanger may have considerable flexibility in the loading direction, it is still generally much stiffer than the piping system in the direction supported.

A hanger is suspended from a structure with the hanger rod pivoting at a fixed point as shown in Fig. 6.3. When the pipe moves due to thermal expansion or other forces, the hanger rod slants creating a horizontal resistance. It can also lift up the piping somewhat. These types of secondary effects are normally not implemented in typical analysis methods or in computer software. However, if the slanting angle is limited to no more than 4 deg., then the horizontal resistance can be ignored. Figure 6.3 shows that at a 4 deg. slanting, the horizontal resistance is about 70 units per 1000 units of support force. To reduce the effect of this horizontal resistance to a minimum, a hanger is normally installed with some initial slanting so that it becomes vertical at the operating condition. When the headroom of the piping is insufficient to achieve this less than 4 deg. slant angle, a trapeze hanger may be used.

- Restraint

Any device that prevents, resists, or limits the movement of the piping is called a restraint. Restraints refer generally to devices other than the weight sustaining devices, or supports defined

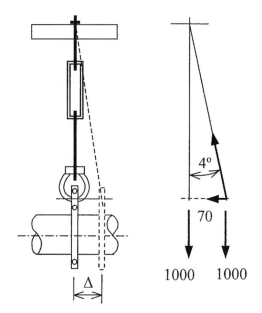

**FIG. 6.3
LATERAL FORCE CREATED BY DISPLACEMENT**

above. Figure 6.4 shows some of the more commonly used restraints. There is no limit as to how many directional restraints can be installed at one point. The guides and stops shown in the figure are all combined with a vertical support. An anchor can also be considered as the combination of six directional restraints.

- Strut

A directional restraint consists of a compressive column with two rotational joints. A strut is subjected to both tensile and compressive loads. It is used when there is no suitable support structure nearby. It can also be used to reduce the restraint friction force. A strut can be conveniently attached to a nearby structure in a generally skewed direction.

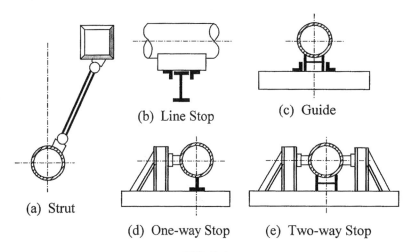

**FIG. 6.4
DIRECTIONAL RESTRAINTS**

- Stop

  This is a device that stops the piping movement in one or more translation directions. Meanwhile, a *line stop* specifically refers to the device that stops the pipe in the axial direction. Unless specified otherwise, a stop acts in both plus and minus directions (also known as double-acting).

- Guide

  The guide stops the pipe from moving in the lateral direction. The guide is double-acting in nature. For a long straight segment of piping, guides are generally provided at every other support span. (This term was used by the Kellogg book for rotational restraint.)

- Space maintenance stop

  The stop details given in Fig. 6.4(b) and (c) are mainly for space maintenance. They hold the piping in the proper space and resist moderate occasional forces. For large stop forces, more solid constructions, such as the ones shown in Fig. 6.4(d) and (e), should be used.

- One-way stop

  This is a directional stop that stops the pipe in either the plus or the minus direction. The pipe is free to move in the opposite direction.

- Double-acting

  A restraint that is active in both plus and minus directions. This requires the stop members to be welded on both sides of the support shoe as in Fig. 6.4(b), (c), and (e).

- Single-acting

  The device is active only in either the plus or minus direction. If one of the stop members in Fig. 6.4(b) and (c) is removed, the stop becomes single-acting. A one-way stop is single-acting.

- Limit stop

  A stop that is active only after the pipe has moved a certain amount. The allowable displacement is controlled by the gap between the pipe and the stop. It should be noted that most stops and guides are installed with a 1/16-in. (1.5 mm) construction gap to prevent binding between the pipe and the stop. Besides those located near equipment, the construction gaps are normally ignored in the analysis.

- Brace

  This device is similar to a strut, but is used specifically in resisting occasional loads and reducing vibrations. A brace is commonly attached with a pre-compressed spring or friction unit. It affords a pre-set amount of initial resistance, after which a constant amount or sloping amount of resistance persists. The device does not need any initial movement to activate the stopping action. Figure 6.5 shows a brace with a pre-compressed spring. Detail (a) shows the force-displacement relation, and detail (b) shows the modeling technique. The brace can be modeled as the combination of one linear spring and one elastic-plastic non-linear restraint. A limit stop can also be superimposed to limit the maximum travel.

- Resting or sliding support

  A device that provides support from beneath the piping but offers no resistance other than frictional to horizontal motion. A *resting support* is a single-acting device that stops the piping from moving only in the downward direction.

- Rigid (solid) support

  A support or restraint that consists of solid structural members only. Although the support structure may possess some inherent flexibility, it is still much stiffer than the piping in the direction of support.

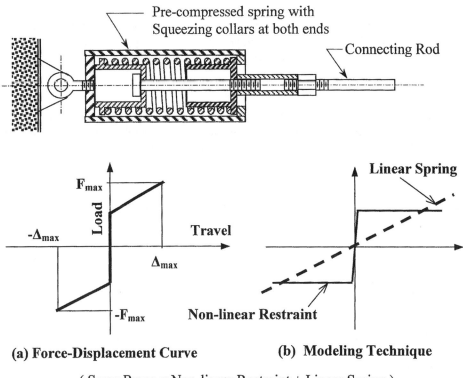

FIG. 6.5
PRE-COMPRESSED SPRING SWAY BRACE

- Spring (resilient) support

  A support that includes a very flexible member such as a spring. A *variable spring support* is much more flexible than the piping in the direction supported. It is called a variable support because the magnitude of its supporting load changes as the pipe moves.

- Constant-effort support

  A support that applies a relatively constant support force to the piping regardless of the pipe movement (e.g., compensating spring or counterweight device).

- Snubber

  A device that resists shock loads such as earthquake and water hammer, but does not resist the slow moving thermal expansion movement. Snubbers can be classified as hydraulic or mechanical as shown in Fig. 6.6. With hydraulic snubbers, the restraint movement pushes a piston inside a fluid filled cylinder. The cylinder has a passage to the reservoir. Inside the passage, there is a spring loaded check valve that allows the fluid to flow through gradually but shuts off the flow when the fluid velocity reaches a threshold limit. The mechanical snubber, on the other hand, uses a ball screw and ball nut assembly to convert the translational restraint movement to rotational movement. When the rotational acceleration reaches a certain level, the rotation of the torque transfer drum is bound still by the capstan spring. Either type of snubber requires a small initial movement before the snubbing action is activated. Therefore, it is not effective in reducing steady-state small amplitude vibrations. In piping stress analyses, the snubber is active for dynamic loads, but not for static loads.

Pipe Supports and Restraints   157

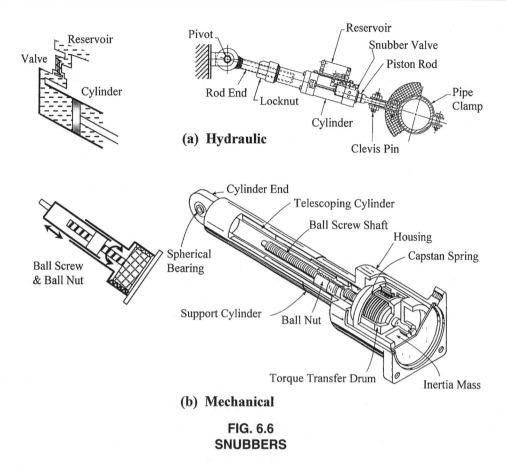

**FIG. 6.6
SNUBBERS**

- Rotational restraint

This is a device that prevents the pipe from rotating about one specific axis. It is sometimes referred to as a *moment restraint*. The rotation can be prevented with a specially designed device or by two coupled parallel stops. If it is done by two parallel stops, it is also modeled as two stops in the analysis.

6.2 SUPPORT SPACING

Adequacy of the support is determined by the sustained stress limitation. If the sustained stress meets the code requirement, then the support system is considered acceptable. This criterion is what computerized stress analyses are based on. If the sustained stress is within the allowable limits, then it does not matter how often the piping is supported. However, there are other concerns that require the setting of the allowable support spans for horizontal lines. These concerns are discussed in the following subsections.

1. Layout of Support Steels and Sleepers

These support locations are normally determined before the piping system is analyzed for acceptable stress. Layout designers will need allowable spans that ensure safe stress to work with.

2. Limit Pipe Displacements

The displacement of the pipe needs to be limited for better appearance and to prevent excessive pocketing. The displacement limitation is also required to avoid interference between pipes. Although excessive displacements will be discovered during stress calculations, it is better to limit them when the piping is being routed.

3. Avoid Support Rotations

Excessive pipe rotations twist support shoes making the supports ineffective. In general, limiting the sagging displacements can prevent excessive rotations.

4. Provide Rigidity

A certain amount of piping system rigidity is needed to reduce undesirable vibration. This, in part, is related to limiting the displacements.

The allowable support spans are determined by the stress and displacement criteria. These can be evaluated by the semi-fixed beam approach as discussed in Section 2.7. Based on the allowable bending stress criterion, the spacing is limited to

$$L_1 = \sqrt{\frac{10ZS}{w}} \qquad (6.1)$$

where S is the design weight bending stress {B31.1 [2] uses $S = 2300$ psi (15.86 MPa), whereas CEN [3] uses $S = 10$ MPa (1450 psi), and Kellogg [1] uses $S = 2750$ psi (18.96 MPa)}; Z is the section modulus of the pipe cross-section; and w is the weight force per unit length of pipe.

Based on the allowable sagging criterion, the spacing is limited to

$$L_2 = \sqrt[4]{\frac{128EI\Delta}{w}} \qquad (6.2)$$

Δ is the design sagging displacement. B31.1 uses $\Delta = 0.1$ in. (2.5 mm) for power plants, whereas Kellogg suggests a Δ of 0.5 in. to 1.0 in. (12.5 mm to 25 mm) for process plants.

The allowable span, L_S, is therefore taken as the smaller of L_1 and L_2. Table 6.1 shows the suggested pipe support spacing by B31.1 for power plants. A similar table can be constructed accordingly for each type of service with different allowable stress and displacement criteria.

The suggested pipe support spacing is applicable only for uniform pipe without any attached concentrated weight, such as a valve or flange. It is not applicable for the overhanging span either. For overhanging spans, the quarter circle approach as shown in Fig. 6.7 can be used. The curve shown is a quadrant of a circle having a radius equal to the allowable spacing. The two coordinates of the point

**TABLE 6.1
SUGGESTED PIPE SUPPORT SPACING
FOR POWER PIPING [2]**

Nominal pipe sizes		Suggested maximum span, L_S			
		Liquid service		Gas service	
in.	mm	ft	m	ft	m
1	25	7	2.1	9	2.7
2	50	10	3.0	13	4.0
3	80	12	3.7	15	4.6
4	100	14	4.3	17	5.2
6	150	17	5.2	21	6.4
8	200	19	5.8	24	7.3
12	300	23	7.0	30	9.1
16	400	27	8.2	35	10.7
20	500	30	9.1	39	11.9
24	600	32	9.8	42	12.8

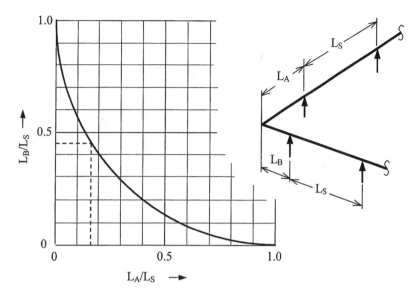

**FIG. 6.7
OVERHANGING PIPING SPANS**

on the curve determine the two allowable overhanging lengths. For instance, if $L_B = 0.45 L_S$, then the allowable L_A is $0.16 L_S$.

6.3 ANALYSIS OF PIPING SYSTEMS RESTING ON SUPPORTS

The most common and economical approach in dealing with the countless piping in a process plant or power plant is to rest the piping on pipe racks or other support structures [4]. The piping is supported either directly on the pipe wall or through pipe shoes. These types of supports are generally called resting supports. These supports are single-acting, because they only stop the pipe from moving downward but allow the pipe to move up freely. Due to the nature of this non-linearity, exact solutions are not expected for piping that goes through various temperature cycles. Therefore, three major schools of thought have developed in the pipe stress software community with regard to resting supports and temperature cycles. The analysis results and the qualities of the system designed differ considerably among the methods used.

A simple example shall be used to demonstrate the merits and pitfalls of some analysis approaches designed to satisfy the code requirements and philosophy. Figure 6.8 shows a typical piping system resting on the support structure. The piping has one end connected to a process tower and the other end connected to another process equipment. The system is supported at three locations. As the temperature of the process fluid rises, the tower grows upward and the pipe expands. With the tower connection moving up gradually, the piping system also goes through the following sequence of changes:

- With a small tower movement, the piping is held down on all supports by the weight of the piping including fluid and attachments. Some thermal expansion (displacement) stress is generated, but the weight stress remains the same as in the cold condition.
- As the temperature rises gradually and the tower movement increases somewhat, the piping will lift from the first support (support 20). A further increase of the tower movement will lift the pipe off support 30, thus making a large portion of the piping unsupported. This substantially increases the sustained weight stress.
- As the system reaches the maximum operating temperature, the tower connection moves up some more, but the pipe is still being held down at support 40. The expansion stress increases

160 Chapter 6

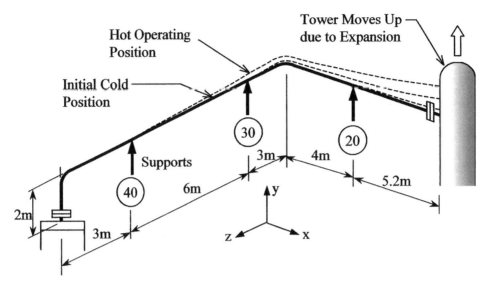

**FIG. 6.8
PIPING RESTING ON SUPPORTS**

due to the higher temperature and larger tower movement. However, the sustained weight stress remains unchanged as no additional supporting location is lifting off its support.
- When this maximum operating temperature is maintained for a certain period, the thermal expansion stress (displacement stress) will be relaxed somewhat. The amount of relaxation depends on stress level and operating temperature. However, the sustained weight plus pressure stress remains the same.
- When the plant cools down, the piping moves back on to its supports. This reduces the weight stress to its initial cold condition weight stress. The system, however, generates some reverse thermal expansion stress due to relaxation at operating temperature.
- If considerable yielding or creep occurs at hot condition, the pipe may return to the support point before the temperature reaches the ambient temperature. A continued cooling down to ambient temperature will cause high thermal stresses and loads due to stoppage by the support that prevents the pipe from moving further down. Generally, this will cause unpredictably high loads and stresses at the tower connection.
- In the next operating cycle, the weight stress goes back to the hot condition stress sustained, but the expansion stress will be reduced to a level corresponding to the relaxed state.

Three main approaches have been adopted by computer software packages in dealing with systems like the one shown in Fig. 6.8.

1. *General straightforward approach.* This is an approach commonly adopted by general-purpose finite element programs. In this approach, sustained stress and expansion stress are calculated separately without checking the situation influenced by the other load. Sustained stress is calculated by considering only the weight and pressure loads at ambient state. All supports are considered active because no temperature and support displacement is involved. The expansion stress range is calculated only with the temperature change. No weight influence is considered. If the pipe lifts off from the support due to temperature only, it is considered inactive for the expansion analysis.

    This approach may mishandle both sustained and expansion stresses. First, the calculated sustained stress is the stress under ambient condition. The most important sustained stress under hot condition is not calculated. Second, expansion stress may be underestimated, because the restraining effects of the supports, over which the piping is held down by weight, are ignored.

2. *Algebraic subtraction approach.* In this approach, sustained stress is calculated considering only the weight and pressure loads under ambient state. All supports are active because no temperature or support displacement is involved. An operating condition including temperature and weight is also analyzed, with the supports that lift off removed from the analysis. The expansion stress range is calculated by algebraically subtracting the hot operating condition by the weight condition at ambient state (initial cold condition).

Three major issues are at stake in this approach. First, the calculated sustained stress is the stress under ambient condition. The most important sustained stress under hot condition is not calculated. Second, by subtracting the initial cold weight load from the operating load, this approach includes the cyclic weight stress range, changing from ambient to hot conditions, in the expansion stress range. This is not consistent with the code philosophy of separating sustained stress from self-limiting expansion stress. Third, the stress for the temperature plus weight condition depends greatly on the signs of the moments of the two loads included. If the moment of the weight change is in the opposite direction of the moment of the temperature change, the calculated expansion stress will be smaller than that calculated by the temperature change alone. This is not correct because yielding and relaxation can change the sign of the thermal expansion moment during the course of operation.

Although the code allowable thermal expansion stress range includes the consideration of sustained stress, the combination of expansion stress range and sustained stress shall be done absolutely. Because the signs of expansion stress and sustained stress are immaterial, their algebraic summation or subtraction is meaningless.

3. *Operating condition approach.* In this approach, all supports are checked at the operating condition, which normally involves temperature plus weight and pressure. If the pipe lifts off from a support at operating condition, that particular support is then treated as inactive for both the sustained weight plus pressure stress calculation and the expansion stress calculation. By the same token, if the pipe is held down on a support at operating condition, that support is treated as active for both sustained and expansion stress calculations. With this method, sustained stress and expansion stress are calculated independently once the activity of the supports is determined.

The sustained weight plus pressure stress calculated with this approach is the true sustained stress at hot operating condition, when the stress is high and the pipe is weak. The one thing that may appear to be improper to some inexperienced analysts is that the calculated weight displacement may show a downward movement at support locations. This downward displacement represents only the movement of pipe from a thermally lifted condition. At support locations, the operational displacements combining weight and temperature will either be zero or in the upward direction. The calculated expansion stress is the potential stress range, recognizing that the sign and the magnitude may change throughout the operating cycles.

From the above discussions, it is obvious that the *operating condition approach* is the only method that meets the code philosophy and requirements. The other two approaches have flaws in calculating the sustained stress and the expansion stress range.

Analysis of piping resting on supports is nothing new. Engineers have routinely analyzed this type of piping for more than two decades. The erroneous concepts of some computerized approaches and the blind acceptance of computer results by engineers, however, are new. Attracted by the glamorous nature of thermal flexibility analysis, many engineers have forgotten that sustained stress is much more important than expansion stress. Sustained stress is the primary stress, whereas expansion stress is a secondary stress. At low temperatures, when the hot allowable stress has the same value as the cold allowable stress, the weight allowable stress limit is only about one-third of the expansion stress allowable limit. At higher operating temperatures in the creep range, the weight allowable stress limit can be as low as only one-tenth of the expansion stress allowable limit. Therefore, it is important to note that the first priority of piping analysis is to accurately determine the sustained weight stress under hot operating condition. This is not to say that expansion stress is unimportant. A good analysis shall calculate both sustained and expansion stresses as accurately as possible.

It should be noted that by calculating weight stress at cold condition, the result is not expected to indicate where a spring support is needed. It is only when the weight stress at hot operating condition is calculated that the engineer will be able to detect where and when a spring support is needed. Spring supports are used to reduce the weight stress at hot operating condition.

The operating condition approach may be somewhat conservative for pipes that only lift up a very small amount from the support. In this case, the rule of thumb is to consider the support double-acting to check both sustained and expansion stresses. If both stresses are within the code allowable limits, then the system should be considered acceptable.

6.4 VARIABLE SPRING AND CONSTANT EFFORT SUPPORTS

As previously noted, the most economical and efficient way to support the piping is to simply rest the piping on a rigid support structure. Being supported with rigid supports, the pipe will either generate a potentially huge upward force when the pipe expands downward or leave the support inactive when the pipe moves upward. One way to maintain the pipe properly supported is to replace the rigid support with a spring support. With spring support, the pipe will always be supported with an appropriate amount of force, regardless of its vertical movement. The magnitude of the supporting force, however, changes somewhat as the pipe moves vertically. Because of this change in support force as the pipe moves up and down, spring support is also called a variable spring support. Although the support force changes throughout the operating cycle, the amount of change is predictable from the spring selected. In the design, the spring is properly selected so that the load variation is within the acceptable limit that will not compromise the integrity of the piping system.

Since the pipe always has the tendency to move in the vertical direction, it takes some engineering judgment to decide when and where the spring support shall be used. A straightforward method often mentioned is the use of the free vertical thermal movement of the piping as the criterion. In this method, the system is first analyzed for thermal expansion, without any support included, to find out the free vertical piping movement at the intended support location. If the free vertical thermal expansion exceeds a certain limit, say 0.5 in. (13 mm), then a spring support is used. If it exceeds 3 in. (75 mm), then a constant-effort support is used. This type of approach may appear to be reasonable, but actually does not serve any real purpose. At the location near a fixed point, such as an anchor or equipment, even a very small vertical displacement is too much to use a rigid support. On the other hand, a large vertical free thermal displacement in a wide-open area can still be supported with a rigid support due to the large flexibility of the piping. With a computerized environment, a good approach is to use double-acting rigid supports as much as possible for a trial. After the trial analysis, the supports that generate huge thermal loads are removed to see if the sustained stress is still within the allowable range. If it is not, then spring supports should be used at some or all of the locations where the trial double-acting rigid supports are eliminated.

In the configuration shown in Fig. 6.8, a 12-in. (300-mm) pipe will lift off and be unsupported at all three supports with a tower movement of 3.2 in. (80 mm). Ideally, all three supports should be replaced with spring hangers. However, with a compact layout as shown, only the mid-support may be required to be spring-supported, whereas the other two are simply removed. This, naturally, needs to be verified with an analysis. It should be also noted that some of the resting supports, although not active under operating condition, might be required for the hydrostatic test. However, a support required for hydrostatic test but not suitable for proper piping operation should be removed after the test.

6.4.1 Variable Spring Hanger Selection Procedure

Spring hangers are selected based mainly on allowable load variation. The load variation is defined as

$$\text{Load variation} = \frac{\text{Cold load} - \text{Hot load}}{\text{Hot load}} = \frac{\text{Displacement} \times \text{Spring rate}}{\text{Hot load}} \quad (6.3)$$

Industry standards [5] limit the load variation to no more than 25%. This means that, regardless of the pipe movement, the spring will carry no more than 125% and no less than 75% of the properly balanced portion of load. The unbalanced load will be shifted to the neighboring supports or equipment. In actual applications, a smaller design load variation may be specified for critical piping systems, such as the ones operating at creep range or connected to rotating equipment.

In the past, due to the unavailability of long springs, it was customary to use a constant-effort support whenever the expected vertical movement exceeded 2 in. or 3 in. depending on the criticality of the service. Nowadays, as extra long spring hangers are widely available from vendors, piping engineers tend to use the variable spring support as much as possible due to its load adjustability. Constant-effort supports are reserved for extra large displacements that exceed the application range of the variable spring support.

For easier application in the spring hanger selection process, Eq. (6.3) can be rearranged around the spring rate as follows:

$$\text{Spring rate} = \frac{\text{Load variation} \times \text{Hot load}}{\text{Displacement}} \quad (6.4)$$

Equation (6.4) is the main engine of the spring hanger selection process, which is summarized into the following steps.

1. Determination of hot load

Hot load is the balanced load that exerts the least weight stress to the system. This load is obtained by analyzing the weight load with all supports considered as rigid. To minimize the load carried by the equipment, occasionally the vertical constraint at the equipment is mathematically released while the balanced load is being calculated. This balanced load is called the hot load because we want the spring to carry this load at operating condition.

2. Calculating the operational displacement

This operational vertical displacement is determined by analyzing the operating condition with weight and thermal expansion combined. The hot load determined in the previous step is applied as an external support force. Spring rates of the spring hangers are ignored, because the spring load applied is the hot load.

3. Selecting allowable load variation

The maximum allowable load variation is 0.25 (25%). Within this maximum variation, other limitations may be used depending on experiences and project specification [6]. A load variation of 0.15 (15%) is generally recommended for hot piping operating at creep range, and a load variation as low as 0.06 (6%) may be specified for piping connected to sensitive equipment.

4. Calculating the maximum spring rate

After determining the hot operating load, vertical pipe movement, and allowable load variation, the maximum spring rate is calculated using Eq. (6.4). A higher load variation results in a higher spring

164 Chapter 6

rate. The actual spring rate used shall be less than the maximum spring rate calculated. The terms spring rate and spring constant are interchangeable.

    5. Selecting the hanger supplier

Spring hangers are selected based on a supplier's catalog. Although a supplier can furnish almost any spring hanger specified, the cost is often prohibitive for specially manufactured hangers. Therefore, it is important to select the hanger that is offered in the supplier's catalog. In case the supplier has not yet been selected during the design analysis stage, a reference supplier's catalog can be used. One popular reference catalog is the Grinnell Pipe Hangers catalog [7]. You can, for instance, select the hanger with the Grinnell catalog, and then specify it as Grinnell Fig-xxx, Size-xxx, Type-xxx, or equal. This approach works, because most suppliers offer very similar products.

    6. Selecting the basic spring

The supplier's catalog provides a selection chart similar to the one presented schematically in Table 6.2. Each supplier offers about 20 to 25 hanger sizes. The chart shows size 0 through size 22. Each hanger size has up to five displacement ranges: normal (N), short (S), long (L), extra long (XL), and double extra long (XXL). Suppliers have different names for these displacement ranges. Grinnell, for instance, calls them Fig-B268, Fig-82, and so forth. Because using the figure number to identify the displacement range has become a tradition, the displacement ranges are also referred to as "Figures." This practice also lessens the confusion with other types of ranges, such as the load range. Each hanger size also has its load range, the same for all figure numbers (displacement ranges). The size 2 hanger in Table 6.2, for instance, has a working load range of 95 lb to 162 lb regardless of the figure number. In addition to the working load range, there are also top and bottom margins to accommo-

TABLE 6.2
VARIABLE SPRING HANGER SELECTION CHART IN (LB),
(IN) UNITS - SCHEMATIC

		Range				Hanger Size					
		XXL	XL	L	N	S	0	1	2	* * * * * * * *	21	22
(Spring Displacement, in)	Top Margin	(0.0)(2.0)	(0.0)(1.5)	(0.0)(1.0)	(0.0)(0.5)	(0.0)(0.25)	43	63	81	* * * * * * * *	18750	25005
	Working Range, in	0	0	0	0	0	50	74	95	* * * * * * * *	21875	29173
		↓	↓	↓	↓	↓	*	*	*		*	*
							* * * *	Support Loads, lbs			* * * *	
		↓	↓	↓	↓	↓	*	*	*		*	*
		10	7.5	5	2.5	1.25	88	126	162	* * * * * * * *	37500	50010
	Bottom Margin	(12)(14)	(9.0)(10.5)	(6.0)(7.0)	(3.0)(3.5)	(1.5)(1.75)	95	137	178	* * * * * * * *	40625	54178
							Spring Rate - (lb/in)					
						S	30	42	54	* * * * * * * *	12500	16670
						N	15	21	27		6250	8335
						L	7.5	10.5	13.5		3125	4167
						XL	5	7	9		2083	2778
						XXL	3.75	5.25	6.75	* * * * * * * *	1563	2084

Notes: 1 lb = 4.4482 N; 1 in = 25.4 mm; 1 lb/in = 0.175 N/mm

date the uncertainty involved with the data and calculation. Beyond the top and bottom margins, the hanger behaves just like a rigid hanger.

The hanger selection process starts with locating the hot load on the support load area to determine the hanger size. Up to two hanger sizes may contain a given hot load. For instance, a load of 110 lb is within the working load ranges of sizes 1 and 2. We shall select the smallest and shortest spring that is suitable for the operation. A smaller, shorter spring is cheaper and also requires less installation space, which can be very critical. Start from the smaller size, and look for the spring rate that is equal to or less than the maximum spring rate determined previously. From this spring rate, trace backward to find the corresponding displacement range, or figure number.

7. Checking the working range

The selected spring has to work within the working range. We already have the hot load located inside the working range; the next step is to determine if the cold load is also located inside the range. This cold load is calculated by:

$$\text{Cold load} = \text{Hot load} + (\text{Displacement} \times \text{Spring rate}) \tag{6.5}$$

In this equation, upward displacement is considered plus and downward displacement is minus. If the cold load is not within the working range, the other hanger size that also contains the hot load may be tried. If neither of the two hanger sizes works with the spring rate, a lower spring rate — that is, a figure with longer displacement range — can be tried. If no figure can accommodate the hot-cold load range, then a constant-effort support shall be used.

The spring hanger selection process is tedious and can be overwhelming for less-experienced engineers. The task, however, has been implemented in most pipe stress analysis computer software packages. The designer only has to instruct the computer where to put the hanger, what is the allowable load variation, and which supplier is to be used. The computer will automatically go through the procedure and print out a selection table as shown in Table 6.3. The table has all the information, except hanger type and overall dimension, needed to order the hanger.

6.4.2 Constant-Effort Supports

A constant-effort support is used when the load variation of a variable spring hanger exceeds its acceptable value or when the vertical pipe displacement exceeds the range of the variable spring hanger.

TABLE 6.3
COMPUTER GENERATED SPRING HANGER SELECTION TABLE

```
■ SIMFLEX REPORT PROCESSING
File  Reports  Help
.SIMFLEXS-PIPE STRESS ANALYSIS (7.0 ) (ASME B31.3) - PENG ENGINEERING, HOUSTON
-------------------------------------------------------------------------------
RESTING SUPPORT EXAMPLE                                            17 SEP 05
12 INCH SCH-40, 750F (400C), WITH 3.2-IN (80-MM) TOWER MOVEMENT    RestSpt.DA

:** SPRING HANGER SELECTION TABLE **
-------------------------------------------------------------------------------
USING --     GRINNELL      -- VARIABLE SPRINGS

(FOR TRAPEZE SPRINGS, THE LOAD SHOWN IS FOR EACH SPRING)
-------------------------------------------------------------------------------
POINT   OPERATN   COLD-SET   VERTICAL-DISPL    SPRING   LOAD    GRINNELL-SPRG NUMBER
NO.     LOAD      LOAD       NORM  MAX   MIN   RATE     VAR     -------------- SPRING
        ( N )     ( N )      (MM)  (MM)  (MM)  (N/MM)   (0/0)   FIGURE   SIZE
-------------------------------------------------------------------------------
 30     11557     13437      53.7  53.7   .0   35.0     16.3    TRIPLE   13      1
```

A constant-effort support consists of a spring with an ingeniously proportioned linkage that offers a constant support force through a fairly large displacement range. Although constant-effort supports are mostly hung from above the pipe, they are traditionally called "supports," leaving the term "hangers" to indicate variable springs.

Selection of the constant-effort support is very straightforward. The information needed is the operating load and travel range. Support vendors suggest that the specified travel range should be equal to the maximum actual operating displacement plus 20%, and in no case should the additional value be less than 1 in. (25 mm). The hanger size is then selected from the vendor catalog. One of the unique characteristics of the constant-effort support is the size-load relationship or, rather, non-relationship. The working load for the same size of support largely depends on the travel. This is somewhat puzzling, but can be easily explained.

Figure 6.9 shows the schematic representation of the constant-effort support. The support consists of a spring and a rocking yoke. The yoke rocks around a pivot pin located at the lower end of the support frame. One end of the yoke is connected to the pre-compressed spring, whereas the other end is connected to the load rod. The load, W, is supported by the pulling force, F, of the spring. By balancing the moment about the pivot axis, we have $FS = WL$. Because FS remains relatively constant for each size, the relation can be written as WL = constant for each size of the support. From the sketch, it is obvious that the longer the travel, d, the wider the L required. Therefore, the greater the travel, the lower the support load that can be sustained by a given size of support.

Although the load of a constant support can be adjusted by about ±10% in the field, it does not offer any indication when the balanced load is reached. Because the load, once set, remains the same regardless of pipe position, the exact balance largely depends on an accurate calculation. To estimate the accurate support load, some potential variations on the weight of pipes and components have to be considered. An ordered pipe normally has some overweight. The overweight is even more common for components such as forged elbows and tees. The weight of pipe clamps and other suspended support components have to be included as concentrated weight for the calculation to be accurate.

6.4.3 Spring Support Types and Installations

The computer-generated spring hanger selection table does not give the support type, which is dictated by the availability of space and support structure. Each spring hanger supplier normally offers about seven or eight different spring hanger types for installing in different situations. Figure 6.10 shows typical applications of hanger types offered by Grinnell Corporation [7].

Regarding the installation of spring hangers, different practices have evolved for different industries. Power plant piping is typically hung from the floor structure above, using rigid hangers. When a spring

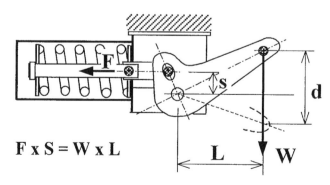

FIG. 6.9
CONSTANT-EFFORT SUPPORT SCHEMATIC SKETCH

Pipe Supports and Restraints   167

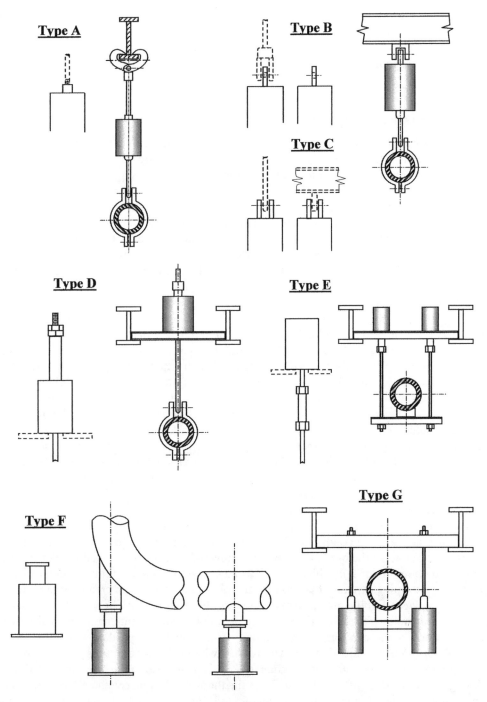

FIG. 6.10
SPRING HANGER TYPES AND APPLICATIONS (BASED ON GRINNELL SPRING [7])

hanger is needed, it is just a matter of replacing the rigid hanger with a spring hanger. The headroom is generally not a problem. In this case, the piping layout seldom requires any change because of the use of spring hanger. On the other hand, process piping is normally resting on the support structure underneath, without any support structure available overhead. In this case, the use of spring hangers or spring supports is an important matter. In many occasions, the piping layout has to be revised to accommodate not only the spring hanger hardware, but also the piping movement. In some process plant projects, the use of spring hanger and sometimes even rigid rod hanger requires prior approval from the project management.

6.4.4 Setting of Loads — Hot Balance and Cold Balance

The load of a spring hanger changes throughout the operating cycle. By setting the pre-operation cold condition load at a certain level, we can predict the load at operating condition based on the expected vertical displacement. Again, load variation is calculated as spring rate times the vertical displacement. The load that balances the weight of the portion of the piping supported is called the balanced load. Intuitively, we might like to set the balanced load at midway of the operating cycle and shift one-half of the load variation to the cold condition and the other half to the hot condition. However, in actual applications, the piping system is either hot balanced or cold balanced.

Hot balance sets the hanger in such a way that the spring load at hot operating condition equals the balanced load. This is the preferred approach that, at least theoretically, minimizes load and sustained stress at the most important hot condition. This is a must for piping operating at creep range, where creep damage depends largely on sustained stress at operating condition. Figure 6.11 shows the scheme of setting the spring given in Table 6.3. The hot load is equal to the balanced load of 11,560 N (2600 lb). The setting of the cold load is then determined by adding the expected load variation to the hot load — that is, the cold load is 13,440 N (3020 lb). The spring is locked at this cold load with a red-painted locking lug or pin. A colored warning label is attached to remind the operator to pull out the pin before operation. After the installation and hydro test is completed, the locking lug or pin is removed. Because at cold condition the hanger is pulling more force, or less force as the case may be, than the balanced load, the piping will experience some jerk or movement when the lug is removed. This can alarm field engineers if the movement is significant. Fortunately, the lugs are removed one spring at a time so a very large movement is rarely observed. If all calculations are exact, the magnitude of the unbalanced load is within the allowable load variation with which the system is designed for. For the spring shown in Fig. 6.11, the spring force reduces gradually as the pipe temperature rises and the piping moves up. The spring force eventually reaches the balanced load of 11,560 N when the piping system reaches the operating temperature.

Cold balance sets the hanger force to balance the weight load at cold condition. This leads to some unbalanced force at operating condition when the piping is at hot operating condition. In theory, this approach is not as good as the hot balance option. However, most experienced field engineers, especially those who represent large machineries operating at low to moderate temperature, insist on using cold balance on the piping systems attached to their equipment. This often creates arguments between the machinery engineer and the piping stress engineer. The personnel friction can be avoided and the job can be better handled if we know the rationale behind the equipment engineer's insistence.

Accurate support loads can only be determined from accurate pipe data and calculations. The mathematical portion of the equation is not a problem when using a computer program, but the data portion is much less reliable. The pipe weight can be 10% heavier than the theoretical value, and the forged fitting can be as much as 30% heavier than the theoretical value calculated from the equivalent pipe. The insulation covering and support clamps and floating attachments are often not included in the calculation. Because some of these items are not known for sure, the calculated support load is naturally less than exact. Furthermore, the calculation model often ignores connecting piping of smaller sizes, and the boundary conditions used are just approximate assumptions. Therefore, it is possible that the cold set load is way off the presumed value, and the resulting hot load is not at all predictable.

The displacement of the low to moderate temperature piping, such as some of the large compressor piping systems at a process plant, is generally small. This means that the actual load variation,

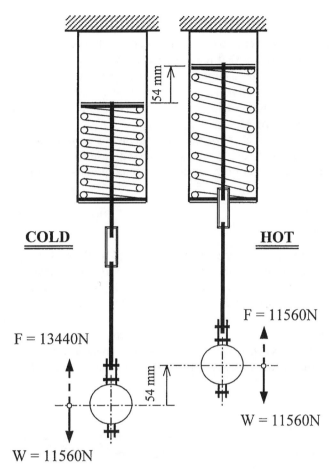

FIG. 6.11
HOT BALANCE SPRING LOAD

calculated as the product of spring rate and displacement, and the percentage of load variation are also generally small. Therefore, if there is a means to achieve the exact balance at cold condition, then the unbalanced load at hot condition is small and predictable. The method used by the equipment engineer is to apply no load at all on the connecting flange or the final field weld when the piping is being connected to the equipment. The springs are unlocked after the hydro test. The piping is then pulled using the support, spring or rigid, to bring the connecting flange to the equipment matching flange without the use of any other external forces. In this way, it is fairly certain that the weight load on the equipment at cold condition is practically zero. The hot load, which results from spring load variation, is small and predictable. This approach of connecting the piping to equipment, however, is only feasible for gas or vapor piping.

Cold balance is very popular in the field, but somehow does not yet capture the attention of stress engineers. To facilitate the field installation process, the spring should be set in such a manner that the cold load is in balance with the weight of the portion of the piping being supported. The vertical restraint at the equipment connection should be released mathematically when calculating the spring load, so the potential weight load at the equipment connection is as small as possible. To achieve the no-load connection, the supports have to be well distributed and positioned throughout the piping system. Without the proper pulling points, the no-load situation can never be achieved.

170 Chapter 6

6.5 SUPPORT OF LONG RISERS

A long vertical line is called either a riser or a down comer, but is generally referred to as a riser. A riser has a few unique characteristics that require special attention. The riser has the potential of producing a large vertical expansion, has a large concentrated weight acting at one point in the horizontal plane, and has a large liquid column if it is a liquid line. These unique characteristics affect the support scheme and analytical process. In the following, we will discuss the support scheme, load calculation, and analytical approach for the risers.

Figure 6.12 shows three typical cases of riser arrangements. In case (a), the riser is hung directly underneath a piece of equipment such as drum or vessel; case (b) shows a riser connected to two considerably long horizontal runs at both top and bottom; and case (c) shows a typical process line running alongside a tall vessel.

6.5.1 Support Schemes

Because of the large vertical displacement, it is a natural tendency for engineers to use constant-effort or long variable spring supports. However, due to the large concentrated weight, a slight amount of load variation or inaccuracy of data leaves a significantly unbalanced weight that can generate a large bending moment at the horizontal portion of the piping. A rigid support is required somewhere nearby to absorb this load variation and uncertainty. In case (a), the entire riser grows down at hot operating condition. There is no choice but to put the spring supports as close to the equipment as possible to minimize the vertical displacement. In case (b), there is always a theoretical zero vertical displacement point located at the riser. Rigid supports or hangers located at close proximity of this

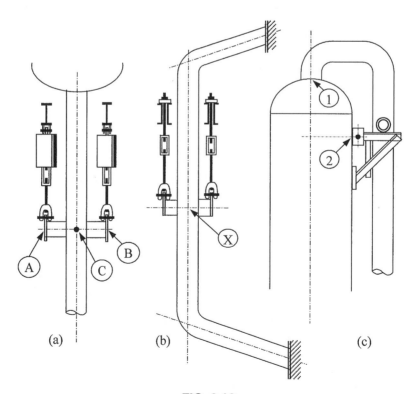

**FIG. 6.12
SUPPORTS AT RISERS**

point should be used. Rigid hangers are generally more convenient as they can be hung from overhead structural steel. Due to the flexibility of the horizontal runs, the location of the zero vertical displacement point needs not be exact. Case (c) is a typical process line connected to a tall vessel. Because the temperature of the part of the vessel at the connection is the same as the pipe temperature, assuming the vessel is not internally insulated, a rigid support can always be used and located close to the connection. For a very hot long riser, it may require supports at multi-levels to reduce the weight stress.

6.5.2 Support Loads

Support loads can be determined by allocating the proper portion of the piping system, including pipes, components, attachments, liquid content, and insulation, to each support. This is done by a mixture of common sense and simple arithmetic. In a computerized design environment, the computer automatically determines the loads. Computerized calculation is quick and accurate, but the determined loads are often neither what we want nor are they the most suitable ones. Several special procedures have to be applied in performing the computerized load determination analysis.

In case (a), a normal computerized analysis will allocate most of the weight to the equipment, as it is the most rigid support. Some spring load is assigned to one of the springs, leaving another spring with no load at all. This, of course, is not what we want. What we want is to have the spring carry most of the weight and to have the weight shared equally by the two springs. To achieve this, first, the vertical translational restraining effect at the equipment has to be released when calculating the weight support load. This anchor release procedure will leave essentially no weight load at the anchor or equipment. Most computer software packages will automatically do this if instructed. To distribute the spring load, a seemingly more accurate analysis scheme is to model trunnions A and B as two short pipes, each connected with a spring. However, this elaborate approach often produces two completely different springs with occasionally no load at all at one of the springs. This is attributed to the theoretical approach of the computer software. When the computer distributes the weight support load, it considers all spring supports as rigid in the calculation. This works just fine if the supports are well separated. With two rigid supports located close by, as in this case, only one support will be active due to the cranking action of the horizontal run located down below. The proper method of analysis is to consider the effect of the two springs as a combined spring located at the center point, C, and then determine the support load at this center point. The load so determined is divided by 2, from which the load of each spring is obtained. Because the two spring loads generate zero combined moment at the center point, the one center spring approach is mathematically equivalent to the two separated, but equal, springs. The trunnions have to be modeled, however, to have their weight counted. The weights of hanger attachments are lumped in the analysis as concentrated weight acting at the center point. In most pipe stress analysis software packages, the two springs can be automatically selected by instructing the computer to select two springs at the center point.

The loads on the rigid hangers in case (b) are difficult to determine both analytically and practically. A computerized analysis will most likely predict that only one hanger will carry the weight load due to the cranking action of the horizontal runs. The other hanger will simply sit there with no load. The roles of the two hangers, however, might be switched when the system is at operating condition. Although the loads on rigid hangers can be adjusted or shifted somewhat by tightening the rods, they are not predictable. Therefore, unlike in case (a), the method of using one-half of the load determined at the center point is not feasible. The computer-calculated loads for the rigid hangers have to be used. This means that the total load is generally carried by only one of the hangers. More realistic loads may be calculated by including the proper stiffness or spring constant of the hanger and support structure in the analysis. The stiffness of the rigid hanger and structure assembly is in the range of 10^5 lb/in. to 10^6 lb/in. (1 lb/in. = 0.175 N/mm).

The situation for case (c) is similar to that of case (b), but with some differences. Here, the riser is generally long and slender and offers some flexibility to seat both sides of the support. The support structure is a combined assembly used for both sides of the support. In general, if the support trunnions are shimmed properly, then the theoretical support situation can be achieved. Even so, some extra margin needs to be provided for the trunnion design load.

6.5.3 Analysis Method

In the analysis, insulation and content are lumped together with the pipe weight. In other words, the analysis is based on total weight, including pipe, insulation, refractory, and content, per unit length of the pipe. That is, a unit length of the pipe is combined with a unit length of insulation and a unit length of liquid to become the total weight per unit length of pipe in the analysis. This may appear to contradict the fact that the liquid weight of the entire riser acts upon the bottom of the riser, rather than on each unit length of the pipe. This convenient approach, however, does not affect the support load distribution calculation, as all weights are still acting on the same vertical centerline. As for the sustained stress calculation, the method is conservative as soon as the design pressure is taken as the pressure existing at the bottom of the riser.

6.6 SIGNIFICANCE OF SUPPORT FRICTION

A piping system may involve several different types of friction force. Flexible joints, such as Dresser couplings and ball joints, need a certain breakup force to start sliding or rotating. These required breakup forces are due to the friction of the tightly packed joint. These friction forces will be discussed in Chapter 7. Another type of friction involving soil pressure will be discussed in Chapter 10. The friction forces discussed here are the ones produced at the sliding surfaces of the supports.

6.6.1 Effects of Support Friction

Support friction in a piping system prevents the pipe from thermal expansion, thus creating higher stress in the piping and higher load on the connecting equipment. However, in certain instances friction can help stabilize the system and reduce potential damage. Even in dealing with pure thermal expansion, friction can serve as guides preventing a large load from transmitting to a piece of sensitive equipment. Therefore, there is no rule of thumb as to whether it is non-conservative to ignore friction. In general, when dealing with dynamic loads, friction tends to reduce the magnitude of both the pipe stress and the equipment load. In this case, the omission of friction in the design analysis is conservative. However, there is no general rule governing the static load. With static loads, the effect of friction needs to be investigated and simulated as closely as possible to the real situation. When designing the support structure, however, it is always required to include the pipe support friction forces.

The effect of support friction is very important in some areas. Analysis of the long transmission pipeline, for instance, focuses entirely on the balancing of friction force against potential expansion force. If friction is not included in the analysis, the analysis would be meaningless. Another area of importance is the piping connected to rotating and other sensitive equipment. Rotating equipment is notorious for its low allowable piping load. Sometimes the friction at one support can completely change the acceptability of the piping system. Take the system shown in Fig. 6.13, for instance, where the restraint at point 25 is installed to protect the compressor at point 10. The effect of friction at point 25 is demonstrated by comparing the analysis results [8] of the case with friction against the case without friction. It is clear that the effect of friction at the restraint is very significant. By applying the American Petroleum Institute Std-617 [9] criteria, only the load calculated with the restraint but without friction is acceptable. Restraint friction has to be either removed or substantially reduced (see also Chapter 9).

The effect of support friction is also very significant for long off-site piping. Figure 6.14 shows a zigzag section of liquefied natural gas piping. From the comparison chart, it is clear that support

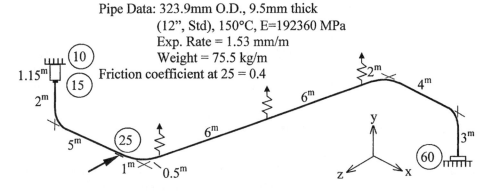

**FIG. 6.13
EFFECT OF SUPPORT FRICTION ON COMPRESSOR PIPING**

friction dominates the response of the system. Without support friction, the axial anchor force stems mainly from the stiffness of the system resisting the thermal expansion. Because the system is very flexible, this thermal resisting force is very small at 16,746 N as compared with the frictional resistance of 222,600 N. The support friction tends to prevent the piping from moving to the flexible direction,

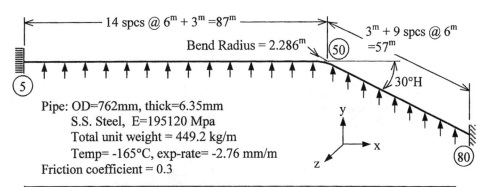

**FIG. 6.14
EFFECT OF SUPPORT FRICTION ON OFF-SITE PIPING**

174 Chapter 6

because it only takes a very small force to stop the movement in the flexible lateral direction. Therefore, support friction also serves as a guide to a long flexible line. This is the reason why the moment at anchor 5 is much smaller with support friction than the moment for the case with support friction ignored. In addition, because of the guide effect, support friction tends to make the system less flexible. This causes the bending moment, thus the stress, at the bend to be almost three times as high when support friction is present as compared with the case when support friction is ignored.

6.6.2 Method of Including Friction in the Analysis

There are a few general steps to follow when including support friction in the analysis. The nature of friction is first idealized as elastic perfect plastic resistance as shown in Fig. 6.15(a). When the pipe is forced to slide on the support surface, initially it does not move because of friction resistance. As the pushing effort increases to a certain limit, the pipe starts to move and follows with a slight snap because of the drop in friction resistance after the pipe has started moving. The initial resistance force is called static friction, and the resistance after the pipe moves is called sliding friction or dynamic friction. Because friction at each support reaches the maximum level of static friction at different times, the lesser dynamic friction is used in the design of piping system. The model of friction force is, therefore, similar to a linear elastic/perfect plastic non-linear restraint. The linear elastic portion represents the flexibility of the support. This linear portion is also needed for stabilizing the mathematical simulation processes.

The analysis is an iteration process involving a trial-and-error approach. There are two main approaches used in piping stress analysis: direct friction force approach and resilient friction restraint approach.

The *direct friction force approach* starts the analysis by first assuming that friction stops the pipe from sliding on the support surface. The resultant stopping force is then calculated to compare with the full friction force, $F\mu$. If the resultant stopping force is less than the full friction force, then the pipe is stopped, as assumed. If the resultant stop force is greater than the full friction force, then

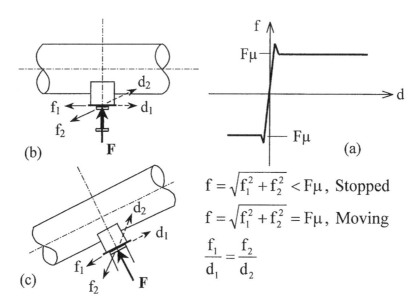

FIG. 6.15
SUPPORT FRICTION COMPONENTS

the pipe slides with a total resistance force equal to the full friction force. The analysis is performed again, assuming that there is no stop to pipe sliding, except the full friction resistance force, at those support locations where the stop force is bigger than the friction force. In the analysis, friction force is applied as external force in the direction against the pipe movement. Because the sliding of the pipe on the support surface involves two degrees of freedom, the friction force is two-dimensional. It comprises two perpendicular forces applied in the direction perpendicular to the support direction. Figure 6.15(b) and (c) shows the schematic directions. As the friction force works directly against the displacement, its components, f_1 and f_2, are proportional to the displacement components, d_1, and d_2, as shown in the figure.

The general analysis procedure, described above, works well for some simple and rather stiff systems, such as the compressor piping given in Fig. 6.13. In this system, the reaction force at point 25 is first calculated without the friction. Based on the calculated restraint reaction force, a full friction force is determined and allocated proportionally to the displacement components in x and y directions. The calculated friction components f_x and f_y are then applied as external forces in a re-analysis. The restraint reaction and pipe displacements will likely change after the re-analysis. A new set of adjusted friction components are again applied as external forces in another trial analysis. Because the system is rather stiff, a convergence will be reached in just a couple of trials. However, this direct friction force as external force approach does not work very well on long flexible piping systems involving many supports.

The *resilient friction restraint approach* is more stable especially for long and flexible systems, such as the one shown in Fig. 6.14, in which the direct friction force approach does not work at all. With the direct friction approach, the trial lateral friction forces can twist the system into a very unpredictable situation, and the trial-and-error iteration can deviate farther and farther away from the balanced position. One workable scheme to deal with this type of flexible system is to model all potential friction forces as resilient restraints. Each component of friction force is represented by a resilient restraint. A resting support in y direction, for instance, will have two resilient friction restraints in x and z directions, respectively. In each trial analysis, the spring rates of these resilient restraints are determined from the potential friction forces and the pipe displacement of the previous trial analysis. Because the simulated friction force is resilient, no extreme pipe displacement will be produced. The analysis generally converges to the balanced state in a few iterations, although some systems may take a couple dozen iterations to reach the balanced state. The resilient restraint approach is summarized as follows:

(1) Each friction support and restraint is assigned with two orthogonal resilient friction restraints perpendicular to the direction of the support or restraint. The resilient restraints are aligned with the local coordinates of the sliding surface, and their reaction forces represent the expected friction forces.
(2) The analysis starts out with no friction effect to calculate the preliminary support reaction force, F, for each support or restraint. It also calculates the preliminary displacements at two perpendicular directions to the support or restraint. The two perpendicular directions, of course, coincide with the directions of the resilient friction restraints. For a resting support in the y direction, for instance, the two displacements needed are d_x and d_z.
(3) Each iteration of the analysis is performed using the updated trial spring rates for the resilient friction restraints. The spring rate is updated with current support reaction force and displacement information as

$$\text{Spring rate} = \frac{F\mu}{\sqrt{d_1^2 + d_2^2}} = \frac{F\mu}{\sqrt{d_x^2 + d_y^2}}$$

The same spring rate is used for both x and y directions at each support or restraint, so the direction of the resultant friction force is directly against the displacement direction.

(4) The results of each iteration are checked to see if variations from the last iteration are all within the range of acceptable values. If the variation between current iteration and last iteration is not acceptable for any support, the new spring rates are calculated and a new iteration is performed.

6.6.3 Application of Friction Force

The procedure outlined is for the analysis including all loads, such as weight, thermal, etc. This is the condition required for evaluating equipment loads. However, in code stress compliance analysis, it is often required to separate the loads into different analyses so stresses can be separated into proper categories. For instance, the weight load is not included when the thermal expansion stress is being calculated. However, without the weight load, friction is mostly non-existent. This becomes a paradox in piping stress analysis, because friction would never get into the act. Friction is irrelevant in the weight load case, yet non-existent in the thermal expansion case.

Weight initial load is the key element for solving the above paradox. The cited inconsequential friction paradox can be solved by maintaining the initial weight load for each support or restraint. An analysis considering only the weight load is performed to determine the magnitude of the weight load that is normally carried by each support or restraint. These nominal weight loads are then recorded and remembered. We will then recognize that these weight initial loads are always present at the supports to affect the friction effect. The friction force can then be calculated with this weight initial load superimposed on the support reaction force produced from thermal expansion and other loads.

Because the friction is mostly the result of weight load, some may argue that friction force is a sustained load simply because weight is a sustained load. This, of course, is not true because friction force is a passive force. It is always in the defensive mode (and is never in an active mode to affect anything). Friction force will tend to reduce the effect of a sustained load, but it does not have the damaging nature of a sustained load. Friction is actually an effect rather than a force.

Friction generally serves as restraints to a piping system. It increases the effective stiffness of the system. Obviously, the stiffer the system, the higher the load, and also the higher the stress, which will result in the piping acting against thermal expansion. Therefore, inclusion of friction in thermal expansion analysis is necessary. On the other hand, friction can help reduce the damage from occasional load, such as earthquake and wind. In these cases, the inclusion of friction is a matter of judgment. It should be clearly specified in the design specification.

6.6.4 Methods of Reducing Friction Force

Support friction is not desirable in some situations. In the examples given in Figs. 6.13 and 6.14, friction has increased anchor reactions and pipe stresses. The necessity of reducing support friction is seldom the consequence of pipe stress. The reduction is mostly required for reducing the equipment load and support structure load. We are all aware of just how small the allowable piping load can be at an equipment connection. However, we often forget that the support structure cannot take a large friction force either. Most support structures are mainly designed for the vertical load. They are generally very weak in resisting the horizontal load.

In the system shown in Fig. 6.13, restraint 25 generates a high x-direction friction force, causing a very large x-direction equipment force and z-direction equipment moment. Because these loads are not acceptable, one way to solve the problem is to reduce the friction force.

The best way to reduce support and restraint friction is to eliminate the sliding surface. This can be achieved by using hangers and struts. The strut as shown in Fig. 6.4(a) can be used here. However, there are situations in the field wherein the use of hangers and struts is not feasible. For example, for off-site and yard piping in a process plant, sliding support is often the only feasible support.

There are two popular means of reducing friction force at support and restraint surfaces: rollers and low-friction sliding plates. Figure 6.16(a) shows a sketch of the roller support, which converts sliding action into a rolling action. Roller supports not only offer less friction than the sliding support, but also help maintain the alignment of the pipe. They serve as guides for the pipe and, at the same time, accommodate slight axial rotation of the pipe. Rollers are popular for piping that is laid in a trench or in a dusty environment.

Figure 6.16(b) shows the general arrangement of low-friction sliding plates. Two major types of materials are used for the sliding plate: graphite and Teflon (poly-tetra-fluoro-ethylene, PTFE). Teflon is used at lower temperatures, from −330°F to 300°F (−200°C to 150°C), whereas graphite is used mainly at higher temperatures. Teflon is often filled with glass and other agents to increase the allowable bearing pressure. The sliding pair assembly may either have two low-friction plates or have one low-friction plate paired with a polished stainless steel plate. Both cases have about the same coefficient of friction, although the one with the stainless steel plate has a somewhat higher coefficient. The assembly with one low-friction plate paired with a stainless steel plate is more popular due to its sturdy construction and resistance to tearing off.

The friction coefficient of the Teflon sliding plate is very consistent at about 0.05. However, at contact pressures less than 300 psi (2000 kPa), the coefficient is high and unpredictable. Therefore, it is important to maintain the contact pressure at above 300 psi. The general guideline is to maintain the pressure at a minimum of 500 psi (3500 kPa) with a 0.1 design friction coefficient. In addition, the allowable bearing pressure of the Teflon sliding plate is rather low at the 2000 psi range. At 100°F, for instance, the allowable pressure is 2000 psi for an unfilled Teflon plate and 4000 psi for a filled Teflon plate. The allowable pressure decreases rapidly with an increase in temperature. These limitations imply the need for a rather precise contact area for each support. The use of the standard size plate is not permitted in many cases. The size of the contact area is likely to change when the pipe moves. One

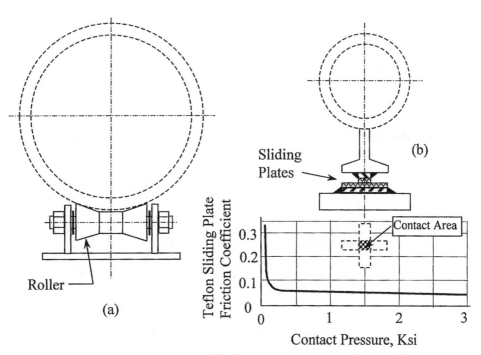

FIG. 6.16
LOW FRICTION SUPPORTS

way to ensure the size of the contact area is to use two perpendicular narrow rectangular strips, one on each side of the contact surface.

There are some difficulties associated with pipe support sliding plates. Because of the small size used in low load situations, the plate often drops off the contact surface when pipe movement is significant. Moreover, because sliding plates are required to be perfectly parallel to the surfaces to work, special installation procedures are required. Slight rotations of the pipe can also substantially reduce the effectiveness of the plates. Overall, low-friction sliding plates have been shown to be less than ideal in piping with large displacements. Movement should be minimized as much as possible by providing more frequent loops and offsets.

6.7 SUPPORT OF LARGE PIPES

Large pipes pose some special requirements in their support and analysis technique, mainly due to their high r/t ratio. Engineers tend to consider shear pipes larger than 24 in. (600 mm) in nominal diameter as large pipes, because some piping components standards cover only up to 24 in. However, the need for special requirements is not very clear-cut. It needs to be investigated on as case-to-case basis.

Large pipes generally tend to have higher r/t ratio due to the generally smaller internal pressure and smaller percentage of surplus thickness. The pipe shell of a large pipe has a higher tendency to deform at support locations than that of a smaller pipe. Once the shell deforms locally, the stress field becomes very complicated. Therefore, the current design procedures on large pipe depend mostly on field experiences and empirical data.

The load of the pipe, weight or otherwise, is carried by the shell with the tangential shear force distributed by a sinusoidal function as shown in Fig. 6.17. The tangential shear force per unit length, V, at location φ degree away from the zenith (top) is

$$V = \frac{W}{\pi r} \sin \varphi \tag{6.6}$$

where W is the total shear force acting at the given cross-section under consideration. The equation shows the tangential shear force distribution at an undisturbed area without any support or restraint located nearby. In this case, the maximum shear force, which is located at the equator, is equal to $W/\pi r$. This maximum value is twice as big as the average value of $W/2\pi r$. This doubling of the average shear force and thus the shear stress is called the shear distribution factor as noted in Chapter 2.

As the shear force transmits to the support location, its distribution gradually changes to a very complex pattern at the support point. However, according to Saint-Venant's principle [10], the distri-

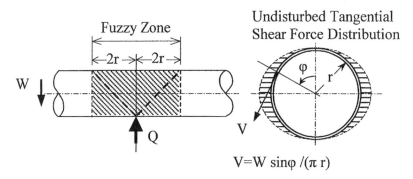

FIG. 6.17
TANGENTIAL SHEAR FORCE DISTRIBUTION

bution pattern will remain relatively undisturbed at about one diameter away from the support point. The area extending one diameter wide from each side of the support is a fuzzy zone where the shear distribution pattern is not well understood. Some type of semi-empirical approach is needed in dealing with the situation. The design of the supports and associated pipes of a large diameter piping system follows mostly one of the three approaches: Roark's saddle, Schorer's ring girder, and Zick's saddle.

6.7.1 Saddle Supports Using Roark's Formula

Saddle supports are used in this design to partially stiffen the pipe shell in order to better support the load. When an un-stiffened pipe rests on saddle supports, high local longitudinal as well as circumferential stresses are generated at the areas of the pipe shell adjacent to the tips of the saddles. The stresses decrease as the saddle angle increases, but they are practically independent of the saddle width. Saddle width is the dimension parallel to the pipe axis. For a pipe that fits the saddle well, the maximum value of these localized stresses will probably not exceed that indicated by the formula [11]

$$S_{max} = k \frac{Q}{t^2} \ln\left(\frac{r}{t}\right) \tag{6.7}$$

where

Q = total saddle reaction
r = pipe radius
t = pipe thickness
k = coefficient given by

$$k = 0.02 - 0.00012(\theta - 90) \tag{6.8}$$

where θ is the saddle angle in deg. It is the arc angle of the contact between saddle and pipe. This maximum stress is almost wholly due to circumferential bending and occurs at points at about 15 deg. above the saddle tips.

The maximum value of Q the pipe can sustain is about 2.25 times the value that will produce a maximum stress, as calculated by Eq. (6.7), equal to the yield point of the pipe material. This formula has been specified by the American Water Works Association (AWWA) in its guidelines [12] for designing steel water pipe since the 1960s.

The coefficient given by Eq. (6.8) is based on the results of tests performed on very thin-walled pipe. The report of Evces and O'Brien [13] involving ductile iron pipe has established that, for thicker pipes, the k values given by

$$k = 0.03 - 0.00017(\theta - 90) \tag{6.9}$$

provide better and excellent correlation between the stresses predicted by Eq. (6.7) and the actual stresses as measured when θ is between 90 deg. and 120 deg. Equation (6.9) is used by Ductile Iron Pipe Research Association (DIPRA) in its design guide [14].

In addition to the circumferential bending stresses discussed above, there are two general stresses that have to be considered: flexural beam bending stress and pressures stress. The empirical stress given by Eq. (6.7) presumably also includes the shear stress and the direct tangential membrane stress due to support load.

The maximum stress given by Eq. (6.7) is mainly circumferential bending, which is considered a secondary stress and is included in the evaluation of fatigue. The stress is also often conservatively treated as belonging to the local primary stress category for the design of pipe supports. The local primary stress is combined with the primary membrane stress due to pressure, weight, and other mechanical loads, to become the total local primary stress. The total local primary stress is generally limited to 1.5 times the basic allowable stress of the material. The pressure stress involves hoop and longitudinal stresses. As discussed in Section 4.5.1, because the longitudinal stress is roughly one-half the magnitude of the hoop stress with the same sign, only the difference representing one-half of the hoop stress is included in the combined stress intensity calculation. Considering that hoop stress is limited to the basic allowable stress, the stress criterion for the total local primary stress excluding pressure stress is 1.0 of the basic allowable stress. Both AWWA and DIPRA adopt this approach and limit the sum of the circumferential bending stress, as calculated by Eq. (6.7), and the beam flexural bending stress to the basic allowable stress, which is equal to, or less than, 2/3 of the yield strength.

The circumferential bending stress, due to support load, reduces substantially [15, 16] with the increase in internal pressure. Therefore, there is a question of combining this circumferential bending stress with the membrane pressure stress. However, because the potential variation of the support load is considerable due to settlement and shifting of the support piers, the conservative approach of including the pressure stress is necessary.

6.7.2 Ring Girder Supports

In this approach, the pipe is stiffened with ring girders, which are supported at two points located at the equator as shown in Fig. 6.18(a). This is a theoretical approach developed by Schorer [17] based on the theory of elasticity on thin shells. The design involves checking the pressure effect on the locally stiffened zone and calculating the force and stress of the stiffening ring due to the support load. It calculates the stresses at the pipe shell and at the stiffening ring.

With the pipe shell stiffened at the support point, the support load is transmitted from the pipe to the support ring by the shear forces distributed by sinusoidal function as shown in Fig. 6.17. This is the same shear distribution as in the portion of the pipe shell remote from the support point. By maintaining sinusoidal shear distribution at the pipe, the support load does not generate localized stress at the pipe. However, the pressure does generate localized pipe shell stress due to the choking effect of the stiffening ring. The pipe shell stresses due to weight and other external loads are calculated based on ordinary beam formulas, and the stresses due to pressure are calculated by choking formulas.

Choking force and moment. Figure 6.19(a) shows the deformation of a locally stiffened shell subjected to an internal pressure. The edge of the pipe at the pipe-ring junction is called the rim. To

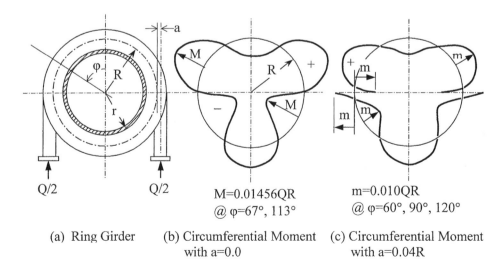

(a) Ring Girder

(b) Circumferential Moment with a=0.0
M=0.01456QR
@ φ=67°, 113°

(c) Circumferential Moment with a=0.04R
m=0.010QR
@ φ=60°, 90°, 120°

FIG. 6.18
CIRCUMFERENTIAL BENDING MOMENT AT RING GIRDER

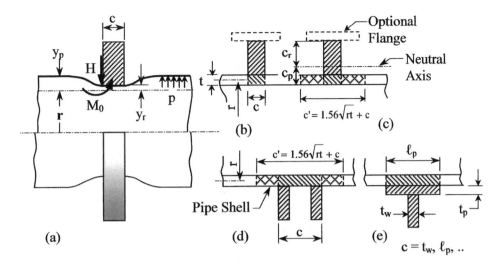

FIG. 6.19
RING GIRDER EFFECT ON SHELL

investigate the effect of the stiffening ring, we take the pipe and the ring as two separate free bodies divided at the junction. Due to internal pressure, p, the radius of the unrestricted pipe shell increases by y_p. However, because the rim of the shell is restricted by the stiffening ring, whose radius increases only by y_r, the shell is choked at the rim by the amount

$$y = y_p - y_r \tag{6.10}$$

This choking is accomplished by the rim force, H, and rim moment, M_0. The magnitudes of H and M_0 are per unit circumference of the rim. This is similar to the situation discussed in Section 5.3 for the semi-infinite beam on elastic foundation. At the rim where $x = 0$, Eqs. (5.19) and (5.20) can be rewritten as (P is replaced with H)

$$y = \frac{2H\beta}{k} - \frac{2M_0\beta^2}{k} \tag{6.11}$$

$$\theta = -\frac{2H\beta^2}{k} + \frac{4M_0\beta^3}{k} \tag{6.12}$$

where $\beta = \dfrac{1.285}{\sqrt{rt}}$, $k = \dfrac{Et}{r^2}$ for Poisson's ratio $\nu = 0.3$.

Due to symmetry and the fact that the pipe is connected to the ring, the slope is zero at the rim. With $\theta = 0$, Eq. (6.12) becomes

$$M_0 = \frac{H}{2\beta} \tag{6.13}$$

Substituting (6.10) and (6.13) to (6.11), we have

$$y_p - y_r = \frac{H\beta}{k} = \frac{r^2 H\beta}{Et} \tag{6.14}$$

The increase in the radius of unrestricted pipe shell due to internal pressure is

182 Chapter 6

$$y_p = \frac{pr^2}{tE} \tag{6.15}$$

The ring receives the internal pressure force over the width, c, and the radial reaction force, H, from the choking of the pipe. Because H is transmitted from both sides of the ring, a total of $2H$ are acting on the ring. The circumferential hoop tension over the ring is

$$N_t = (pc + 2H)r \tag{6.16}$$

This hoop tension generates a hoop stress of $S_t = N_t/A_r$, which results in an increase of the radius by

$$y_r = S_t \frac{r}{E} = (pc + 2H)\frac{r^2}{A_r E} \tag{6.17}$$

where A_r is the ring cross-section area. Substituting Eqs. (6.15) and (6.17) to (6.14), we have

$$\frac{pr^2}{tE} - (pc + 2H)\frac{r^2}{A_r E} = \frac{r^2 H \beta}{Et}$$

or

$$H = \frac{p(A_r - ct)}{\beta A_r + 2t} = \frac{p}{\beta} \frac{(A_r - ct)}{\left(A_r + 2t\frac{1}{\beta}\right)} \tag{6.18}$$

In addition, from (6.13), we have

$$M_0 = \frac{p}{2\beta^2} \frac{(A_r - ct)}{\left(A_r + 2t\frac{1}{\beta}\right)} \tag{6.19}$$

Substituting $\beta = 1.285/\sqrt{rt}$, (6.18) and (6.19) become

$$H = 0.78p\sqrt{rt}\,\frac{(A_r - ct)}{(A_r + 1.56t\sqrt{rt})} \tag{6.20}$$

$$M_0 = 0.303prt\,\frac{(A_r - ct)}{(A_r + 1.56t\sqrt{rt})} \tag{6.21}$$

The choking force and moment are determined at the edge of the ring defined by the c width. The ring area, A_r, is the total ring area combined with the shell located inside the c width. When the pipe is locally thickened or padded as in case (e), the c width is defined based on the length of the thickened pipe, ℓ_p. If this length is equal to or longer than $\ell_c = 4.0\sqrt{r(t + t_p)} + t_w$ (see Eq. (5.28)), the locally thickened pipe or the padded pipe will be considered as the base pipe with ring width $c = t_w$, and the pipe thickness equal to $(t + t_p)$. If ℓ_p is less than ℓ_c, then $c = \ell_p$ with the maximum c limited to $= 1.56\sqrt{r(t + t_p)} + t_w$. The pipe thickness in this latter case is the un-thickened base pipe thickness.

Stresses at pipe shell. The H and M_0 discussed above are for each unit circumferential width of the pipe shell. The longitudinal shell bending stress is equal to M_0 divided by the section modulus $t^2/6$ of the shell plate. That is,

$$S_\ell = \frac{M_0}{t^2/6} = 1.82\frac{(A_r - ct)}{(A_r + 1.56t\sqrt{rt})}\frac{pr}{t} \tag{6.22}$$

The longitudinal bending stress in (6.22) is expressed in terms of pr/t, the hoop stress of the unrestricted shell. The stress is higher for a bigger ring cross-sectional area, A_r, and a smaller ring width, c. For a very rigid ring, the stress can reach up to 1.82 times of the unrestricted pipe hoop pressure stress.

The choking of the shell generates the longitudinal bending stress, and at the same time decreases the pressure hoop stress. The pressure hoop stress at the rim is reduced by the y_r/y_p factor. In addition to the choking stress, the shell is also subject to a longitudinal pressure stress of $pr/2t$, unless the pipe is fully restrained, in which case the Poisson effect due to hoop stress governs.

The longitudinal shell bending stress is a secondary stress. Its evaluation requires the calculation of the total primary plus secondary stresses. The combined stress intensity requires the calculation of both longitudinal and circumferential stresses. The total longitudinal stress is calculated by combining the above shell bending stress due to pressure, beam bending stress due to weight and other mechanical loads, beam bending as well as direct axial stress due to thermal expansion, and longitudinal pressure stress. Locally or across the whole cross-section, the longitudinal stress will have tensile portion and compressive portion, due to bending. The compressive portion has to be combined with the circumferential pressure hoop stress, which is in tension. The combination of longitudinal stress with circumferential stress is required only when the stresses of the two directions are in opposite signs. Because the longitudinal shell bending moment also creates a Poisson stress with the same sign in the circumferential direction, it can be subtracted from the multi-directional combination.

The limit of the total primary plus secondary stress is three times of the basic allowable stress. Subtracting 1.5 times of the basic allowable for primary membrane plus local primary stress, the allowable for the secondary stress including choking bending and thermal expansion is 1.5 times of the basic allowable stress.

When the shell bending stress becomes excessive, a ring pad (Fig. 6.19(e)) or a locally thickened pipe at the support location can be used. It is interesting to note that, although the longitudinal choking stress is produced at the support, it is independent of the support load. Therefore, a reduction of the support span will not reduce the stress. This choking stress receives considerably different treatment in different applications. It may be treated as a primary stress in some cases [12], or simply ignored in other cases [16].

The internal or external pressure generally determines the pipe wall thickness. Besides pressure, the water or liquid weight can also dictate the wall thickness in low-pressure large liquid piping. The sum of the longitudinal beam bending stress and pressure stress is limited to 1.0 times the basic allowable stress. Subtracting half the basic allowable stress for the longitudinal pressure stress, the flexural beam bending stress, which is controlled by the span length, should be limited to no more than one-half of the basic allowable stress.

Because the weight of the pipe and content may control the wall thickness required, the half-full condition, such as when filling the system, also needs to be investigated. The longitudinal beam flexural stress at the half-full condition can be higher than the stress at the full condition. The stress ratio depends on the parameter, k, and is determined as follows:

$$\frac{S_\ell}{S_L} = \frac{1}{\sqrt{k}}, \quad \text{where } k = \frac{L}{r}\sqrt{\frac{t}{r}} \tag{6.23}$$

where S_ℓ and S_L are the longitudinal beam bending stresses at the half-full and full conditions, respectively, and L is the span of the supports. The half-full stress is higher than the full stress when k is less than 1.0. At the half-full condition, the allowable beam bending stress is the same as the basic allowable stress. This higher allowable is permitted mainly due to the absence of the internal pressure at the half-full condition.

For low-pressure liquid pipe, the possibility of instability or shell wrinkling due to compressive stress is also a concern. The compressive longitudinal stress, $S_{\ell c}$, is limited to one-half of the compressive yield strength of the pipe material or the value given by (see also Eq. (1.4))

$$S_{\ell c} \geq 0.06E\left(\frac{t}{r}\right) \quad (6.24)$$

which is derived from the theoretical tube column critical stress $S_c = 0.605E(t/r)$ [11] by applying a 0.4 imperfection factor and dividing by a safety factor of 4. The American Society of Mechanical Engineers (ASME) B&PV Code, Section VIII, Division 1 [18] should be consulted for a more accurate allowable compressive stress.

Stresses at stiffening ring. Support load is transferred from pipe to support by the shear forces distributed in sinusoidal form as shown in Fig. 6.17. These shear forces produce circumferential bending moments at the ring when the ring is fixed at certain points. For the ring supported at two diametrical locations, the bending moment at any given point located φ angle from the top can be calculated as follows:

- For the first quadrant, $0 < \varphi < \pi/2$

$$M_{\varphi,1} = \frac{QR}{2\pi}\left[\varphi\sin\varphi + \frac{3}{2}\cos\varphi - \frac{\pi}{2} + \frac{2a}{R}\left(\cos\varphi - \frac{\pi}{4}\right)\right] \quad (6.25)$$

- For the second quadrant, $\pi/2 < \varphi < \pi$

$$M_{\varphi,2} = -\frac{QR}{2\pi}\left[(\pi-\varphi)\sin\varphi - \frac{3}{2}\cos\varphi - \frac{\pi}{2} - \frac{2a}{R}\left(\cos\varphi + \frac{\pi}{4}\right)\right] \quad (6.26)$$

where R is the radius of the ring girder neutral axis, and a is the eccentricity of the support point from the neutral axis of the ring as shown in Fig. 6.18(a).

Equation (6.26) is identical to (6.25), except that the sign is reversed, if the φ angle in (6.25) is replaced with $(\pi - \varphi)$. Therefore, except for the opposite sign, the moment curves as shown in Figure 6.18(b) and (c) are symmetric about the horizontal diameter. The moments in the third and fourth quadrants are identical with those of the second and first quadrants due to symmetry in loading. These equations can also be obtained from Roark's [11] ring formulas.

When supports are applied at the neutral axis with zero offset, the ring circumferential bending moment is zero at the equator. Figure 6.18(b) shows the moment distribution with $a = 0$. The maximum moment in this case is $0.01456QR$ occurring at $\varphi = 67$ deg. (+) and $\varphi = 113$ deg. (−). This maximum moment can be reduced considerably by applying the support reactions outside the neutral axis of the ring. When applying the support reactions outside the neutral axis it generates an external moment equal to $Qa/2$ at each side of the equator. This moment will then divide into two halves with one-half transmitting through the upper portion and the other half to the lower portion of the ring. Although the ring circumferential bending moment at the equator has increased to $Qa/4$ from zero, the maximum bending moments at the other locations are reduced. The optimum offset is achieved with $a = 0.04R$, which gives the maximum (+) and minimum (−) bending moment of $0.010QR$. This represents a 45% reduction of the maximum circumferential bending moment from the situation without a support offset. The ring moment distribution with this optimum support location is shown in Fig. 6.18(c).

Some words of caution are in order for the optimized bending moment. Because the moment distribution is based on the assumption that support columns offer only a pure vertical support function without any other restraining effect, the effect of the offset support forces is not at all certain. Furthermore, the $0.04R$ optimum offset for an 80-in. pipe, for instance, is only 1.6 in. It would require considerable quality control for both constructing and locating the actual support point at the base to achieve this required offset. In addition, support columns normally need some type of gussets to resist lateral loads due to friction, wind, earthquake, and other factors. Once the columns are laterally stiffened to the shell, the ideal moment distribution of the ring changes. See Zick's approach given in Section 6.7.3 for further discussions.

The maximum bending moments, $M_{\varphi,\max} = 0.01456QR$ or $0.010QR$, depending on the design specification, are used to calculate the ring bending stress.

$$S_{b,p} = \frac{M_{\varphi,\max}c_p}{I}, \quad S_{b,r} = \frac{M_{\varphi,\max}c_r}{I} \qquad (6.27)$$

where $S_{b,p}$ is the stress at the pipe; $S_{b,r}$ is the stress at the outer rim of the ring; I is the combined effective moment of inertia of the ring-shell combination; and c_p and c_r are the distances from the neutral axis to the pipe inside surface and ring outside rim, respectively, as shown in Fig. 6.19(c). At (+) moment locations, the stress at the pipe is compressive and the stress at the outer rim of the ring is tensile. The signs of the stresses reverse at (−) moment locations. The combined effective moment of inertia of the ring is calculated by including the actual ring section plus an effective width of pipe shell extending $0.78\sqrt{rt}$ from each side of the ring as shown in the cross-hatched section in Fig. 6.19.

In addition to the circumferential bending moment, the shear forces also result in a direct ring force, which is the maximum at $\varphi = \pi/2$. The direct force and the resulting stress are

$$N = \frac{Q}{4} \quad \text{and} \quad S_n = \frac{Q}{4A_r} \qquad (6.28)$$

The stress is compressive above the support, and tensile below the support.

The ring also produces hoop stress due to direct pressure and the choking reaction from the pipe. By converting the choking reaction from both sides of the ring to an equivalent pressure of $p_H = 2H/c$, the ring can be considered as subjected to a total equivalent pressure of $p_e = p + p_H = p + 2H/c$. Substituting H from Eq. (6.20), the hoop stress at the ring is calculated with a simple formula

$$S_p = \frac{cp_e r}{A_r} = \frac{pr}{A_r}\left[c + 1.56\sqrt{rt}\,\frac{(A_r - ct)}{(A_r + 1.56t\sqrt{rt})}\right] \qquad (6.29)$$

The above three stresses, S_b, S_n, and S_p, are added absolutely at the pipe and at the outer rim of the ring to become the maximum combined stresses at those locations. The total stress is generally limited to 1.0 times of the basic allowable stress. The ring should be appropriately proportioned to avoid the possibility of warping. Allowances for support load shifting should also be considered.

The ring girder is generally supported by two short columns. This type of construction is popular due to the relatively long support span and high underneath clearance allowed. The long span reduces the number of supports required and the related substructure costs. The high clearance is ideal for crossing streams, swamps, and marshes. However, the support columns and the support points have to be properly designed to ensure good distribution of the support load and to reduce the support friction force. Rockers, rollers, and low-friction sliding plates may be used for the support points. Confined neoprene support pillows can also be used. Gussets and lateral bracings may be required at the columns to resist the lateral loads from the friction force and occasional loads, such as wind and earthquake. Schorer's approach has been adopted and outlined by AWWA in its design guide for steel water pipe [12].

6.7.3 Saddle Supports by Zick's Method

Zick [16] has developed a set of formulas and procedures for analyzing and designing large horizontal vessels supported by two saddles. This method has been widely used due to its ease of application. It is an ASME Section VIII [18] recommended procedure for the design of supports for horizontal vessels and is the basis for the relevant section of the British Standard BS-5500 [19]. It is also widely used in the design of supports for large diameter pipes. However, due to the rather small diameter and r/t ratio of the piping, as compared with the vessels, some adjustments are required for piping applications.

Beam bending stress and reduction of pipe section modulus. Based on strain gauge studies, Zick has discovered that when the vessel is supported by a saddle, only the portion of the shell in contact with saddle plus a small portion nearby is effective in resisting the beam bending moment. The hatched top portion of the shell cross-section, as shown in Figure 6.20, is ineffective in resisting the bending moment. The section modulus of the effective portion, which is the saddle angle plus one-sixth of the non-contact angle, is only a small fraction of the normal section modulus with the entire circumference of the shell active. Table 6.4 shows the section modulus reduction factors for different saddle angles. Take a 120-deg. saddle, for instance, the effective pipe section modulus is only 0.1066 of the full-pipe section modulus. This reduction factor reaches 0.0655 for the 90-deg. saddles normally used in piping. Therefore, to adopt this effective section modulus, an equivalent stress intensification factor (SIF) of 15.27 has to be applied for 90-deg. saddles in piping stress analyses. This appears to be inconsistent with field experiences on piping. The strength of the pipe does not suddenly decrease to 10% of the original strength by adding a saddle. This is the impression that many engineers have about this reduced section modulus. The saddle, if anything, should have reinforced the pipe instead.

Because the section modulus reduction is independent of the support load, the behavior of the ring supporting its own weight may reveal some connection. A large thin vessel shell often does not have enough ring stiffness to hold its own weight. Thus, the shell requires the use of saddles and stiffening rings to maintain its roundness before any load is applied. For piping systems with support types as tentatively suggested by Table 6.6, the section modulus reduction is most likely insignificant. However, this needs to be further investigated either in the filed or in the laboratory. In any case, the section modulus reduction is applicable only to shells for very low pressure applications.

For beam loading, we also have to take into account the fact that a reduction in section modulus also means a reduction in moment of inertia, thus a reduction in stiffness. This reduction in moment of inertia serves as a hinge, which considerably reduces the moment carrying capacity. That is, a very small beam bending moment will be produced at the support location of a continuous piping when the section modulus of the pipe over the support point is reduced. The increase in beam bending stress at the support is, therefore, not as much as the supposed reduction in the section modulus.

Circumferential stress at horn of saddles. The pipe load sustained by the pipe located far away from the support is distributed as shear forces in a sinusoidal function such that no circumferential bending stress is generated. As the load transfers from the pipe to saddle, the shear forces redistribute and, at the same time, generate circumferential bending moments on the pipe shell. As previously discussed, the exact shear forces distribution at the fuzzy zone in the proximity of the support is not known. Calculating the exact bending moments resulting from the shear and the support forces is not possible. Zick's method uses a semi-empirical approach.

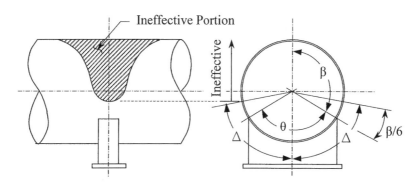

FIG. 6.20
EFFECTIVE SECTION MODULUS OF LARGE VESSEL OVER A SADDLE AGAINST BEAM BENDING

TABLE 6.4
SECTION MODULUS REDUCTION OVER SADDLE SUPPORT OF THIN CYLINDRICAL SHELL [16]

Saddle angle (θ), deg.	Δ, rad	Section modulus reduction factor, K_z	Equivalent stress intensification factor, $1/K_z$
90	1.178	0.0655	15.27
120	1.396	0.1066	9.38
140	1.541	0.1412	7.08
160	1.687	0.1817	5.50
180	1.833	0.2286	4.37

$$K_z = \frac{\Delta + \sin\Delta\cos\Delta - 2\frac{\sin^2\Delta}{\Delta}}{\pi\left(\frac{\sin\Delta}{\Delta} - \cos\Delta\right)}$$

Stiffening ring over the saddle. Although the saddle may or may not have a stiffening ring, the analytical model with a stiffening ring over the saddle is our first working model to deal with the saddle support problems. The semi-empirical approach will extend the application of this model to non-stiffened saddles. When there is a stiffening ring over the saddle, the shell is maintained in fair roundness, so the shear forces distribution is considered the same as the distribution of the undisturbed pipe section as shown in Fig. 6.17. If the ring is fixed at the horns of the saddle, the circumferential bending moment, M_φ, at any point, A, located φ angle from the top is given by:

$$M_\varphi = \frac{QR}{\pi}\left\{\cos\varphi + \frac{\varphi}{2}\sin\varphi - \frac{3}{2}\frac{\sin\beta}{\beta} + \frac{\cos\beta}{2} - \frac{1}{4}\left(\cos\varphi - \frac{\sin\beta}{\beta}\right)\right.$$
$$\left.\times\left[9 - \frac{4 - 6(\sin\beta/\beta)^2 + 2\cos^2\beta}{(\sin\beta/\beta)\cos\beta + 1 - 2(\sin\beta/\beta)^2}\right]\right\} \quad (6.30)$$

where the angles φ and β are in radians. The moment distribution is shown schematically in Fig. 6.21. For un-stiffened shells, to be discussed later, the ring radius, R, shall be replaced with the pipe radius, r.

The maximum moment occurs at the horns of the saddle and is a function of the saddle angle, θ. Therefore, the magnitude of the maximum moment is obtained by substituting $\varphi = \beta = \pi - \theta/2$ in the equation. The corresponding maximum moments are: $M_{\varphi,\max} = -0.0827QR$ for $\theta = 90$ deg.; $= -0.0529QR$ for $\theta = 120$ deg.; $= -0.0317QR$ for $\theta = 150$ deg.; and $= -0.0174QR$ for $\theta = 180$ deg. For piping, the saddle angles used are mostly 90 deg. and 120 deg. However, the vessel codes require that the saddle angle shall not be less than 120 deg. for vessels.

When the saddle encloses the whole bottom half of the shell with $\theta = 180$ deg., the situation is similar to Schorer's ring discussed in Section 6.7.2. However, in Schorer's case, the bending moment at the equator is zero due to the absence of the saddle. This is an interesting contrast to Zick's maximum bending moment of $0.0174QR$ at the equator for $\theta = 180$ deg. Zick's maximum bending moment is considerably greater than that of Schorer's, even without the support offset. We should use our own judgments, based on the actual support leg configuration, to select the proper design moment.

The circumferential bending stress at the ring is calculated by dividing the moment with the section modulus of the ring. The section modulus is calculated based on the combined ring shell effect, including a part of the effective shell as shown in Fig. 6.19(c) and (d). If two or more rings are used, then the maximum bending moment is divided equally by the number of rings to become the moment for each ring. The effective shell width of each ring for the multi-ring situation should not overlap each other.

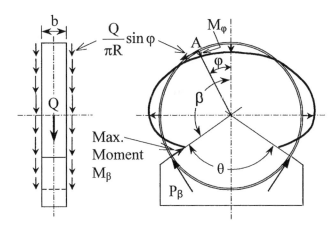

**FIG. 6.21
CIRCUMFERENTIAL BENDING MOMENT AND DIRECT FORCE, RING IN PLANE OF SADDLE**

In addition to the circumferential bending moment, the ring is also subject to a direct tangential force, P. The direct force at the saddle horn location is equal to

$$P_\beta = \frac{Q}{\pi}\left[\frac{\beta \sin\beta}{2(1-\cos\beta)} - \cos\beta\right] - \frac{\cos\beta}{R(1-\cos\beta)}(M_\beta - M_t) \tag{6.31}$$

where M_β is the maximum moment at the horn and M_t is the moment at the top of the ring. M_t is obtained by setting $\varphi = 0$ in Eq. (6.30). The P_β values are: $0.359Q$ for $\theta = 90$ deg., $0.340Q$ for $\theta = 120$ deg.; $0.302Q$ for $\theta = 150$ deg., and $0.250Q$ for $\theta = 180$ deg. The circumferential stress due to the direct force is equal to P_β divided by the cross-sectional area of the ring-shell combination, including the effective width of the shell.

The total circumferential stress consists of the bending stress due to circumferential moment, the direct compressive stress due to direct tangential force, and the hoop stress due to pressure, p. Due to flow requirements, the ring is always installed outside the shell in piping. Therefore, for a ring over the saddle, the maximum circumferential bending stress is tensile at the pipe and compressive at the outer edge of the ring. For an external ring over the saddle, the maximum circumferential stresses at the pipe shell and outer edge of the ring are

$$S_{c,max} = S_{c,\beta} = -\frac{P_\beta}{A_r} + \frac{c_p M_\beta}{I_r} + \frac{c'rp}{A_r} \quad \text{at the pipe} \tag{6.32a}$$

$$S_{c,max} = S_{c,\beta} = -\frac{P_\beta}{A_r} - \frac{c_r M_\beta}{I_r} + \frac{c'rp}{A_r} \quad \text{at ring} \tag{6.32b}$$

where A_r and I_r are the cross-sectional area and moment of inertia of the effective ring, including the actual ring and the effective shell width as given in Fig. 6.19, which also shows the definition of c_r, c_p, and c'.

Saddle without stiffening ring. When the shell is supported on a saddle without any stiffening ring installed, the shear forces tend to concentrate near the horns of the saddle. The shear distribution is, therefore, different from the ideal sinusoidal form. Because the exact shear forces distribution is not known, Zick solved the problem with a two-step semi-empirical approach.

First, by regarding the shear distribution as unchanged from the undisturbed portion, the circumferential bending moment and direct shear force are the same as in the case when a stiffening ring is installed over the saddle. Therefore, the same Eqs. (6.30) and (6.31) are used for the un-stiffened shell over the saddle by replacing ring radius, R, with pipe radius, r. However, the values given by these equations are artificial ones. They are just for reference purposes. Our main purpose is to find the stress, so we hope that there is a direct correlation between the stress and this reference bending moment.

Second, based on Zick's observations, the stresses calculated on the assumption that a certain width of shell is effective in resisting the hypothetical moment, M_β, agreed conservatively with the results of strain gauge surveys. It was found that this effective width of shell should be equal to 4 times the shell radius. It is interesting to note that this effective width coincides with the fuzzy zone width given in Fig. 6.17. However, this is just pure coincidence, because no relation is postulated. Saddle width is not a controlling factor in the calculation of this circumferential stress. Zick also suggested that internal pressure stresses do not add directly to local bending stresses, because the shell rounds up under pressure.

The direct tangential force is also smaller than the P_β calculated by Eq. (6.31) due to a change in shear distribution. Zick used $Q/4$ as the reasonable direct force. This force will be resisted by the shell width directly over the saddle plus an effective width as shown in Fig. 6.19.

The total circumferential stress at shell is the summation of the direct shear stress due to tangential shear force and shell bending stress due to circumferential moment. The pressure hoop stress is not included. That is,

$$S_{c,\beta} = -\frac{Q}{4t(b+1.56\sqrt{rt})} \pm \frac{M_\beta}{4r(t^2/6)} \tag{6.33}$$

The stress due to circumferential bending moment is a local secondary stress. For the combined stress, Zick suggested an allowable value of 1.5 times the basic tension allowable stress, provided the compressive strength of the material equals the tensile strength.

Stiffeners added adjacent to the saddle. When stiffening rings are added adjacent of the saddle, the circumferential bending moment distribution is different from that of the case when the ring is added directly over the saddle. Zick considered the ring/saddle combination as a combined ring with sinusoidal distribution of the shear forces. The combined ring is assumed to be supported at saddle horn locations with two tangential forces as shown in Fig. 6.22. With the same shear force distribution as the undisturbed section, the circumferential bending moment on the combined ring at a point φ angle from the top is

$$M_\varphi = -\frac{QR}{2\pi}\left\{\frac{\pi-\beta}{\sin\beta} - \varphi\sin\varphi - \cos\varphi[3/2 + (\pi-\beta)\cot\beta]\right\} \tag{6.34}$$

The equation is applicable only to the portion above the saddle. A negative sign is applied to Zick's original formula to maintain a consistent sign convention. With the revised sign convention, the same Eq. (6.31) can be used to calculate the direct force.

The maximum bending moment occurs at $\varphi = \rho$ near the equator for the range of common saddle angles from 90 deg. to 150 deg. The maximum bending moments and corresponding direct forces for the common saddle angles are as follows: For $\theta = 90$ deg., $M_\rho = 0.077QR$, $P_\rho = 0.304Q$, occurs at $\rho = 99$ deg.; for $\theta = 120$ deg., $M_\rho = 0.0581QR$, $P_\rho = 0.273Q$, occurs at $\rho = 94$ deg.; for $\theta = 150$ deg., $M_\rho = 0.0355QR$, $P_\rho = 0.218Q$, occurs at $\rho = 84$ deg.

When the saddle angle equals 180 deg., the case becomes identical to Schorer's ring without the support offset discussed previously in this chapter. With $\theta = 180$ deg., the bending moment at the

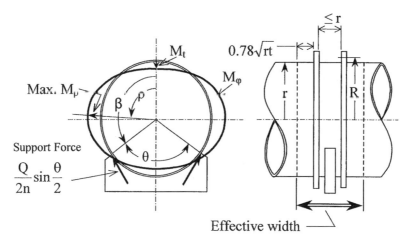

FIG. 6.22
STIFFENING RINGS ADJACENT TO SADDLE

equator is zero, which is quite in contrast to the near-maximum value that occurred at the equator with smaller saddle angles.

The combined stress is calculated as done earlier using an equation similar to (6.33), except that the effective width is used for both tangential shear stress and circumferential bending stress. The moment and force given are the total for the combined ring or rings. If more than one ring are used, and they are widely separated by more than $1.56\sqrt{rt}$, then the moment and the force are divided equally among the rings for calculating the stress of each individual ring.

6.7.4 Support Types

When dealing with large pipe supports, a tough question is when to use saddles or other support types. Although some large operating companies and engineering companies do provide some guidelines on the support types, the industry as a whole lacks a consistent approach to this problem. In the following, we will try to establish some basis with the hope of finding a reasonable applicable approach.

Obviously, the easiest and most economical support type is to support the pipe directly on its bare shell. However, not every pipe can be supported on its shell because it produces the largest support stress on the pipe among all support types. The potential pipe stress due to the support load has to be estimated somehow, for the direct bare shell supported situation, before the proper support type can be determined.

Following Roark's and Zick's approaches, a circular ring model is used to investigate the pipe stress due to support load. Support load is transmitted from the pipe to the support through a circular ring resting on the support structure. The sinusoidal form of the tangential shear forces distribution is again assumed to transmit the load from the pipe to the support. Figure 6.23 shows the circumferential bending moment distribution at the pipe shell. We know the maximum bending moment occurs at the support point and is equal to $0.239Qr$ on paper, but we do not exactly know how to use it for the following reasons: (1) because it is an artificial moment due to an artificial shear distribution; and (2) because the effective width of the ring is not known. The approaches by Roark and Zick, again, are the likeliest solutions.

Zick approach. In treating the saddle support, Zick uses the artificial moment paired with an effective ring width of four times the shell radius. If the same approach is used, then the stress at the shell will be

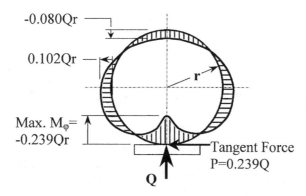

**FIG. 6.23
CIRCUMFERENTIAL MOMENT ON A RING WITH CONCENTRATED LOAD**

$$S_{b,\max} = \frac{M_{\varphi,\max}}{Z} = \frac{0.239Qr}{bt^2/6} = \frac{0.239Qr}{4rt^2/6} = 0.3585\frac{Q}{t^2}$$

The fact that the stress is independent of the pipe diameter makes this approach unfavorable. The pipe diameter always has some role in determining the stress.

Roark approach. Roark's formula as given in Eq. (6.7) contains a $\ln(r/t)$ factor that makes it more suitable for this application. The question is how to determine the k value. We have k values determined by Eq. (6.8) for large r/t ($r/t > 100$) pipes and k values determined by Eq. (6.9) for small r/t pipes, but they are for saddle supports with saddle angles ranging from 90 deg. to 150 deg. They should not be considered directly applicable to the case of 0-deg. saddles, that is, no saddle. However, a conservative k value can probably be derived from the general trend of the data.

Pipe shell stress is determined by the circumferential bending moment and the direct tangential force. The stress due to the direct tangential force is generally small and will be ignored for the time being. The stress due to the circumferential bending moment is calculated by dividing the moment with the section modulus, if we know the moment and section modulus. Unfortunately, we know neither. One reasonable assumption we can use is that the stress is proportional to the artificial moment determined with the sinusoidal distribution of the tangential shear forces as in Zick's approach. By comparing Zick's saddle tip moment and Roark's pipe stress of saddle supported pipe, a relation between Zick's moment and Roark's k value may be established. Table 6.5 shows the k values, for both high r/t and low r/t, at various saddle angles together with the corresponding Zick's saddle tip bending moments. It is obvious that the $k/(M/Qr)$ ratio decreases as the saddle angle decreases. Because stress is proportional to k, smaller $k/(M/Qr)$ values will have less stress for a given M/Qr. Therefore, it is reasonable to consider it conservative to use the same $k/(M/Qr)$ ratio of a 70-deg. saddle for saddles with smaller angles. Here, we have extrapolated the Roark's formula a little to cover 70-deg. saddles.

By setting the $k/(M/Qr)$ ratio to 0.3093 (a larger value is more conservative), we have the k value for the bare pipe support as

$$\frac{k}{M/Qr} = 0.309, \quad \text{i.e., } k = 0.309\frac{M}{Qr} = 0.309\frac{0.2387Qr}{Qr} = 0.072$$

Using $k = 0.072$ in. Eq. (6.7), we have

$$S_{b,\theta=0} = 0.072\frac{Q}{t^2}\ln\left(\frac{r}{t}\right) \tag{6.35}$$

TABLE 6.5
PIPE CIRCUMFERENTIAL BENDING STRESS NEAR THE HORNS OF SADDLES

Saddle angle (θ), deg.	Roark's k values		Moment at horn (M/Qr)	k/(M/Qr)	
	High (r/t)	Low (r/t)		High (r/t)	Low (r/t)
170	0.0104	0.0164	0.0215	0.4837	0.7627
150	0.0128	0.0198	0.0317	0.4038	0.6246
120	0.0164	0.0249	0.0528	0.3106	0.4716
90	0.0200	0.0300	0.0826	0.2421	0.3631
70	0.0224	0.0334	0.1080	0.2074	0.3093
0	(0.0308)[1]	(0.0453)[1]	0.2387	[2]	[2]

[1] Beyond the applicable range of the formula, probably un-conservative.
[2] Can be conservatively set at the same values as those of the saddle with 70-deg. saddle angle.

The maximum value of Q that the pipe can sustain is about 2.25 times the value that will produce a bending stress, calculated by the above formula, equal to the yield point of the pipe material. That is, by setting an allowable bending stress to the yield point of the material, the safety factor based on stress is 2.25. This is comparable to the B31 code fatigue safety factor for local skin bending stress. It is also comparable to the safety factor for local primary stress. This limit on shell bending stress can be used to determine the required support type. Using a yield strength of 30,000 psi and a support span of 40 ft, the tentative support attachment types for some pipe sizes are tabulated in Table 6.6. The 40-ft span is not necessarily the actual span. It is an adjusted span considering the effect of uneven settlements between supports. The table is termed as tentative because the rationale is not solid and there is room for improvement.

The allowable spans given in Table 6.6 are for the bare pipe without any attachment. When the allowable span equals or exceeds 40 ft, bare pipe support is possible. With a pad of the same thickness as the pipe, the section modulus due to shell bending increases 2 to 4 times depending on the pad-pipe bonding condition. Assuming conservatively that it is doubled, the allowable span will also be doubled. Pad support can be used for the bare pipe allowable span greater than or equal to 20 ft. The saddle used in piping is generally the 120-deg. ones. With 120-deg. saddles, the k value is 0.0164 for high r/t and 0.0249 for low r/t as given in Table 6.5. These values are about ¼ of the k value used in Eq. (6.35) for constructing Table 6.6. Therefore, a saddle will increase the allowable span by four times.

Shell bending stress is not required in equilibrium with the applied load. It is secondary and concentrated in a local area. Therefore, it is combined, if cyclic, with thermal expansion stress and other displacement stresses for fatigue evaluation. In general, pipe weight is not cyclic, but water weight is, if drained frequently. The portion of stress, which is cyclic, should be added to the thermal expansion stress or alternately subtracted from the allowable stress. It should be noted, however, that B31 evaluation uses one-half of the theoretical secondary stress as the basis. The stress given by Eq. (6.35) shall be considered a theoretical stress.

In addition to the shell bending stress, there is the direct membrane stress due to direct tangential force. This stress is determined by dividing the tangential force $P = 0.239Q$ with the effective cross-sectional area. That is,

$$S_t = \frac{0.239Q}{(b+1.56\sqrt{rt})t} \quad (6.36)$$

where b is the width of the support. This is a local membrane stress and is checked together with pressure hoop stress. The combined stress shall not exceed 1.5 times the basic allowable stress of the pipe material. Because this stress is compressive, it is not additive to the pressure hoop stress of the pipe with internal pressure.

TABLE 6.6
TENTATIVE SUPPORT TYPES BASED ON EXTENDED ROARK'S FORMULA
(INCLUDING 0.05 IN. CORROSION ALLOWANCE)

Pipe Dia. (in)	Nominal Thick (in)	r/t	Weight with Water lbs / ft of pipe	Bending Stress due to Weight psi / ft of pipe	Allowable Span (ft)	Support Attachment Type
16	0.25	32	123	767	39	Pad
16	0.375	21	141	293	103	None
16	0.50	16	161	149	202	None
20	0.25	40	178	1182	25	Pad
20	0.375	27	204	458	65	None
20	0.50	20	225	240	125	None
24	0.25	48	245	1707	18	Saddle
24	0.375	32	277	654	46	None
24	0.50	24	308	348	86	None
30	0.25	60	370	2727	11	Saddle
30	0.375	40	408	1026	29	Pad
30	0.50	30	447	541	55	None
36	0.25	72	517	3980	8	Ring
36	0.375	48	564	1488	20	Pad
36	0.50	36	610	777	39	Pad
42	0.25	84	689	5495	5	Ring
42	0.375	56	744	2041	15	Saddle
42	0.50	42	798	1061	28	Pad
48	0.25	96	886	7279	4	Ring
48	0.375	64	948	2688	11	Saddle
48	0.50	48	1010	1390	22	Pad
54	0.25	108	1107	9330	3	Ring
54	0.375	72	1177	3431	9	Ring
54	0.50	54	1247	1769	17	Saddle
60	0.25	120	1352	11651	3	Ring
60	0.375	80	1431	4274	7	Ring
60	0.50	60	1509	2197	14	Saddle
70	0.25	140	1816	16153	2	Ring
70	0.375	93	1908	5895	5	Ring
70	0.50	70	1999	3020	10	Saddle

Based on 30,000 psi circumferential bending stress.
Support types selected based on the following: (a) Direct bare pipe support with allowable bare pipe span >40 ft. (b) Pad having the same thickness as pipe nominal thickness with 12 in. length and 120-deg. encirclement.
Applicable for allowable bare pipe span >20 ft. (c) Saddle required for allowable bare pipe span <20 ft. (d) Ring stiffener required for allowable bare pipe span <10 ft.

6.8 PIPE STRESSES AT INTEGRAL SUPPORT ATTACHMENTS

Integral support attachments are attached to the pipe by welds. The supporting loads are transmitted through the attachment then to the pipe. Complex stress fields are created at the interface of the attachment and the pipe. These stresses can be calculated readily by using the finite element method (FEM) implemented in some specialized computer software packages available in the market. FEM, although powerful, requires special techniques in modeling and stress interpretation. It is used frequently in the study of special cases, but rarely in routine piping design activities.

For routine attachment design activities, the cookbook method is preferred. Some of the popular methods are:

(1) The ASME Power Boiler Code [20, 21] approach. The boiler code provides a set of formulas and charts for the design of rectangular attachments welded in the longitudinal direction of the pipe. The applied load is a radial force, longitudinal moment, or both.
(2) Welding Research Council Bulletin #107 (WRC-107) [22]. This bulletin provides methods and data for calculating the local stresses at rectangular, as well as round, attachments to both spherical and cylindrical shells.
(3) Welding Research Council Bulletin #198 [23, 24]. This bulletin provides formulas for calculating stress indices for attachment loads including radial force, shear force, longitudinal moment, circumferential moment, and torsional moment. Local stresses due to pressure deformation and temperature gradient are also discussed. The stress indices are mainly used in Class 1 nuclear piping.
(4) British Standard BS-5500 [19]. Appendix G of BS-5500 provides methods of calculating stresses due to local loads on some common local attachments on pressure vessels.
(5) Kellogg's choking model [1]. M.W. Kellogg Company derived a set of design formulas based on the behavior of the pipe under circumferential uniformly distributed radial line loading. This model is generally referred to as the choking model. The design formulas so derived are considered conservative.

More detailed discussions on the power boiler formula, Kellogg method, and WRC-107 are given in the following subsections.

6.8.1 Power Boiler Formulas for Lug Stresses

Special notes. The power boiler formulas are formulated in inches/pounds units. They are not in consistent units. Therefore, all other units are converted to in./lb first before the allowable loads and stresses are calculated. After the loads and stress are calculated, they can be converted back to the units desired.

The power boiler code has listed the design procedure for the support lugs attached to boiler tubes. The main stress concerned is the shell bending stress that is additive to the pressure hoop stress. The procedure uses the model of a circular ring of unit width subjected to a radial force. This is the same model shown in Fig. 6.23, except that the force now is the applied force instead of a reaction force. The maximum shell bending moment occurs at the forcing point, and is equal to $0.239Fr$. F is the applied force per unit width of the ring. The nominal bending stress, disregarding the curvature of the ring, equals to

$$S_b = \frac{M}{Z} = \frac{0.239Fr}{t^2/6} = 0.717F\frac{D}{t^2}$$

or

$$F = 1.395 S_b \frac{t^2}{D} \tag{6.37}$$

The old version of the code, issued before 1992, used an allowable bending stress of 20,000 psi. This resulted in an allowable force roughly equals to

$$F_{\text{allow}} \cong 28{,}000 \frac{t^2}{D} \quad \text{(in inch pound units)} \tag{6.38}$$

Equation (6.38) was further adjusted by the following factors: (1) the tensile stress is the controlling quantity when combining with the pressure hoop stress; and (2) both bending moment and internal pressure produce higher stress at the inside surface of the pipe wall as shown in Fig. 6.24. Because the compression loading, the one pushing the shell, produces tensile stress at the inside surface, it has lower allowable values. The allowable loads are plotted against D/t^2 in a composite chart. The shortcomings of the old code are: (1) it did not consider the width of the lug; (2) it did not allow the credit of the surplus strength of the pipe over the design pressure; and (3) it did not distinguish the pipe material.

The procedure was revised in 1992 to take into account the lug width and the surplus strength of the pipe. The allowable load is also based on the basic allowable stress of the pipe material. The allowable loads are curve fit into two load factor curves given as follows:

- For compression loading

$$L_{f,c} = 1.618 X^{[-1.020 - 0.014(\log X) + 0.005(\log X)^2]} \tag{6.39}$$

- For tension loading

$$L_{f,t} = 49.937 X^{[-2.978 + 0.898(\log X) - 0.139(\log X)^2]} \tag{6.40}$$

where $X = D/t^2$. The log terms are the logarithms to the base 10. These load factors are not used to calculate the stresses; rather, they are used to calculate the allowable loads. The allowable loads are also dependent on the allowable stress, $S_{b,\text{allow}}$, set as

$$S_{b,\text{allow}} = 2.0 S_a - S \tag{6.41}$$

where S_a is the basic allowable stress of the pipe material and S is the pressure hoop stress due to design pressure. This is equivalent to an allowable of $2S_a$ for the primary plus secondary stresses $(S_b + S)$.

The effect of the lug thickness is adjusted by the ratio of the potential bending moment as compared to the maximum bending moment of $0.23873 Fr$ produced with lug width $b = 0$. By considering the lug as a saddle, the potential bending moment at the edge of the lug can be found from Zick's equation (Eq. (6.30)) by substituting $\varphi = \beta = \pi - \theta/2$ to the equation. The adjustment factor, K, is equal to

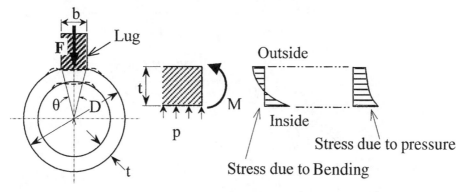

FIG. 6.24
CIRCUMFERENTIAL STRESSES AT LUG ATTACHMENT

TABLE 6.7
K VALUES (ADJUSTMENT FOR LUG THICKNESS)

Angle (θ) deg.	Edge moment $M\varphi/Fr$	K value $0.239/(M\varphi/Fr)$
0	0.23873	1.000
2	0.23402	1.021
4	0.22936	1.041
6	0.22476	1.062
8	0.22022	1.084
10	0.21574	1.107
12	0.21132	1.130
14	0.20696	1.154
16	0.20266	1.178
20	0.19424	1.229
30	0.17421	1.370
40	0.15561	1.534

$0.23873 Fr$ divided by the potential bending moment at the edge of the lug. The factors for some of the lug angles are tabulated in Table 6.7. This table is consistent with Table PW-43.1 of the boiler code.

The allowable loads per unit length of the lug are calculated separately for tension and compression loads by

$$L_a = K(L_f)S_{b,\text{allow}} \qquad (6.42)$$

L_f is the load factor. A tension load uses tension load factor, $L_{f,t}$, and a compression load uses compression load factor, $L_{f,c}$. The applied loads are then checked with the allowable loads. The allowable for the tension or pulling load is higher than that of the compression load.

The power boiler approach is only applicable to loads acting in the plane of the lug plate as shown in Fig. 6.25(b). The applied loads are decomposed and combined into a radial load acting at the center point of the lug plus a bending moment acting longitudinally. The shear force acting parallel to the lug

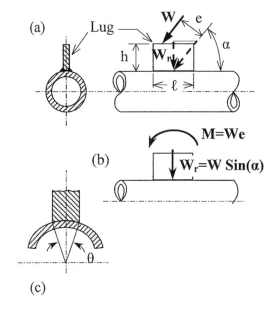

FIG. 6.25
LOADS AT LUG

is ignored. The lug is not capable of resisting the loads and moments acting perpendicular to the lug plate. The center force and the moment have to be converted into the radial force per unit length of the lug. With the force and moment as shown, the maximum radial force per unit length is calculated as

$$F = -\frac{W_r}{\ell} \pm \frac{6M}{\ell^2} \qquad (6.43)$$

where ℓ is the length of the lug. The negative sign implies compressive force. The moment is mainly resulting from the eccentricity of the applied force. The radial force per unit length due to the moment can be considered as the bending stress multiplied with the thickness of the lug. Because of the moment, there are maximum and minimum forces at the edges of the lug. If W_r is compressive, then only the maximum compressive unit force needs to be evaluated. Otherwise, evaluations are required for both compression and tension. A compressive force has less allowable value.

As shown in Fig. 6.25(a), most of the loads are not acting normal to the pipe, nor are they acting at the center of the lug. The applied force has to be converted to a central radial force, W_r, and a moment equal to the force W multiplied with the arm of the eccentricity, e. That is, $M = W(e)$. Shear force at the lug-pipe interface is ignored.

Example

A 1000-lb load applied at the center of the outer edge of the lug with an angle of 50 deg. as shown in Fig. 6.25(a). (The force shown is a generic force acting off center of the lug's outer edge.) The tube has an OD of 4 in. and thickness of 0.300 in. The 0.300-in. thickness is the net thickness after subtracting corrosion and other allowances. The lug is 3 in. long, 2 in. high, and 0.5 in. thick. The pipe design pressure is 2000 psi, and the allowable stress is 15,000 psi.

To check the acceptability of the load, the load is first converted to standard force and moment as shown in Fig. 6.25(b), that is,

$$W_r = W \sin \alpha = 1000 \sin(50) = 766 \text{ lb}$$

$$M = We = Wh\cos \alpha = 1000 * 2 * \cos(50) = 1286 \text{ lb-in.}$$

Substituting the above to Eq. (6.43), we have applied force per unit length as

$$F = -\frac{766}{3} \pm \frac{6 \times 1286}{3^2}$$
$$= +602 \text{ lb/in. (tension)}$$
$$= -1112 \text{ lb/in. (compression)}$$

These forces shall be checked against the allowable. The first step to figuring out the allowable is to calculate the X factor, which is

$$X = D/t^2 = 4.0/0.3^2 = 44.44$$

From which the load factors are calculated from Eqs. (6.39) and (6.40) as

$$L_{f,c} = 0.03255; \text{ and } L_{f,t} = 0.04052$$

The pressure hoop stress due to 2000 psi design pressure is

$$S = \frac{PD}{2t} - yP = \frac{2000(4)}{2(0.3)} - 0.4(2000) = 13{,}333 - 800 = 12{,}533 \text{ psi}$$

The allowable lug bending stress, from Eq. (6.41), is

$$S_{b,\text{allow}} = 2.0 S_a - S = 2.0 \, (15{,}000) - 12{,}533 = 17{,}467 \text{ psi}$$

We also need a lug thickness correction factor to calculate the allowable loads. With 0.5-in. thickness, $\theta = \sin^{-1}(0.5/2) = 14.5$ deg. The K factor from Table 6.7 is 1.161. The allowable loads are calculated from Eq. (6.42) as

$$\text{Compression: } L_{a,c} = K(L_{f,c})S_{b,\text{allow}} = 1.161\,(0.03255)\,17{,}467 = 660 \text{ lb/in.}$$

$$\text{Tension: } L_{a,t} = K(L_{f,t})S_{b,\text{allow}} = 1.161\,(0.04052)\,17{,}467 = 821 \text{ lb/in.}$$

The allowable compression load of 660 lb/in. is smaller than the applied compression force of 1131 lb/in. Therefore, the 3-in.-long lug is not acceptable mainly due to the eccentric moment. To spread the load to the acceptable range, a 4.5-in.-long lug is required. This can be checked by substituting $\ell = 4.5$ to Eq. (6.43). Thus for $\ell = 4.5$, we have

$$\begin{aligned} F &= -\frac{766}{4.5} \pm \frac{6 \times 1286}{4.5^2} \\ &= 211 \text{ lb/in. (tension)} \\ &= -551 \text{ lb/in. (compression)} \end{aligned}$$

which is within the allowable range.

For a general piping system, in addition to pressure stress, there are other stresses beside the one due to lug loads. These stresses, such as thermal expansion and weight stresses, have to be evaluated in combination with the lug stress. Lug stresses, rather than allowable lug loads, are required for this combination. The shell bending stresses due to lug loads can be calculated from Eq. (6.42) as

$$S_{b,c} = \frac{F_c}{K(L_{f,c})} = \frac{551}{1.161(0.03255)} = 14{,}580 \text{ psi} \quad \text{due to compression load}$$

$$S_{b,t} = \frac{F_t}{K(L_{f,t})} = \frac{211}{1.161(0.04052)} = 4485 \text{ psi} \quad \text{due to tension load}$$

The greater of the two, that is, 14,580 psi, has to be used. If the lug load is cyclic, then this stress has to be added to the thermal expansion stress, due to through-run moment, in the secondary stress evaluation.

In addition to shell bending stress, there is also membrane stress due to direct tangential force. This membrane stress is roughly equal to $0.25F/t$, which is only about a few percent of the bending stress. Therefore, this membrane stress is ignored in the above application. It should be noted, however, that membrane stress is a load-carrying stress, which carries more burden than the bending stress. Because this membrane stress is tensile for a tension load, it should have been added directly to the internal pressure stress for the tension load. After adding this membrane stress, the allowable tension load will be reduced somewhat, whereas the allowable compression load will be increased somewhat. This membrane stress due to tangential force might be enough to blur the distinction between the allowable for the tensile and compression loads, so one uniform load curve may be used for both compression and tension loads.

6.8.2 Kellogg's Choking Model

In previous discussions, all support and lug design procedures use a circular ring model to simulate a section of the pipe. The ring model works well for an actual ring or a line load applied in the longitudinal direction. It does not fit very well on line loads applied in the circumferential direction. Instead of the ring model, Kellogg's approach uses a choking model based on the beam on elastic foundation principle discussed in Section 5.6. The choking model works better than the ring model, especially with trunnions, because it automatically includes the bridging effect of the surrounding pipe body. A trunnion, generally, is a welded attachment made from a piece of pipe.

Unlike a lug that takes only in-plane loading, a trunnion takes both in-plane and out-plane loads. With the choking model, the stresses created by the radial line load distributed in the circumferential direction were derived in Section 5.6. Equations (5.29) and (5.30) are rewritten as follows:

$$S_b = 1.167 \frac{\sqrt{R_m}}{t^{1.5}} f \quad \text{Longitudinal bending stress} \tag{6.44}$$

$$S_c = 0.643 \frac{\sqrt{R_m}}{t^{1.5}} f \quad \text{Circumferential membrane stress} \tag{6.45}$$

where f is the circumferential line load per unit length. From these equations, it is clear that there is a major difference between the choking model and the ring model. In the choking model, membrane stress is significant compared with bending stress. Membrane stress is negligible in the ring model.

Equations (6.44) and (6.45) are applicable to the line load, f, generated from either a rectangular lug or circular trunnion, although a correction factor may be required for certain orientations of the load and the lug. We will start the discussion with the trunnion.

In attachment stress calculations, it is a common practice to consider only the radial force, longitudinal moment, and circumferential moment. The shear forces and the torsional moment are ignored. The orientations of these forces and moments are given with respect to the axis of the trunnion as shown in Fig. 6.26.

The Kellogg approach simulates the trunnion load as a uniform circumferential radial line load whose magnitude per unit width is the same as the maximum trunnion edge reaction per unit edge width. The attachment load is assumed as line loads distributed along the outer envelope of the contact cross-section of the attachment. The stresses are readily calculated by the above equations once the maximum edge reaction, f, is obtained. This turns out to be not as simple as we would have wished. The varying elasticity characteristic around the pipe circumference makes the calculation very complex. The solution is to calculate the reaction based on a uniform flat foundation and apply a correction factor afterward, if necessary. With a uniform flat foundation, the edge reaction to an axial load is uniformly distributed. In this case, the nominal edge reaction equals the applied force divided by the trunnion circumferential perimeter. For a moment load, the edge reaction is linearly proportional to the distance from the centerline as in a bending stress, and the maximum reaction occurs at extreme

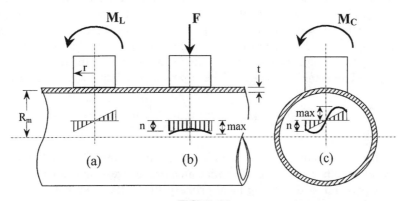

FIG. 6.26
TRUNNION REACTION FORCE DISTRIBUTION

apexes of the trunnion. That is, with a uniform flat foundation, we have the nominal reaction for moment and force as

$$n_1 = \frac{M}{\pi r^2} \qquad (6.46)$$

$$n_2 = \frac{F}{2\pi r} \qquad (6.47)$$

where r is the outside radius of the trunnion. These edge reaction forces can be considered as the stresses multiplied with the wall thickness of a very thin cylindrical shell.

Since the pipe is not a uniform flat foundation, the actual edge reaction is somewhat different from the nominal reaction given by the above equations. Looking from the axial direction of the pipe, the pipe has a ridge on the top that is the most rigid portion in the pipe cross-section. Being the stiffest, this ridge will absorb the greatest amount of force from the trunnion. Figure 6.26 shows the likely distribution of the edge reactions for various loads.

For the longitudinal moment, M_L, the maximum nominal reaction occurs at the ridge, the high force point. Therefore, we can consider the actual distribution of the reaction to be pretty much the same as the nominal reaction distribution. The maximum edge force to be used in the stress calculation is the same force as calculated by Eq. (6.46) without any adjustment.

For the radial load, F, the actual reaction has the maximum value at the ridge, making the distribution non-uniform. The maximum reaction is, therefore, greater than the average nominal value. The maximum edge reaction force to be used in the stress calculation is the nominal value calculated by Eq. (6.47) multiplied with a factor of 1.5.

For the circumferential moment, M_C, the actual maximum edge reaction is shifted away from the extreme apex point. This shifting greatly reduces the effective section modulus and increases the bending stress by several times over the nominal stress for some cases. However, in piping design, with the diameter ratio of the pipe and the trunnion ranging from 2 to 4, the actual maximum edge reaction force is taken as the nominal reaction force calculated by Eq. (6.46) multiplied with a factor of 1.5. This 1.5 factor can be considered as the ratio of the SIFs between out-plane and in-plane bending moments. Judging from the ASME B31 code stipulation that the in-plane SIF of a branch connection equals three-quarters of the out-plane SIF plus 0.25, a reverse out-plane and in-plane SIF ratio may be established for trunnion application. By rearranging the code formula, we have an out-plane SIF of $i_o = (i_i - 0.25)/0.75 < 1.33 i_i$. This shows that the 1.5 ratio suggested above appears to be adequate. i_i and i_o are in-plane and out-plane SIF, respectively. Of course, the rearrangement of code SIF relation is not exactly valid as extra conservatism may have been included in the in-plane SIF. Nevertheless, the use of 1.5 as the adjustment factor on the circumferential moment is generally considered not too far off from reality.

Membrane and bending stresses are calculated by substituting the maximum edge reaction force to Eqs. (6.44) and (6.45). The combined effect of these three loads can be calculated by combining either the edge reaction forces or the stresses. It appears that the combination of the edge forces is easier. Because the edge force created by the radial load distributes uniformly all around the trunnion circumference, it shall be added directly to the edge force resulting from the moment. For the two bending moments, the resultant of the adjusted moments can be used. The combined maximum edge force becomes

$$f = \frac{1}{\pi r^2}\sqrt{(M_L)^2 + (1.5 M_C)^2} + \frac{1}{2\pi r}|1.5 F| \qquad (6.48)$$

The absolute value of the radial force is used in the combination. The combined maximum edge force is then used to calculate the circumferential membrane stress and longitudinal shell bending stress. Membrane stress is a load carrying local primary stress. Its combination with pressure hoop stress is limited to 1.5 times the allowable stress of the pipe material at design temperature. The longitudinal shell bending stress is a secondary stress, which has an allowable value as given by Eq. (4.24).

By conservatively assuming that the sustained longitudinal stress takes up $1.0 S_h$, the allowable secondary stress becomes $1.25 S_c + 0.25 S_h$, where S_c is the allowable stress at ambient temperature (cold) and S_h is the allowable stress at operating temperature (hot). In this secondary stress evaluation, the load range is calculated using the cold modulus of elasticity.

It may be necessary to use separate edge forces to calculate membrane and bending stresses. For membrane, only the loads that have a sustained nature need to be included. These include weight, pressure, and possibly hot thermal reaction, but not the thermal reaction range. Thermal reaction has some sustained nature when dealing with areas of concentrated strain. On the other hand, secondary stress requires the inclusion of the thermal reaction range plus the stresses due to pressure and, possibly, the live weight variation.

When stresses at the connection are excessive, a pad can be used to reduce the stress. With a sufficient pad width to spread the load, the effective thickness for stress calculations is the sum of the pipe thickness and the pad thickness. As discussed in Section 5.5, the required width of the pad to spread the load is $2.0\sqrt{R_m t}$, where R_m is the mean radius of the pipe, not of the trunnion, and t is the effective thickness.

The trunnion connection can also be evaluated by treating the trunnion as a branch pipe with a fabricated branch connection in a routine piping stress analysis. This is conservative for the moment loading because the trunnion connection is stronger than an actual branch connection due to the un-perforated pipe shell. However, it becomes non-conservative for the radial load, because the code does not have an SIF for the radial load. In fact, most piping codes completely ignore the radial load, which is considered as a local axial force. Because axial load is often the main load of a support trunnion, the use of a fabricated branch connection approach is not recommended.

Example

A 10-in. trunnion (10.75 in., 273 mm OD), on a 24-in. (610 mm OD) standard weight pipe (0.375 in., 9.53 mm thick), has an internal pressure of 400 psi (2.76 MPa), axial load of 3000 lb (13,344 N), longitudinal moment of 12,000 lb-in (1,355,811 N mm), circumferential bending moment of 8000 lb-in. (903,874 N mm). Design temperature is 700°F (371°C). The allowable stress, S_c, is 15,000 psi (103.4 MPa) and S_h is 14,400 psi (99.29 MPa). A corrosion allowance of 0.05 in. (1.27 mm) is included in the primary stress calculation.

The maximum edge reaction is calculated by Eq. (6.48) as

$$f = \frac{1}{\pi(5.375)^2}\sqrt{12{,}000^2 + (1.5\times 8000)^2} + \frac{1.5}{2\pi(5.375)}3000 = 320 \text{ lb/in. } (56.0 \text{ N/mm})$$

The pressure hoop stress is

$$S_p = \frac{12\times 400}{0.375 - 0.05} = 12{,}973 \text{ psi } (89.4 \text{ MPa})$$

The circumferential membrane stress due to the trunnion load is

$$S_{c,m} = 0.643 \frac{\sqrt{12}}{(0.375 - 0.05)^{1.5}} 320 = 3847 \text{ psi } (26.52 \text{ MPa})$$

which is within the allowable value of $1.5 S_h - S_p$. Note that subscript m is added besides subscript c to distinguish this stress from the cold allowable stress S_c.

The secondary bending stress is

$$S_b = 1.167 \frac{\sqrt{12}}{(0.375)^{1.5}} 320 = 5633 \text{ psi } (38.84 \text{ MPa})$$

which is within the allowable value of $(1.25 S_c + 0.25 S_h)$.

Occasionally, it is desired to calculate the total primary plus secondary stress intensity due to external loads. This requires the combination of bending stress with membrane stress, which is oriented in the perpendicular direction. Because bending stress also generates Poisson stress at a perpendicular direction with the same sign, this Poisson stress is subtracted from bending stress before being combined with membrane stress (see also Section 5.7.1). That is, the maximum primary plus secondary stress intensity is 5633 (1 − 0.3) + 3847 = 7790 psi (53.71 MPa). This does not include the stress from other loads such as pressure, etc.

In this example, the three loads are chosen in such proportions that their maximum edge loads, after applying the 1.5 adjustment factor for axial force and circumferential moment as in Eq. (6.48), are similar at roughly 133 lb/in. for each load. This helps the comparison between the choking model results and the results given by WRC-107, which will be discussed next.

6.8.3 WRC-107 Stress Evaluation

WRC-107 [22] is a very widely used "cookbook"-style technical guide for calculating local stresses in vessel shells due to external loadings. It is based on theoretical investigations conducted primarily by Prof. P. P. Bijlaard and has become an indispensable tool in the design of pipe and vessel attachments, and also nozzle connections.

WRC-107 is invaluable due to its combined consideration of theory, experimental data, and engineering judgment. In some instances, the theoretical values have been adjusted several hundred percent upward to match the available experimental results. The bulletin assembles 40 data figures, each containing up to a dozen curves. Therefore, it is not possible to present enough information in this book to allow the direct application of the bulletin. In the discussions that follow, only commentary notes regarding the general application of the bulletin will be presented.

Updates and corrections. WRC-107 was first published in 1965, and has later gone through a few updates and corrections. Because stresses are mostly calculated by using a computer program, it is important that the program used covers the latest revision.

Stress locations. The bulletin calculates stresses at four major corners around the attachment. They are labeled as A, B, C, D, as shown in Fig. 6.27. Each point also associates with outside and inside surfaces. They are labeled with upper (U) and lower (L) subscripts. At point C, the outside surface is labeled C_U, and the inside surface is labeled C_L, and so forth. Therefore, a total of eight locations are

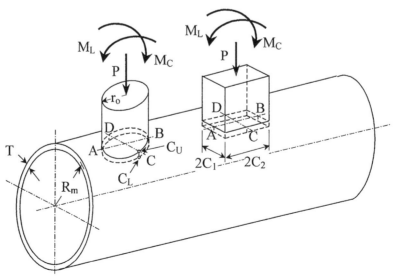

FIG. 6.27
WRC-107 STRESS LOCATIONS

evaluated. The stresses calculated for these locations, however, also contain the stress at off-axis location when the off-axis stress is higher than the stress at the axis.

Combining the overturning moments. The three major loads, which generate significant local stresses, are the axial (radial) force, P, and two mutually perpendicular bending moments, M_L and M_C. The axial force generates stresses at all eight benchmark locations. However, the overturning moments generate stresses only at four locations on two extreme points. For instance, the longitudinal moment, M_L, produces stresses only at points A and B located on the longitudinal axis, and the circumferential moment, M_C, produces stresses only at points C and D located on the traverse axis. Therefore, the stress due to axial force is combined with the stresses due to moments, but the stresses due to the two moments are not combined with each other. This is un-conservative when both longitudinal and circumferential moments are significant [25].

WRC-107 has suggested combining the two orthogonal moments into a resultant moment in calculating the stresses for spherical vessels. This is possible due to the all-around symmetric nature of the spherical vessel. Even with spherical vessels, because the layout of the calculation form suggests the separation of the moments, the combined effects are often overlooked. For cylindrical vessels, because of the unique characteristics of each orientation of the moment, the combination needs to be done at the local stress level. Without this combination, the stress calculated may be as much as 40% un-conservative. Some type of combination should be provided.

Membrane and secondary bending stresses. The calculation forms outlined by WRC-107 calculate the total membrane plus bending stresses. In actual applications, membrane stress has to be separated for local primary stress evaluation. The separation can be readily done with the stresses tabulated in the form, but is often not provided by some computer programs. When internal pressure is included, the stresses calculated with the form are primary plus secondary stresses. Stress concentration factors for the fillets and welds are applied when peak stresses are to be calculated.

Example

As a comparison with Kellogg's choking model, the example calculation given in the choking model is recalculated with WRC-107. The results as given by the PENG-LOCALS [26] computer program are shown in Table 6.8. In the choking model example, the three main loads, P, M_L, and M_C are so chosen that they contribute roughly the same proportion of stress. By comparing Table 6.8 results against the stresses calculated by the choking model, we know that the overall combined membrane stresses are very consistent with each other (3847 versus 3290), but the combined WRC-107 bending stress is much higher (7790 by the choking model versus 14,650 by WRC-107) partly due to the 7086-psi fictitious stress to be explained later.

With the edge force of 133 lb/in. for each component of P, M_L, and M_C loads, the choking model predicts a membrane stress of 1600 psi and a bending stress of 2341 psi for each component of loads. The membrane stresses from WRC-107 are: 946 psi to 2054 psi for P, 608 psi to 1078 psi for M_L, and 402 psi to 1210 psi for M_C. They are fairly consistent with each other. The bending stresses from WRC-107 are: 1522 to 7086 in the circumferential direction for P, 3469 to 3537 in the longitudinal direction for P, 1353 to 2687 for M_L, and 1990 to 4772 for M_C. Besides the 7086-psi stress in the circumferential direction due to P, the other bending stresses are fairly consistent with the choking model, although a 2.0 adjustment factor, instead of 1.5, for P and M_C will produce a better comparison in this case.

The 7086-psi bending stress appears to be a fictitious one due to the lack of WRC-107 data in the region. This stress is calculated by Fig. 1C of WRC-107, which is schematically expressed as Fig. 6.28 in this book. Due to scattering and divergence of the experimental data in the zone with $(r_o/R_m)\sqrt{2R_m/T}$ above 2.0, WRC-107 has the curves completely deleted in this zone. However, without the curves in this zone, the applicable scope of the bulletin is considerably reduced. Therefore, most calculations and computer software packages conservatively use the stress value at $(r_o/R_m)\sqrt{2R_m/T} = 2.0$ for the entire zone. This, in effect, creates a black hole in this zone, where once in this zone the stress will not reduce at all no matter how large the attachment size is increased. A similar situation occurs at Fig. 2C also. This is a big concern of engineers using WRC-107.

TABLE 6.8
WRC-107 STRESS REPORT EXAMPLE

```
PENG.LOCALS- LOCAL STRESSES IN SHELLS (WRC-107,1981) - PENG ENGINEERING, TEXAS

PIPE STRESS ENGINEERING,  Trunnion Example

COMPUTATION SHEET FOR LOCAL STRESS IN CYLINDRICAL SHELL (V-7.0) DATE:00 JAN 06
-------------------------------------------------------------------------------
1.APPLIED LOAD (LB, FT-LB)              3.GEOMETRIC PARAMETERS
    RADIAL FORCE,    P =    3000.0
    CIRCMF. MOMENT, MC=     667.0          GAMMA= 31.50
    LONGIT. MOMENT, ML=    1000.0
    TORSION MOMENT, MT=        .0          BETA=  .398
    CIRC SHEAR FORCE,VC=       .0
    LONG SHEAR FORCE,VL=       .0
2.GEOMETRY (IN) ------------------      4.STRESS CONCENTRATION FACTORS -------
    VESSEL THICKNESS,   T=     .38          DUE TO:
    ATTACHMT O. RADIUS, SRO=  5.38             MEMBRANE LOAD, KN= 1.00
    VESSEL MEAN RADIUS, RM=  11.81             BENDING  LOAD, KB= 1.00
    ATTACHMT CIR WIDTH, 2C1=   .00
    ATTACHMT LON WIDTH, 2C2=   .00      5.ATTACHMT TYPE: ROUND
6.INTERNAL PRESSURE,(PSI) =    .00
-------------------------------------------------------------------------------
----CURVE----   STRESS   -------------------- STRESS (PSI) ---------------------
---READING---  CATEGORY    AU     AL     BU     BL     CU     CL     DU     DL
-------------------------------------------------------------------------------
NO*RM/P= 3.033 S(NO)P    -2054  -2054  -2054  -2054   -946   -946   -946   -946
MO/P   =  .012 S(MO)P    -1522   1522  -1522   1522  -7086   7086  -7086   7086
NO*B/MC= 1.046 S(NO)MC       0      0      0      0   -402   -402    402    402
MO*A/MC=  .066 S(MO)MC       0      0      0      0  -4772   4772   4772  -4772
NO*B/ML= 1.873 S(NO)ML   -1078  -1078   1078   1078      0      0      0      0
MO*A/ML=  .012 S(MO)ML   -1353   1353   1353  -1353      0      0      0      0
STRESS DUE TO PRESSURE       0      0      0      0      0      0      0      0
-------------------------------------------------------------------------------
SUM CIRCMF STRESS SO:    -6006   -256  -1144   -806 -13205  10510  -2857   1770
-------------------------------------------------------------------------------
NX*RM/P= 1.397 S(NX)P     -946   -946   -946   -946  -2054  -2054  -2054  -2054
MX/P   =  .027 S(MX)P    -3469   3469  -3469   3469  -3537   3537  -3537   3537
NX*B/MC= 3.152 S(NX)MC       0      0      0      0  -1210  -1210   1210   1210
MX*A/MC=  .027 S(MX)MC       0      0      0      0  -1990   1990   1990  -1990
NX*B/ML= 1.056 S(NX)ML    -608   -608    608    608      0      0      0      0
MX*A/ML=  .025 S(MX)ML   -2687   2687   2687  -2687      0      0      0      0
STRESS DUE TO PRESSURE       0      0      0      0      0      0      0      0
-------------------------------------------------------------------------------
SUM LONGIT STRESS SX:    -7709   4602  -1119    444  -8790   2263  -2390    703
-------------------------------------------------------------------------------
SHEAR STRESS DUE TO MT       0      0      0      0      0      0      0      0
SHEAR STRESS DUE TO VC       0      0      0      0      0      0      0      0
SHEAR STRESS DUE TO VL       0      0      0      0      0      0      0      0
-------------------------------------------------------------------------------
SUM  SHEAR STRESS SS:        0      0      0      0      0      0      0      0
-------------------------------------------------------------------------------
COMBINED STRESS/ MEMBR    3132   3132    976    976   3264   3264    844    844
INTENSITY,  S  / TOTAL    7709   4858   1144   1250  13205  10510   2857   1770
-------------------------------------------------------------------------------
MAXIMUM COMBINED STRESS INTENSITIES (PSI) :     (+)-O  (+)-I   (-)-O   (-)-I
A. MEMBRANE ----------------------------------   3290   3290     581     581
B. TOTAL SKIN --------------------------------  14652   9017    2431    1814
-------------------------------------------------------------------------------
NOTES: 1. A=RM*BETA,  B=RM*RM*BETA, O=PHI - SIGN CONVENTION FOLLOWS WRC-107
       2. S(Q)F IS THE STRESS IN Q-DIRECTION DUE TO F-LOAD
       3. NO*RM/P= 1.397,  MO/P=  .055 AT POINTS C AND D
          NX*RM/P= 3.033,  MX/P=  .028 AT POINTS C AND D
```

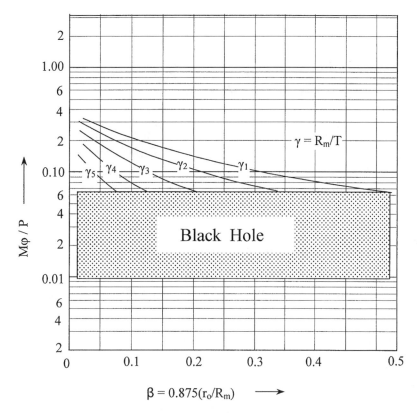

FIG. 6.28
SCHEMATIC CHART FOR CIRCUMFERENTIAL BENDING STRESS AT TRAVERSE AXIS, DUE TO RADIAL LOAD (BASED ON WRC-107, FIG. 1C)

6.9 TREATMENT OF SUPPORT STIFFNESS AND DISPLACEMENT

In a process or power plant, the sizes of structural members, such as columns and beams, are comparable to, or smaller than, the sizes of many of the pipes. The effectiveness of the structure members as pipe supports can be an issue on the integrity of the piping system. Sometimes it is even hard to say which is supporting which. Therefore, in the analysis of a large-diameter piping system, it is important to include the stiffness of the support structure. The stiffness of the support structure can be included either by modeling the structure directly with the piping, or by considering the structural stiffness as the support spring rate. In the latter case, the spring rate has to be estimated first using whatever methods available. The approach of using the support spring rate is more popular than the direct modeling approach.

Most support structures also experience various movements, including thermal expansion, earthquake/wind displacement, and so forth. These movements impose loads on the piping and have to be included in the analysis. When the support is rigid, the support structure movement will directly cause the pipe to move the same amount as the support. However, when the support structure has significant flexibility, the movement at the pipe is generally different from that of the support structure as shown in Fig. 6.29. Pipe movement is determined by the support force and the support spring rate. It is not known before the analysis is performed; therefore, it is not part of the input data. The data input is the support structure movement. It should be noted that the pipe may also move in the opposite direction of the support movement if the structure is flexible.

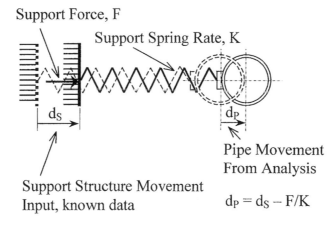

FIG. 6.29
SUPPORT AND PIPE MOVEMENTS

One puzzling effect of including the support stiffness is the settlement displacement due to weight. The weight load will cause the pipe to move downward on flexible supports. However, as the pipe is installed and adjusted with zero initial displacement, the movements that might show on the computer output are just paper movements and should be ignored.

REFERENCES

[1] M. W. Kellogg Company, 1956, *Design of Piping Systems*, revised 2nd edition, Chapter 8, John Wiley & Sons, Inc., New York.

[2] ASME B31.1, *Power Piping*, 2001 Edition, American Society of Mechanical Engineers, New York.

[3] European Standard EN-13480-4, 2002, *Metallic Industrial Piping — Part 3: Design and Calculation*, European Committee for Standardization.

[4] Peng, L. C., 2006, "Stress Analysis for Piping Systems Resting on Supports," *Chemical Engineering*, 113(2), pp. 48–51.

[5] MSS SP-58, 1975, *Pipe Hangers and Supports — Materials, Design, and Manufacture*, Manufacturers Standardization Society of the Valves and Fittings Industry, Arlington, VA.

[6] MSS SP-69, 1966, *Pipe Hangers and Supports — Selection and Application*, Manufacturers Standardization Society of Valves and Fittings Industry, Arlington, VA.

[7] Grinnell Corporation, 1990, *Grinnell Pipe Hangers*, Catalog PH-90, Grinnell Corporation, Pipe Support Division, Cranston, RI.

[8] Peng, L. C., 1989, "Treatment of Support Friction in Pipe Stress Analysis," *PVP-Vol.169 Design and Analysis of Piping and Components*, Book No. H00484, ASME, New York.

[9] API Standard 617, *Centrifugal Compressors for General Refinery Service*, American Petroleum Institute, Washington, D. C.

[10] Den Hartog, J. P. 1952, *Advanced Strength of Materials*, McGraw-Hill Book Co., New York. (discussed the principle in several topics).

[11] Young, W. C., 1989, *Roark's Formulas for Stress and Strain*, 6th ed., McGraw-Hill Book Co., New York.

[12] AWWA 1989, *Steel Water Pipe — A Guide for Design and Installation (M11)*, 4th ed., American Water Works Association.

[13] Evces, C. R. and O'Brien, J. M., 1984, "Stresses in Saddle Supported Ductile Iron Pipe," *Journal of the AWWA*, 79(11), pp. 49–54.

[14] Bonds, R. W., 1995, *Design of Ductile Iron Pipe on Supports*, Ductile Iron Pipe Research Association, Birmingham, AL.
[15] Stokes, R. D., 1965, "Stresses in Steel Pipelines at Saddle Support," *Civil Engineering Transactions*, October 1965, The Institution of Engineers, Australia. (Discussed in *Tubular Steel Structures — Theory and Design*, Troitsky, M. S., 1982, The James F. Lincoln Arc Welding Foundation, Cleveland, OH.)
[16] Zick, L. P., 1951, "Stresses in Large Horizontal Cylindrical Pressure Vessels on Two Saddle Supports," *Welding Journal Research Supplement*, 30(9), pp. 435–445, and revision of January 1971.
[17] Schorer, H., 1933, "Design of Large Pipelines," *Transaction of the ASCE*, 98, pp. 101–119.
[18] ASME, 1989, "Boiler and Pressure Vessel Codes, Section VIII," *Rules for Construction of Pressure Vessels*, Division 1, American Society of Mechanical Engineers, New York.
[19] British Standard BS-5500, *Specification for Fusion Welded Pressure Vessels (Advanced Design and Construction) for Use in the Chemical, Petroleum, and Allied Industries*, British Standard Institute, UK.
[20] ASME, 1992, "Boiler and Pressure Vessel Codes, Section I," *Power Boilers*, ASME, New York.
[21] Bernstein, M. D. and Yoder, L. W., 1998, *Power Boilers — A Guide to Section I of ASME Boiler and Pressure Vessel Code*, ASME Press, New York.
[22] Wichman, K. R., Hopper, A. G. and Mershon, J. L., 1979, "Local Stresses in Spherical and Cylindrical Shells due to External Loadings," *Welding Research Council (WRC) Bulletin 107*, 1981 Reprint, The Welding Research Council, New York.
[23] Dodge, A. G., 1974, "Secondary Stresses Indices for Integral Structural Attachment to Straight Pipe," *Welding Research Council (WRC) Bulletin 198, Part 1*, The Welding Research Council, New York.
[24] Rodabaugh, E. C., Dodge, W. G., and Moore, S. E., 1974, "Stress Indices at Lug Supports on Piping Systems," *WRC-198, Part2*, The Welding Research Council, New York.
[25] Peng, L. C., 1988, "Local Stresses in Vessels — Notes on the Application of WRC-107 and WRC-297," *Transaction of ASME, Journal of Pressure Vessel Technology*, 110, pp. 106–109.
[26] Peng Engineering, *PENG.LOCALS, User's Manual*, Peng Engineering, Houston, TX.

CHAPTER 7

FLEXIBLE CONNECTIONS

In Chapter 3, we discussed thermal expansion and piping flexibility. A piping system has to be flexible enough to absorb the thermal expansion displacement, without creating unacceptable stresses in the pipe or excessive reaction loads in the connecting equipment. The piping may have enough flexibility just from the turns and offsets created by the natural layout. Additional loops and offsets can be provided if the flexibility from the natural layout is not enough. There are circumstances, however, when flexible connections are needed for economic or practical reasons.

Aside from several concerns, which will be discussed later, the flexible joint is an easy solution to the piping flexibility problem. This can be demonstrated by the simple system shown in Fig. 7.1. The system is an L-shaped 8-in. (200 mm), schedule-40 (8.18 mm thick) piping operating at 600°F (316°C) from an ambient installation condition of 70°F (21°C). The system, with its natural layout, generates an anchor force of 1135 lb, an anchor moment of 13,381 lb-ft, and an elbow stress of 14,300 psi. The stress is well within the code allowable range, but the anchor loading, especially the moment, is too high for most connecting equipment. A complex loop with cleverly placed restraints would be required to reduce the anchor load to an acceptable level. The problem, on the other hand, can be easily solved with flexible joints.

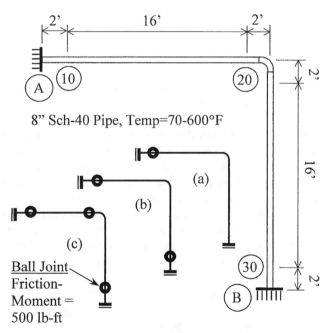

FIG. 7.1
ARTICULATION OF PIPING SYSTEM

209

TABLE 7.1
EFFECTS OF FLEXIBLE CONNECTIONS (SEE FIGURE 7.1 FOR ARTICULATION SCHEMES)

Articulation Schemes	Anchor Force, Same at A/B		Anchor A Moment, lbs-ft	Anchor B Moment, lbs-ft	Stress at Bend, psi
	Horz.-Force, lbs	Vert.-Force, lbs			
None	1135	1135	13381	13381	14300
(a)	1010	455	1413	12523	12600
(b)	373	373	1250	1250	10200
(c)	62	62	624	624	1000

Let us say point A is an equipment connection that cannot take the loading from the piping as laid out. In this case, a ball joint can be placed at point 10 near the anchor to effectively remove the moment. In a two-dimensional planar system like this one, a hinge joint can be used instead of the ball joint. However, ball joints are used as examples in this and subsequent discussions on this system. The data given below is obtained from analyses made with a computer program. Scheme (a) is the situation with one ball joint located at point 10. The moment at anchor A has been reduced to 1413 lb-ft. This is not exactly zero, due to the 500 lb-ft friction resistance on the joint and 455 lb of vertical force acting with a 2-ft moment arm from the anchor to the joint. If the load at anchor B is likewise too high, then a second joint can be placed at point 30 near the anchor, as shown in (b). Ultimately, three joints can be placed in a plane creating a complete flexible system as shown in (c). Results of these schemes are tabulated in Table 7.1 for comparison. Theoretically, if the joints were frictionless, then there would be no force, moment, and stress in a three-joint system. The small forces, moments, and stresses are all due to the sliding friction of the ball joints.

Problems with flexible joints. Flexible joints appear to be efficient, simple, and clean devices for reducing pipe stress and equipment load due to thermal expansion, yet they are used very sparingly in actual constructions. In fact, many design specifications stipulate that an expansion or flexible joint should not be used unless approved beforehand by the owner of the plant or its representatives. Some of the reasons are as follows:

- Flexible joints are specialty items requiring special engineering. Special installation procedures are also needed.
- An extra item besides the piping proper means extra cost, although this may not be true after taking into account the extra cost of the expansion loop system that would have been required otherwise. An accurate cost comparison is difficult to come by without detailed analysis, which in itself can be costly.
- Most of the flexible joints need periodic inspection and maintenance. Because the joints are not running, they do not generate noise, vibration, and other vital signals. Due to this quietness, the required periodic maintenance activities are easily forgotten by the plant personnel.
- It requires expertise in engineering to accurately estimate the effect. It also requires ingenuity in managing the pressure thrust force for some types of joints.
- It is difficult to provide insulation covering to accommodate the flexing requirement.
- A horizontal joint tends to trap liquid and sediment creating a very corrosive environment.
- The joints, especially the bellow type, are normally stressed beyond the yield point to provide the flexing capability. The thin highly stressed parts are susceptible to corrosion damage and stress corrosion cracking.
- Potential sticking or packing degradation in slip type joints may lead to loss of flexing capability or leakage.
- Special supports, guides, and anchors are normally needed.
- The fact that the flexible joint poses a weaker link in the system is enough to discourage most decision makers.

Situations that favor flexible joint. As much as we dislike the problems associated with a flexible joint, there are occasions when the flexible joint is not only cost-effective but also a necessity. In large-diameter high-temperature piping, for instance, a workable layout is very expensive and difficult to come by without using some expansion joints. The following are some areas that favor the use of flexible joints:

- Available space is not enough to provide a conventional expansion loop. The connection between two pieces of nearby equipment is one example of this situation.
- Equipment allowable load is so low that it is impossible to meet no matter how large and extensive is the expansion loop that is provided. Most of the exhaust piping of a steam turbine drive belongs to this category.
- Large high-temperature, low-pressure piping would have required a huge expansion loop that simply would have been too expensive and occupy too much space to install. The flue gas line from a heater or from a cat-cracking unit of an oil refinery is an example of this case.
- A vacuum line that cannot tolerate the pressure drop from the loop. The condensate line from the power plant steam condenser is a typical example.
- The process requires smooth straight flow, such as the catalytic cracking riser in an oil refinery.
- To isolate vibration from a process machine to the piping and vice versa.
- For the fit-up of a large-diameter piping to a machinery. This is for construction and maintenance use. It does not serve as an expansion stress reduction device.

7.1 BASIC FLEXIBLE JOINT ELEMENTS AND ANALYTICAL TOOLS

Similar to the situation of other standard catalog components, such as valves and pumps, the stress concerns of the flexible joints proper are handled mainly by the manufacturers. The bellow elements and joint assemblies, for instance, are designed and manufactured by the manufacturers based on applicable standards [1] and codes [2, 3]. The standard generally used is the *Standards of the Expansion Joint Manufacturers Association* commonly referred to as the EJMA standard. The duty of piping engineers is to ensure that the joint is fit for the application, and the interface of the piping satisfies the manufacturer's allowable load and deformation requirements. The interface between piping and the flexible joint is somewhat more complicated than the interface between piping and other components. Because the characteristics of the flexible joint dictate the piping load and stress, they have to be included in the stress analysis of the piping system. However, when handled properly, the flexible joint seldom poses any allowable load difficulty to the piping. This is very much unlike the case of rotating equipment, whose allowable load always poses a problem for the piping. In the following, the basic flexible joint elements and their analytical tools are discussed.

7.1.1 Generic Flexible Connections

A general or generic flexible connection represents any concentrated additional flexibility that exists at a certain point of the piping system. In analytical terms, each point of the piping is considered as the junction of two pipe beam elements. The two end points that join together will always move in unison with equal displacements and rotations. Occasionally, there is a device inserted between these two ends to provide a special added flexing capability between them. This special device can be located geometrically at a point or included as a short length. In any case, the added flexibility can be simulated as concentrating at the center point of the device.

Figure 7.2 shows the schematic function of the flexible joint. The joint is considered as located at a small zone centered at point N. Mathematically, point N is separated into two end points, N_A and N_B. Although these two end points are located at the same geometrical space, they are two independent

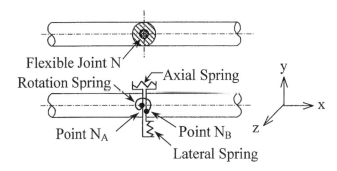

FIG. 7.2
GENERAL FLEXIBLE JOINT

points each with six degrees of freedom. The relative movements between these two points are defined by the interacting characteristics of the flexible joint. In the axial direction, the interaction is characterized as an axial translational spring with a spring rate of K_X, and a twisting rotational spring with a spring rate of K_{RX}. For the lateral directions, there are two mutually orthogonal axes perpendicular to the pipe axis. Each of these two axes also has translational and rotational springs. Overall, there are six characteristic springs at a general flexible joint. To simplify the discussions that follow, a set of working keywords will be used to represent the characteristics of the joint. The format of the keywords is FT* for Flexible joint in Translation in *-axis, and FR* for Flexible joint in Rotation around *-axis. Each joint has six spring components as follows.

$FTX = K_X$; $FTY = K_Y$; $FTZ = K_Z$. These are translational flexible springs in the x, y, and z directions, respectively. K_X, K_Y, and K_Z are the corresponding spring rates (typically in lb/in. or N/mm). If any of the springs is present, the movements at point N_A will be different from that of point N_B. The differential magnitudes depend on the internal forces and the spring rates. Take the x direction, for instance; here, the internal x force equals the differential x displacements between N_A and N_B, multiplied by the x direction spring rate K_X. That is, $F_x = (D_{xB} - D_{xA})K_X$, and so forth. The flexible joint spring should not be confused with the support spring, which generates an external support force due to pipe movement relative to the support structure.

$FRX = K_{RX}$; $FRY = K_{RY}$; $FRZ = K_{RZ}$. These are rotational flexible springs about the x, y, and z directions, respectively. K_{RX}, K_{RY}, and K_{RZ} are the corresponding spring rates (typically in lb-in/rad or N-mm/rad). These springs will allow differential rotations between points N_A and N_B. The differential rotation depends on the spring rate and the internal moment about that particular direction interested. Take the axial direction, for instance, where the internal twisting torsional moment equals the differential rotation of the two end points multiplied by x-rotational spring rate. That is, $M_X = (R_{xB} - R_{xA})K_{RX}$. Here, the x axis is assumed to lie in the axial direction of the given pipe segment.

The above keywords are only applicable to the joints aligned with the common global coordinate axes of the system. They are not directly applicable for joints aligned in a skewed direction, in which case a set of orientation parameters is also needed. Because it is quite complex to furnish the orientation parameters correctly, another set of more specialized keywords are preferred. They are called local coordinate flexible joints, defined as FTA, FTV, FTH, FRA, FRV, and FRH, which are specified with corresponding spring rates. FTA is a translational flexible spring in the axial direction; FTV is a translational flexible spring in the vertical plane perpendicular to the pipe axis; and FTH is the translational flexible spring in the horizontally oriented direction perpendicular to both FTA and FTV. FRA, FRV, and FRH are defined in the same manner for rotational flexible springs. This set of local coordinate flexible joints requires clearly defined pipe directions.

7.1.2 Bellow Elements

Most flexible joints are constructed with bellows serving as pressure containing flexible elements. Bellow elements are generally constructed either with metal or with synthetic rubber. The following discussions focus mainly on the metallic bellow elements. For rubber-type bellows, the manufacturer's data sheet should be consulted. Furthermore, because this book deals mainly with the design and analysis of a piping system, only the general bellow behaviors related to piping applications will be discussed. With that in mind, the most common double-U type metal bellow is used as the typical example.

Deformation of individual convolution. The most fundamental characteristics of the bellow element are the force-deformation relationships of the individual convolution. These fundamental characteristics serve as the basis for the engineering of the bellow flexible joints. The design of the bellow is similar to the design of the piping system. In piping, the pipe has to be thick enough and stiff enough to resist the pressure and occasional load, but also has to be flexible enough to absorb the thermal expansion. Similarly, the bellow also has to be thick enough to resist the internal pressure, yet thin enough to absorb the deformation without generating excessive forces and stresses. These mutually conflicting requirements pose a great challenge to the design of the bellow. It has to be thick, but not too thick.

The membrane stress due to pressure in the bellow is limited to the basic allowable stress of the material at the design temperature. This primary membrane stress is always less than the yield strength of the material. This is the universal rule in the design of piping components. However, to provide the flexing capability, the bellow is generally stressed beyond the yield strength of the material. Because flexing capability is provided for the displacement, its corresponding stress is self-limiting. Therefore, the flexing stress at the bellow is evaluated by the fatigue criterion. For economical and practical reasons, manufacturers often rate their commercial bellows to a safe operating limit of 3000 cycles. This can differ from manufacturer to manufacturer and is generally identified in the catalog.

The fatigue of the bellow depends on the deformation of the convolution. A bellow is generally an assembly of multiple convolutions. The deformation at each convolution in a bellow is different. The fatigue life of the bellow is determined by the most severely deformed individual convolution. Therefore, finding and accessing the most severely deformed, in terms of overall combined effect, individual convolution is the main engineering procedure when it comes to bellow applications.

The deformation of the individual bellow can be classified into two categories. One is from the axial force and the other from the bending moment. Figure 7.3 shows the relationship of these two deformations. Figure 7.3(a) shows the deformation due to an axial force applied at the centerline of the bellow element. If the force is not applied at the centerline, it needs to be decomposed into one at the centerline combined with a bending moment. With this axial force, the convolution deforms uniformly around the entire circumference of the convolution. The deformation is expressed as

$$e_x = q_a - q \quad \text{(Negative is compression)} \tag{7.1}$$

This axial deformation per convolution is the most fundamental characteristic quantity of the bellow. A bellow is designed based on a certain uniform axial deformation per convolution. Once the bellow is so designed, no deformation at any point along the entire circumference of any convolution of the bellow can exceed this design axial deformation per convolution. In other words, if the allowable deformation per convolution is set as $e_{x,\text{allow}}$, then no deformation, uniform or otherwise, at any point on any convolution should exceed this limit.

The deformation due to bending follows the same rule as the flexural strain on a beam cross-section due to a bending moment. The deformation varies linearly as shown in Fig. 7.3(b). Bellow convolution is viewed as a flexible cylinder with a diameter the same as the pitch diameter of the bellow. The deformation is measured at this pitch diameter. From the figure, we have the convolution deformation due to rotation as

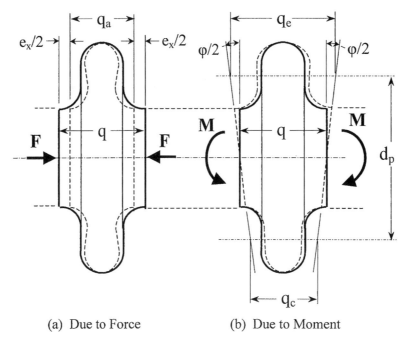

(a) Due to Force (b) Due to Moment

FIG. 7.3
DEFORMATION OF INDIVIDUAL CONVOLUTION

$$e_\theta = q_e - q = q - q_c \tag{7.2}$$

Because $\varphi = (q_e - q)/(d_p/2) = 2e_\theta/d_p$, we have

$$e_\theta = \frac{\varphi d_p}{2} \quad (\varphi \text{ is rotation per convolution}) \tag{7.3}$$

To determine the relationship between the force and deformation, a convolution spring rate due to axial load is defined as

$$f_w = \frac{F}{e_x} \tag{7.4}$$

EJMA calls f_w the working spring rate. Because normally the bellow is designed to function beyond the elastic limit, an accurate spring rate is not available. Based on piping flexibility analysis requirements, we know that the modulus of elasticity at ambient temperature should be used. As for the effect of stretching beyond the yield point, the values used differ slightly among manufacturers [1]. However, to be consistent with the piping analysis and also to be on the safe side, the modulus of the linear elastic range is recommended.

The relation between e_θ and M can also be expressed in terms of f_w. This is done by considering the bellow as a very flexible thin tube with a mean diameter d_p. The spring rate per unit width of circumference can be found from Eq. (7.4) as

$$f_{w,u} = \frac{f_w}{\pi d_p} \tag{7.5}$$

The maximum force per unit width of circumference due to φ rotation is $e_\theta(f_{w,u})$, and the equivalent bending stress is obtained by dividing the unit width force by the thickness as

$$S_b = e_\theta \left(\frac{f_w}{\pi d_p}\right) \frac{1}{t} = \left(\frac{\varphi d_p}{2}\right)\left(\frac{f_w}{\pi d_p}\right)\frac{1}{t} = \frac{f_w \varphi}{2\pi t} \tag{7.6}$$

Because the maximum bending stress due to the bending moment on a tube is the moment divided by the section modulus, we have

$$S_b = \frac{M}{Z} = \frac{M}{\pi (d_p/2)^2 t} \tag{7.7}$$

Equating Eqs. (7.6) and (7.7), the relation between ϕ and M becomes

$$M = \left(\frac{f_w \varphi}{2\pi t}\right) \frac{\pi d_p^2 t}{4} = \frac{1}{8} f_w d_p^2 \varphi \tag{7.8}$$

This is the moment and rotation relation, thus rotational spring rate, expressed in term of the axial spring rate.

The per-convolution deformation due to the combined axial force and moment is

$$e_{com} = e_x \pm e_\theta = \frac{F}{f_w} \pm \left(\frac{\varphi d_p}{2}\right) = \frac{F}{f_w} \pm \frac{4M}{f_w d_p} \tag{7.9}$$

Equation (7.9) is the basis for evaluating the combined force and moment effect on a bellow. e_{com} at any convolution of the bellow should not exceed the allowable axial deformation per convolution, $e_{x,allow}$.

Deformation of bellow element. The single convolution deformation discussed above can be used to investigate some basic functions of the bellow element consisting of multiple convolutions. The force-displacement relationship so established can be used to implement the bellow structural element for analyzing the piping system containing bellows. With the bellow structural element included in the computer program, a piping system can be routinely analyzed, whether with bellow or without bellow. The use of the bellow element is much more convenient than the generic flexible joint approach that is often used. Figure 7.4 shows three basic modes of deformation of a bellow element. The following force-displacement relationships are derived based on EJMA formulas.

Figure 7.4(a) shows the *axial deformation*. This is the most basic deformation mode. It is also the basic reference data for all other deformation modes. Axial deformations are uniform throughout all convolutions. From Eq. (7.4), we have

$$F = f_w e_x = f_w \frac{x}{N} = (K_x)x \quad \text{where } K_x = \left(\frac{f_w}{N}\right) \tag{7.10}$$

Figure 7.4(b) shows a *pure rotation*. Deformations are similar for all convolutions. From Eq. (7.8) and noting that $\theta = N\varphi$, we have

$$M_\theta = \frac{1}{8} f_w d_p^2 \varphi = \frac{1}{8}\left(\frac{f_w}{N}\right) d_p^2 \theta = \frac{d_p^2}{8} K_x \theta = (K_{R\theta})\theta$$
$$\text{where } K_{R\theta} = \frac{d_p^2}{8} K_x \tag{7.11}$$

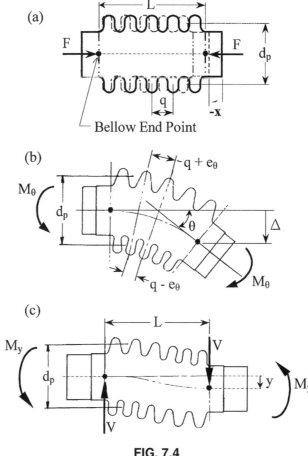

FIG. 7.4
BELLOW ELEMENT

$K_{R\theta}$ is the rotational spring rate due to θ rotation. This is the EJMA rotational spring constant formula, which accompanies some free lateral displacement, Δ, at the end. This formula is applicable to the generic rotational flexible joint located at the center of the bellow. Equation (7.11) has to be adjusted for the free end displacement, Δ, when implemented in a bellow structural element. The effect of this free end displacement will be discussed later.

Figure 7.4(c) is the deformation pattern due to *lateral displacement*. The situation is treated based on displacement, rather than on force as in the other modes. Distribution of the deformation is not uniform throughout the bellow. The maximum convolution deformation occurs at both ends. It is in a localized rotational mode and has both tension and compression at the same convolution. At both ends, the EJMA formula for deformation per convolution is

$$e_y = \frac{3d_p y}{NL} \tag{7.12}$$

The lateral displacement creates not only the lateral force but also the bending moment at the ends. The relationships between the force, moment, and lateral displacement are given by EJMA as

$$M_y = \frac{f_w d_p e_y}{4} = \frac{3}{4}\left(\frac{f_w}{N}\right)\frac{d_p^2}{L}y = \frac{3d_p^2}{4L}K_x y = (K_{Ry})y$$

$$\text{where } K_{ry} = \frac{3d_p^2}{4L}K_x \tag{7.13}$$

$$V = \frac{f_w d_p e_y}{2L} = \frac{3}{2}\left(\frac{f_w}{N}\right)\frac{d_p^2}{L^2}y = \frac{3}{2}\left(\frac{d_p}{L}\right)^2 K_x y = (K_V)y \qquad (7.14)$$

$$\text{where } K_V = \frac{3}{2}\left(\frac{d_p}{L}\right)^2 K_x$$

where L is the bellow effective length, which is the free natural length adjusted by the axial deformation, if any. K_{ry} is the rotational spring rate due to the y lateral displacement. K_V is the lateral spring rate. The lateral-rotational spring rate, K_{ry}, is inversely proportional to the bellow effective length, and the lateral spring rate, K_V, is inversely proportional to the square of the bellow effective length. Because the potential axial deformation can constitute a significant portion of the bellow length, it needs to be included in the calculation of spring rates. To be on the safe side, the potential minimum length, after subtracting the potential axial deformation, should be used.

The spring rates as given by Eqs. (7.10), (7.11), (7.13), and (7.14) are functions of the axial spring rate, K_x; effective bellow length, L; and bellow pitch diameter, d_p. Once K_x, L, and d_p are known, the spring rates of all directions and thus the stiffness matrix of the bellow element can be constructed.

The *EJMA rotational spring rate*, the one given by the manufacturer, implies that there is an accompanying free lateral movement, Δ, at the end as shown in Fig. 7.5(a). This free lateral displacement, sometimes called the relief displacement, can be calculated by

$$\Delta = c\sin\theta = R\tan(\theta/2)\sin\theta = \frac{L}{\theta}\tan(\theta/2)\sin\theta \cong \frac{L}{2}\theta \qquad (7.15)$$

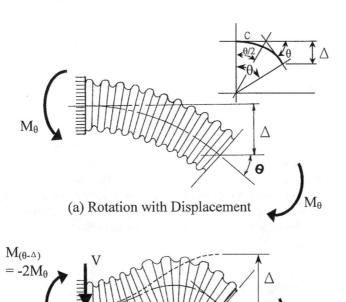

(a) Rotation with Displacement

(b) Rotation without Displacement

FIG. 7.5
BELLOW ROTATION WITH NATURAL DISPLACEMENT

The above approximate equation, using $\tan(\theta/2) = \theta/2$ and $\sin\theta = \theta$, is applicable for small rotations. For $\theta = 10$ deg., the common maximum allowable, the deviation of the approximation is about 0.2%.

Without this free displacement, the bellow has to be laterally moved back a distance of Δ. This creates an additional moment and force as shown in Fig. 7.5(b). The increased moment and force can be calculated from Eqs. (7.13) and (7.14) as

$$M_\Delta = \frac{3d_p^2}{4L}K_x\Delta = \frac{3d_p^2}{4L}K_x\frac{L}{2}\theta = \frac{3d_p^2}{8}K_x\theta \tag{7.16}$$

$$V_\Delta = \frac{3}{2}\left(\frac{d_p}{L}\right)^2 K_x\Delta = \frac{3}{2}\frac{d_p^2}{L^2}K_x\frac{L}{2}\theta = \frac{3}{4}\frac{d_p^2}{L}K_x\theta \tag{7.17}$$

Adding this additional moment to the EJMA moment as given by Eq. (7.11), we have the total moment for the no-displacement condition as

$$M_{(\theta+\Delta)} = \frac{3d_p^2}{8}K_x\theta + \frac{d_p^2}{8}K_x\theta = \frac{1}{2}d_p^2 K_x\theta \tag{7.18}$$

This moment is four times as big as the moment with the end of the bellow free to move laterally in the bending plane. The moment at the other end of the bellow is only half of this amount, but in the reversed direction of the original moment. It is obvious that the combined effect of the rotation and the lateral displacement is not a simple matter of just adding their individual numbers together. A combination of the rotation with some properly oriented displacement may produce much less bellow stress than the rotation alone. This simple fact is often overlooked even by experienced engineers.

7.2 USING CATALOG DATA

As the design and manufacture of bellow expansion joints is still pretty much an art mastered through years of practice, there have been few hard formulas for determining the strength and characteristics of the bellow from its appearance provided. Each manufacturer adopts its own unique bellow shape, material, fabrication method, testing scope, and qualification procedure to produce a safe bellow for applications within their published limitations. What the piping engineer has to do is to ensure that the characteristics such as spring rate, pressure thrust force, and so forth are included in the piping analysis, and the resulting bellow deformations from relative piping movements are within the manufacturer's limitations. We shall rely on the manufacturer's experience and integrity to provide the required margin of safety.

7.2.1 Background of Catalog Data

The catalog data is based mainly on the characteristics of the single bellow convolution, which differs from manufacturer to manufacturer. It may also differ from size to size and from pressure class to pressure class. For each basic convolution, it is designed to be strong enough to resist the pressure rated. The convolution is also checked for lateral stability (squirm) based on the rated pressure and the maximum bellow length offered for such convolution. Additional thickness over that required for the bursting strength may be needed. The convolution is then evaluated for axial spring rate and the allowable axial deformation for the reference fatigue life. The reference fatigue life used also differs from manufacturer to manufacturer. The popular ones are 3000, 4000, and 5000 cycles. The reference fatigue life of the joint is also generally listed in the catalog. The pressure rating, spring rate, and fatigue life are often confirmed by the available theoretical formulas and are generally backed by experiments. The experiments, however, are generally performed on multi-convolution units. The characteristics of the bellow element are derived mathematically by stacking the required number of basic convolutions together.

The basic data provided for a basic convolution comprise the spring rate, f_w, and the allowable axial deformation, $e_{x,\text{allow}}$, for the reference fatigue life. Axial deformation is determined by assuming that deformation is uniform around the whole circumference. For non-uniform distributions, such as the linear distribution in bending, no deformation at any point is allowed to exceed this amount. From these two basic data, we can calculate the following spring rates and allowable movements.

$$K_x = \frac{f_w}{N} \quad \text{(Axial spring rate, from Eq. (7.10))}$$

$$K_R = \frac{d_p^2}{8} K_x \quad \text{(Rotation spring rate, from Eq. (7.11))} \quad (7.19)$$

$$K_L = \frac{3}{2}\left(\frac{d_p}{L}\right)^2 K_x \quad \text{(Lateral spring rate, from Eq. (7.14))}$$

K_L is given in Eq. (7.14) as K_V. The bellow length, L, used in calculating the lateral spring rate is the modified potential minimum length. To construct the allowable chart in the catalog, each manufacturer uses its own criteria for determining the effective length. One popular approach is to take the effective length as the natural length minus one-half of the allowable axial movement. A simplified approach that considers the effective length equal to the natural length subtracted by 0.5 in. (12.7 mm) has also been used. This modification of the length is done by most manufacturers to estimate the potential maximum lateral spring rate. Again, it should be noted that the rotational spring rate implies an accompanying lateral displacement, and the lateral movement also creates the bending moment as discussed previously in this chapter. EJMA and some manufacturers also provide some torsional sprint rate. However, because it is very large, the torsional spring rate is generally ignored or, rather, considered rigid.

The allowable non-concurrent movements are calculated by

$$x = N e_x \quad \text{(Axial)} \quad (7.20)$$

$$\theta = N\varphi = N\frac{2}{d_p}e_\theta = N\frac{2}{d_p}e_x \quad (\text{Rotation}, e_\theta \leq e_x) \quad (7.21)$$

$$y = \frac{NL}{3d_p}e_y = N\frac{L}{3d_p}e_x \quad (\text{Lateral}, e_y \leq e_x) \quad (7.22)$$

where $e_x = e_{x,\text{allow}}$. Again, the bellow length used is the natural length subtracted by the expected axial movement. The allowable rotation is generally further limited to no more than 10 deg., regardless of the number of convolutions, by some manufacturers. This additional rotational limitation is adopted to avoid column instability due to the offset caused by the end rotation.

7.2.2 Using the Catalog

Before using the catalog, it is important to note that the data found in the catalog differs greatly from manufacturer to manufacturer. The catalog of the specific expansion joint in mind should be used.

From the catalog, a preliminary selection should be made based on pressure rating and the expected movements. Alternatively, this selection process can be left to the manufacturer by supplying the design parameters such as pressure, temperature, expected movements, and so forth. The designer shall then incorporate the pressure thrust force and spring rates, either from the catalog or from the manufacturer, into the mathematical model of the piping system for analysis. If the equipment reaction, pipe stress, or the movement exceeds the allowable value, a longer bellow or other arrangement should be tried. The process repeats until all reaction forces, pipe stresses, and bellow movements are within the allowable range. The following are some notes regarding the catalog data.

Pressure rating. The pressure rating given is for the temperature below the benchmark temperature listed. The benchmark temperature also differs among manufacturers. The two popular benchmark temperatures adopted are 650°F and 800°F. The pressure rating is for the bellow element itself. For flange and other connecting components, the standard rating for each component should apply. Pressure rating is based on both rupture and stability criteria. If the system design temperature is higher than the manufacturer's benchmark temperature, the rating has to be adjusted. For the rapture criterion, the rating is proportional to the allowable stress. On the other hand, the rating is proportional to the modulus of elasticity when the bellow is governed by the stability criterion. Because the reduction in the modulus of elasticity is not as rapid as the reduction in the allowable stress as the temperature increases, the adjustment of the rating for temperature above the benchmark temperature can always be based on the allowable stress ratio. That is,

$$\text{The adjusted rating} = \text{Catalog rating} \times \frac{\text{Allowable stress at higher temperature}}{\text{Allowable stress at benchmark temperature}}$$

The rating should only be adjusted downward. It should not be increased for temperatures lower than the benchmark temperature. The rating adjustment at lower than benchmark temperature should also be checked by the ratio of the modulus of elasticity to cover the stability limitation. Because the ratio of the modulus of elasticity is very close to unity at lower temperatures, the use of the unadjusted rating for a lower than benchmark temperature is not overly conservative. When in doubt, the manufacturer of the bellow should be consulted.

Effect of rotation. The end rotational deformation of the bellow initiates the eccentricity of the element. This eccentricity can promote premature instability. Therefore, the pressure rating may be reduced by the increase in end rotational deformation [4], or it can only be used in conjunction with a further limitation on rotational deformation. Some manufacturers [5, 6] put a 10-deg. limitation on the allowable rotation regardless of the bellow length.

Bellow effective area. This effective area is also referred to as the pressure thrust area. This area is used to calculate the pressure thrust force, which will be resisted by the anchor, tie-rod, or the equipment. The effective area is often used to calculate the pitch diameter of the bellow. Piping analysis has to include the pressure thrust force either by applying the forces explicitly or by giving the effective area or pitch diameter. The pressure thrust force acting at both ends of the bellow toward the piping is a major problem in a piping system installed with an expansion joint.

Spring rates. The spring rates given in the catalog have to be incorporated in the piping analysis. To obtain correct results, the spring rates in all directions have to be included at the same time. These rates include the axial translation rate, two mutually perpendicular but equal lateral spring rates, and two mutually perpendicular but equal bending spring rates. The torsional spring rate can be taken as rigid.

There are two general approaches for including the bellow spring rates in the analysis. One is to consider the bellow as a generic flexible joint located at the center point of the bellow. The center point is mathematically separated into two ends, which are connected with springs having the spring rates as given by the catalog. From Eq. (7.19), and using the keywords given in Section 7.1.1, we have the flexible joints as FTA = K_x, FTH = K_L, FTV = K_L, FRH = K_R, FRV = K_R. The other approach considers the bellow as a general structural beam element with a stiffness matrix constructed from the spring rates given by or derived from the catalog. The latter approach is generally referred to as the bellow element method. In this method, the data for the axial spring rate, bellow effective length, and bellow pitch or effective diameter is generally required.

Non-concurrent movements. The movements are classified into three categories: axial, lateral, and rotational. More accurately, these movements are the bellow deformations produced by the relative movements of the pipe connected to the ends of the bellow. It is very important to remember, however, that when the movements are taken as the differential quantities of the two ends, the free lateral displacement due to rotation has to be subtracted. The allowable movements listed in the catalog are non-concurrent movements with only one movement existing at a time. For concurrent movements, they are evaluated with the usage factor defined as

$$\text{Usage factor} = \frac{x}{x_a} + \frac{y}{y_a} + \frac{\theta}{\theta_a} \leq 1.0 \quad (7.23)$$

where x, y, and θ are concurrent axial, lateral, and rotational movements, respectively. The terms x_a, y_a, and θ_a are the non-concurrent allowable movements listed in the catalog for axial, lateral, and rotational, respectively. The usage factor is limited to 1.0.

Equation (7.23) assumes that x, y, and θ are the total effective or normalized movements and are in the same base as x_a, y_a, and θ_a, all of which are based on a reference fatigue life. Because x, y, and θ are generally associated with different components having different operating cycles, some adjustments and combinations as described in the following are generally required before the equation is evaluated.

Adjustment for operating cycles. As noted previously, the non-concurrent allowable movements given in the catalog are based on a reference fatigue life. Therefore, to compare with the catalog data, the actual movements have to be adjusted to the same base as the catalog data. Two approaches have been used for this adjustment.

One method [4] is to assign a dominant operating mode as the base mode with its operating movements as the base movements. The movements of all other operating modes are considered the same as the base movements with equivalent additional operating cycles calculated as

$$N = \sum (r_i^4 N_i) \quad \text{for } i = 1,\ldots,n \quad (7.24)$$

where n is the number of operating modes, N_i is the number of operating cycles for mode i, and r_i is the ratio of the deformation of mode i to that of the base mode. The fourth power over the deformation ratio is used instead of the fifth power used by American Society of Mechanical Engineers B31 in the evaluation of piping components. This fourth power arrived from the results of tests on stainless steel bellow elements [7]. Equation (7.24) essentially converted the deformations of all operating modes into one base deformation, which is generally the maximum deformation, with N- equivalent number of operating cycles. Bellow selection is then based on the base movements and the total equivalent number of operation cycles.

The other approach [5, 6] favors a more simplified method. In this approach, the movements of each operating mode are converted to normalized movements corresponding to the reference fatigue life. A chart similar to Fig. 7.6 is published in the catalog for this conversion. The normalized movements for all postulated operating modes are summed absolutely to become the total normalized movements in each movement category. These normalized movements are then substituted into Eq. (7.23) to calculate the fatigue usage factor. The usage factor is limited to equal or less than unity.

Figure 7.6 shows the schematic chart for adjusting the operational movements. The curve is an example for a reference fatigue life of 3000 cycles. The manufacturer who rates its allowable movements based on 3000 operating cycles would use a similar chart. The adjustment factor, C, for 3000 cycles is 1.0. For an operating mode with 1000 expected cycles, the adjustment factor is 0.8. This means a 1.0-in. movement will be adjusted to become a 0.8-in. normalized movement corresponding to 3000 operating cycles. We shall use an example to demonstrate this idea.

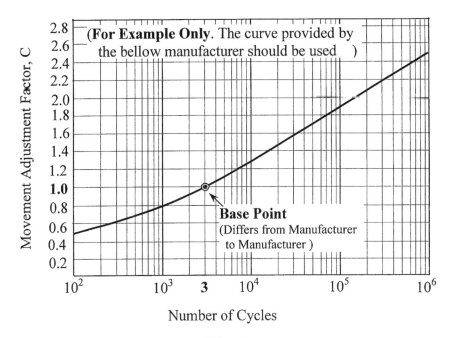

FIG. 7.6
MOVEMENT ADJUSTMENT FOR DIFFERENT OPERATING CYCLES

As an example, consider a bellow element that has non-concurrent allowable movements of 3.86 in. axial, 1.1 in. lateral, and 10 deg. rotational, rated for 3000 operation cycles. This bellow is subjected to three postulated operating modes with expected movements as shown in Table 7.2. The table shows one cycle of installation mode due to misalignment, 200 cycles of normal startup and shutdown during the life of the plant, and 10,000 cycles of lifetime major fluctuations due to load swing and so forth. For the installation mode the adjustment factor is 0.5, so the expected movements 0.42 in., 0.08 in., and 0 deg. are converted to the normalized movement by multiplying them with this 0.5 factor. It should be noted that unless given specifically by the manufacturer, an adjustment factor less than 0.5 should not be used. For the fluctuation mode, the adjustment factor is 1.29. Therefore, the expected movements should be increased by this 1.29 factor to become the normalized movements. By the same token, the normalized movements for each operating mode can be calculated and added together category by category. The total normalized movements from all three operating modes are 1.41 in., 0.37 in., and 1.63 deg. for axial, lateral, and rotational, respectively. Applying these values to Eq. (7.23), we have

TABLE 7.2
CALCULATION OF NORMALIZED MOVEMENTS

Movement Category	Installation Mode (1 Cycle)			Start-up/Shut-down (200 Cycles)			Fluctuation Mode (10000 Cycles)		
	Actual	"C"	Norm.	Actual	"C"	Norm.	Actual	"C"	Norm.
Axial	0.42	0.50	0.21	1.20	0.57	0.68	0.40	1.29	0.52
Lateral	0.08	0.50	0.04	0.25	0.57	0.14	0.15	1.29	0.19
Rotation	0.00	0.50	0.00	1.50	0.57	0.86	0.60	1.29	0.77

Total Normalized Movements: Axial = 0.21 + 0.68 + 0.52 = 1.41"; Lateral = 0.37"; Rotation = 1.63°.

$$\text{Usage factor} = \frac{1.41}{3.86} + \frac{0.37}{1.1} + \frac{1.63}{10}$$
$$= 0.365 + 0.336 + 0.163$$
$$= 0.864 \le 1.0$$

Because the fatigue usage factor is less than 1.0, the bellow is suitable for the expected operational movements.

Torsional moment. Most manufacturers insist that the bellow should not be subjected to any torsional moment. This "zero load" syndrome is very familiar to piping engineers. However, we all know that no matter how well you have designed the piping system the bellow will still experience some torsional moment. EJMA does have a criterion for this torsion moment. It is based on the shear stress generated at the cylindrical tangent. The shear stress should not exceed $0.25 S_h$ of the bellow material. That is,

$$S_s = \frac{2 M_t}{\pi d^2 t} \le 0.25 S_h \tag{7.25}$$

where S_s is the shear stress at the cylindrical tangent portion of the bellow, d is the outside diameter of the cylindrical tangent, t is the wall thickness of the bellow, and S_h is the code basic allowable stress at design temperature. The values of d and t are generally not given in the catalog. They have to be obtained from the manufacturer. If required, the value of d can be taken as the outside diameter of the connecting pipe.

7.2.3 Calculating Operational Movements

Because EJMA rotational movement implies an accompanying lateral movement, the calculation of the operational movements is not as straightforward as one might think. The calculation of the lateral movement, the one with the least allowable range, requires some adjustments. It needs to distinguish between the portion that is pure lateral movement and the portion that comes from the natural flexing of the rotational movement.

Figure 7.7(a) shows the end displacements of a bellow element. These end displacements are calculated directly from an analysis of the piping system. Many engineers have wrongly assumed that

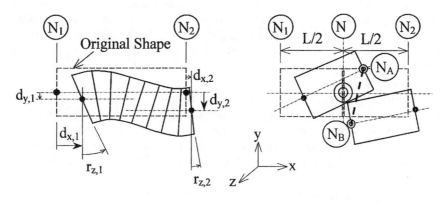

(a) End Displacements (b) Joint Displacements

FIG. 7.7
BELLOW DEFORMATION

bellow movements are the differences of the end displacements. For instance, the axial movement or deformation equals $(d_{x,2} - d_{x,1})$, and lateral movement equals $(d_{y,2} - d_{y,1})$, and so forth. This is correct for axial displacement, but wrong for lateral movement when there is an end rotation involved. Natural flexing due to rotation has to be subtracted from the differential lateral movement.

There are two main approaches for calculating the correct EJMA lateral movements. One is to use the differential displacements at the joint location when the bellow is modeled as generic flexible joints located at the center of the bellow as shown in Fig. 7.7(b). This center flexible joint approach eliminates lateral flexing due to rotation. Bellow movements can be calculated directly from differential displacements and rotations between points N_A and N_B. N_A and N_B are mathematically located at the same location, but are connected with springs in between them. In this particular case, the figure shows that the EJMA lateral movement so obtained is much larger than the apparent lateral movement obtained by differential displacements of the bellow ends. However, in general, the EJMA lateral movement is smaller than the apparent movement in most cases.

When the analysis uses the bellow beam element, bellow deformations are calculated directly from the forces and moments generated. It can automatically calculate the total equivalent axial deformation due to the combined axial, lateral, and rotational movements. This total equivalent axial deformation can be compared directly with the allowable axial deformation of the bellow.

7.2.4 Cold Spring of Expansion Joint

For reasons to be discussed later, an expansion joint is generally cold sprung for 50% or more of the expected movements. This cold spring has to be incorporated in the piping layout to accommodate it. The axial cold spring is generally pre-set in the factory and fixed with shipping bars. The lateral and rotational cold springing are done in the field as the joint is being connected to the piping system. In the expansion joint community, interestingly, the cold spring done in the factory is called pre-setting, and only the one done in the field is referred to as cold spring. To effect the cold spring, the piping is laid out with the proper gap for the final pull. However, cold springing is mainly done on the expansion joint with the piping held stationary. In other words, the piping is not pulled. The shipping rods holding the joint in the axial direction remain connected, whereas the attachments guarding the lateral movement are removed before the cold spring. The following are the advantages of cold spring [1].

Force reduction. With a 50% cold spring, for example, the bellow at ambient installation condition is stretched to 50%, but opposite the expected movements. The bellow is subjected to an initial force equal to one-half of the full potential bellow force expected if the bellow were not cold sprung. The initial force decreases as the system gradually warms up. The force reduces to zero at the mid-point of the operating temperature, then reverses gradually to reach one-half of the full potential bellow force at the full operating temperature. Therefore, instead of facing the full potential force, the bellow experiences only one-half of the full potential force. Although the reduction of the force acting on the connecting equipment may not be as much as the reduction of the bellow force, substantial reduction of the equipment force is also expected.

Stability. As discussed previously, the stability of the bellow is greatly reduced by the end rotation of the bellow. With a 50% cold spring, the effective end rotation is reduced by 50% of the potential full rotation. This reduces the bellow body curvature by half, thus substantially increasing stability against internal pressure. The bellow end rotation generally occurs at places where two or more bellow elements are assembled in a functional unit, such as the universal joint to be discussed in the next section.

Bellow clearance. With the rated movements, the deformation of all or some of the convolutions can represent a large percentage of the bellow pitch, resulting in a considerably reduced clearance. A severely out-of-shape convolution also behaves quite differently from the original shape. Again, with a 50% cold spring, the potential movement is divided into two halves, one-half in the extension and the other in compression. Therefore, the clearance is reduced only by 50% as much.

Component clearance. When an expansion joint is furnished with internal sleeves, external covers, or tie devices, they require adequate clearance to accommodate the movements. The clearances between these components are generally constructed uniformly around the circumference. With movements tilted to one side, the clearance should be large enough to cover the full potential movements. A large clearance makes the joint shaky. With a 50% cold spring, the clearance needed is reduced by half.

7.3 APPLICATIONS OF BELLOW EXPANSION JOINTS

EJMA [1] and most of the manufacturers' catalogs have an extensive collection of examples on the application of bellow expansion joints. The applications can be classified based on the direction of the main movement as applied on the bellows. The most direct applications use the axial deformation of the bellow to absorb pipe expansion. The expansion is handled directly through bellow axial displacement. Other applications may use the bellow's lateral deformation and rotation.

7.3.1 Application of Axial Deformation

The most basic characteristic of the bellow is its axial flexibility. Comparatively, a very small force is needed to deform the bellow axially a significant amount than that is needed to squeeze the same amount of deformation in the piping system. Therefore, a bellow is especially suited for absorbing the expansion of a long straight line. Because a bellow is not capable of transmitting the longitudinal pressure stress, it results in a large pressure thrust force acting toward the piping at both ends of the bellow. Taking care of these pressure thrust forces is the main concern of bellow expansion joint applications. Extensive anchors and guides placed at strategic locations are always required.

Figure 7.8 shows the typical anchor and guide requirements at a bellow expansion joint. To start with, two anchors located at each side of the bellow are needed to stop the pressure thrust force from pushing the piping system and its connecting equipments. The pressure thrust force is calculated by multiplying the maximum operating pressure with the effective bellow area given by the manufacturer. The effective area is generally the same as the circular area of the pitch diameter of the bellow. To

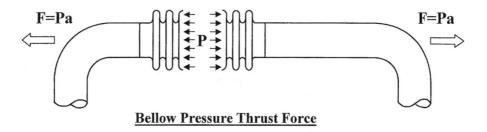

Bellow Pressure Thrust Force

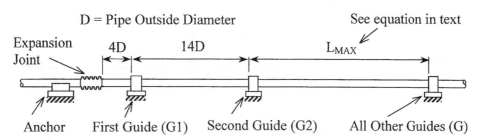

FIG. 7.8
ANCHOR AND GUIDE SCHEME OF AXIAL EXPANSION JOINT

reduce the number of guides required, the bellow is generally located close to one of the anchors as shown. The other anchor is located on the other end of the straight section whose expansion is being absorbed. The anchors facing the full pressure thrust force are called the main anchors. In addition to the pressure thrust force, p_a, the main anchor also faces the friction loads from supports and guides, F_f, fluid flow momentum force, F_ρ, and bellow resistance, $K_x x$. That is,

$$F_A = p_a + F_f + F_\rho + K_x x \tag{7.26}$$

The pressure thrust forces are isolated to within the anchors, but still work on the piping located within. In an internal pressure environment, the pressure thrust force acts as a compressing load on the piping located within the main anchors. This can create column-buckling problems for the piping as well as for the bellow. Lateral guides located at strategic locations are needed to ensure the stability of the system. Because the stability of both the bellow and pipe column is very sensitive to end rotation, a pair of coupling guides very close to the bellow are required. EJMA recommends that the first guide should be located at a distance of 4 pipe diameters away from the edge of the bellow, and a second guide placed at 14- pipe diameters away from the first guide as shown in Fig. 7.8. These two guides, called G1 and G2, respectively, are mainly for the protection of the bellow and to limit the end rotation of the piping. Besides G_1 and G_2, general guide or guides, G, are required to protect the main piping from buckling. The maximum distance between general guides is given by EJMA as

$$L_{MAX} = 0.131 \sqrt{\frac{EI}{p_a + K_x x}} \tag{7.27}$$

Figure 7.9 shows several typical arrangements of the bellow expansion joints in absorbing axial movements. Figure 7.9(a) shows a single bellow to absorb the expansion of a relatively short straight

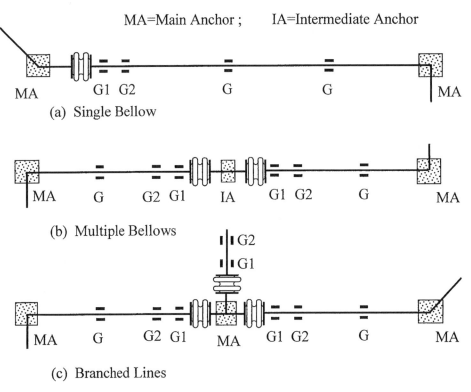

FIG. 7.9
TYPICAL ANCHOR AND GUIDE ARRANGEMENTS ON STRAIGHT LINES

piping run. Two anchors at the ends are needed to resist the pressure thrust force. The bellow is placed near one of the anchors to save the amount of guides required. A set of G1 and G2 is placed near the other end of the bellow. Based on the L_{max} criteria given in Eq. (7.27), the rest of the piping run requires two additional general guides. In addition to the guides, the piping also needs proper weight supports, which are not shown in the figure.

Figure 7.9(b) represents the typical arrangement of a long straight run. Due to limitations of the bellow capacity and of the support shoe movement, the piping is divided into multiple sections. The length of the section is determined by the allowable bellow axial movement and allowable pipe travel on supports. Depending on the designer's preference, the bellow may be placed either at the end or at the middle of the section. Generally, it is considered more economical to place the bellow at the end of the section as shown in the figure. In this case, the bellows of two adjacent sections are generally grouped together to reduce the number of guides required. However, this end bellow arrangement produces larger support shoe movements, which may be critical for some cases. For a long piping run, many pairs of bellows may be required. The anchor placed in between the bellows of two adjacent sections is called the intermediate anchor (IA). The load of the intermediate anchor comes from the differential pressure thrust force of the two adjacent bellows and from the support friction and bellow deformation forces of the two connected pipe sections.

The load at the intermediate anchor is theoretically zero when the two bellows are identical in size and pressure, and the bellows are symmetrically located. In actual applications, significant forces might be produced in some situations that should be checked. Figure 7.10 shows some of the potential loads generated due to the pressure effect. Case 1 has two identical bellows and thus theoretically zero pressure thrust force on the IA. In the actual design, 10% to 20% of the full pressure force may be used depending on the design specification. This uncertainty factor is to compensate for the uncertainty due to pipe misalignment and bellow end rotation. If considerable bellow end rotation is expected, such as in piping with stratified flow, the higher factor should be used. A non-uniform temperature distribution under stratified flow conditions bows the pipe, creating significant end rotations. In case 2, the size changes in between two bellows. Theoretically, only the pressure force at the smaller bellow is balanced. The IA design force will be the unbalanced differential force plus a certain percentage of the balanced pressure force. Case 3 presents the situation where a valve is placed between the bellows. In this case, the IA load is greatest when the valve is closed with zero pressure at one of the bellows. The differential bellow force, K, is the force resulting from differential deformation of the two bellows.

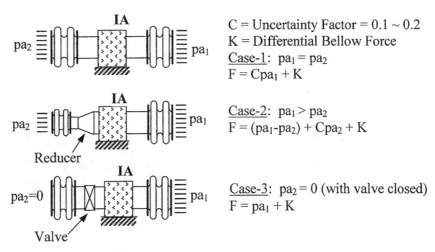

**FIG. 7.10
INTERMEDIATE ANCHOR DESIGN LOADS**

The uncertain IA load should be considered as acting in both axial and lateral directions. The friction force and bellow deformation force at the IA are generally very small when the system is properly supported, making the forces from both sides of the IA balance each other. However, because the piping at both sides of the IA does not necessarily move in synch, EJMA recommends taking the friction force and bellow deformation force from one side only as the design load for the IA. The design specification shall give the specific design load for each specific situation. The intermediate anchor placed close to the bellow, such as the ones shown in Figs. 7.9 and 7.10, is not subject to friction force, which is absorbed by the main anchor or the IA located away from the bellow.

Figure 7.9(c) shows a branch connection with the bellow placed at the branch line, as well as at each of the through runs. The anchor is a main anchor due to the unbalanced pressure thrust force applied at the branch line.

Alignment guides. The guides required in expansion joint installations have to be effective in both up-and-down and horizontal lateral directions. It is popular to use catalog items, termed as alignment guides as shown in Fig. 7.11, for this purpose. However, it should be noted that the alignment guide as offered by the manufacturer's catalog is not capable of taking the weight load. Therefore, the alignment guides are installed in addition to the normal weight supports. The guides do not replace the weight supports. Some designers use a special support detail that combines the weight support with the guide function. In this case, the guide function should include both vertical and horizontal directions. For some liquid lines, the weight hold-down effect is enough to act as a guide in the vertical direction, thus only the horizontal direction guide needs to be added. This can be done by placing guide lugs against both sides of the sole plate of the support shoe, provided the support shoe is strong enough. However, hold-down stops are always recommended for expansion joint installations due to the difficulty of assessing the adequacy of the weight hold-down effect.

In-line pressure balanced expansion joint. From the above discussions, it is clear that an extensive anchor and guide system is required when using an expansion joint to accommodate axial movement. This can be very expensive or difficult to install in some cases. For instance, to provide a main anchor at piping located several stories high in the air would require beefing up of the entire structure from the ground level and up. Because the problem stems mainly from the pressure thrust force acting at the bellow, an expansion joint without this pressure thrust force will make the application much simpler. A joint without this pressure thrust force effect is called a pressure balanced joint. The one that works on axial deformation is called an in-line pressure balanced expansion joint. As shown in Fig. 7.12, the in-line pressure balanced joint uses an ingenious bellow and linkage arrangement, so the pressure thrust force at one bellow is canceled by that of another bellow. Figure 7.12(a) shows the basic arrangement.

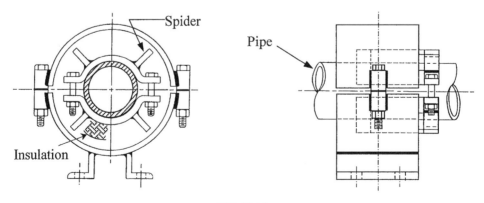

**FIG. 7.11
ALIGNMENT GUIDE**

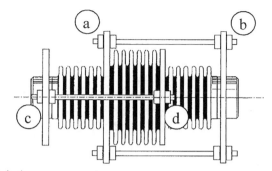

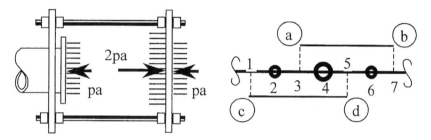

(a) Arrangement

(b) Balancing the Pressure Forces   (c) Modeling Technique

FIG. 7.12
IN-LINE PRESSURE BALANCED EXPANSION JOINT

In addition to the two basic bellows corresponding to the size of the pipe, a bigger bellow is placed in between these two. The larger bellow has twice the effective pressure thrust area of the basic bellow. The larger bellow is also called the balancing bellow. By interconnecting these bellows with linkages as shown in Fig. 7.12(b), the pipe at each end of the joint receives zero pressure thrust force. Each pipe end is subjected to two smaller pressure thrust forces (from the two basic bellows) that is balanced by one big pressure thrust force from the balancing bellow. Although the exact effective area of the bellow is not at all exact, a fairly good balance is achieved by fine-tuning the manufacturer's experimental data.

The in-line pressure balanced joint can be analyzed as one combined equivalent flexible joint with its axial spring rate equal to the sum of the spring rates of the three bellows. By squeezing the overall joint one unit length, it squeezes one unit length on each of the basic bellows, and extends one unit length from the large bellow. Therefore, the spring rate of the overall joint is the sum of the spring rates of the three individual bellows. This combined equivalent joint works well for pure axial movement, but does not offer any clue on lateral and rotational movements. To accommodate the small lateral and rotational movements expected on most joints, the in-line pressure balanced joint can be analyzed using the model shown in Fig. 7.12(c). The two basic bellows can be modeled as bellow elements or generic flexible joints at points 2 and 6. The large bellow is modeled likewise at point 4. The three joints are connected with spools of the main process pipe. The two sets of linkages can be simplified as single rods located at the centerline of the pipe. Linkage a-b is connected to points 3 and 7 at the pipe, whereas linkage c-d is connected to points 1 and 5 at the pipe. Pressure thrust forces are either specified explicitly or by giving the bellow effective diameters.

Corner pressure balanced expansion joint. To absorb axial movement, a pressure-balanced joint can be easily installed at the corner of a change of direction. Instead of using a large balancing bellow as in

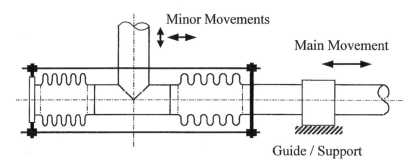

**FIG. 7.13
CORNER PRESSURE BALANCED EXPANSION JOINT**

the in-line joint, the corner type uses double bellows as shown in Fig. 7.13. The pressure thrust force from one bellow is balanced by that of the other bellow of the same size, transmitted through tie-rods. For a unit axial movement from the main pipe, one bellow faces one unit of extension, whereas the other bellow experiences the same amount of compression. The overall axial spring rate is the sum of the spring rates of the two bellows. The corner pressure balanced joint also absorbs some lateral and rotational movements, thus broadening its applicability. It is often used at connections to rotating equipment such as steam turbines and pumps.

7.3.2 Lateral Movement and Angular Rotation

In accommodating lateral and rotational movements, the bellows are generally restrained with tie-rods. This eliminates the requirement for main anchors because the pressure thrust forces are resisted and isolated by the rods. As discussed previously, the bellow rotation always comes with a lateral relief movement, and the bellow lateral movement always generates a bending moment. Therefore, it is not easy to find a pure lateral or pure rotation in actual applications. The lateral movement and angular rotation generally work together.

Single tied expansion joint. An expansion joint is often assembled with tie-rods to prevent the pressure thrust force from pushing the piping system. However, once the joint is tied, it loses its ability to flex axially, except to absorb the expansion of the pipe located between the tie-rod lugs. Therefore, only lateral and rotational flexing capabilities are available with tied expansion joints. Figure 7.14 shows a typical case of using the lateral flexing of a tied joint to absorb the pipe expansion. Although the movement is in the axial direction of the main piping run, the joint is placed in the perpendicular direction of the run. Because the lateral deformation of the bellow generates a rotational bending moment as well as a lateral force, the end of the joint is expected to rotate somewhat. This calls for a guide that allows some vertical piping movement. This type of guide for keeping the system from moving out of plane is called a planar guide (PG). No main anchor is needed in these types of applications.

In the analysis of the piping system, the bellow is again modeled either as a bellow element or as a generic flexible joint located at the center of the bellow. The pressure thrust forces, which are resisted by the tied rods, are either applied explicitly as external forces or specified by giving the effective bellow diameter. The tie-rod combination can be modeled either as four equally spaced rods tied to four spider-shaped lugs, or by simplifying them as one combined rod located at the centerline of the pipe and connected to the two ends of the bellow. The end connections of the tie-rods are connected to generic flexible joints having very small rotational spring constants.

Universal expansion joint. The allowable lateral movement of the bellow is roughly proportional to the square of the bellow length. A slight increase in bellow length can substantially increase the allowable movement. However, due to stability concerns, bellow length can only be increased by a certain

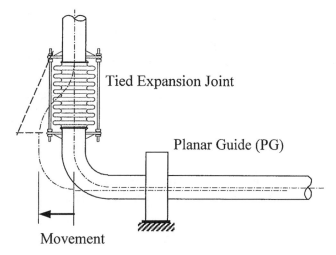

FIG. 7.14
USING BELLOW LATERAL MOVEMENT

amount. Therefore, a single bellow joint can accommodate a relatively small amount of lateral movement. For large movements, a double bellow arrangement is generally used. Figure 7.15 shows an arrangement wherein two small bellows are connected with a spool pipe in between. The stability of the bellow is determined by the total length of the two bellows and is fairly unaffected by the length of the spool pipe. By lengthening the spool pipe, this arrangement can take a very large lateral movement. However, it should be noted that a long spool piece itself might have a column-buckling problem that needs to be checked. Because this type of joint can also sustain a fair amount of rotation and absorb the axial expansion in between the tie-rod connections, it is generally referred to as a universal expansion joint.

The data sheet provided by the universal expansion joint manufacturer generally gives only the allowable lateral movement and overall lateral spring rate of the assembly, K_L. This, of course, is in response to the purchase specification, which generally makes a request for these two items. However,

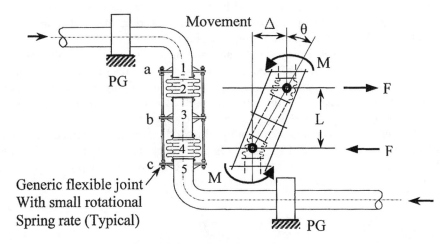

FIG. 7.15
UNIVERSAL EXPANSION JOINT

the data on the overall assembly cannot be used readily in the analysis of the piping system. Information regarding each bellow is required for the analysis. This sometimes calls for the reverse conversion of the overall bellow data into individual bellow data. The universal expansion joint utilizes mainly the rotational flexibility of the bellow to accommodate the lateral movement. Axial flexibility is needed only to absorb the axial expansion of the assembly located inside the tie-rod boundary. As the pipe moves laterally a Δ distance, both bellows rotate an angle θ, which is equal to Δ/L. The lateral movement produces the lateral force F and a bending moment M, which is equal to

$$M = K_R \theta = K_R \frac{\Delta}{L} \tag{7.28}$$

where K_R is the rotational spring constant of the individual bellow. By taking the moment around the center of one of the bellows, we have the relation between the force and moment as $2M = FL$. That is,

$$F = \frac{2M}{L} = \frac{2K_R \Delta}{L^2} \tag{7.29}$$

From Eq. (7.29) and the definition of the overall lateral spring rate as $K_L = F/\Delta$, we have

$$K_R = \frac{L^2}{2}\frac{F}{\Delta} = \frac{L^2}{2}K_L \tag{7.30}$$

This K_R can be used for piping analysis. This is just an approximate method. To be more precise, the manufacturer should be consulted for the individual bellow data, which should include axial, lateral, and rotational spring rates. The allowable movements for the individual bellow may also be obtained for reference.

In most constructions, the movement of the spool pipe in between the bellows is controlled by the lugs located at the center of the spool. These lugs are connected to tie-rods to support the spool and also to help distribute the movement evenly to both bellows. In the analysis, the bellows at points 2 and 4 in Fig. 7.15 can be modeled either as bellow elements or as generic flexible joints located at the center of the bellows. An elaborate analysis would have modeled all the tie-rods in the locations as constructed. However, because in actual operations not all rods are active and only one of the mid lugs is active due to differential rotation between the rod and the pipe, this elaborate modeling is not as precise as one might expect. In a simplified but realistic approach, the tie-rods are grouped as one rod located at the centerline of the pipe. It is connected to the lug locations with generic flexible joints with very small rotational spring rates. It should be noted that for most universal joints, the overall lateral spring rate is very small. The magnitude of the rotational spring rate used at tie-rod connections can have a significant effect on the overall lateral resistance.

Pressure balanced universal expansion joint. Axially, the universal expansion joint can absorb axial deformation only from the expansion of the pipe assembly located inside the tie-rod boundary. It is unable to absorb the axial movement from the expansion of the piping outside the expansion joint assembly. When the system has both large lateral as well as axial movements, a pressure balanced universal joint can be used. Figure 7.16 shows a typical arrangement.

The pressure-balanced universal joint is a combination of the pressure-balanced joint shown in Fig. 7.13 and the universal joint shown in Fig. 7.15. It is generally analyzed with a simplified model that combines all the tie-rods into one large rod located at the centerline of the pipe. The necessary data points, needed for the analysis, are also shown in the figure.

7.3.3 Hinges and Gimbals

To eliminate the bellow pressure thrust force, a bellow expansion joint is often tied with tie-rods. However, the tied expansion joint does not work very well for absorbing rotational movement. The

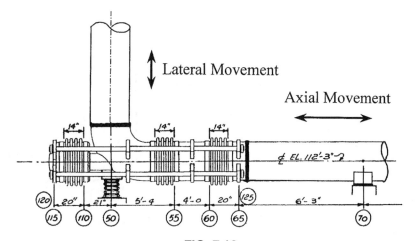

FIG. 7.16
PRESSURE BALANCED UNIVERSAL EXPANSION

effectiveness of tied expansion joints is also questionable for shorter joints due to the hindering mobility of short rods. Hinge and gimbal bellow joints, as shown in Fig. 7.17, work more smoothly and effectively in absorbing rotational movements.

A *hinged joint* allows the pipe ends at the joint to rotate in a plane. Due to its planar motion, it is mainly used to absorb the expansion of piping lying essentially in a plane. The system shown in Fig. 7.1 is an example of good locations for hinge joints.

Hinge joints are generally installed in pairs or in triplets. Figure 7.18 shows a couple of typical applications. In Fig. 7.18(a), two hinges are used to absorb the expansion in the x direction. To ensure smooth movement of the hinges, the orientations of the hinges have to be installed precisely. Because of the rotation and the expansion of the y leg, some y movements are expected in the piping. Planar guides (PGs) are needed to maintain the piping in the plane and to protect the hinge plates from warping. PGs also allow the pipe to move in the y direction. The distance between the hinges determines the allowable movement. The longer the distance, the larger the movement allowed. When the y expansion is significant, it may require the use of three hinges. A three-hinge system theoretically eliminates all the forces and stresses due to thermal expansion. However, the bellow spring rate and hinge pin friction do produce significant forces and stresses in actual applications. In the analysis, the hinge is modeled as a generic rotational flexible joint with a spring rate the same as the bellow

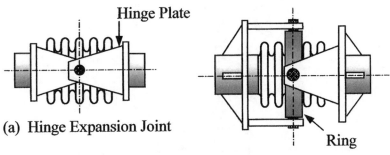

FIG. 7.17
HINGE AND GIMBAL EXPANSION JOINTS

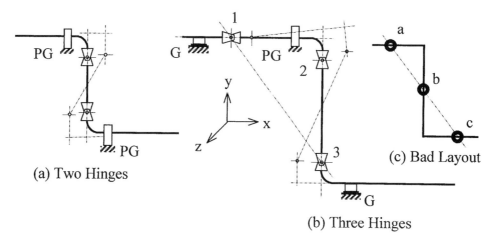

**FIG. 7.18
HINGED SYSTEMS**

rotational spring rate. This, however, does not include the hinge pin friction. A more precise analysis would manually increase the spring rate to cover the friction.

A *bad layout* of a three-hinge system is occasionally observed in the field. For a three-hinge system, the second hinge should be located as far as possible away from the line connecting the first and the third hinges. This facilitates a smooth movement of all joints. Occasionally, a less-experienced designer would line up all three hinges, locating the second hinge on the line connecting first and third hinges as shown in Fig. 7.18(c). In this case, the movement of the second hinge is not well defined because it can move to either side of the connecting line. Therefore, the movement is very difficult to initiate in this case. Once it does start moving, it can exhibit a jerking action with large sudden movements.

Gimbal joints can rotate around two axes perpendicular to the joint axis. They are used in situations when there are two principal movements not lying in the same plane as shown in Fig. 7.19. Because hinges, as well as gimbals, do not have the capacity to absorb axial movement, some vertical flexibility of the piping is required. When the piping does not have enough flexibility to absorb vertical expansion, a third flexible joint is needed. This third joint often uses a hinge instead of a gimbal.

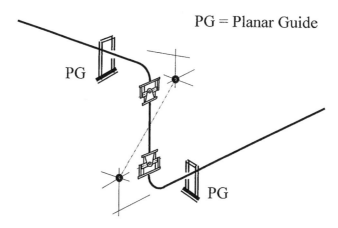

**FIG. 7.19
APPLICATION OF GIMBAL JOINTS**

7.4 SLIP JOINTS

Slip joints allow the contact surfaces of the pipes joined together to slip away from each other. Depending on the modes of the sliding motion, slip joints are divided into two main types: axial slip joint and rotational slip joint. The slip joint has a rugged construction that makes it suitable for hostile environments, such as inside a ditch, underwater, or underground. However, as the sliding surfaces are not perfect seals, seepage may develop along the surface. This leakage concern has prevented this type of joint from being used in hazardous materials. To ensure the tightness of the joint, considerable force has to be constantly maintained on the gasket or packing. This results in a fairly large internal friction force resisting the slipping movement. In some cases, this internal friction force is so huge that the joint simply losses its flexing capability.

An *axial slip joint* allows the pipe to slide into it axially and at the same time allows the pipe to also rotate axially. Axial slip joints come in different styles. For low temperature and low-pressure lines, the compression sleeve as shown in Fig. 7.20(b) and other clamp-on couplings are commonly used for accommodating some moderate amount of expected pipe thermal expansion. The compression sleeve, often referred to as Dresser Coupling, is very popular in water distribution systems where temperature change is mainly attributed to climate change. For large movements at higher temperatures and pressures, an internally guided construction with a secure packing gland as shown in Fig. 7.20(c) is used. Similar to bellow joints, all axial slip joints need main anchors to resist the pressure thrust force. Properly located guides are also needed to ensure column stability. However, in joints with internal guides, the G1 guide as given in Fig. 7.8 is not required.

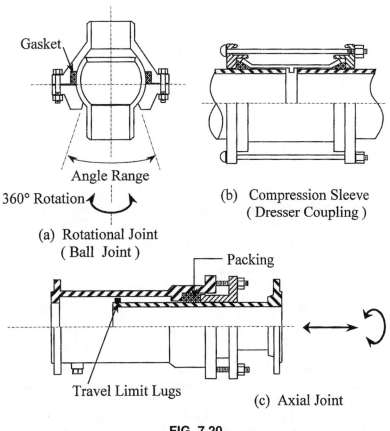

**FIG. 7.20
SLIP JOINTS**

The internal friction force of an axial slip joint is very significant and should be taken into consideration in the design and analysis of piping and supports. Because an axial joint is normally installed to accommodate the expansion of a straight piping run, the anchor design load can be accurately estimated by simple addition of the pressure thrust force plus the joint internal friction force plus the support external friction forces. A computerized analysis may be required for piping systems that are not entirely straight. In this case, internal friction force is simulated as the spring force of the axial flexible joint. Because internal friction force is the product of movement and spring constant, it varies with the actual movement. The analysis may require a couple of iterations to match the spring force with the internal friction force provided by the manufacturer of the joint. For instance, with an expected movement of x and an internal friction force of F, the spring rate of the flexible joint will be F/x. However, once the internal friction force is included, the actual movement would be changed to x'. A revised spring rate, F/x', will have to be used for a re-analysis. The iteration should converge very quickly to reach the exact internal friction force given by the manufacturer.

The axial slip joint can also accommodate axial rotation of the pipe. Occasionally, a joint is specially constructed to accommodate only the axial rotation of the pipe. Because no axial movement is allowed, internal lugs can be installed to resist the pressure thrust force. Therefore, no anchor is needed for this type of axial rotation joint. By strategically locating a couple of these types of rotational joints, the resulting piping system, much like a multi-hinged system, can accommodate very large movements.

A *rotational slip joint* is generally referred to as a ball joint. It is also occasionally called a ball-and-socket joint. As shown in Fig. 7.20(a), the common type of ball joint is constructed with three main pieces. It has an inner ball shaped adapter enclosed by a two-piece, dome-shaped housing. The gasket is placed at the junction of the two housing pieces. The joint is capable of rocking at the specific range permitted by the opening of the outer housing. It is also capable of rotating 360 deg. axially. In the practical sense, it is considered capable of rotating in any direction.

The ball joint is a very rugged component that is suitable to hostile environments, such as offshore and loading dock applications. It can sustain considerable abuse by piping and equipment operators. Again, to maintain the tightness of the joint, sufficient force has to be applied and maintained at the gasket or packing. This results in a considerable friction force creating a fairly large resisting moment against the rotation of the joint. The moment required to rotate the joint is called the break-off moment, whose magnitude is available from the manufacturer of the joint. This break-off moment can have a very significant effect on the flexibility of the piping and has to be taken into consideration in the design and analysis of the piping system.

In the analysis, the ball joint is treated as a generic flexible joint with three directions of rotational flexibility. This, by using the keywords given in Section 7.1.1, is equivalent to FRX, FRY, and FRZ. The spring rates of the joints are determined by the break-off moment and the expected resultant relative rotational movement. Assuming the break-off moment is M_b, and the resultant relative rotational movement is θ, then the spring rates for all directions of the joint are the same at $K_R = M_b/\theta$. That is,

$$\theta = \sqrt{(r_{xa} - r_{xb})^2 + (r_{ya} - r_{yb})^2 + (r_{za} - r_{zb})^2}$$

$$K_R = M_b/\theta$$

$$\text{FRX} = K_R; \quad \text{FRY} = K_R; \quad \text{FRZ} = K_R$$

where r_{xa} is the x rotation at side a, and r_{xb} is the x rotation at side b, and so forth. Side a can be considered as the housing, and side b the ball adapter. Both sides are located mathematically at the same location, which is the center of the joint.

The relative rotational movement is likely to change after the joint frictional moment is included in the analysis. If the calculated relative rotational movement is different than the one assumed in the calculation of the spring rate, the analysis has to be repeated with an adjusted spring rate calculated based on the recently calculated relative rotational movement. The analysis should converge after a couple of iterations.

The above analysis approach is applicable only to single load case static analyses. For multi-load case and dynamic analyses, a suitable approach should be devised on a case-to-case basis.

7.5 FLEXIBLE HOSES

Hoses are used to connect two distant points with considerable relative movements. They are often used to transfer fluid or fluidized material from the storage area to a car, truck, or boat, and vice versa. At a gas station, we often use a hose to fill the gasoline into the car. Most loading and unloading hoses are made of plastic or synthetic rubber. However, the hoses used to accommodate piping movements are generally made of metal. Although the principle of application is the same for both metallic and non-metallic hoses, this section discusses only the metallic hoses.

7.5.1 Types of Metallic Hoses

There are two basic types of metallic hoses: corrugated hoses and interlocked hoses. Figure 7.21 shows the general construction of these two types.

Corrugated hose is constructed with a bellow of very long length. Its behavior is fundamentally the same as the bellow expansion joint. The hose will resist the hoop pressure stress, but cannot sustain the longitudinal pressure stress. As discussed previously, the long bellow also has a tendency to squirm under internal pressure. To resist the longitudinal pressure stress and prevent squirm, corrugated hoses are often constructed with braids wrapping around the outside surface as shown in Fig. 7.22. The braid cover also protects the corrugation from scratch and wear. The braided hose, similar to a

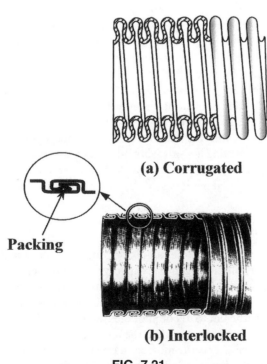

FIG. 7.21
BASIC TYPES OF METAL HOSES

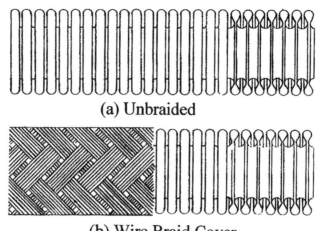

**FIG. 7.22
BRAID COVER OF CORRUGATED HOSE**

tied expansion joint, cannot accommodate any axial movement. On the other hand, the un-braided hose can tolerate only a very small internal pressure.

A corrugated hose is prone to abuse due to the lack of a limiting mechanism. It can be bent beyond the acceptable range without even being noticed. The situation is even more critical for braided hoses. Because its corrugations are invisible from outside, a braided hose does not show readily when it is damaged. Therefore, corrugated hose is not suitable for manual handling in such situations as loading/unloading and switching operations.

Corrugated hose has a continuous metal wall thus making it pressure-tight. It is suitable for handling any type of gas and liquid as long as it is compatible with the hose material.

Interlocked hose is constructed with links that are kept tight with packing material. There are clearances provided between the links that afford the capability of accommodating some axial movement. As the hose is being bent, the clearances gradually close. At a certain point when the clearances are completely closed, the hose becomes stiff and cannot bend any further. This sudden stiffening effect serves as a warning to the handler, preventing the interlocked hose from being over bent. This automatic warning feature makes the interlocked hose especially suitable for manual handling.

The packing mechanism at the interlocked links does not offer a perfect seal. Therefore, the interlocked hose is satisfactory for carrying low-pressure air, steam, and water, but is generally not suitable for conveying gases and "searching" liquids such as kerosene and alcohol. The outside of the interlocked hose is relatively smooth, making it easy to handle without any covering.

7.5.2 Application and Analysis of Flexible Hoses

Hoses can be used to accommodate a wide range of piping and equipment movements. As the hoses are extremely flexible, their installations generally involve very little effort. The force required is just a little more than what is required to carry the weight of the hose. However, there are a few general precautions that need to be exercised. First, the hose should not be subjected to twisting; second, the length of the hose should be sufficient to accommodate the offset and movement; and third the installation space should be adequate to accommodate the length.

The allowable *minimum bend radius* is the most fundamental limitation on the installation of flexible hoses. For interlocked hoses, the limiting radius depends largely on the clearances between links. It has less to do with the stress and fatigue, so it generally has only one limiting radius for all applications. For corrugated hoses, on the other hand, the limiting radius depends on the stress at the corrugations. For pressure hoses with braided reinforcement, the corrugation stress comes mainly from the bending of the hose. Therefore, setting a limitation on the magnitude of the bending will control the corrugation stress. In other words, the installation is acceptable if the hose is not bent beyond the limiting radius. Similar to the situation discussed in the bellow expansion joint, the mode of failure of the hose corrugation is due to fatigue. Therefore, the bend radius limitation depends also on the number of operating cycles expected. Most manufacturers provide two limiting radii, one for static application involving a one time fit-up installation, and the other for operational movement involving many cycles of intermittent flexing. The whole design and installation process involves ensuring that this minimum radius is maintained under the initial layout condition and throughout the expected operation modes.

Figure 7.23 shows an example installation for accommodating a lateral movement of Δ. The design process starts with the determination of the minimum bend radius. This radius is available from the manufacturer's catalog, but we should remember that different operating modes have different limiting bend radii. Assuming that the limiting bend radius is R, the minimum length of the hose is calculated by Eq. (7.32). The required hose length consists of a fitting length, A, transition tangent length, D, and flexing length, 2ℓ. The flexing length is also called the live length, which is calculated as

$$2\ell = 2R\theta = 2R\cos^{-1}\left(\frac{R - \Delta/2}{R}\right) = 2R\cos^{-1}\left(1 - \frac{\Delta}{2R}\right) \tag{7.31}$$

where θ and $\cos^{-1}(\ )$ are expressed in radians. The transition tangent length is normally taken to be the same as the hose diameter, d. The required minimum total develop length of the hose is

$$L = 2A + 2D + 2\ell = 2A + 2d + 2R\cos^{-1}\left(1 - \frac{\Delta}{2R}\right) \tag{7.32}$$

This is a theoretical length to satisfy the expected lateral movement. Once the length is determined, the theoretical design is completed. However, the real problem may have just begun. Normally, in an engineering company, the stress engineer determines the hose length requirement, but the piping designer does the actual layout. The installation space provided by the layout of the piping will generally be somewhere between the projected length and the developed length. It will also require

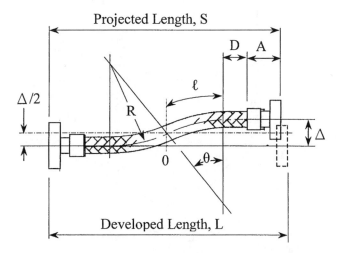

FIG. 7.23
HOSE LENGTH REQUIRED FOR OFFSET

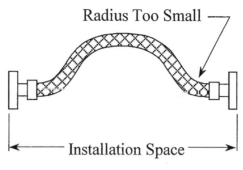

**FIG. 7.24
INSUFFICIENT SPACE**

some flexibility of the piping in the axial direction. Occasionally, a designer may provide too narrow an installation space, thinking that the flexible hose can be easily squeezed into the space. Although a slight squeeze is acceptable to most hose lengths, the necessary squeezing is impossible in many cases. This insufficient installation space has created a considerable amount of unacceptable installations. Figure 7.24 shows the typical situation. The radii at the transition corners are too small and generate excessive stresses. This type of unusually sharp turns can fail after subjecting to just a few cycles of the operational movement.

For a general loop used to accommodate in-plane movements, as shown in Fig. 7.25, the fundamental design factors are still length and space. A sufficient hose length is needed to maintain the minimum radius throughout all the postulated operating modes. At the same time, the space has to be wide enough to accommodate the length without creating too small a radius in any part of the hose. The design in this case may start with the space required. The first step is to determine the closest

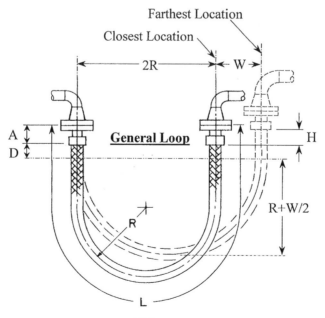

**FIG. 7.25
HOSE LENGTH FOR GENERAL LOOP**

location from all potential movements. This can be either the installed location or the location occurring at some point of time during operation. Obviously, the distance between the two elbows has to be greater than $2R$ or twice of the minimum radius permitted by the hose. To determine the hose length, the potential maximum distance between two elbows also has to be determined. Using the notations given by Fig. 7.25, we find the hose length as

$$L = \pi(R + W/2) + H + 2A + 2D \tag{7.33}$$

where D is the tangent transition length generally set to be the same as the hose diameter. Installation of the hose loop is straightforward. As long as the hose length and piping space are sufficient, no undue bending on the hose is expected. The elbows at the ends are very important. Without them, sharp bending is required to force the hose into the connections.

7.5.3 Analysis of Hose Assembly

The hose assembly is not generally analyzed for forces and stresses involved. As long as the hose length and installation space are determined, no significant resisting force is expected besides the weight of the hose and its content. This is the general practice, partially due to the difficulty of the analysis involving large displacements. However, by considering the layout shape as the neutral baseline, an analysis can be performed to determine the resistance and deformation of the hose assembly subject to end movements and body forces.

The analysis procedure follows a few basic steps. (1) The hose is divided into a series of elements along the layout shape. (2) Each element is considered a bellow having spring rates as given by the manufacturer or determined as bellow elements discussed previously. (3) Because the hose is braided, the axial spring rate is very large and shall be considered inflexible; thus, only lateral spring rates and bending spring rates are considered. (4) At the center of each element a generic flexible joint is assigned and given two lateral springs and two bending springs. This is similar to the gimbal joint with some added lateral flexibility. (5) The elements are considered rigid members, having the same unit weight as the hose and its content. The system is then analyzed in the same manner as a regular piping system.

7.6 EXAMPLES OF IMPROPER INSTALLATION OF EXPANSION JOINTS

One of the reasons that many operating companies and engineering companies shun an expansion joint is due to its unpredictability. Very often the joint is just not functioning as intended. Very frustratingly, nobody seems to know the real reason for some of these problems. We have seen some seemingly very well designed installations that simply refuse to work. In the following, a few case histories are listed to demonstrate that many of the failures are actually attributable to a lack of common sense. Unfortunately, this lack of common sense can ultimately shut down a whole plant.

7.6.1 Direction of Anchor Force

The direction of the anchor force is an important factor in the design of the main pressure thrust anchor. To achieve the safest and most economical design, the strongest orientation of the anchor is always aligned with the direction of the anchor force. The direction of the anchor force is apparent in most cases; yet, occasionally, it can vary according to the modes of operation. Figure 7.26 shows an anchor that failed due to a change of the operating mode.

In Fig. 7.10, we have noted that when there is a valve placed in between the bellow and the anchor, the situation with the valve closed should be considered as one of the design conditions. In other words, the intermediate anchor shall be designed with the same load as for the main anchor. The full pressure thrust force has to be used.

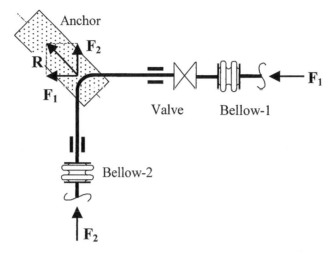

FIG. 7.26
POTENTIAL DIRECTIONS OF ANCHOR FORCE

The situation shown in Fig. 7.26 is not a problem of force magnitude. Rather, it is a problem of force direction. The anchor is designed for the combination of two full bellow pressure thrust forces, F_1, and F_2, with $F_1 = F_2$. The resultant design load is R, which is 1.414 times as high as the full bellow pressure thrust force. This resultant force, however, is acting in the 45-deg. oblique direction from the direction of the individual bellow pressure thrust force. A narrow rectangular anchor block oriented in this 45-deg. direction was constructed for this service, with the loading point located at the tip. The anchor failed when it was required to shut off the valve.

When the valve is closed, F_2 becomes zero, as bellow-2 has no internal pressure. The anchor load is actually reduced to just 1.0 times the full bellow pressure thrust force. However, the smaller force is now acting in an eccentric oblique direction to the anchor block. Although the anchor is strong enough for the larger force acting in the intended direction, it is not strong enough to resist the smaller force acting eccentrically from the design direction.

There are two possible scenarios in the problem. One is that the engineer simply overlooked the operating mode involving the shutoff of the valve. The other is that the engineer transmitted only the largest force with its direction to the foundation designer. Even experienced engineers can often commit the latter mistake.

7.6.2 Tie-Rods and Limit Rods

Expansion joints are often installed with tie-rods and limit rods. Tie-rods are used to resist the longitudinal pressure thrust force, thereby preventing the bellow from being overstretched by pressure during normal operation. Limit rods, on the other hand, are used to protect the bellow from being overstretched during the events of anchor failure or pipe buckling. A misjudgment of the purpose of tie-rods can lead to piping failure or equipment operational difficulty.

Figure 7.27(a) shows a bellow fitted with tie-rods. As discussed previously in this chapter, this type of expansion joint is used to accommodate lateral movement and rotation. It cannot accommodate axial movement, except the absorption of the expansion of the piping assembly located between the two tie-rod connections. These are tie-rods, as can be judged from the fact that no anchor is placed to resist the bellow pressure thrust force. In this case, the nuts on the tie-rods are close against the collar of the attachment. If the nuts are turned away from the collar to leave a gap either during operating or at idle condition, the gap will be closed as soon as the line is pressurized. Uninterrupted, this will

Flexible Connections 243

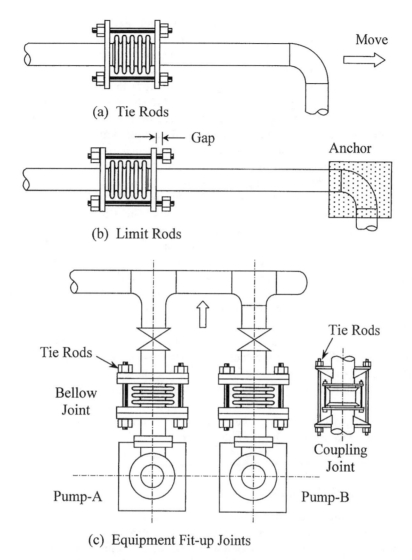

FIG. 7.27
PROBLEMS ASSOCIATED WITH TIE-RODS AND LIMIT-RODS

eventually stretch the bellow to beyond its allowable safe limit. The nut locations, thus the length of the joint, are set by the manufacturer according to the design specification. They should not be moved under any operation or idle condition. If they are to be moved for installation or maintenance purposes, they should only be moved by a minimal amount, enough for dismantling the joint. Bellow deformation should always be maintained within the allowable working limit. The nuts shall be returned to their proper location before the piping is put back into operation. It is recommended that nut positions be locked by double nuts.

Figure 7.27(b) shows the limit rods. In most piping systems handling hazardous materials, the design specification often calls for limit rods to be installed on all bellow expansion joints designed for absorbing axial movements. These types of installations require comprehensive anchor and guide systems as shown in Figs. 7.8 and 7.9. To guard against the failure of the main anchor due to washout or other situations, limit rods are provided to prevent the bellow from being excessively overstretched during the event. Proper gaps are provided between the nut and the collar of the rod attachments. During operation, the nuts are easily turned in and out changing the gap size. If an operator mistakenly thinks that the rods are tie-rods for resisting operation pressure, the bolts might be improperly tied, leaving no gaps at the nuts. If this occurs, the anchor and the connecting equipment could be pulled to failure as the system contracts while cooling down to the ambient temperature. Because the nuts are so easy to turn, it is not unusual to see field and operating personnel turn the nuts in a little one day and out a little the next day. A proper warning label may be attached to the joint for this case, but it is not foolproof because the label can get lost or worn out in the long run. Locking devices using double nuts or other means are recommended. Frequent training of field personnel may also be required.

Figure 7.27(c) shows the joints used in the fit-up of multi-machinery installations. This involves mostly large pipes operating at close to ambient temperature. Cooling water and water supply systems often require multiple pumps operating in parallel with one idling as backup. Spaces for connections are very tight requiring flexible connectors for practical fit-up. The flexible joints used range from synthetic rubber joints, to slip couplings, to metal bellow joints. These joints have generated more than their fair share of operational problems to the connecting equipment, mostly in the case of pumps.

Because the piping operates mostly at ambient temperature and the branch legs are generally not very long, the thermal expansion in general is not a problem. The problem is the unbalanced pressure thrust force acting on the equipment. The loosening of the tie-rod nuts in some of the units is the source of this difficulty. In many cases, the nuts are not checked to the specified tightness after the installation. When a structure operates with loose nuts, the bellow thrust force is not balanced. The unbalanced pressure thrust force can push the equipment out of alignment, causing vibration, overheating, and other problems on the equipment. The bolts should be tight, but not overly tight as to pull the equipment out of alignment. More comprehensive discussions on equipment loads are given in a separate chapter.

Requirements of the tie-rods are generally easy to grasp for bellow-type flexible joints. However, it is not as obvious in coupling-type joints. Because the slipping surface of the coupling joint is not visible from the outside, engineers often mistake the packing stuffing bolts as the tie-rods and forget the need for a separate set of tie-rods as shown in the figure.

7.6.3 Improperly Installed Anchors

The anchor is so important in the application of bellow expansion joints that many engineers seem to automatically place an anchor or anchors in an expansion joint installation. The anchor is required in most bellow expansion joints that are not tied with tie-rods. An anchor placed at the piping that has a tied expansion joint is an invitation to an operational problem. Figure 7.28 shows one such installation that might have crippled the operation of an entire plant.

As discussed in earlier chapters and will be more fully discussed in a later chapter, piping loads are generally very critical to rotating equipment. A steam turbine drive, for instance, can only sustain a very small piping load such that a practical piping layout without using a flexible joint can hardly meet the load limitation. An expansion joint is often needed to solve the equipment load problem. The layout shown could have accomplished this task, except that the installation of the anchor completely spoils the attempt.

With the anchor installed, the pipe will not move at the anchor point. Because the anchor point does not move, the pipe located between the anchor and the turbine expands into the joint. This leaves the nuts loose with gaps. When tie-rods become loose, they no longer resist the pressure thrust force. The thrust force is thereby absorbed by the anchor and by the equipment. Although the anchor might have been designed to handle the force, the equipment is generally not able to resist this type of force without creating shaft misalignment. This misalignment results in the vibration of the machinery and overheating of the bearings and seals.

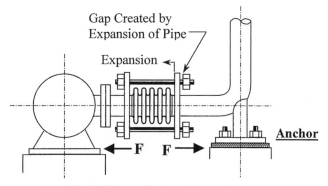

(a) **WRONG** Installation with an **Anchor**

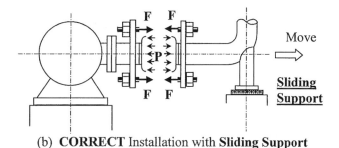

(b) **CORRECT** Installation with **Sliding Support**

FIG. 7.28
A SITUATION WHEN ANCHOR IS THE PROBLEM

The correct installation is to allow the support point to slide as shown. The movement of the elbow ensures the tightness of the tie-rods, because the pressure keeps pushing the elbow until the rods are tight. The tightened tie-rods can then absorb the bellow pressure thrust force, leaving no unbalanced pressure thrust force on the machine.

REFERENCES

[1] EJMA, *Standards of the Expansion Joint Manufacturers Association, Inc.*, Expansion Joint Manufacturers Association, Inc., Tarrytown, NY.
[2] "Appendix X, Metallic Bellows Expansion Joints," ASME B31.3, *Process Piping*, American Society of Mechanical Engineers, New York, New York.
[3] Becht IV, C., 2004, *Process Piping — The Complete Guide to ASME B31.3, 2nd ed.*, ASME Press, New York, NY.
[4] Tube-Turn, 1964, *Bellows Expansion Joints*, Catalog from Tube Turns, Div. of Chemetron Corporation, Louisville, KY.
[5] Pathway, 1987, *Flexway and X-Press, Round Metal Expansion Joints Manual 185G*, Catalog from Pathway, A subsidiary of Dover Corporation, El Cajon, CA.
[6] PT&P, 1997, *Expansion Joint Catalog*, Piping Technology & Products, Inc., Houston, TX.
[7] Markl, A. R. C., 1964, "On the Design of Bellows Elements," *Piping Engineering*, 1969, Tube Turns Division of Chemetron Corporation, Loisville, KY.

CHAPTER 8

INTERFACE WITH STATIONARY EQUIPMENT

The main purpose of piping stress analysis is to ensure the structural integrity of the piping and to maintain the operability of the system. The latter function is mainly to ensure that the piping forces and moments applied to connecting equipment are not excessive. Excessive piping loads may hinder the proper functionality of the equipment. The function of maintaining system operability requires the investigation of the interface effects with connecting equipment. There are three main interface effects between piping and connecting equipment:

(a) *The loads imposed on piping from equipment*. This involves mainly the expansion of the equipment. The expansion of a vessel, for instance, can be large enough to have a considerable effect on piping stress. Occasionally, the vibration of equipment, although not detrimental to the equipment itself, can amplify through the piping to create a problem.
(b) *Flexibility of equipment*. The equipment is generally considered a rigid member in the analysis of piping. However, the flexibility, either from equipment supports or from the equipment itself, may significantly affect the results of the analysis.
(c) *Effect of piping loads on equipment*. The equipment is generally designed for loads based on the function of the equipment. It may or may not consider the potential piping loads that it might eventually have to take. It is important to ensure that the piping loads are acceptable to the equipment whether the equipment is designed for any piping loads or not.

This chapter discusses the interfaces of piping with stationary equipment, such as valves, heaters, and pressure vessels. The interfaces with rotating equipment will be discussed in the next chapter.

8.1 FLANGE LEAKAGE CONCERN

The possibility of flange leakage occurring well before the failure of the pipe or the flange is a major concern of piping engineers when the allowable piping expansion stress-range runs well over the yield strength of the piping material. Even with the structural integrity of the flange intact, the system is still not functional if the flange tightness is not maintained. Flange leakage is a very complex problem involving many factors. Inadequate pressure rating, poor gasket selection, insufficient bolt loading, temperature gradient, bolt stress relaxation, piping forces and moments, and so forth, can all cause leakage at a flange. In this chapter, we will limit our discussion to the effects of piping forces and moments.

The pipe loads applied to the flange have received considerable attention from piping engineers. Many new construction projects require that these piping loads be evaluated to assure the tightness as well as the structural integrity of the flanges. This problem has been broadly investigated. Blick [1], Koves [2], Tschiersch and Blach [3] have presented some theoretical treatments, and Markl and

George [4] have conducted a large number of experiments on the subject. From the extensive tests made on 4-in. Class 300 American Society of Mechanical Engineers (ASME) B16.5 [5] flanges, Markl and George have found that "even under unusually severe bending stresses, flange assemblies did not fail in the flange proper, or by fracture of the bolts, or by leakage across the joint face. Structural failure occurred almost invariably in pipe adjacent to the flange, and in rare instance, across an unusually weak attachment weld; leakage well in advance of failure was observed only in the case of threaded flanges." These test results pretty much confirm the expectation of the conventional flange selection process. In the conventional design process, the flange is selected based on the pressure rating. If the design pressure is smaller than the pressure rating of the flange, then the flange is expected to function just as well as the pipe of the same pressure rating.

The test results by Markl and George are so convincing that, for B16.5 flanges, as long as the design or service pressure is within the pressure rating of the flange, the flanges are considered capable of taking the same pipe load as that of the pipe-to-flange welds with appropriate stress intensification factors applied [6]. The stress intensification factor for welding neck flanges is 1.0. However, due to the large varieties of flanges involved in many different types of industries, the question remains as to whether the 4-in. Class 300 data is applicable to all flanges.

In this "design by analysis" era, engineers are generally not satisfied with "rule of thumb" approaches. A more definite evaluation approach is preferred. That is when the confusion starts. Several methods have been proposed and used, yet they give widely different results. To determine the merits of each method, the standard flange design procedure will be briefly discussed first.

8.1.1 Standard Flange Design Procedure

The standard flange design procedure was first developed in the 1930s and was adopted by the ASME Code for Unfired Pressure Vessel in 1934 [7, 8]. Through decades of practices and refinements, the current ASME Boiler and Pressure Vessel Code, Section VIII, Division 1, Appendix 2 design rules, generally referred to as Appendix 2 rules, have been universally adopted for the design of flanges subject to internal pressure. Countless flanges have been designed by this simple and easy-to-use cookbook approach.

Using an integral type flange as an example, the flange is idealized into three ring sections as shown in Fig. 8.1. It consists of the flange ring, hub ring, and an effective length of pipe section. The stresses at each ring section are calculated by applying a circumferentially uniform bending moment at the face of the flange ring. This bending moment represents the total loading applied at the flange. Once

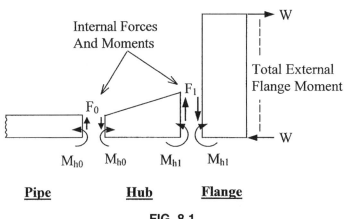

FIG. 8.1
IDEALIZED FLANGE ANALYTICAL MODEL

this moment is known, the stresses at the flange and at the hub are calculated with code formulas and charts. The main task of the designer is to determine the magnitude of this bending moment under seating and operating conditions. Appendix 2 rules do not cover the pipe force and moment.

Figure 8.2 shows the loadings on a flange subject to internal pressure. Their magnitudes are calculated as follows:

$$\begin{aligned}
H_D &= \frac{\pi B^2}{4} P, \quad \text{Pressure end force from pipe (operating condition)} \\
H_T &= \frac{\pi (G^2 - B^2)}{4} P, \quad \text{Pressure force at flange face (operating condition)} \\
H_{G,2} &= b\pi G y, \quad \text{Gasket force for seating (seating condition)} \\
H_{G,1} &= m(2b)\pi G P, \quad \text{Gasket force for sealing (operating condition)}
\end{aligned} \quad (8.1)$$

where b is the effective gasket seating width, m is the gasket factor, and y is the minimum required gasket seating stress. These b, m, and y values are given by the Appendix 2 rules. Due to the rotational deformation (cupping) of the flange, gasket stress is not uniform. In general, the outer edge receives a much higher stress than average, thus making the outer rim portion of the gasket the effective area. The effective gasket seating width is roughly equal to one-half of the gasket contact width, and the gasket load diameter, G, is located between the outer contact diameter and the mean contact diameter of the gasket.

The bolt load, W, is the summation of H_D, H_T, and H_G under each of the two conditions. However, to guard against overstressing the flange from the actual tightening of the bolts, the design bolt load is further adjusted by the total allowable bolt force. The total allowable bolt force is calculated by

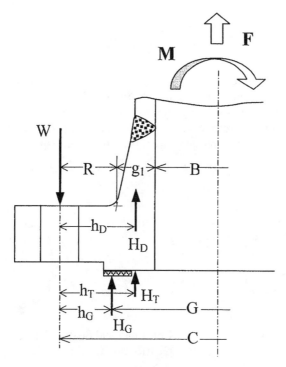

FIG. 8.2
LOADINGS AT FLANGE

multiplying the total bolt root areas by the bolt allowable stress. The design bolt load is taken as the average of the total allowable bolt force and the bolt force required by the above H_D, H_T, and H_G. This additional bolt force, over what is required for pressure and seating, balances with H_G, thus causing H_G to also increase by the same amount.

The total flange moment is calculated by using the bolt centerline as the pivot. The moment arms are taken as

$$\begin{aligned} h_D &= R + g_1/2 \\ h_G &= (C - G)/2 \\ h_T &= h_G + (G - B)/4 \end{aligned} \tag{8.2}$$

The total flange moment becomes

$$\begin{aligned} M_O &= H_D h_D + H_T h_T + H_{G,1} h_G \quad \text{Operating condition} \\ M_G &= (H_{G,2} + H_B) h_G \quad \text{Seating condition} \end{aligned} \tag{8.3}$$

H_B is the additional bolt force of the actual design bolt force used over the required bolt force based on H_D, H_T, and H_G. Once M_O and M_G are determined, flange stress and hub stress are calculated using the formulas and charts given. Three stresses — longitudinal hub stress, radial flange stress, and tangential flange stress — are calculated. The actual stress calculation is beyond the scope of the book. We will only use the procedure outlined to discuss the various methods proposed for evaluating the pipe force and moment.

8.1.2 Unofficial Position of B31.3

ASME B31 Mechanical Design Committee Report [9] issued for B31.3 [10] has stipulated that the moment, M_L, to "produce leakage" of a flanged joint with a gasket inside the bolt circle can be estimated by

$$M_L = \frac{C}{4}(S_b A_b - P A_p) \tag{8.4}$$

where A_p is the area of the circle to the outside of gasket contact, S_b is bolt stress, and A_b is total root area of flange bolts. This equation uses a rigid flange model having the pipe moment resisted by the bolt and gasket combination. By idealizing the bolt force and also the sealing force as distributed line loads located around the bolt circle as shown in Fig. 8.3(b), the residual sealing force per unit circumference, after subtracting the pressure force, is uniform and equal to $(S_b A_b - P A_p)/(\pi C)$. With a

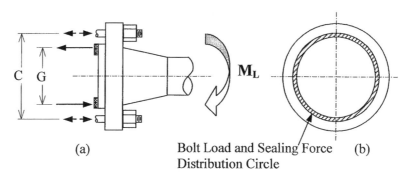

FIG. 8.3
GASKET AND BOLT FORCES DUE TO BENDING MOMENT

bending moment applied, the bolt force and thus the sealing force will be linearly redistributed across the diametrical direction. The maximum and minimum forces per unit circumference due to the moment occur at two extreme points and equal to $M_L/(\pi C^2/4)$. The moment will cause the sealing force at one end to increase and at the other end to decrease. The flange is assumed to leak when the sealing force, after subtracting pressure and moment forces, at any point of the circumference is zero. In other words, the flange will leak when $(S_b A_b - PA_p)/(\pi C) - M_L/(\pi C^2/4) = 0$.

This equation is not officially included in the code, but is being used by some engineers. However, because it is not a design formula, it is occasionally misapplied. In using this formula, there are a few things that need to be noted.

(1) Many engineers have mistaken the M_L as the allowable moment, which is not correct. Equation (8.4) is used to predict the moment to produce leakage. We do not normally design a flange to leak. Therefore, to use the formula, some type of margin has to be included. This can be done by applying a proper safety factor or to set aside some residual gasket loading for maintaining tightness. The latter is given by Appendix 2 rules [8] as $H_{G,1} = m(2b)\pi GP$, which is also given in Eq. (8.1). In other words, the total bolt force has to be subtracted by $H_{G,1}$ as well as PA_p to become the surplus sealing force before being converted to the allowable bending moment.

(2) Bolt stress, S_b, can be taken as the basic allowable stress of the bolt. However, it should be noted that the flange might not be designed for the full allowable stress of the bolt. Assuming the flange is designed for the full basic allowable stress of the bolt, then the use of $S_b = (S_c + S_h)/2$ may be considered as having a safety factor of 1.5. This is because the allowable stress is generally increased by 50% when dealing with piping loads. (See also item (B) in Section 8.1.3). Occasionally, engineers might even use the empirical stress [11] actually applied on the bolt while tightening. This stress can be as high as twice the allowable stress for 1-in. (25 mm) bolts as an example. This single application tightening stress shall not be considered as available for repetitive piping loads.

(3) Figure 8.3 shows the relationship of gasket force, bolt force, and the pipe moment. Theory and experiments have shown that with a given direction of moment, the load at each particular bolt can either increase or reduce [1], depending on the relative stiffness of the flange, gasket, and bolt. The bolt load in response to the pipe moment changes very little. The pipe moment is resisted mainly by the gasket. Therefore, it appears to be logical to use the gasket loading diameter, G, instead of the bolt circle diameter, C, in Eq. (8.4).

Equation (8.4) is simple and its intent clear, but is not very easy to apply. Judging from the above listed concerns, clear specifications are needed for using the equation.

8.1.3 Equivalent Pressure Method

The fact that we have a set of very reliable formulas and rules for designing flanges subject to internal pressure provides the incentive for evaluating the pipe force and moment based on these rules. To use the pressure design procedure, the first thing needed is to find the relationship between the internal pressure and the pipe force and moment. The M. W. Kellogg [12] has suggested an equivalent pressure approach. The approach assumes that the action of the moment and force is equivalent to the action of the pressure, which produces a gasket stress that is the same as the gasket stress produced by the force and the moment. Figure 8.4 shows this equivalence relation. By letting S_F represent the gasket stress due to force, S_M the maximum gasket stress due to moment, and S_P the gasket stress due to equivalent pressure, we have

$$S_F + S_M = S_P$$

or

$$\frac{F}{\pi Gb} + \frac{M}{\pi G^2 b/4} = \frac{\pi G^2 P_e/4}{\pi Gb}$$

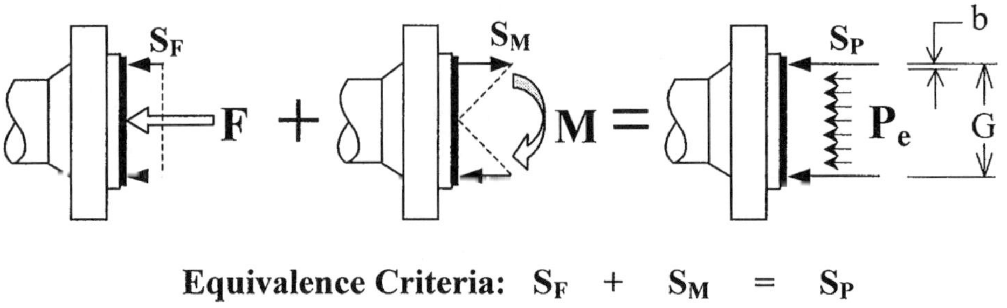

Equivalence Criteria: $S_F + S_M = S_P$

FIG. 8.4
EQUIVALENT PRESSURE DUE TO FORCE AND MOMENT

$$\text{i.e.,} \quad P_e = \frac{4F}{\pi G^2} + \frac{16M}{\pi G^3} \tag{8.5}$$

Once equivalent pressure P_e is determined, flange stress can be calculated using Appendix 2 formulas and rules. The above equivalent pressure formula is considered conservative because the maximum gasket stress due to moment occurs only at the small areas near two diametrically extreme points. The bending moment is the dominant loading.

In practice, equivalent pressure is combined with design pressure to become the total equivalent pressure. That is,

$$P = P_d + P_e \tag{8.6}$$

The equivalent pressure approach is very popular among piping engineers. There is little doubt about the usefulness of this approach, but its applications are still not uniform. The following are two major diversities:

(A) Rating table lookup. This approach has been adopted by a number of computer software packages. It uses the rating table as the sole evaluation tool. Once the total equivalent pressure is calculated, the rating table is checked for this total pressure. The pipe force and moment is considered acceptable to the flange if the total equivalent pressure is within the flange rating pressure. (See Table 1.1 for pressure ratings.) This method is simple and conservative, but has very limited practical use. It can be used for quick checking the acceptability of the piping load. However, because it is exceptionally conservative when used as a definite evaluation rule, it usually creates — rather than solves — problems.

In designing a piping system, the designer generally selects the flanges and valves based on the pressure rating. For a given design pressure, the flange selected will have a rating pressure equal to or greater than the design pressure. It is not unusual to select the flange having the rated pressure the same as or close to the design pressure. In this case, the flanges are still expected to perform satisfactorily under moderate piping forces and moments, in addition to the design pressure. Experiences do confirm this expectation. Had the piping system in this case been evaluated for leakage using equivalent pressure with the rating table lookup procedure, most of the flanges would have been disqualified even with minimal piping loads. The total equivalent pressure, including design pressure and pipe forces and moments, is generally greater than the rating pressure, which is almost entirely taken up by the design pressure in this case. The flanges would have to be replaced with ones of higher rating. This unusual requirement can generate a shocking impact to the plant involved. Some systems that have gone through this type of leakage evaluation have installed different classes of flanges at different locations on the same piping system. This creates not only waste and confusion, but also the quality assurance problem of installing the right flange at the right location.

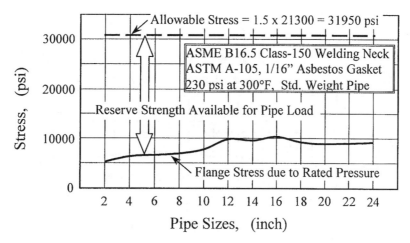

FIG. 8.5
ASME B16.5 CLASS 150 FLANGES

It is to be emphasized that the rating table lookup approach is a quick and conservative evaluation method. However, a flange connection is still most likely satisfactory even if it fails to pass the rating table look up evaluation.

(B) Flange stress calculation. In addition to providing sufficient strength against the rated pressure, the rating of the standard flanges [5, 13–15] also provides some reserve strength to cover other potential loads, including pipe forces and moments. The reserve strength, as judged by ASME Section-VIII Appendix 2 rules, is not uniform across sizes and classes. Smaller sizes and lower pressure classes generally have higher reserve strengths. Figures 8.5, 8.6, and 8.7 show these trends [16]. The exact amount of reserve strength can only be determined by calculating the flange stress.

The stresses given in Figs. 8.5, 8.6, and 8.7 are calculated according to ASME Section-VIII Appendix 2 procedures. Three stresses are calculated: longitudinal hub stress (S_H), radial flange stress (S_R), and tangential flange stress (S_T). The allowable value for these calculated stresses depends on the nature of the loading. For the pressure design, the allowable for S_H is $1.5S_h$, and the allowable for

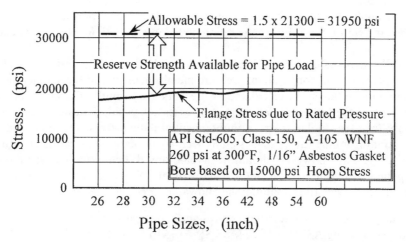

FIG. 8.6
API STD-605 CLASS 150 FLANGES

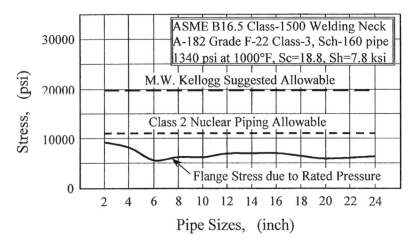

**FIG. 8.7
ASME B16.5 CLASS 1500 FLANGES**

each of S_R and S_T is $1.0S_h$. S_h is the allowable stress of the flange material at design temperature. When the piping load is included, the allowable stresses are increased to $1.5S_h$ for each of the three stresses. These same allowable stresses are applicable to the flanges subject to steady static loads as well as thermal expansion loads. These allowable values, however, are considered too small when dealing with thermal expansion load ranges at high operating temperatures. Based on experiences, M. W. Kellogg [12] suggested an allowable of $1.5(S_c + S_h)/2$ for bolt stress and for each of the three calculated flange stresses, for loads including thermal expansion. This is 1.5 times of the average of hot and cold allowable stresses. S_c is the allowable stress at ambient temperature.

8.1.4 Class 2 Nuclear Piping Rules

The Class 2 nuclear piping code [6] has comprehensive rules regarding the evaluation of flanged connections subject to piping loads. Based on this code, the flanged connections can be evaluated by any of the following methods.

1. Any flanged joint

This method is applicable to any type of flanged joint. It is based on the equivalent pressure approach described in Section 8.1.3. Either the rating table lookup method or stress calculation method can be used.

The rules are the same as in Section 8.1.3, except that both bending and torsional moments are evaluated, but separately. The longitudinal pressure membrane stress at the smaller end of the hub is added to the longitudinal hub stress calculated by the standard formula. The allowable stresses are all set to $1.5S_h$ for each of the three calculated flange stresses. As usual, the calculated three stresses are longitudinal hub stress, radial flange stress, and tangential flange stress. These allowable stresses are significantly different from the values suggested by Kellogg on flanges operating at high temperatures as shown in Fig. 8.7. This, however, should not create any difficulty for nuclear piping due to the moderate temperature environment in water reactor plants. When dynamic loads are included, the equivalent pressure is halved, thus doubling the allowable forces and moments.

2. Standard flanged joints at moderate pressures and temperatures

Flanged joints confirming ASME B16.5, MSS SP-44, API Std-605, or AWWA C207 Class E (275 psi), with design and service pressures below 100 psi, and design and service temperatures below 200°F, have an allowable piping moment of

$$M_A \leq \frac{A_b S_b C}{4} \tag{8.7}$$

This formula is similar to Eq. (8.4) by ignoring the pressure term. By ignoring the pressure term, it may appear to be very un-conservative. However, the formula is based on two prerequisites that make it sufficiently conservative: (1) the flange has to be manufactured by an established standard; (2) the operating pressure is less than 100 psi, and the operating temperature does not exceed 200°F. For a flange made by one of the standards listed above, the minimum rating pressure is 235 psi for Class 150 or equal. This means that the 100-psi limiting pressure is less than one-half of the rated pressure. From Fig. 8.5, it is clear that flange stress due to one-half of the rated pressure for Class 150 ASME B16.5 flange is negligible compared with the allowable stress. By taking into account the inherent conservatism of the standard flange, it is obvious that Eq. (8.7) is still fairly conservative. The allowable moment is doubled when the dynamic loads are included.

One thing that the code does not mention is the special case with API Std.-605 Class 75 flanges. API-605 Class 75 has a rating pressure of only 140 psi, which is very close to the 100-psi limiting pressure. Therefore, there is a concern that Eq. (8.7) may not be sufficiently conservative for API-605 Class 75 flanges.

3. ASME B16.5 flanged joints with high strength bolting

ASME B16.5 was available before the standard flange design procedure was finalized. Without clear-cut procedures, the design tends to be more conservative. This makes B16.5 more conservative than other standards. This evaluation is applicable only to B16.5 flanges using bolting material having an allowable stress of not less than 20,000 psi at 100°F. For flanged joints in this category, the allowable bending or torsional moment (considered separately) is

$$M_A \leq \frac{C}{4} S_{bm} A_b \frac{S_y}{S_{yn}} \tag{8.8}$$

where S_{bm} = 12,500 psi (86.2 MPa) is fixed for the reserve bolt stress available for pipe moment and S_{yn} = 36,000 psi (248.2 MPa) is fixed for the nominal yield strength of the flange material. S_y is the yield strength of the flange material at design temperature. The allowable moment is doubled when the dynamic loads are included.

8.2 SENSITIVE VALVES

There are many valves required in a plant to effectively manipulate and control the process flows. Some of these valves are sensitive to piping loads due to either their limited stroking forces or their weaker cross-sections. A safety relief valve, for instance, operates upon the balance of the pressure force and the set spring force. In a sense, the set spring force determines at what pressure the valve shall pop open to safeguard the system. The pipe force, if large enough, may strain the valve body so much as to create binding between the valve stem and its guide. This will create an obstruction or generate an excessive friction force, which is not included in the original balance equation of pressure force and spring force. As a result, the valve will open at a higher pressure than originally set. The same phenomenon is also applicable to some of the control valves.

Normally, valves are made as strong as the connecting pipe of the same size and thickness. However, due to control characteristics requirements and possibly economic reasons, safety relief valves and control valves are often one size smaller than the main connecting pipe as shown in Fig. 8.8. The inlet of a safety relief valve is also generally one size smaller than the outlet. This makes the inlet connection two sizes smaller than the main outlet piping.

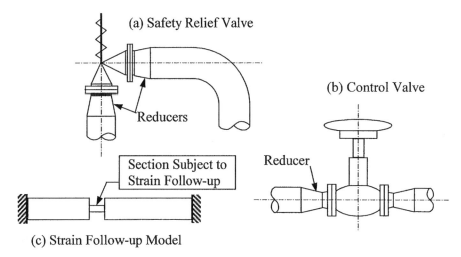

**FIG. 8.8
SENSITIVE VALVE CONNECTIONS**

The smaller size of the valve creates a local weakness in the system that promotes a strain follow-up or elastic follow-up when the stress exceeds the yield point. As shown in Fig. 8.8(c), when a short weak piece is connected to longer and larger pieces, the stress at the weak piece may reach the yield point, whereas the stress at the larger piece could still be well within the elastic limit. Because the yielding weak piece can take no more or little additional force, it ends up absorbing all the additional expansion of the larger pieces, which require additional forces to produce additional deformation. This strain follow-up can eventually overstrain the weak piece, which corresponds to the relief valve or the control valve. Therefore, special attention is needed regarding the pipe loads on these sensitive valves. The codes do caution about the strain follow-up phenomena, but no definite limitation criterion is given. To mitigate the problem, there are several approaches adopted by experienced engineers, some of which are outlined below.

(1) Reduce the allowable stress by 50%. Due to the shake-down phenomenon, the allowable stress for the primary plus thermal expansion range, not including local peak stress, is set as high as twice the yield strength. By reducing the allowable stress by half, the maximum primary plus thermal expansion stress is limited to below the yield point thus preventing the strain follow-up.
(2) By considering thermal expansion as a primary load. By evaluating the thermal expansion as a primary load, the stress is automatically limited to within the yield strength.
(3) Using a minimum of Class 300 valves. In addition to limiting the stress of the connecting pipe, some design specifications require that the control valves and safety relief valves have a minimum flange class of 300. Although lighter classes of valves may be satisfactory stress-wise, they may not be stiff enough to prevent excessive deformation.

Some engineers may consider the above design approaches too conservative. However, the strain follow-up and excessive deformation at those sensitive valve locations are very likely to occur. Engineers should at least be aware of the problem.

8.3 PRESSURE VESSEL CONNECTIONS

Very high percentages of the piping are either starting from or ending at a vessel. To ensure the integrity of these connections, three things need to be considered in the design analysis of the piping system: (1) loading imposed to the piping by the vessel; (2) flexibility of the vessel; and (3) effect of the piping load on the vessel.

8.3.1 Loadings Imposed to Piping from Vessel

The most common load imposed by the vessel is displacement of the vessel nozzle. The displacement can come from the thermal expansion of the vessel, the settlement of the vessel foundation, the movement caused by earthquake and wave motion, and so forth. Except for thermal expansion displacement, the other displacements are generally given in the design specification. Generally, it is the piping engineer's duty to determine the thermal expansion displacement of the nozzle connection and take it into account in the design analysis.

Figure 8.9 shows the general layout of a horizontal vessel. There are four major nozzles, numbered 1 through 4, for connecting process piping. The thermal expansion movement at each nozzle has to be figured out before the piping is analyzed for flexibility and stress. To calculate the thermal expansion at the nozzles, the fixed point from which the vessel expands has to be determined first. This starts with the selection of the fixed support.

A horizontal vessel is generally supported with two saddled supports. One of the supports is fixed, whereas the other can slide within slotted bolt holes. The fixed end is selected strategically so the expansion of the vessel cancels a portion of the expansion of the piping. This arrangement results in the minimum relative expansion, thus minimizing the expansion force and stress. The fixed end has to be selected at a very early stage in the project so the vessel can be fabricated at the same time the piping is being designed. In some organizations, a planning drawing is used for these types of early comments from all engineering disciplines.

The fixed support determines the zero expansion point in the horizontal direction. Vertically, the vessel is generally assumed as fixed at the bottom of the vessel as shown in the figure. This assumption

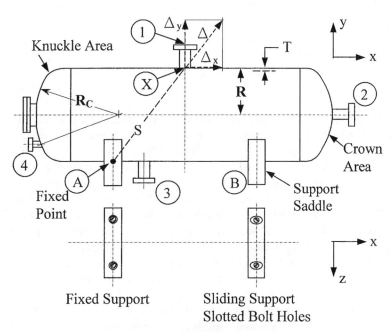

FIG. 8.9
VESSEL CONNECTIONS

ignores the vertical expansion of the support saddle. A more accurate fixing position can be determined from the actual temperature distribution of the saddle. Once the fixed point is determined, the thermal expansion at each nozzle is calculated by the expansion rate times the distance between the nozzle and the fixed point.

The official interface with the vessel is the face of the nozzle flange. However, for the piping design analysis, the piping model is extended to the nozzle-shell intersection, such as point X at nozzle 1. This is to include the flexibility of the nozzle neck and also to incorporate the shell flexibility. It is customary to calculate the thermal expansions for all nozzles at the same time and put them on the vessel drawing for use by all related piping engineers.

The expansion of point X is calculated by multiplying the distance, S, with the thermal expansion rate of the vessel shell. In case the vessel shell temperature is not uniform, an average expansion rate can be used. This expansion is Δ in the A-X direction. However, for piping analysis, this expansion is decomposed into directions coinciding with the global coordinates adopted for the analysis. In this case, it is decomposed into Δ_x and Δ_y in the x and y coordinates, respectively. Of course, Δ_x can also be directly calculated by multiplying the x offset between A and X by the expansion rate, and so forth. Expansions for all other nozzles are determined in a similar manner.

8.3.2 Vessel Shell Flexibility

The vessel shell has significant flexibility that can significantly affect the result of the piping analysis. However, due to mostly psychological concerns, the inclusion of vessel flexibility in the piping analysis is still not universal. Some vessel engineers worry that the inclusion of shell flexibility will ultimately result in a stiffer piping system that might cause damage to the vessel. This is partly true, but it mostly has an adverse effect on the quality of the plant. More often than not, we would see flimsy piping hanging off the vessel just for the sake of flexibility. This is mainly due to an inconsistency of design practices. On one hand, the piping is designed by piping engineers who consider the shell as inflexible (rigid), and on the other, the piping load is evaluated by vessel engineers who consider the shell as flexible. This type of double standard approach results in a very low allowable piping load for a thinner vessel wall, whose flexible nature cannot be counted in the calculation of the piping load.

Treating the vessel connection as rigid, although conservative for thermal expansion analysis, can result in a very uneconomical design. Sometimes, it can also become very un-conservative in a high occasional load environment. The system given in Fig. 8.10 can be used to explain these effects.

The system was analyzed by Stevens et al [17] using theoretical flexibility parameters developed by Bijlaard [18, 19] and correction factors based on the work of Cranch [20]. In the figure, the analysis results are tabulated for the case where the shell is considered rigid and the case that takes shell flexibility into account. When the shell connection is considered rigid, the piping moment increases from 883,608 lb-in to 3,654,876 lb-in at the compressor nozzle, and the moment increases from 92,604 lb-in to 2,701,620 lb-in at the shell nozzle. The x-direction piping force increases from 12,454 lb to 72,512 lb at both connections with the shell considered rigid. By treating the shell connection as rigid, the bending moment at the shell nozzle is almost 30 times as high as the one calculated by including shell flexibility. If shell flexibility is not included, the resulting artificially high piping forces and moments may necessitate the use of a very large expansion loop or loops to reduce them to within acceptable limits of both the vessel and the compressor. The big loop, resulting from the rigid connection analysis, is not only expensive, but also becomes a potential source of operational problems. The loop is prone to vibration and is more likely to suffer from damage due to earthquake and other occasional loads. Proper inclusion of shell flexibility in the analysis is not only economical, but also safer. The system shown in Fig. 8.10 is somewhat on the high end of the flexibility spectrum. Most systems would have the piping forces and moments increased by two or three times when the connections are considered rigid, as compared with that calculated with flexible connections.

Among the six degrees of freedom at a vessel connection, the flexibilities in the directions of the two bending moments and the direct axial force are considered significant. Figure 8.11 shows these

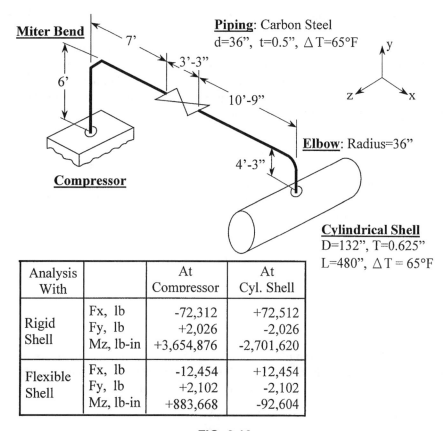

FIG. 8.10
EFFECT OF VESSEL FLEXIBILITY ON PIPING FORCES AND MOMENTS [17]

directions on the cylindrical and spherical shells. Flexibilities at torsion and direct shear directions are generally ignored and considered rigid. The flexibility of the vessel connection is still not very exact. The practical approach would combine theoretical backgrounds with experimental data and field experience. The following are some of the practical approaches used by piping engineers.

Kellogg's choking model for bending on cylindrical shell nozzle. As discussed previously in Section 6.8.2, M. W. Kellogg [12] has used the choking model for calculating support trunnion stresses, which

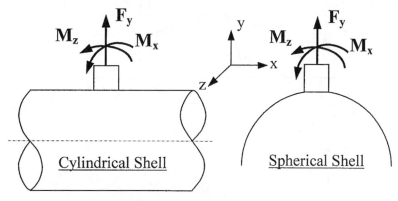

FIG. 8.11
DIRECTIONS WITH SIGNIFICANT FLEXIBILITY

is also applicable in calculating the vessel local stress, as will be discussed later in this section. Kellogg also used the same choking model to calculate the flexibility of the vessel connection. Figure 8.12 shows a cylindrical shell nozzle subject to a longitudinal bending moment, M, under this choking model. With this moment applied, a rotation on the nozzle base is expected. Corner a pulls up a distance y, whereas corner b pushes down the same distance y. The choking model is mainly used to determine the magnitude of y. From the theory of beam on elastic foundation, the choking model as discussed in Section 5.6 gives the displacement as

$$y = \Delta_y = 0.643 \left(\frac{R}{T}\right)^{1.5} \frac{f}{E} \quad (8.9)$$

where f is the load per unit length around the shell circumference. In the choking model, f is uniform around the circumference. Here, we assume that it is the maximum unit load resulting from the moment. From Eq. (6.45), we have

$$f = \frac{M}{\pi r^2} \quad (8.10)$$

Combining Eqs. (8.9) and (8.10) and noting that $\theta = y/r$, we have

$$\theta = \frac{y}{r} = \frac{0.643}{r}\left(\frac{R}{T}\right)^{1.5} \frac{1}{E}\left(\frac{M}{\pi r^2}\right) = \frac{0.2047 M}{E r^3}\left(\frac{R}{T}\right)^{1.5} \quad (8.11)$$

The Kellogg formula uses a non-consistent unit of M resulting in a different constant, which has a factor of 12 over the value given in Eq. (8.11). When including the shell connection flexibility in the analysis, a flexible joint has to be assigned with the spring constant calculated as

$$K_R = \frac{M}{\theta} = \frac{E r^3}{0.2047}\left(\frac{T}{R}\right)^{1.5} \quad (8.12)$$

As noted in Chapter 6, the choking model is better suited for longitudinal bending. For circumferential bending, the flexibility may be several times higher. That is, the spring constant for circumferential bending may be several times smaller than the K_R value given by Eq. (8.12). Kellogg suggested using

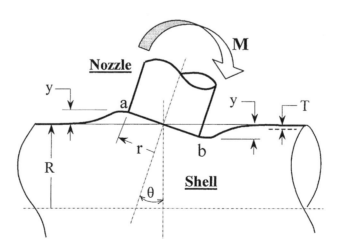

FIG. 8.12
CHOKING MODEL ON SHELL FLEXIBILITY

the same spring constant for both longitudinal and circumferential bending when the connection stress is also calculated with the choking model as given in Section 6.8.2. The idea is to self-compensate for the over estimated circumferential bending moment with the underestimated stress calculation of the circumferential bending. Nowadays, the thinking is a little different, because the magnitude of the bending moment and the magnitude of the stress are both important. In a system connected to sensitive equipment, the bending moment, rather than the stress, is generally used as the acceptance criterion. Therefore, an accurate estimate of the bending moment is required. In this case, as a rule of thumb, the spring rate for the circumferential bending is taken as one-third of the value given by Eq. (8.12). This is consistent with the factor used by the nuclear piping in pre-1983 editions of the code [21]. The Kellogg approach may appear too simple to more theoretically minded engineers, yet many companies have used it for more than 50 years with quite satisfactory results.

Nuclear piping branch flexibility. The nuclear piping code [21] has a set of rules for calculating the flexibility of branch connections. This set of rules is based on the report prepared by Rodabaugh and Moore [22]. The rules adopt the piping practice of assigning a flexibility factor, k, to the component. The flexibility factor is the ratio of the deformation or rotation of the pipe component to that of the plain pipe of the same length. Because the junction flexibility is concentrated at the joint, the length is actually zero. The rules use a reference length equal to the outside diameter of the nozzle or branch pipe. That is, the relation of the flexibility factor and rotation, θ, is defined as

$$\theta = k\left(\frac{Md}{EI_b}\right) \tag{8.13}$$

I_b is the moment of inertia of the nozzle. The flexibility factors are calculated by the following two equations:

$$k = 0.1\left(\frac{D}{T}\right)^{1.5} \sqrt{(T/t_n)(d/D)} \left(\frac{t}{T}\right) \quad \text{for circumferential bending} \tag{8.14}$$

$$k = 0.2\left(\frac{D}{T}\right) \sqrt{(T/t_n)(d/D)} \left(\frac{t}{T}\right) \quad \text{for longitudinal bending} \tag{8.15}$$

For each branch construction detail, the rules have an adjusted value for the effective branch thickness or nozzle thickness, t_n. For normal stub-in nozzle connections used at the vessels, the effective thickness t_n is the same as the nozzle thickness, t. To apply these flexibility factors in the piping analysis, a flexible joint is placed at the nozzle-shell intersection with the joint spring constant calculated by

$$\text{Spring constant, } K_R = \frac{M}{\theta} = \frac{1}{k}\left(\frac{EI_b}{d}\right) \tag{8.16}$$

To get an idea of the difference between circumferential bending and longitudinal bending flexibility, two D/T parameters are checked. For $D/T = 100$, the ratio of flexibility between circumferential and longitudinal bending is $100/20 = 5$; and for $D/T = 50$, the ratio is $35.4/10 = 3.54$.

Axial flexibility of nozzle connection on cylindrical vessel. Flexibility in the axial direction of the nozzle has very little effect on average piping systems. It counts only as an equivalent expansion of a small short piece of pipe. However, it can determine the design strategy of the short connections between two vessels or between a vessel and a piece of sensitive equipment.

The axial flexibility of the nozzle connection is generally taken from Bijlaard's original data [18] supplemented by British Standard BS-5500 [23]. This flexibility data involves a few key parameters such as $\alpha = L/R$, $\gamma = R/T$, $\beta = r/R$. Here, vessel length, L, plays a significant role in axial flexibility. This is quite different from bending flexibility, which is independent of the vessel length. Based on

BS-5500, the flexibility of a given α and β can be represented as a straight line in a log-log scale. A general equation, therefore, can be constructed as

$$\frac{y}{F}ER = A_n \left(\frac{R}{T}\right)^m, \quad \text{where } A_n = f(\alpha,\beta), \quad m = f(\beta)$$

Based on Bijlaard's data, the following three equations for three β values can be constructed as

$$\frac{y}{F}ER = A_0 \left(\frac{R}{T}\right)^{2.22} \quad \text{for } \beta = 0 \text{ (Point Load)} \tag{8.17}$$

$$\frac{y}{F}ER = A_8 \left(\frac{R}{T}\right)^{2.125} \quad \text{for } \beta = 1/8 \tag{8.18}$$

$$\frac{y}{F}ER = A_4 \left(\frac{R}{T}\right)^{2.10}, \quad \text{for } \beta = 1/4 \tag{8.19}$$

(A_n) values vary with L/R of the vessel, and are given in Fig. 8.13 for the three β benchmark values. The flexibility, y/F, is generally found by interpolating between that of two β benchmark values. The first step in determining the flexibility is to select the two equations that encompass the given β value. The flexibility of each β benchmark value is then calculated. The final flexibility is found by interpolating between the flexibilities of the two β benchmarks. The spring rate, which is F/y, is the inverse of the flexibility. Due to the limited available data, the above equations are neither exact nor in standard form. The approaches given by other literature sources may be different in form and values.

Vessel length L is the effective length between two stiffening rings or heads. The curves in Fig. 8.13 assume that the nozzle is located at the middle of the effective length. When the nozzle is located off-center of the effective vessel length, Bijlaard suggested an equivalent length as $L_{eq} = 4 x (L - x)/L$, where x is the distance between the nozzle and the nearest head or stiffening ring. When using the curves, this L_{eq} is used instead of L in calculating α. That is, $\alpha = L_{eq}/R$.

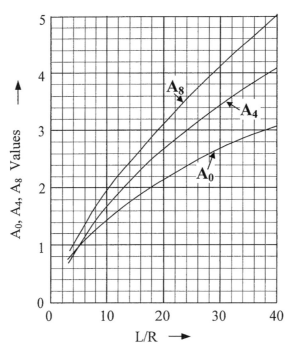

FIG. 8.13
VESSEL LENGTH COEFFICIENT

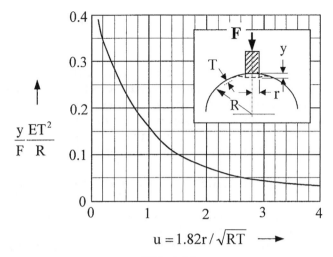

FIG. 8.14
NOZZLE AXIAL DISPLACEMENT AT SPHERICAL SHELL

Flexibility of nozzle connection at spherical shell. Due to the symmetrical nature of the spherical shell, its theoretical treatments derive better results than that for the cylindrical shell. The theoretical data given by Bijlaard [24] can be used directly for practical engineering [25]. However, because Bijlaard's theoretical data consistently over-predict the flexibility of the nozzle connection, the rigid insert model is generally used [23] for engineering applications as a way of compensating for this over-prediction. The flexibility of a rigid insert is less than that of a nozzle connection with a hollow pipe.

Bijlaard's theoretical deformations at rigid inserts due to axial forces and bending moments are summarized into two charts. Figure 8.14 shows the axial deformations of the shell due to axial forces. The deformation given is at the edge of the insert. The axial spring rate of the shell at the nozzle connection is F/y, which is extracted from the chart value $(y/F)(ET^2/R)$.

For the bending moment loading, the edge deformation is given as shown in Fig. 8.15. The chart value has to be converted to a rotational spring rate as follows:

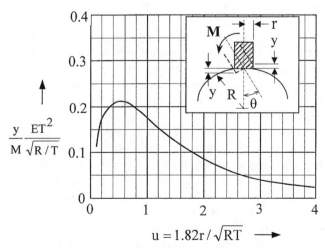

FIG. 8.15
NOZZLE ROTATIONAL DISPLACEMENT AT SPHERICAL SHELL

$$K_{RX} = K_{RZ} = \frac{M}{\theta} = \frac{M}{y/r} = r\frac{M}{y} = r\frac{ET^2}{\sqrt{R/T}} \bigg/ \left[\frac{y}{M}\frac{ET^2}{\sqrt{R/T}}\right] \qquad (8.20)$$

The value inside the brackets is taken from the chart. Again, once the spring rate at each direction is obtained, a flexible joint is placed at the shell nozzle junction for analyzing the piping.

The above flexibility relations are for radial nozzles only. For nozzle connections not in the radial direction, like nozzle 4 in Fig. 8.9, a special arrangement is needed to prevent it from using the wrong flexibility in the piping analysis. In this case, a short artificial radial piece is placed at the junction to properly orient the vessel before the piping is run in its proper direction.

Effect of reinforcing pad. When the reinforcing pad is wide enough, the nozzle connection flexibility is calculated by assuming that the shell has a thickness equal to the sum of the shell and pad thickness. Based on the discussion given in Section 5.5, a pad width of $2.0[R(t + t_p)]^{1/2}$ is required to dilute the junction effect to a negligible level at the edge of the pad. Therefore, with a pad width of $2.0[R(t + t_p)]^{1/2}$ or greater, the connection flexibility is calculated the same as a non-reinforced shell with a thickness of $T = (t + t_p)$. For narrower reinforcing pads, the flexibility is calculated by [26]

$$y = y_{rp} + (y_u - y_{rp})\left(\frac{t}{t + t_p}\right)^2 \qquad (8.21)$$

where y_{rp} is the deformation calculated by assuming that the whole pad as the rigid insert, and y_u is the deformation calculated by considering the pipe as a rigid insert on the un-reinforced original shell. Rotation on the reinforced connection is obtained via the same method.

8.3.3 Allowable Piping Load at Vessel Connections

In typical engineering practice, the vessel engineer is the one who evaluates the acceptability of the piping load at vessel nozzle connections. The piping engineer calculates the load and then submits it to the vessel engineer for evaluation. The procedure is somewhat cumbersome and sometimes unpleasant for both piping and vessel engineers. Some preliminary evaluation by the piping engineer can save considerable time spent in back-and-forth communications.

Rule of thumb approach. In the earlier days, one approach of limiting the pipe stress to 6000 psi (41.37 MPa) was often used. This rule of thumb approach is still used nowadays by some engineers. However, there are two items that need to be noted regarding this approach. First, stress is based on the section modulus of the pipe, not that of the nozzle, which can be heavier than the pipe. Second, pipe stress is calculated assuming the vessel connection is rigid. This rigid assumption is required to compensate for the fact that a thinner wall vessel offers more flexibility but resists less of the load. If the vessel connection is considered rigid, then the calculated piping load will be proportionally much greater than what actually exists for a thinner vessel. On the other hand, fixing the allowable piping load, thus the piping stress, to the vessel will apply a lower safety factor to a thin vessel than to a thick vessel. However, combining the two together, a rather uniform safety factor is afforded to all vessels, thin or thick. The 6000-psi limiting stress is generally considered sufficiently conservative.

This approach requires mutual understanding between the piping engineer and the vessel engineer. If the piping engineer adopts this approach but vessel engineer does not accept it, a problem may arise due to the non-compatibility of methods used by the two parties. For instance, if the piping load is calculated based on the assumption that the vessel connection is rigid, it most likely will not pass the evaluation based on the vessel connection stress produced by such a piping load. It should be reiterated that the piping load calculated by the rigid vessel connection assumption is fictitious, and thus should be treated accordingly.

Based on local stress calculation. The best means to evaluate the piping load is to calculate the vessel stresses created at the vessel connection. The calculations are exactly the same as those discussed in Section 6.8 for integral support attachments. The power boiler [27] formula, Kellogg [12] approach, and the WRC-107 [28] method have all been used. Two supplemental notes are discussed here.

(1) The stresses shall be calculated based on actual piping loads. Therefore, vessel connection flexibility has to be properly estimated and included in the piping analysis. If connection flexibility is not included, the piping load calculated is just a reference value that is not suitable for calculating the actual stress.
(2) The piping load, whether from weight or thermal expansion, may have to be considered as a primary load to a vessel connection. This is due to the strain follow-up discussed in Section 8.2. The vessel connection area is very small compared to the whole piping system. A yield at the connection is not likely to relax the piping load, thus prompting a continuous yielding at the connection to an eventual failure. Therefore, besides the short connecting piping between two pieces of equipment, the loads from all other piping systems are generally considered as sustained to the vessel connections. The resulting vessel connection stresses should also be evaluated by the rules of primary stress.

8.3.4 Heat Exchanger Connections

Heat exchanger connections are also considered vessel connections. However, they deviate from vessel connections in two instances. One is that the heat exchanger shell is generally small compared to an ordinal vessel. Therefore, the heat exchanger can be modeled as pipe and be directly included in the piping analysis. The stress intensification factor for the fabricated branch connection, reinforced or un-reinforced, can be used to calculate the junction stress.

The other deviation is that the heat exchanger is a factory-manufactured item, which means that the manufacturer is responsible for the piping and shell interface. The manufacturer will normally specify the allowable piping loads that the heat exchanger nozzle can sustain. The piping engineer has to ensure that the piping load on the nozzle is within the allowable limits. Because of this division of responsibility, the analysis with the model including the heat exchanger directly as a pipe element can only serve as a reference. The only thing that is certain is the allowable load given by the manufacturer. The manufacturer's allowable loads are often derived from the analysis using WRC-107 [28].

8.4 POWER BOILER AND PROCESS HEATER CONNECTIONS

Power boilers, or steam boilers, and process heaters all involve many internal tubes, which are then connected to the external piping. The characteristics of the interface with the piping depend on the type of equipment.

In a power boiler, internal tubes are generally assembled into a drum or header. Because both drum and header are located externally and are easily supported from surrounding structures, the interface is generally considered rigid. Only the expansion of the drum or header is included in piping analysis. The boiler manufacturer specifies the allowable piping load at each of the nozzle connections. The piping load has to be within the allowable limit, which generally can be met without too much difficulty.

If the power boiler is located within a process plant, the connecting piping is required to be designed and analyzed according to B31.1 [29] power piping code, instead of B31.3 [10] process piping code required by the process plant piping. This is because the jurisdiction of the process piping code starts from the first shut-off valve of the boiler. The piping before the first shut-off valve belongs to the power boiler code [27], which requires its piping be designed by B31.1. Because piping analysis can only be separated at a point that is physically anchored, an analysis has to cover both before and after shut-off valve sections. Some engineers may conduct two analyses, one with B31.1 and the other with B31.3. The B31.1 analysis checks only the portion before the shut-off valve, and the B31.3 analysis checks the rest. However, because B31.1 code is generally regarded as more conservative than B31.3 in steam and water applications, a single run with B31.1 is deemed as acceptable. B31.1 and B31.3 differ only in the calculation of stress and its allowable. The piping load at the nozzle connection is the same regardless of the code applied.

The situation at a process heater is quite different. Internal tubes in a process heater generally connect directly to the external piping. This requires the piping analysis to include the effect of heater tubes. A more elaborate analysis would model the whole heater internals with the connecting piping. However, this would require the inclusion of all tube supports and associated friction effects to achieve reasonable analysis results. In most cases, only a few coils are included to provide some flexibility. The rest of the tubes are considered rigid by placing an anchor or some restraints at the end of the analytical model.

For the allowable piping loads on heater connections, API Std-560 [30] states that "Heater terminals shall be deigned to accept the moments and forces, or the movements listed in Table 7, unless otherwise specified by the purchaser." Table 7 is rearranged as Fig. 8.16 in this book. The table may be interpreted as providing either the allowable forces and moments, or the allowable movements. But

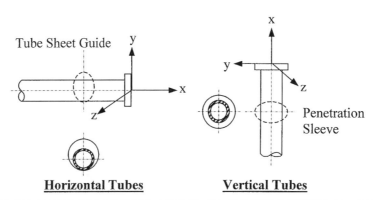

Allowable Forces and Moments [30]								
Pipe Size (in)	Axial Forces Fx		Lateral Force Fy and Fz		Torsion Moment Mx		Bending Moment My and Mz	
	N	Lbs	N	Lbs	N-m	Lbs-ft	N-m	Lbs-ft
2	445	100	890	200	475	350	339	250
3	667	150	1334	300	610	450	475	350
4	890	200	1779	400	813	600	610	450
5	1001	225	2002	450	895	660	678	500
6	1112	250	2224	500	990	730	746	550
8	1334	300	2669	600	1166	860	881	650
10	1557	350	2891	650	1261	930	949	700
12	1779	400	3114	700	1356	1000	1017	750
Allowable Movements, mm (in) [30]								
		Horizontal Tubes			Vertical Tubes			
		Δx	Δy	Δz	Δx	Δy	Δz	
Radiant Terminals		0	+25^{mm} (1^{in})	±25^{mm} (1^{in})	0	±25^{mm} (1^{in})	±25^{mm} (1^{in})	
Convection Terminals		0	+13^{mm} (1/2^{in})	±13^{mm} (1/2^{in})	--	--	--	

FIG. 8.16
ALLOWABLE FORCES, MOMENTS, AND MOVEMENTS AT HEATER TERMINALS

the general understanding is that the heater will allow the movements, yet still tolerate the forces and moments listed. The listed allowable movements are to accommodate the piping expansion in certain directions. However, there is also the movement of the heater connection that is to be accommodated by the piping system. This is the movement of the heater nozzle due to the tube expanding from its internal support point. This movement is generally in the axial direction of the nozzle and is in the outward direction for hot tubes.

In the analysis, the allowable movements are modeled as limit stops having gaps equal to the allowable movements in the corresponding directions. These limit stops tend to reduce the thermal expansion forces of the piping. In the axial direction, the allowable movement is given as zero by API Std-560. In reality, it generally has a movement that requires the piping to accommodate. This movement is generally given by the heater manufacturer and is treated as the displacement that pushes toward the piping system. This movement is automatically included if the piping analytical model includes the heater internal tubes.

The allowable movement is generally provided through the use of an oversized penetration sleeve or tube sheet support hole as shown in Fig. 8.17. The penetration holes and tube sheet holes are generally circular shaped. This requires the use of skewed restraints to guide the pipe to move around the hole on horizontal tube connections. This skewed restraint is needed in addition to the limit stops required for limiting the maximum movement.

On the horizontal tube arrangement, the tube is initially sitting on the bottom of the hole with the contacting point 1, and the tube centerline located at point a. As the tube moves in the lateral z direction, it is forced to move up along the circumference of the support hole. When the tube reaches the maximum z allowable movement, the tube contact point shifts to point 2 with centerline point located at point b. That is, the tube will move along the a-b path instead of in the pure z direction. In this case, the a-b path is at a 45-deg. angle from the horizontal z axis. This is the direction of the skewed restraint to be used in the analysis. In addition to the two skewed restraints located in the +45 deg. and −45 deg directions, two side limit stops and one top limit stop are also needed to limit the overall allowable movements. For a circular tube sheet hole, the allowable vertical movement Δy can be twice as large as the allowable horizontal movement Δz, if the pipe does not move sideward. Because the pipe is allowed to move sideward, the vertical movement is allowed the same amount of movement as the lateral movement.

On the vertical tube arrangement, the allowable movement is generally uniform around the whole circumference. It has equal Δy and Δz, but the magnitudes decrease when the tube moves in the skewed direction. As a tube moves in a 45-deg. direction, for instance, the allowable movements are reduced to 0.707 times the maximum allowed values. The resultant movement equals the maximum movement given. In cases when the external piping has a tendency to move to one side, a shifting of the allowable movement may be arranged with the manufacturer.

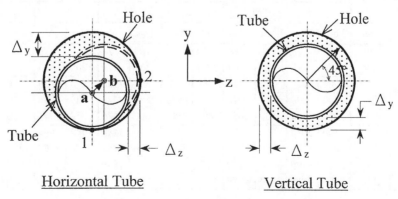

FIG. 8.17
MODELING OF ALLOWABLE MOVEMENTS

The allowable forces and moments are very small compared to the piping stress. This is especially true as the pipe sizes become bigger. This is the main reason why many analyses include the heater internal tubes in the analytical model in order to get some relief on the piping loads. The low allowable forces and moments are mainly attributable to the weak support systems of the tubes and the low allowable stress of the tube material in a high temperature environment. Therefore, the first priority of the external piping is the support system, which should essentially leave no piping dead load at the connecting nozzles. For thermal expansion of the piping, the more realistic connecting load can be obtained by including the heater internal tubes. However, it should be noted that the support of the heater tube generally does not have very much extra margin for resisting external piping load. In the analysis, some artificial restraints and anchors can be assumed, but the connecting loads should be strictly limited to the allowable value given. One should not assume that the piping load is satisfactory just because the heater tube stress under this combined model is within the allowable. This is not correct, because the analytical model with assumed restraints and anchors obscures the stresses at supports and at other small connections, such as pigtails.

8.5 AIR-COOLED HEAT EXCHANGER CONNECTIONS

Air is generally not as good a coolant as water because of its higher ambient temperature and lower heat transfer coefficient. However, the use of air can save significant costs over the containment and

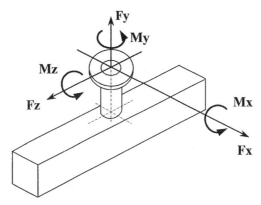

Nozzle Sizes NPS		Allowable Moments						Allowable Forces					
		Mx		My		Mz		Fx		Fy		Fz	
in	mm	Lb-ft	N-m	Lb-ft	N-m	Lb-ft	N-m	Lbs	N	Lbs	N	Lbs	N
1.5	40	50	70	70	90	50	70	100	440	150	670	100	440
2	50	70	90	120	160	70	90	150	670	200	890	150	670
3	75	200	270	300	410	200	270	300	1330	250	1110	300	1330
4	100	400	540	600	810	400	540	500	2220	400	1780	500	2220
6	150	1050	1420	1500	2030	800	1080	600	2670	750	3340	750	3340
8	200	1500	2030	3000	4070	1100	1490	850	3780	2000	8900	1200	5340
10	250	2000	2710	3000	4070	1250	1690	1000	4450	2000	8900	1500	6670
12	300	2500	3390	3000	4070	1500	2030	1250	5560	2000	8900	2000	8900
14	350	3000	4070	3500	4750	1750	2370	1500	6670	2500	11120	2500	11120

FIG. 8.18
ALLOWABLE PIPING LOADS AT AIR COOLER NOZZLES [31]

infrastructure required for water cooling. Air-cooled heat exchangers are popular for condensing petroleum vapors. They are also often used for cooling natural gas to reduce the volume for easy transmission. Most mediums cooled by air-cooled heat exchanger are at rather low temperatures. Some of them are also at very low pressures. Because of its low temperature and low pressure nature, an air-cooled heat exchanger poses some unique problems to connecting piping. The first concern of piping engineers is the allowable forces and moments that an air-cooled heat exchanger can sustain.

Figure 8.18 shows the allowable forces and moments as given by API Standard 661 [31] for air-cooled heat exchanger nozzles. The numbers have very little to do with the actual piping stress. They are just the commonly accepted and agreed-upon values determined from the practicality and economics of the heat exchanger. The allowable values have a sudden jump for 8-in. nozzles. After that, they remain relatively constant for all bigger nozzles. As usual, these allowable values are very difficult to meet, especially at bigger nozzle sizes. Special treatments are required to achieve the satisfactory design of the piping system.

There are some unique layouts of the connecting piping for low pressure applications. Because of the low pressure, a slight change in the inlet or outlet pressure can greatly affect the flow distribution through the tube bundles. To ensure even distribution of the flow through all heater sections, a cascading piping layout or a large header construction as given in Fig. 8.19 is required. This ensures that the inlet and outlet pressures at all tube bundles are the same, thus achieving the same flow and cooling effect at all bundles. In other words, this type of cascading layout enables all exchanger sections to deliver the full cooling effect. The requirement of delivering equal inlet and outlet pressure at all exchanger sections makes it very difficult to implement any practical expansion loop. Any deviation in the piping layout from the ideal configuration will no doubt change the inlet and outlet pressure distribution. To forgo the expansion loop, the piping analyst is required to explore all possible relief scenarios to come up with a good design that can accommodate the allowable nozzle loads.

Most air-cooled heat exchangers have a floating header construction as shown in Fig. 8.19. This type of construction allows the header to be shifted sidewise and also lift up somewhat. These facts have to be incorporated in the piping analysis to have realistic piping loads. This means that the header and possibly the tube bundle flexibility have to be included in the analytical model of the piping. When the header is connected to either the inlet or outlet, only the connected piping is modeled in the analysis. However, if the header is connected to both inlet and outlet piping, then both inlet piping and outlet piping have to be analyzed together. The header can be regarded as a piece of pipe with a corresponding cross-sectional moment of inertia of the header or, alternatively, it can be simply considered as a rigid member. The header is supported at both ends with single acting supports. The

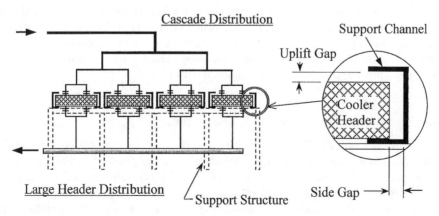

FIG. 8.19
TYPICAL LEAD PIPING AND SUPPORT SCHEME OF AIR COOLER

ends are also restrained in the sidewise and upward directions with gaps as given by the manufacturer. The header weight and part of the tube bundle weight also have to be included in the analysis. The resistance of the tube bundle to the upward header movement can be simulated with a proper spring rate, but is often ignored due to a lack of spring rate data. The resistance to the z direction rotational movement is generally considered rigid.

Because the piping is generally inflexible in the z direction that is parallel to the header, the most basic and important requirement of the air-cooled heat exchanger is to provide enough side-gap to accommodate this z direction expansion of the piping. This side gap should be at least equal to one-half of the expansion of the total width of the cooler assembly subjected to the highest temperature of the piping. A modification should be requested if the side gap is not enough to accommodate this pipe expansion.

8.6 LOW-TYPE TANK CONNECTIONS

A tank is a vessel, so its connections can be treated in the same manner as the vessel connections discussed in Section 8.3. However, the tank connections located at the very lower end of the tank cannot be treated in the same manner as ordinary vessel connections due to some unique characteristics associated with the region. These "low-type" tank connections are located within a distance of 1.25 $\sqrt{Rt}$ from the tank bottom. Connections of this type generate considerable rotations to the connecting piping due to the bulging of the tank shell under the product head, the hydrostatic pressure of the liquid contained. The surroundings of the connection is also much more complex than that of a vessel connection. Therefore, the flexibility of these low-type tank connections cannot be readily estimated from the vessel connection formulas. A set of unique approaches is needed to handle the interaction between the tank and the piping. Based on their working company's experiences and information generated by their proprietary computer program, Billimoria and Hagstrom [32] have developed a set of stiffness coefficients and allowable load factors for analyzing these low-type tank connections. These data and formulas were field verified by Billimoria and Tam [33] and were later adopted by API Standard 650 [34] for the design of steel storage tanks. API-650 puts this evaluation approach in its "Appendix P – Allowable External Loads on Tank Shell Openings."

The discussions that follow are based on Billimoria and Hagstrom's approach. Because the data involves interpolation between widely separated benchmark values, the data so obtained for each specific connection parameter is not very precise. Therefore, it is necessary to provide a higher margin on the allowable. It is also necessary to note that API Standard 650 does not mandate the use of this evaluation. Its use is left to the mutual agreement of the individual manufacturer and the purchaser. Standard 650 also states: "It is not intended that this appendix (evaluation method) necessarily be applied to piping connections similar in size and configuration to those on tanks of similar size and thickness for which satisfactory service experience is available." Therefore, the piping engineers may or may not use the exact approach outlined here. However, the phenomena given here are significant and should be dealt with properly. In particular, the displacement and rotation of the tank shell have created some known problems and should be handled with care.

8.6.1 Displacement and Rotation of Tank Connection

There are two sources of shell displacements. One is the thermal expansion of the tank shell, and the other is the diametrical elongation due to pressure from the product head. The rotation occurs only at the low type connections. The rotation is produced by the choking of the tank bottom. The general behaviors of tank displacement and rotation have been discussed in Section 5.8. Because the tank is simply resting on the foundation, the rotational resistance is small. Therefore, the working formulas pertinent to tank connections are based on a zero-moment restraint at the shell-bottom plate junction.

Figure 8.20 shows the relationship of the displacement and rotation at the lower portion of the tank. As the tank receives a temperature increase, either from the ambient temperature or the product, the shell expands increasing the radius by the amount,

$$\Delta_T = R\alpha(T_2 - T_1) \tag{8.22}$$

where α is the expansion rate. In addition to thermal expansion, the shell also elongates due to internal pressure. The pressure elongation is calculated by

$$\Delta_P = \frac{S_{hp}}{E}R = \frac{P}{Et}R^2 \tag{8.23}$$

where P is internal pressure. In most cases, P is due to the liquid head only, that is, $P = \rho g H$, where g is the gravitational acceleration and ρ is the density of the liquid. $\rho = 1.94$ slugs/ft^3, or 1000 kg/m^3 for water.

This increase in radius reduces to almost zero at the tank bottom due to the choking of the tank bottom plate. The choking effect reduces the radial displacement, but creates the rotation of the tank shell.

Calculation of the displacement and rotation at the nozzle location requires an estimate of the thermal expansion of the tank bottom plate. For internal pressure, it is pretty much agreed that the tank bottom plate is rigid and does not move outwardly. However, for thermal expansion, the situation is different. One school of thought is that the tank bottom expands at the same amount as the bulk of the shell. Another theory is that the tank bottom does not move in response to the change in tank shell temperature. This is based on a couple of factors: (1) the tank bottom generally has lower temperature due to conduction from the foundation, and (2) the friction force at the bottom plate is enough to suppress the thermal expansion. To cover different situations, a base expansion factor, B, is introduced. Tank bottom expansion is assumed to be B fraction of the shell thermal expansion. That is, the bottom plate expands $B\Delta_T$ in the radial direction. B ranges from 0 to 1.0. Billimoria and Hagstrom [32] use 0.0, whereas API-650 uses 1.0 — meaning that the tank bottom plate expands the same amount as the shell. It should be noted that a smaller B results in a smaller nozzle displacement, but generates

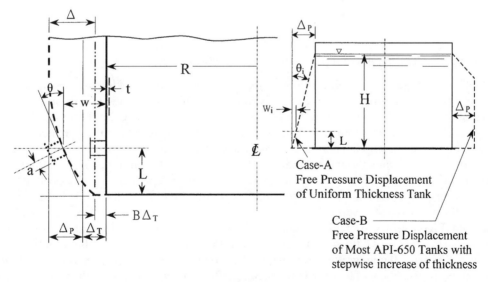

FIG. 8.20
DISPLACEMENT AND ROTATION AT LOW-TYPE TANK CONNECTION

a greater nozzle rotation. In general, the rotation has a much larger impact on the piping than the displacement. The amount of free radial displacement choked at the bottom is

$$\Delta = \Delta_P + (1-B)\Delta_T \tag{8.24}$$

From Eq. (5.41), the displacement and rotation at the point located L distance away from the bottom are determined as follows:

$$w = \Delta[1 - f_1(\beta L)] + B\Delta_T = [\Delta_P + (1-B)\Delta_T]\left(1 - e^{-\beta L}\cos\beta L\right) + B\Delta_T \tag{8.25}$$

$$\theta = \Delta\beta f_3(\beta L) = [\Delta_P + (1-B)\Delta_T]\beta e^{-\beta L}(\cos\beta L + \sin\beta L) \tag{8.26}$$

where $\beta = \dfrac{1.285}{\sqrt{Rt}}$

Equations (8.25) and (8.26) are for tanks with a uniform free displacement at the lower portion of the tank shell. For tanks with uniform thickness, the free displacement due to liquid pressure is sloped from zero at the top to the maximum at the bottom as in case A of Fig. 8.20. This is attributable to the varying liquid pressure, which produces zero hoop stress at the top and the full maximum stress at the bottom. Therefore, for a tank with a uniform thickness, the displacement and rotation at the nozzle connection have to be adjusted with the initial free displacement and rotation whose values are determined by

$$w_i = -\Delta_P \frac{L}{H}, \quad \theta_i = -\frac{\Delta_P}{H} \tag{8.27}$$

The above initial free displacement and rotation have to be added to the displacement and rotation calculated by Eqs. (8.25) and (8.26), respectively, if the thank thickness is uniform.

However, because most of the large tanks constructed under API Std-650 have a stepping increase of thickness, the thickness at each elevation or course is selected to have a hoop stress the same as or close to the allowable stress of the plate. Therefore, the free displacement due to pressure head is the same throughout the tank, except at the top portion where the displacement is smaller due to the limitation on minimum thickness. In this case (shown as case B in Fig. 8.20), the adjustment with the initial free displacement and rotation as given by Eq. (8.27) is not required. Adjusting with these non-existing initial displacement and rotation will generally result in an un-conservative analysis.

8.6.2 Stiffness Coefficients of Tank Nozzle Connection

The tank nozzle is generally subject to three force and three moments of loading, but only one force and two moments are considered critical. They are the radial force, longitudinal moment, and circumferential moment. This is similar to the situation of the vessel connection discussed previously in this chapter. One thing that is different from the vessel connection is the highly asymmetric geometry at the tank bottom connection. Therefore, unlike the vessel connection where an axial force creates only axial displacement and a moment loading only creates the rotation in the moment direction, the stiffness or spring rate of the low-type tank connection is not independent. In the asymmetric vertical direction, the spring rates for radial force and longitudinal moment are coupled. They cannot be included in the analysis as the support spring rate in the usual manner. A coupled approach has to be implemented. Figure 8.21 shows this coupling effect of the stiffness coefficients and displacements.

In the radial load (F_R) case, the load not only creates the radial displacement, but also the longitudinal rotation. Similarly, the longitudinal moment, M_L, generates not only the longitudinal rotation but also the radial displacement. The only independent stiffness is the one due to circumferential bending, M_C, which generates only the circumferential rotation. These force displacement relations can be summarized as

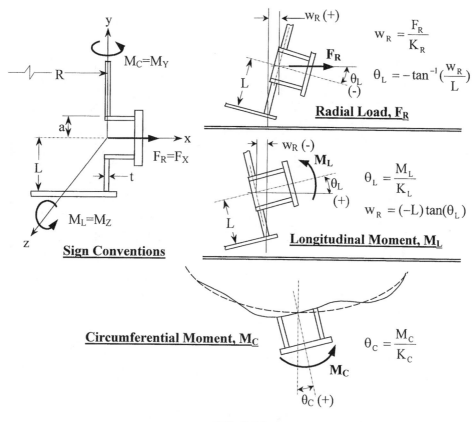

FIG. 8.21
LOADS CONSIDERED ON TANK AND CORRESPONDING DEFORMATIONS

$$\begin{vmatrix} w_R \\ \theta_L \\ \theta_C \end{vmatrix} = \begin{vmatrix} 1/K_R & -L/K_L & * \\ -1/LK_R & 1/K_L & * \\ * & * & 1/K_C \end{vmatrix} \begin{vmatrix} F_R \\ M_L \\ M_C \end{vmatrix} \quad (8.28)$$

The above equation uses $\tan(\theta) = \tan^{-1}(\theta) = \theta$ for the small angles expected. Because the flexibility matrix has to be symmetric for elastic structures, $-L/K_L$ and $-1/LK_R$ are equal. That is,

$$-\frac{L}{K_L} = -\frac{1}{LK_R} \quad \text{or} \quad K_R = \frac{K_L}{L^2} \quad (8.29)$$

Therefore, both K_R and K_L are defined once either one of them is known. Based on their working company's proprietary shell analysis software, Billimoria and Hagstrom [32] have come up with charts for K_R, K_L, and K_C. Their K_R and K_L charts do not exactly show the relationship expressed by Eq. (8.29), likely because the rigid body relationships as given by Fig. 8.21 are not exactly correct. However, because the symmetric flexibility and stiffness matrices are required for the elastic structural analysis of the piping, only one of them needs to be used. Tests [33] have shown that the deviation of the chart value and the experimental value differs more in K_R than in K_L. Therefore, K_L is a better choice to use. Figures 8.22 and 8.23 show the reformatted charts. Figure 8.22 represents the case with reinforcement on the shell, and Figure 8.23 is for the case with reinforcement at the nozzle only. Two benchmark nozzle positions are used. They are at $L/2a = 1.0$ and $L/2a = 1.5$. The charts are

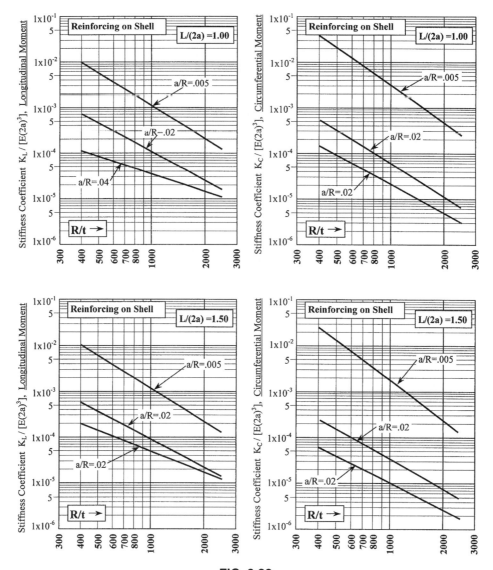

FIG. 8.22
TANK NOZZLE STIFFNESS COEFFICIENTS WITH REINFORCING ON SHELL [32, 34]

constructed for three a/R values: $a/R = 0.005$, $a/R = 0.02$, and $a/R = 0.04$. The interpolation between nozzle locations can be done linearly, but the interpolation between a/R ratios is generally done by logarithm scales.

8.6.3 Allowable Piping Loads at Tank Connections

Again, based on their proprietary shell analysis software, Billimoria and Hagstrom developed a unique cookbook method for evaluating piping loads. The evaluation criteria are based on the fact that the hoop stress at the un-choked shell due to liquid head is the same as the basic allowable stress of the material. The total membrane stress including the piping load is allowed to 110% of the basic allowable stress. This 110% factor is a conservative approach considering the wide influence of the stress field. Most piping and vessel codes allow 150% of the basic allowable stress for total local mem-

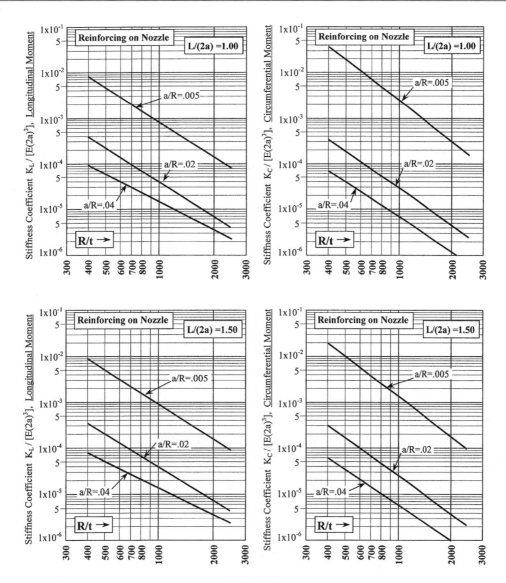

FIG. 8.23
TANK NOZZLE STIFFNESS COEFFICIENTS WITH REINFORCING ON NOZZLE ONLY [32, 34]

brane stress. The total surface stress is limited to three times the basic allowable stress. However, the surface stress evaluation is omitted due to the conservatism of the membrane stress evaluation criteria. Once the membrane stress is satisfied, the surface stress is also considered satisfied.

The piping load evaluation method involves two nomograms and a set of allowable coefficient charts. The method is straightforward, but is somewhat mysterious to most piping engineers. The method only tells whether the load is acceptable. It gives very little clue as to how bad the loads are and how to rectify the situation if the loads are unacceptable. The nomogram is also somewhat difficult to implement in a computer program. An explanation based on the allowable stress values may give engineers a better understanding of the procedure.

Piping loads are evaluated based on the stresses at four major corners, A, B, C, and C', as shown in Fig. 8.24. The stresses are represented by dimensionless parameters whose allowable values are equal to 1.0. These stress parameters are proportional to the loads F_R, M_L, and M_C, and are defined as

276 Chapter 8

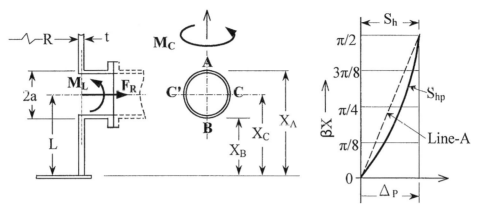

(a) Stress Locations and Sign Convention   (b) Pressure Hoop Stress

FIG. 8.24
STRESSES AT TANK NOZZLE CONNECTION

$$\frac{\lambda}{2Y_F}\left(\frac{F_R}{F_P}\right), \quad \frac{\lambda}{aY_L}\left(\frac{M_L}{F_P}\right), \quad \text{and} \quad \frac{\lambda}{aY_C}\left(\frac{M_C}{F_P}\right), \tag{8.30}$$

respectively, where

$\lambda = a/\sqrt{Rt}$, $F_P = \pi a^2 p$ = pressure end load on nozzle
Y_F, Y_L, Y_C = allowable coefficients for F_R, M_L, and M_C, respectively

To get a better idea of the method, a closer look at the nature of these dimensionless stress parameters is necessary. Because the limiting value of these dimensionless parameters is 1.0, these parameters correspond to a stress value of $1.1S_h$ — that is, 110% of the basic allowable stress at the operating temperature. Furthermore, the tank shell hoop stress at the nozzle location is assumed to be $1.0S_h$, as is usually the case. The stress due to M_L, for example, can be determined from the dimensionless parameters by setting

$$\frac{\lambda}{aY_L}\left(\frac{M_L}{F_P}\right) = \frac{a/\sqrt{Rt}}{a}\left(\frac{M_L}{Y_L}\right)\frac{1}{\pi a^2 p} = 1.0 = \frac{S_{ML}}{1.1 S_h} = \frac{S_{ML}}{1.1 * pR/t}$$

or

$$S_{ML} = \left(\frac{1.1}{Y_L}\right)\frac{\sqrt{R}}{t^{1.5}}\left(\frac{M_L}{\pi a^2}\right) \tag{8.31}$$

Similarly, for M_C and F_R, we have

$$S_{MC} = \left(\frac{1.1}{Y_C}\right)\frac{\sqrt{R}}{t^{1.5}}\left(\frac{M_C}{\pi a^2}\right) \tag{8.32}$$

$$S_F = \left(\frac{1.1}{Y_F}\right)\frac{\sqrt{R}}{t^{1.5}}\left(\frac{F_R}{2\pi a}\right) \tag{8.33}$$

The above stress formulas are in exactly the same form as given by Kellogg's choking model discussed in Section 6.8.2. $(1/Y_L)$, $(1/Y_C)$, and $(1/Y_F)$, times 1.1 are the stress coefficients for M_L, M_C, and F_R, respectively. Y_L, Y_C, and Y_F are called allowable coefficients whose values are given in Fig. 8.25. A larger Y means a lesser stress is generated per unit load. The original Y_C values are shown as a

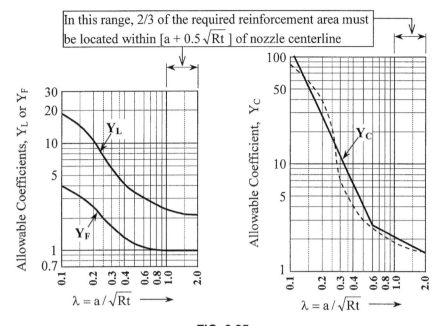

**FIG. 8.25
ALLOWABLE TANK NOZZLE LOAD COEFFICIENTS [32, 34]**

dotted line in the figure. Because it is too sensitive in the mid portion, it is revised to two straight lines in the API Standard-650.

In an unrestrained section of the shell, Kellogg's choking formula predicts a membrane stress coefficient of 0.643 for longitudinal bending as given by Eq. (6.44). An even higher stress coefficient is considered for circumferential bending. However, the situation at a low-type tank connection is quite different. Due to the reinforcing effect of the tank bottom, the stress is generally smaller than the unrestrained section. The stress due to circumferential bending is very small because of a complete stiffening ring effect at the tank bottom. Nevertheless, the Kellogg stress coefficient of 0.643 corresponds to an $Y_L = 1.1/0.643 = 1.71$, which is close to the chart value at the higher (a/R) ratio, the case that is normally encountered in piping branch connections. Furthermore, the allowable coefficients given by Fig. 8.25 are applicable to nozzles with or without a reinforcing pad. The verification tests [33] were performed on the case with a reinforcing pad. In this case, the shell thickness in Kellogg's model should use $(t + t_p)$ instead of t. Using an effective thickness $t_e = (t_p + t) = 2t$ reduces the Kellogg stress coefficient by a factor of $(2)^{1.5} = 2.83$, to 0.227. This brings the values in Fig. 8.25 closer to the ones given by the Kellogg model.

In evaluating the piping loads, in addition to combining the effects of F_R, M_L, and M_C, pressure stress also needs to be considered. The shell hoop stress is normally equal to the basic allowable stress by design. Therefore, in an unrestricted portion of the shell, the allowable stress left over for the piping load is only $1.1 - 1.0 = 0.1$ of the basic allowable stress if the stress due to piping load is in tension. This assumes that the allowable for the total membrane stress, including pressure and piping load, is 110% of the basic allowable stress. In a restricted shell such as the bottom portion of the tank, the pressure hoop stress is smaller due to the restriction on radial displacement. The pressure hoop stress distribution is shown in Fig. 8.24(b). In this bottom portion, the allowable tensile stress due to the piping load is $1.1S_h - S_{hp}$. The pressure hoop stress S_{hp} is proportional to the choked radial displacement given by Eq. (8.25). That is,

$$S_{hp} = S_h[1 - f_1(\beta x)] = S_h\left[1 - e^{-\beta x}\cos\beta x\right] \tag{8.34a}$$

Thus, the allowable tensile stress for the piping load is equal to

$$S_{t,allow} = 1.1 S_h - S_{hp} = \left[1.0 - 0.91\left(1.0 - e^{-\beta x}\cos\beta x\right)\right](1.1 S_h) \tag{8.35a}$$

In the following, the evaluation of the piping load will be based on dimensionless stress coefficients with reference to $1.1 S_h$ as unity. Letting $1.1 S_h = 1.0$, we have

$$S'_{t,allow} = 0.09 + 0.91 e^{-\beta x}\cos\beta x \quad \text{(based on } 1.1 S_h = 1.0\text{)} \tag{8.36a}$$

To simplify the calculation, Billimoria and Hagstrom used a linear line approach to calculate the pressure hoop stress. From Fig. 8.24(b), the pressure hoop stress at the bottom portion of the tank can be represented approximately as the straight line A, assuming the stress reaches the maximum at $\beta x = \pi/2$. Using this straight line, the hoop stress can be calculated as

$$S_{hp} = S_h \frac{\beta x}{\pi/2} = S_h \frac{1.285/\sqrt{Rt}}{\pi/2} x = 0.818 \frac{x}{\sqrt{Rt}} S_h \tag{8.34b}$$

In this case, the allowable tensile stress for the piping load is equal to

$$S_{t,allow} = 1.1 S_h - S_{hp} = \left[1.0 - \frac{1}{1.1} 0.818 \frac{x}{\sqrt{Rt}}\right](1.1 S_h) \tag{8.35b}$$

Letting $1.1 S_h = 1.0$, we have

$$S'_{t,allow} = 1.0 - 0.744 \frac{x}{\sqrt{Rt}} \quad \text{(based on } 1.1 S_h = 1.0\text{)} \tag{8.36b}$$

That is, the allowable stress depends on the location of the stress interested. The allowable tensile stress needs to be subtracted with the pressure hoop stress at the location. The adjusted allowable stress in dimensionless presentation is given by Eq. (8.36b). The minimum tensile allowable stress in dimensionless term is $1.1 - 1.0 = 0.1$ when the pressure hoop stress reaches the basic allowable stress of the material. The allowable compressive stress is not increased by the pressure hoop stress, because the full pressure hoop stress does not exist when tank is empty or less than full. Furthermore, the compressive stress might consist of the component in the meridian direction that is additive to the hoop stress by the maximum shear stress failure theory. In this case, the reduction of the circumferential stress is offset by the addition of the meridian stress.

In combining the effects of forces and moments, the radial force is considered to generate equal stresses at all four corner points. The longitudinal bending is assumed to generate stresses only at points A and B (see Fig. 8.24), and the circumferential moment generates stresses only at points C and C'. Therefore, in addition to the pressure hoop stress, the effect of the longitudinal moment has to be combined with the effect of the radial force. The circumferential moment also needs to be combined with the radial force. The combination of the longitudinal moment and circumferential moment is not required.

By expressing the piping loads as dimensionless quantities, as in Eq. (8.30), the allowable for each of them is 1.0. For combined loads, the sum of them including the pressure hoop stress should not exceed 1.0. This can be expressed by the nomograms as shown in Fig. 8.26. The nomogram consists of two orthogonal axes, each representing the type of loads to be combined. Pressure hoop stress aside, M_L only combines with F_R and M_C only combines with F_R. The axes are all equally scaled in both directions. The 1.0 point is the allowable of the positive load for which the axis represents. The −1.0

Interface with Stationary Equipment 279

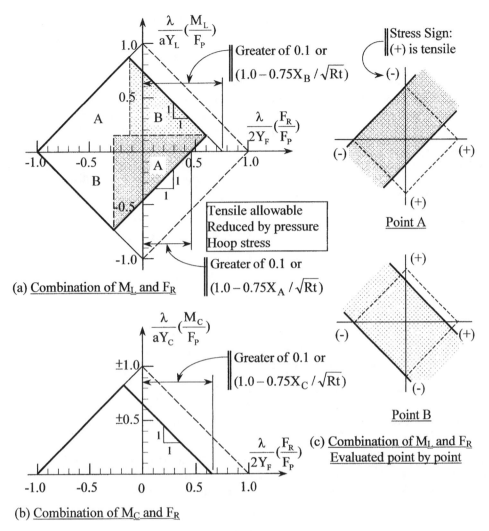

**FIG. 8.26
EVALUATION OF PIPING LOADS AT TANK NOZZLE**

point is the allowable for the negative load the axis represents. The diagonal lines connecting those unit points are the boundary of the combined allowable values. For tensile stresses, the allowable is subtracted with the dimensionless pressure effect, as calculated by Eq. (8.36b), at the point to be evaluated. The construction details of the nomograms are as given in the figure.

The evaluation procedures for the four corner points are combined into two nomograms. Points A and B are combined into Fig. 8.26(a). In this case, the allowable zone is confined by the rectangle marked in bold lines. The nomogram is further divided into four zones with the governing point identified. Because it is the combination of two points, the logic behind the figure is not easy to understand. A separated evaluation of each point as shown in Fig. 8.26(c) should help in this regard. At point A, the stress is tensile for $+F_R$ and $-M_L$. Therefore, the allowable values in $+F_R$ and $-M_L$ axes are each reduced by the effect of the pressure hoop stress corresponding to point A. Similarly, for point B, the allowable value of $+F_R$ and $+M_L$ are each reduced by the effect of the pressure hoop stress corresponding to point B. The allowable zone of each point is shaded. By combing the allowable zones of the two points and by taking the lesser allowable of the two, the combined nomogram is constructed as shown in Fig. 8.26(a). The combined effect of M_C and F_R is simpler due to the symmetric nature of points C and C'. The allowable zones for points C and C' are grouped into one monogram as shown in Fig. 8.26(b).

The nomogram grouping two points together is convenient to apply on graph paper. However, for computerized applications, the point-by point approach as given by Fig. 8.26(c) works better. Furthermore, for computerized applications, the digital or numerical evaluation is more convenient than the nomogram method. The following is the numerical method used by the SIMFLEX [35] computer program.

The numerical evaluation of piping loads on tank nozzles is a process of simply summing up the stress parameters defined by Eqs. (8.30) and (8.36b). For the convenience of discussions, the stress parameters are defined with simpler notations as

$$S_F = \frac{\lambda}{2Y_F}\left(\frac{F_R}{F_P}\right), \quad S_{ML} = \frac{\lambda}{aY_L}\left(\frac{M_L}{F_P}\right), \quad S_{MC} = \frac{\lambda}{aY_C}\left(\frac{M_C}{F_P}\right)$$

$$S_{PA} = 0.75\frac{X_A}{\sqrt{Rt}}, \quad S_{PB} = 0.75\frac{X_B}{\sqrt{Rt}}, \quad S_{PC} = 0.75\frac{X_C}{\sqrt{Rt}} \quad (\max S_{P*} = 0.9)$$

With these notations and sign convention as given by Fig. 8.24, the pipe loads can be evaluated at the four major corner locations of the tank nozzle by a simple additive equation. The allowable value of this evaluation is 1.0. The loads are not acceptable at that particular location if the combined value is greater than 1.0. The combine value is calculated by

At point A: The greater of $\{|S_F - S_{ML} + S_{PA}|$ and $|S_F - S_{ML}|\}$
At point B: The greater of $\{|S_F + S_{ML} + S_{PB}|$ and $|S_F + S_{ML}|\}$
At point C: The greater of $\{|S_F - S_{MC} + S_{PC}|$ and $|S_F - S_{MC}|\}$
At point C': The greater of $\{|S_F + S_{MC} + S_{PC}|$ and $|S_F + S_{MC}|\}$

The parallel columns || signify that the value computed from the expression is taken absolutely. At each point, the first absolute value takes control when the dominant stress is in tension. The second absolute value takes control when the dominant stress is in compression.

8.6.4 Practical Considerations of Tank Piping

Because of the tank shell displacement and rotation, very high piping forces and moments might be generated at the tank nozzle connection. These high piping loads are produced by the combined effect of the product head and thermal expansion. The product head is generally the dominant effect on shell rotation, and is therefore the main promoter of high piping loads. However, these potentially high piping loads are realized only through accurate analysis of the piping system including the interface with the tank. Without interfacing with the tank, especially the shell rotation, the analysis results are meaningless. Figure 8.27 shows the results of a piping analyzed with different tank interface scenarios. The example piping is taken from Billimoria and Hagstrom's paper [32], and the results are obtained by using the SIMFLEX [35] pipe stress analysis computer program.

The example assumes that the tank bottom plate does not expand by temperature change at the shell. This is equivalent to having a zero B factor in Eq. (8.24). For this particular example, the critical load is the longitudinal moment, M_z ($=M_L$). In a correct analysis including tank connection flexibility and both shell displacement and rotation, $M_z = 850,000$ lb-in. Without considering the flexibility of the tank nozzle connection, M_z increases to 1,180,000 lb-in. This is almost a 40% increase over the one that includes tank shell flexibility. Ignoring the connection flexibility is a conservative analysis if the piping can be reasonably designed for it. However, the analyses that do not include the shell rotation under-predict the M_z by a 7 to 1 ratio, or more. In this example, the radial displacement does not have

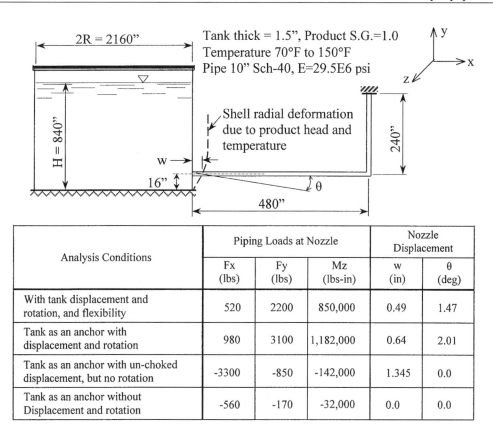

FIG. 8.27
RESULTS OF PIPING ANALYSES WITH DIFFERENT TANK MODELS

a large influence on piping loads for the piping layout given. However, this is not the case for most piping configurations. Radial displacement is also critical in most layouts.

In addition to the tank shell displacement and rotation, a tank is also generally subject to a fairly large amount of settlement due to poor soil conditions. The piping, of course, also has to accommodate this tank settlement. Tank settlement is generally classified into three categories. One is the compaction settlement through the first fill-up or hydrostatic test of the tank. The water weight or product weight causes additional soil compaction resulting in some settlement. The second category is the long-term settlement expected through the usage life of the tank. This is sometimes called plastic settlement. It increases with time, but at a somewhat gradually slow rate. The third category is the elastic settlement caused by the full-empty cycle of the tank. The compaction and elastic settlement can be partially taken care of by connecting the piping after the tank has been hydrostatic tested and the water discharged to half-empty. This connecting at half-empty procedure also reduces the effect of tank shell displacement and rotation. The treatment of long-term settlement varies greatly among operators and conditions. For an expected large long-term settlement, it may not be practical to design the piping system to absorb all settlement at once. Step adjustments of supports at fixed periods may be required. At any rate, in order to achieve a practical design, the settlement profile should be obtained so the proper differential settlements between tank nozzle and pipe supports at various locations can be included in the design analysis. If the settlement profile is not available, then the designer is forced to consider that only the tank settles whereas all the supports do not. This type of design can be very costly and difficult to achieve.

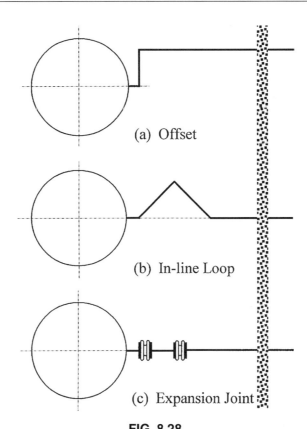

**FIG. 8.28
BASIC TANK PIPING LAYOUTS FOR ACCOMMODATING SHELL ROTATION AS WELL AS LINE EXPANSION**

With all that has been stated about piping loads on tank nozzle, many existing tanks all over the world seem to work just fine without those considerations. In the "Scope" of Appendix P, API Standard-650, judging from past experiences, emphasizes that the method need not be applied to piping connections similar in size and configuration to those on tanks of similar size and thickness for which satisfactory service experience is available. However, engineers are cautioned about this similarity argument. There is no guaranty that just because the tank connection operates satisfactorily for now, it will not fail eventually. The tank shell displacement and rotation swing in cycle for each fill-empty operation. The failure process is accumulated through each operating cycle. It can break when the proper time comes. Therefore, an understanding of the behavior of the connection and its theoretical evaluation is very valuable. At the very least, the piping layout should provide some basic elements for absorbing the displacement and rotation expected from the connection. Fig. 8.28 shows some of the layouts used for providing these basic elements.

REFERENCES

[1] Blick, R. G., 1950, "Bending Moments and Leakages at Flanged Joints," *Petroleum Refiner*, **29**, pp. 129–133. Included in ASME, 1960, *Pressure Vessel and Piping Design – Collected Papers 1927–1959*, pp. 382–393.
[2] Koves, W. J., 1996, "Analysis of Flange Joints Under External Loads," *Journal of Pressure Vessel Technology*, **119**, pp. 59–63.

[3] Tschiersch, R., and Blach, A. E., 1996, "Gasket Loadings in Bolted Flanged Connections Subjected to External Bending Moments," *Proceedings of the 8th International Conference on Pressure Vessel Technology*, **1**, pp. 169–182.

[4] Markl, A. R. C., and George, H. H., 1950, "Fatigue Tests on Flanged Assemblies," *Transactions of the ASME*, **72**(1), pp. 77–87.

[5] ASME/ANSI B16.5, *Pipe Flanges and Flanged Fittings*, Formerly ASA B16e, ASME, New York, NY.

[6] ASME B&PV Code, Section-III, *Rules for Construction of Nuclear Power Plant Components*, Subsection NC, Class 2 Components, subparagraph NC-3658.3, 1989 edition, ASME, New York, NY.

[7] Waters, E. O., Wesstrom, D. B., and Williams, F. S. G., 1934, "Design of Bolted Flanged Connections," *Mechanical Engineering*, vol. 56, May, pp. 311–313.

[8] ASME B&PV Code, Section-VIII, Division 1, *Pressure Vessels*, Formerly *Unfired Pressure Vessels*, Appendix 2, "Rules for Bolted Flange Connections with Ring Type Gaskets," ASME, New York, NY.

[9] ASME, 1958, *ASME B31 Mechanical Design Committee Report*, Oct., Revised April 1984.

[10] ASME Code for Pressure Piping, B31.3, *Process Piping*, ASME, New York.

[11] Wesstrom, D. B., and Bergh, S. E., 1951, "Effect of Internal Pressure on Stresses and Strains in Bolted-Flanged Connections," *Transactions of the ASME*, **73**(5), pp. 553–568. Included in ASME, 1960, *Pressure Vessel and Piping Design — Collected Papers 1927–1959*, pp. 121–136.

[12] The M. W. Kellogg Company, 1956, *Design of Piping Systems*, revised 2nd ed., John Wiley & Sons, New York, NY.

[13] API Standard 605, *Large-Diameter Carbon Steel Flanges*, American Petroleum Institute, Washington, D. C.

[14] MSS SP-44, *Steel Pipe Line Flanges*, Manufacturers Standardization Society of the Valve and Fitting Industry, Inc., Vienna, VA.

[15] ASME B16.47, *Large Diameter Steel Flanges*, ASME, New York, NY.

[16] McKeehan, D. L., and Peng, L. C., 1981, "Evaluation of Flange Connections due to Piping Load," ASME PVP-Vol. 53, *Current Topics in Piping and Pipe Support Design*, ASME, New York.

[17] Stevens, P. G., Groth, V. J., and Bell, R. B., 1962, "Vessel Nozzles and Piping Flexibility Analysis," *Transactions of the ASME, Journal of Engineering for Industry*, **84**, p. 225.

[18] Bijlaard, P. P., 1954, "Stresses From Radial Loads in Cylindrical Pressure Vessels," *The Welding Journal*, **33**(Research Supplement), pp. 615-s–623-s.

[19] Bijlaard, P. P., 1955, "Stresses From Radial Loads and External Moments in Cylindrical Pressure Vessels," *The Welding Journal*, 34(Research Supplement), pp. 608-s–617-s.

[20] Cranch, E. T., 1960, "An Experimental Investigation of Stresses in the Neighborhood of Attachments to a Cylindrical Shell," *Welding Research Council Bulletin*, **60**, pp. 3–44.

[21] ASME B&PV Code, Section-III, *Rules for Construction of Nuclear Power Plant Components*, Subsection-NB, Class 1 Components, Subparagraph NB-3686.5, ASME, New York, NY.

[22] Rodabaugh, E. C., and Moore, S. E., 1979, "Stress Indices and Flexibility Factors for Nozzles in Pressure Vessels and Piping," NUREG/CR-0778, U.S. D.O.C., National Technical Information Service, VA 22161.

[23] British Standard BS-5500, *Specification for Fusion Welded Pressure Vessels for Use in the Chemical, Petroleum, and Allied Industries*, British Standard Institute, UK.

[24] Bijlaard, P. P., 1957, "Computation of the Stresses from Local Loads in Spherical Pressure Vessels or Pressure Vessel Heads," *Welding Research Council Bulletin*, **34**, March.

[25] Rodabaugh, E. C., and Atterbury, T. J., "Flexibility of Nozzles in Spherical Shells," Phase Report No. 3 to U. S. Atomic Energy Commission, TID-24342, June 28, 1966.

[26] Bijlarrd, P. P., 1959, "Stresses in Spherical Vessels from Local Loads Transferred by a Pipe," *Welding Research Council Bulletin*, **50**, pp. 1–9.

[27] ASME B&PV Code, Section-I, Power Boiler, ASME, New York.

[28] Wichman, K. R., Hopper, A. G., and Mershon, J. L., 1979, "Local Stresses in Spherical and Cylindrical Shells due to External Loadings," *Welding Research Council (WRC) Bulletin*, **107**, 1981 Reprint.
[29] ASME Code for Pressure Piping, B31.1, *Power Piping*, ASME, New York.
[30] API Standard 560, *Fired Heaters for General Refinery Service*, American Petroleum Institute, Washington, D. C.
[31] API Standard 661, *Air-Cooled Heat Exchangers for General Refinery Services*, American Petroleum Institute, Washington, D. C.
[32] Billimoria, H. D., and Hagstrom, J., 1978, "Stiffness Coefficients and Allowable Loads for Nozzles in Flat Bottom Storage Tank," *Transactions of the ASME, Journal of Pressure Vessel Technology*, p. 389.
[33] Billimoria, H. D., and Tam, K. K., 1980, "Experimental Investigation of Stiffness Coefficients and Allowable Loads for Nozzle in a Flat Bottom Storage Tank," ASME Paper 80-C2/PVP-59, presented at ASME Century2 Pressure Vessel & Piping Conference, San Francisco, CA, August 12–15, 1980.
[34] API Standard 650, Welded Steel Tanks for Oil Storage, American Petroleum Institute, Washington, D. C.
[35] Peng Engineering, 1980, *SIMFLEX, Pipe Stress Analysis Program — User's Manual*, Houston, TX, website: www.pipestress.com.

CHAPTER 9

INTERFACE WITH ROTATING EQUIPMENT

Rotating machineries are delicate equipment that depend on perfect shaft alignment, balanced rotating parts, and proper clearance for smooth operation. Excessive piping loads and stresses imposed on the equipment can deform machine parts to the point that they may considerably affect the reliability of the equipment. These loads, either from the expansion of piping or weight of the system, can cause shaft misalignment and casing deformation that interferes with the internal moving parts. Therefore, it is important to design the piping system to impose as little load as possible on the equipment. Ideally, it is preferred to have no piping load imposed on the equipment, but this is not possible. The common practice is for the manufacturer to specify a reasonable allowable piping load to which the piping engineers design the piping system to meet. There is no problem with this procedure, except that the allowable loads given by the manufacturers, generally through the standards of their own associations, are too small to be practical. This low allowable piping load is given partially due to the actual design of the machine and partially to protect the manufacturer's interests. The actual load that can be taken could be many times higher for most machines. Nevertheless, most equipment purchasing contracts have a clause stating that the manufacturer will guarantee the machine's performance and integrity only when the piping load is within the allowable limit.

Although the low allowable piping load may have improved the reliability of the machine on paper, it has actually created many unsuspected problems caused by the unusual piping layouts required for reducing the piping loads. To reduce the piping load to the allowable limit, many laborious piping layouts and ingenious restraint schemes are used. These unusual arrangements may show on paper that the allowable load is met, but the reliability of the system is actually compromised. Some of these pitfalls will be discussed later in this chapter.

As mentioned in the introductory chapter, the effort required for designing a piping system connected to a rotating equipment can easily be several times the effort needed for designing a comparable system not connected to a rotating equipment. This can be better explained with Fig. 9.1. This figure shows the allowable piping moments for American Petroleum Institute (API) pumps and National Electrical Manufacturers Association (NEMA) turbines. These two are the most common types of rotating equipment and are covered by well-established standards. To get an idea of the significance of these allowable moments, the moments that generate a pipe stress of 6000 psi on the pipes with standard wall thickness are also given. This value (6000 psi) is one of the reference piping stresses used by some companies for designing the piping system connected to a pump.

At pipe sizes 3 in. or smaller, the allowable moment corresponds to a pipe stress of about 8000 psi. For an 8-in. pipe, the pump allowable moment corresponds to 4000 psi at the pipe, and the turbine allowable moment corresponds to a pipe stress of only 1500 psi. For 12-in. nozzles, the pump allowable moment corresponds to only 2500 psi pipe stress, and the turbine allowable moment corresponds to only about 1000 psi pipe stress. This means that for a 12-in. (300 mm) piping system, the pipe stress has to be reduced to 1000 psi (7000 kPa) in order for the piping load to meet the turbine allowable. In

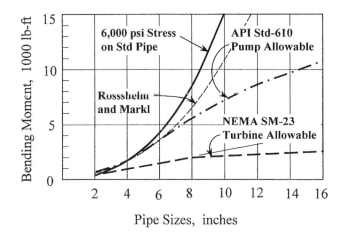

**FIG. 9.1
ALLOWABLE PIPING LOADS**

comparison, a piping system that does not connect to any rotating equipment can have an expansion stress plus sustained stress of about 30,000 psi for most of the common piping materials. Rotating equipment allowable piping loads are based on the normal operating condition that includes pressure, weight, and thermal expansion loads. This stress comparison pretty much conveys the difficulty of designing a piping system connected to rotating equipment.

9.1 BRIEF BACKGROUND OF ALLOWABLE PIPING LOAD ON ROTATING EQUIPMENT

9.1.1 When Nobody Knew What to Do

Most industrial plants built before the 1950s did not follow any specific requirements on the magnitude of piping loads applied to rotating equipment. The design calculation was also not sophisticated enough to routinely estimate the loads applied. There were no specific allowable loads given at that time. The only guideline was a manufacturer's insistence on putting no piping load on the equipment. Later, based on calculations made on successful operating installations, Rossheim and Markl [1] found that the average piping reaction acting on the pumps were:

$$\text{Vertical force, lb} = 3.25(D + 3)^3$$
$$\text{Lateral forces, lb} = 1.50(D + 3)^3 \quad \text{(Inconsistent units)} \quad (9.1)$$
$$\text{Bending and torsion moments, lb-in.} = 60(D + 3)^3$$

where D is the outside diameter of the connecting pipe in inches. The moment values are also charted in Fig. 9.1. These were the actual loads placed on the pump nozzles by sensibly designed piping systems. This also meant that, although no load was permitted, the lowest loads the engineers could come up with were those shown in Eq. (9.1). Equation (9.1) serves only as a reference. It is not a working formula. However, it could serve as a guideline when a new standard is being developed. One thing that is basic to Eq. (9.1) is that the piping loads are proportional to the cube of the adjusted pipe diameter. Although this relationship is often taken for granted by piping engineers, it is quite in contrast to popular industrial standards that have allowable loads increased to less than proportional to the pipe size. Therefore, the gap of understanding between piping engineers and equipment manufacturers grows wider as the pipe size gets bigger.

9.1.2 First Official Set of Allowable Piping Loads

In 1958, a new era began when NEMA published the allowable piping forces and moments on its SM-20 for mechanical-drive steam turbines [2]. This was the first set of official requirements on piping load applied to rotating equipment. When NEMA SM-20 was first published, it was widely rejected by the piping community as well as plant operators. By looking at Fig. 9.1 again, it is easy to understand this sentiment. For small piping, whose flexibility is very easy to increase, the NEMA allowable is quite consistent with established experience. However, as the pipe size gets bigger, the NEMA allowable is way too low compared to the prevailing piping load criteria used at that time. This initial rejection subsided gradually through a period of about 15 years. The standard eventually became de facto when piping analysis was fully computerized. This realization was partially promoted by developers of piping stress analysis software.

The allowable piping load of the NEMA steam turbine standard, currently SM-23, is very stringent (and will be fully discussed later). Nevertheless, it was adopted as the model scheme by other standards, by applying a somewhat larger factor to partially reduce the conservatism. The API centrifugal compressor standard [3] uses a 1.85 factor and industrial pump standard [4] vendors often use a multiplication factor of 1.3 over the NEMA allowable. However, as will be noted later, the NEMA SM-23 allowable does have a couple of illogical elements that make it not a very good example to follow.

9.1.3 Factors Behind the Low Allowable Piping Load

Piping engineers have been wondering if rotating equipment could be manufactured to take a higher piping load without grossly increasing the cost. The answer depends on the type of equipment and the environment it serves. The factors behind this low allowable piping load will be explored first.

Piping engineers often think that manufacturers give a low allowable load to protect their own interests. This notion is not necessarily true, because many units of the equipment indeed cannot take very much load. The main problem is that there is a weak link that is often overlooked in the design of the equipment. Figure 9.2 shows a typical pump installation, which can be divided into three main parts: the pump body, the foundation, and the pedestal/baseplate. Without special input from piping or equipment engineers, the routine design of the pump assembly places a different significance on

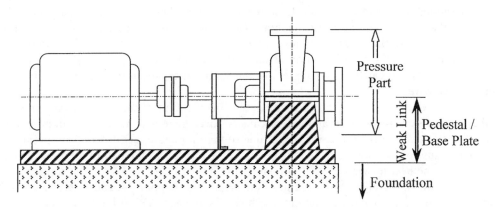

FIG. 9.2
THE WEAK LINK AT ROTATING EQUIPMENT

different parts of the pump. The pump body is designed to be as strong, if not stronger, than the piping so that the body can resist the same internal pressure as the piping. The foundation, normally designed with the combined pump and motor effect, is also massive and stiff due to the limitation of the soil bearing capacity. It is also designed for the piping load and shaft torque as requested or specified. However, the pedestal/baseplate is a different story. Without considering any piping load, the pedestal/baseplate is designed only by shaft torque and pump weight. That is the main reason why earlier versions of the pump allowable piping loads were based on the weight of the pump only [5]. This weak link is the main reason why some units of the equipment do indeed can take very low piping load.

By understanding this situation, the problem can actually be relieved rather easily. Improvements have already been seen in pump applications. Pump application engineers, who long realized the low allowable piping load problem, customarily specified double (2×) or triple (3×) baseplates to increase the allowable piping load by two or three times, respectively. Surprisingly to most, the cost of a 2× or 3× pump was only marginally more than that of a regular pump. Actually, it should not have been surprising, because all the vendor needs to do is add a couple of braces or stiffeners to the pedestal/baseplate. Recognizing the popularity of 2× and 3× baseplates, API formally adopted the philosophy to its pump standard. Since the sixth edition of API Std-610 [5], the pipe load allowable has been increased to a level that makes the 2× and 3× baseplate specification no longer necessary. In other words, the strength of the whole pump assembly has become fairly uniform that no additional allowable load can be squeezed out without adding a substantial cost. Figure 9.1 also shows that the API Std-610 allowable values are comparable to the loads that have been applied by sensible designs. Regrettably, this is only true for API Std-610 pumps. For other pumps, the 2× and 3× specification is still recommended.

9.2 EVALUATION OF PIPING LOAD ON ROTATING EQUIPMENT

Each type of rotating equipment has its own piping load allowable and evaluation method. However, the general procedures are all about the same. Although most modern pipe stress analysis computer software packages automatically, when requested, evaluate the piping loads on most common types of equipment, there are still procedures that need to be followed. This section discusses the general background and procedures. The detailed background and procedures for each type of specific equipment will be discussed in separate sections that follow.

9.2.1 Effect of Piping Loads

Normally, piping loads are evaluated for two categories of effect on the equipment. The first effect is the loads applied at an individual nozzle connection. The other is the combined effect of all the loads applied at all nozzles on the entire machine. The limitation of the load at an individual nozzle is to ensure the integrity of the nozzle and prevent local deformation of the casing. The combined effect is limited to prevent excessive deformation of the pedestal and baseplate. Excessive deformation on the pedestal or baseplate causes misalignment of the shaft either within the machine or between the machine and its drive.

The evaluation of the combined effect requires the specification of the moment resolving point at which the moments are resolved or applied. Because a force generates moment at all points that are not in the line of action of the force, the magnitude of the moment effect differs from point to point. Because the evaluation of combined effect is mainly done to prevent the deformation of the machine support system, the best measure of the moment is the one applied at or near the support location. The resolving point is, therefore, best specified at or near these support locations. NEMA SM-23 has the resolving point specified at the center of the exhaust flange face. API Std-610 uses the center of the pump as the resolving point. The forces and moments at each nozzle connection are resolved to the resolving point by the following relations using the notations given by Fig. 9.3:

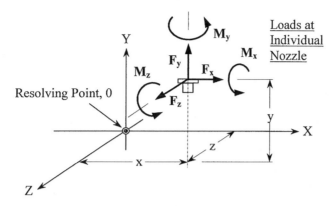

FIG. 9.3
TRANSFERRING LOADS TO RESOLVING POINT

$$F_{x,0} = \sum_{i=1}^{n} F_{x,i}; \quad F_{y,0} = \sum_{i=1}^{n} F_{y,i}; \quad F_{z,0} = \sum_{i=1}^{n} F_{z,i}$$

$$M_{x,0} = \sum_{i=1}^{n} (M_{x,i} - F_{y,i}z_i + F_{z,i}y_i) \quad i = 1,\ldots,n \text{ is the nozzle number}$$

$$M_{y,0} = \sum_{i=1}^{n} (M_{y,i} - F_{z,i}x_i + F_{x,i}z_i) \quad 0 \text{ is the resolving point}$$

$$M_{z,0} = \sum_{i=1}^{n} (M_{z,i} - F_{x,i}y_i + F_{y,i}x_i)$$

(9.2)

The resultant force is just the summation of forces from all individual nozzles. The resultant moment in each direction is the combination of moments of the same direction at all individual nozzles plus the moments due to the effects of forces from all individual nozzles. For an inlet/outlet two-nozzle turbine, the combination is easier, because the exhaust nozzle is also the resolving point whose forces do not contribute any additional moment. Remember that forces at the inlet nozzle still contribute to the resulting moment at the resolving point. Some piping engineers occasionally forget this force effect on the moment.

9.2.2 Movements of Nozzle Connection Point

The movements of the nozzle can greatly affect the analysis results, and thus, affect the requirements of an expansion loop. The movements come from the thermal expansion of the equipment and the support movements due to earthquake, wind, wave, and so forth. The foundation movement is also generally classified as a support movement. Unless given specifically, support movements are not considered. Only thermal expansion movement is considered in the design analysis. Because the casing and support temperature is not uniform throughout the machine, estimation of the movement is not accurate. This is especially true for a compressor or turbine whose casing temperature varies greatly across the machine. Therefore, the data given by the manufacturer should be used if available. Accurate estimation of the nozzle movement is critical when supports are needed at nearby locations.

Depending on the analytical approach used, the movement can be either directly entered in the analysis as an anchor displacement or calculated by the computer from the expansion of the rigid casing member applied with proper temperature. With nozzle movements given by the equipment

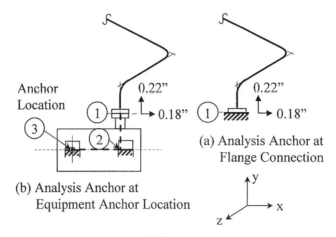

FIG. 9.4
NOZZLE MOVEMENT APPLICATION

manufacturer as $D_x = 0.18$ in. and $D_y = 0.22$ in., Fig. 9.4 shows the two methods of describing them in the analysis.

Method (a) is the typical approach used for analyzing a single piping system. The piping is anchored at the face of the connecting flange. The nozzle displacements specified by the manufacturer are entered directly as anchor movements. This method calculates the loads applied at an individual nozzle connection, but does not give any idea of the load contribution to the combined loading. When the combined resultant load exceeds the limit, this method gives very little clue as to which line is the major contributing factor.

Method (b) is used when the user wants to find out the contribution of the line to the resultant forces and moments at an equipment anchor point or moment resolving point. In this case, the analysis model extends into the equipment with two rigid members, 1-2 and 2-3, representing the casing. The manufacturer-specified nozzle movements can be applied in two ways. Method (b)-1 considers the rigid casing member at ambient temperature and applies the nozzle movements at anchor point 3, as anchor movements. Method (b)-2 considers no movement at anchor 3 and applies proper temperature at members 1-2 and 2-3 to produce the specified nozzle movement at point 1, the flange connection. Nozzle movement, in this case, cannot be directly applied at point 1 as anchor displacement, because that would cause huge fictitious forces and moments at the rigid members and at anchor 3.

Method (b)-1 is straightforward and does not require any calculation of the temperatures at the rigid members, because the temperatures are ambient. However, this method is only good for analyzing systems connected to one nozzle. If multiple piping systems involving more than one nozzle are to be analyzed, the application of anchor movement at point 3 is not possible. Several sets of nozzle movements could be involved in this case, and the anchor can only be specified with just one set of movements. Although it requires the calculation of the proper temperatures at rigid members 1-2 and 2-3, *method (b)-2* is the better approach.

9.2.3 Analysis Approach

In rotating equipment, piping loads are evaluated at individual nozzles first, after which they are evaluated again for the combined effect with loads from all the nozzles connected to the equipment. The combined loads are calculated at the specified resolving point on the equipment. This calls for special schemes on analyzing the connecting piping so the proper resultant loads can be calculated either automatically with computer software or manually with hand calculations. Figure 9.5 shows two of the schemes.

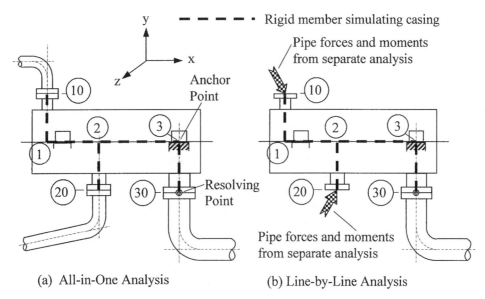

FIG. 9.5
ANALYSIS APPROACHES ON ROTATING EQUIPMENT PIPING

All-in-one analysis. In this scheme (Fig. 9.5a), all piping systems connected to the equipment are analyzed at the same time in one combined model. Each piping system is connected to the equipment with rigid members whose temperatures are properly selected to match the nozzle movements given by the manufacturer. For the anchor location as shown, the y movement at nozzle 30 determines the temperature of rigid member 3-30. The x movement at nozzle 20 determines the temperature of rigid member 2-3, and the y movement of the same nozzle determines the temperature of rigid member 2-20, and so forth. The anchor point may also have some movement, in which case the anchor movements are directly applied at the anchor, and the temperatures of the rigid members are determined based on differential movements between the anchor and the nozzle. If nozzle movements are not given, then the rigid member temperatures are taken as an average of the actual temperatures expected at the portion of casing to which the rigid member represents.

The main benefit of the all-in-one analysis is that the data is available for the computer software to automatically evaluate the combined resultant loads, as well as the individual nozzle loads. The drawback is the complexity of the analysis and the lack of clear indication of each nozzle's contribution to the combined loads. If everything is within the allowable, then there is no problem. However, when combined loads exceed the allowable, it is very difficult to determine which line or lines need(s) to be fixed.

Another inconvenience of the all-in one approach is the difficulty of automatically selecting the proper spring hangers to minimize the weight load at the nozzle. Piping engineers, as well as manufacturers, agree that although reducing the thermal expansion load is generally not easy, at least the weight loads should be reduced to the minimum possible at the nozzle. This can be done by proper selection of spring hangers. A specialized pipe stress analysis computer program normally can do this automatically by releasing the vertical restraint at an anchor when selecting the spring hangers. However, when two or more nozzles share an anchor point, as in all-in-one analysis, this automatic weight load minimizing process does not work.

Line-by-line analysis. In this scheme, the piping systems are analyzed one nozzle at a time. The piping connected to each nozzle is analyzed based on method (a) discussed in Section 9.2.2. This scheme evaluates the loads at each nozzle connection either directly (by computer software) or manually (by comparing with the allowable).

Line-by-line analysis permits a group of engineers to work independently on the same major equipment, such as the main compressor at a process plant or the power turbine in a power plant. Each unit of major equipment has several main process piping systems connected to it. With each engineer or a group of engineers working independently on each line connected to the equipment, the overall turnaround time can be greatly reduced.

Figure 9.5(b) shows the analysis of one piping system combined with the results of two sets of independently analyzed nozzle loads from two other connected piping systems. The nozzle loads of the independently analyzed piping are applied as external forces and moments at the nozzles 10 and 20. It should be emphasized here that the signs of forces and moments should be those corresponding to the loads that are applied to the equipment. A pipe stress computer program has its own unique sign conventions. The loads shown in the member force and moment table may be the ones applied at the piping element, whereas the force and moment given in the support load table generally are the ones applied to the support. Therefore, depending on which table the loads are taken, the signs may have to be reversed before being applied to the analysis.

The example in Fig. 9.5(b) still has one piping system connected to the equipment. This system is generally analyzed by the lead engineer, who takes charge of combining the results from all connected piping systems and ensuring the compliance of all loads applied on the equipment. If the combined loads are not acceptable, he will coordinate the procedure to fix the problem. This assumes that the computer program used can automatically combine the results and tabulate the load compliance result. In any case, all lines can be analyzed separately with loads combined with the formulas given in Eq. (9.2). Results are then compared with the allowable values.

9.2.4 Selecting the Spring Hangers to Minimize the Weight Load

Unlike the thermal expansion load, which is very difficult to control, the weight load is much more manageable. That is the main reason why NEMA SM-23 insists that dead weight of the piping should be entirely supported by pipe hangers or supports. To support the weight, there are two main types of hardware that can be used: rigid support and resilient or spring support. A rigid support can be used only at a location that has negligible vertical thermal movement. If the situation permits, rigid supports should be used whenever it is feasible to do so. This is because rigid supports are more economical and more stable. They can easily absorb any load fluctuation expected from operational condition changes. However, there are many locations where rigid support is not permitted due to the vertical movement of the piping.

Figure 9.6 shows the discharge piping of an overhung volute pump. Except for piping systems operating at ambient temperature, the first support at the discharge piping is generally the resilient type such as a variable spring hanger. From the drawing, it is easy to understand why the pump can take so little piping load. Any vertical force will create a large overturning moment in this type of overhanging construction. The weight of discharge piping plus valves and fittings has to be fully supported by the piping supports.

When there is a support at the proper location as shown in Fig. 9.6(b), the spring can be selected and adjusted to take any weight away from the nozzle connection. In the past, when the spring supports were selected manually, this spring load was simply the sum of the weight of the portion of piping and the components it was assigned to carry. The weight load at the nozzle connection was reduced to zero or close to zero very easily. However, in a computerized analysis environment, one needs a trick for the computer to do the right thing — that is, to have the computer mathematically release the vertical translation restraint when sizing the spring. After the spring selection process is completed, the vertical translation restraint is put back, together with the selected spring force applied, to analyze the piping system for various load cases. This procedure is generally referred to as the anchor release option for spring sizing. Without this anchor release procedure, the computer program would equally divide the weight into two portions. One-half of the weight would go to the spring and the other to the anchor

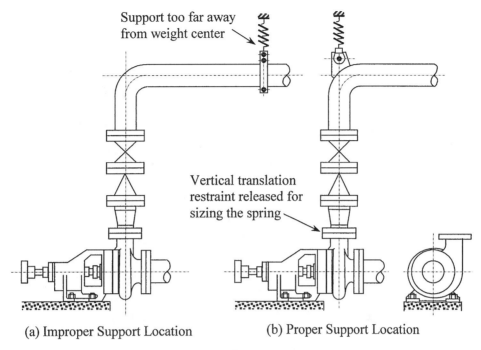

**FIG. 9.6
TAKING CARE OF WEIGHT LOAD**

or nozzle. Then the pump will still carry one-half of the weight when the spring is selected without an anchor release. Anchor release means the anchor cannot take any load in the direction released.

The anchor release procedure works only when there is a spring (or springs) located at proper locations for carrying the weight. If the support is located far away from the weight center as in Fig. 9.6 (a), then the procedure will not work. In this case, the weight force may be reduced to zero or near zero, but the coupling of the spring force and weight creates a huge bending moment. This is due to the large distance between the spring and the weight center. This moment can be more damaging than the weight itself.

9.2.5 Multi-Unit Installation

One of the common omissions by piping stress engineers in analyzing a two-unit installation is skipping the operating mode where one unit of the machines is operating while the other unit is on standby. The schedule-rushed engineer often considers only the situation where both units are operating. In fact, the often-analyzed condition with both units operating is most likely the least critical case. The condition with one unit operating while the other is idling is generally more critical. Take the two-pump system as shown in Fig. 9.7, for instance; theoretically, three operating modes need to be evaluated. These modes include both units operating, unit A operating while unit B is idling at ambient temperature, and the third mode with unit B operating while unit-A is idling. However, an experienced engineer should be able to analyze just the most critical case. In this example, the operating mode with unit A operating while unit-B is idling is the most critical one. Nevertheless, the analysis for the case with both pumps operating is always needed, so the springs, if required, are selected with the potential maximum displacements. In other words, at least two operating modes need to be analyzed.

In a two-pump installation, the most basic requirement is providing enough flexibility between the two pumps. If the flexibility between the two pumps is not enough, then the system will not work no

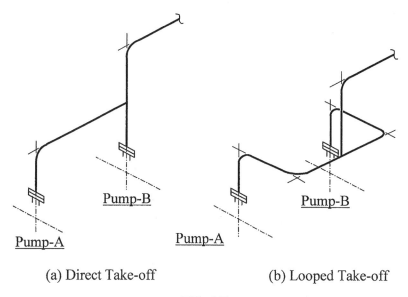

(a) Direct Take-off (b) Looped Take-off

FIG. 9.7
ANALYSIS OF MULTI-UNIT INSTALLATION

matter how flexible the rest of the piping is. Inexperienced engineers often overlook this basic requirement. The flexibility between the two pumps is provided by the take-off piping. Figure 9.7(a) shows direct take-off, which is acceptable for small or low-temperature systems. For large or high-temperature piping systems, a looped take-off as shown in Fig. 9.7(b) is required.

One question often asked concerns the allowable piping load at idling equipment. Because the machine is not operating, it can take a somewhat higher deformation and shaft misalignment. One option that is often used is to increase the allowable by 50% for idling equipment. However, this practice is not universal and needs to be put into the design specification if one wants to use it.

9.2.6 Fit-up the Connection

The final fit-up of piping at a rotating equipment is an act of precise manipulation. It involves the correct dimension and face orientation of the piping together with a proper pull-up procedure so the fit-up load is maintained to a minimum. In some cases, this also involves a cold spring designed to reduce the piping thermal expansion load at operating condition (see also Section 3.6).

To provide the correct dimension and face orientation, a common practice is to make the final piece of pipe slightly longer so it can be trimmed to the exact dimension in the field. Some organizations also have a standard field practice of using localized heating and quenching [6] to bring the piping system into position for bolting up. However, this type of thermal bending is only for very experienced installers. Besides the bending effect, the potentially damaging effects of heating and quenching should also be studied before using this approach.

When a multi-unit installation is operating at near-ambient temperatures, a very tight assembly with a common header is generally used. Under this situation, for both fit-up and maintenance purposes, short bellow expansion joints or flexible couplings are used for the fit-up. In this case, the joints are generally required to be solidified with tie-rods to resist the pressure thrust force. If the joint is not solidly locked, the machine could face operational problems, such as excessive machine vibration. See Section 7.6 and Fig. 7.27(c) for examples of improper installation of expansion joints.

To ensure that no dead weight from the piping and piping components is imposed on the equipment, field engineers from the equipment manufacturer and the operating company will try to use hangers, and hangers only, to bring the piping end flange to alignment with the equipment flange. This requires that all springs be unlocked for adjusting the loads if required. This is contradictory to the common practice of locking the spring hangers during installation and unlocking the spring just before the startup operation. The common practice of making the spring hanger balance the weight at operating condition, although preferred for reducing creep damage, might produce an unpredictable unbalanced force due to uncertainty of the weight and geometry. This unpredictability is not acceptable for sensitive equipment. This is especially true when the piping is expected to operate far below the creep range (see Section 6.4.4). However, this weight load minimizing procedure is not applicable to the system that is to be cold sprung. A cold spring, which requires the proper application of initial forces and moments at the final connection, is used mainly to reduce the thermal expansion load at operating condition. In this case, we can only depend on theoretical calculations to minimize the weight load effect on equipment.

9.3 STEAM POWER TURBINE

The steam turbine used in a power plant is likely the largest rotating equipment that the piping engineer has to deal with. Each turbine is generally connected to main steam, cold and hot reheat steam lines, and roughly ten extraction lines. The lines generally spread out into several independently supported casings. The main steam lines are often connected to the inlet header, rather than directly to the casing. Figure 9.8 shows the schematic piping layout.

For the piping interface, each turbine has a data sheet provided by the manufacturer detailing the thermal movements and allowable piping loads for each piping connection of the turbine. The allowable piping load is not lenient, but is manageable. Besides the extraction steam lines for low-pressure feed water heaters, whose lines are generally fitted with expansion joints, no other connecting piping seems to require a special flexible connection. The extraction steam piping for a low-pressure feed water heater, which is generally located inside the neck of the steam condenser, is connected almost fitting-to-fitting to the turbine. An expansion joint is the only means of providing the flexibility required for absorbing any differential thermal expansion.

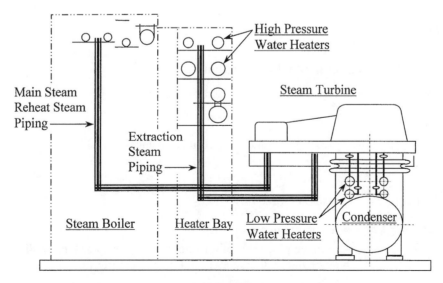

FIG. 9.8
SCHEMATIC LAYOUT OF POWER TURBINE PIPING

Besides the extraction steam to low-pressure feed water heater, all other piping systems have wide spread areas for providing the required flexibility. Take the main steam and reheat steam lines, for instance; here, the connection points of each line is separated by more than 100 ft (30 m) horizontally and another 100 ft vertically. The flexibility of this big L-shape itself is quite large, but may still not be enough to reduce the connecting load to within the allowable limit. However, the space is ample for providing a couple of extra turns, which generally are all that is needed to reduce the piping connecting load to within the allowable.

One very unique practice in power turbine piping is the practice of cold springing. Manufacturers, operators, and engineering companies all agree that cold spring is beneficial to the plant. Most of the main steam and reheat steam lines are either 50% or 100% cold sprung. The extraction steam lines are somewhat more selective in the practice of cold spring mainly due to their sub-creep operating temperature and smaller pipe sizes. See Chapter 3 for general discussions on cold spring.

9.4 MECHANICAL DRIVE STEAM TURBINES

Mechanical drive steam turbines include single-stage and multi-stage turbines intended to drive pumps, fans, compressors, and so forth. The power turbines used in power plants are not included.

NEMA was the first organization that formally specified the allowable piping loads on its rotating equipment. Its NEMA SM-23 [2] standard for steam turbine drives is very well known in the piping community for its stringent allowable to piping loads. To satisfy the NEMA SM-23 allowable for piping loads, two sets of piping loads have to be evaluated. They are the loads at individual nozzles and the combined resultant from all nozzle loads applied at the whole machine.

The individual nozzle load is limited by the strength of the nozzle and casing. An excessive load generates excessive stress or deformation on the nozzle and casing. Excessive stress can cause the nozzle or casing to fail, and high deformation can create interference between rotating and stationary parts. The resultant loads are limited to check the support system of the machine. The combined resultant load is the main cause of shaft misalignment.

Because the resultant moment changes from point to point due to the change in the moment arm of the applying force, a resolving point is specified for the calculation. Since the combined load is used mainly to check the support system, the resolving point should be specified at or near the support point. For a steam turbine drive, the support point is generally near the low-temperature side of the large exhaust nozzle. NEMA SM-23 requires that resultant loads be resolved at the center of the exhaust flange face.

9.4.1 Allowable Loads at Individual Connection

The resultant force and resultant moment imposed on the turbine at any connection shall satisfy the following

$$3F_R + M_R \leq 500 D_e \quad \text{(Inconsistent units)} \tag{9.3}$$

where

$F_R = \sqrt{F_x^2 + F_y^2 + F_z^2}$ = resultant force (lb)
$M_R = \sqrt{M_x^2 + M_y^2 + M_z^2}$ = resultant moment (lb-ft)
D_e = nominal pipe sizes of the connection, for up to 8 in. in diameter (in.)
= (16 + nominal diameter)/3, for sizes greater than 8 in. (in.)

The left-hand side ($3F_R + M_R$) is the applied load, and the right-hand side ($500 D_e$) is the allowable. Equation (9.3), although not a very good example to follow, is also adopted by some other standards.

API Standard-617 [3] for centrifugal compressor is one of them. The equation has three terms representing, force, moment, and diameter, but all of them have different units. This gives an impression that we are adding three oranges with an apple and comparing them with 500 pears. Because of this inconsistency, engineers all over the world are confused as to how to convert it to the metric system. Most pipe stress analysis software packages convert the calculated metric units force and moment to the units given by the equation and then evaluate them with the equation. This sounds inconceivable, but it is the only way to get the job done. To make the equation more universally appealing, it can be converted by multiplying the entire equation with 12 as

$$36F_R + M_R' \leq 6000D_e \quad \text{(Inconsistent units)} \tag{9.3a}$$

where $M_R' = 12M_R$ is the resultant moment in inch-pounds. In this new equation, the number 36 represents 36 in. By the same token, 6000 represents 6000 lb. With 36 and 6000 properly defined, they can be converted to any unit desired. Eventually, a more consistent formula may be forthcoming.

From Fig. 9.1, we get a general idea of the low allowable piping load on turbine drives. This low allowable load dictates the fundamental layout of the connecting piping system, which will be discussed later.

9.4.2 Allowable for Combined Resultant Loads

The combined resultants of forces and moments of the inlet, extraction, and exhaust connections, resolved at the centerlines of the exhaust connection, should meet the following relations for the resultant and for individual components.

(1) The *resultant loads* should meet:

$$2F_c + M_c \leq 250D_c \quad \text{(Inconsistent units)} \tag{9.4}$$

where

$F_C = \sqrt{F_{cx}^2 + F_{cy}^2 + F_{cz}^2}$ = combined resultant force (lb)

$M_c = \sqrt{M_{cx}^2 + M_{cy}^2 + M_{cz}^2}$ = combined resultant moment at resolving point (lb-ft)

$D_{ce} = \sqrt{\sum D_i^2}$ = equivalent diameter of circle opening equal to the total areas of the inlet, extraction, exhaust openings (in.)

D_i = diameter of ith opening (in.)

D_c = D_{ce}, when D_{ce} is 9 in. or smaller

D_c = $(18 + D_{ce})/3$ when D_{ce} is greater than 9 in.

F_{cx}, F_{cy}, F_{cz} = combined resultant force components in x, y, z directions, respectively (lb)

M_{cx}, M_{cy}, M_{cz} = combined resultant moment components in x, y, z directions, respectively (lb-ft)

The resultant moments are calculated by resolving at the center of exhaust flange face. The calculations follow Fig. 9.3 and Eq. (9.2).

(2) The *individual resultant components* should not exceed:

$$\begin{aligned} F_{cx} &= 50D_c & M_{cx} &= 250D_c \quad \text{(Inconsistent unit)} \\ F_{cy} &= 125D_c & M_{cy} &= 125D_c \\ F_{cz} &= 100D_c & M_{cz} &= 125D_c \end{aligned} \tag{9.5}$$

Figure 9.9 shows the coordinate orientation and sign convention. By comparing Eq. (9.3) with Eq. (9.4), it is clear that the allowable value for the combined resultant load is more stringent than that for

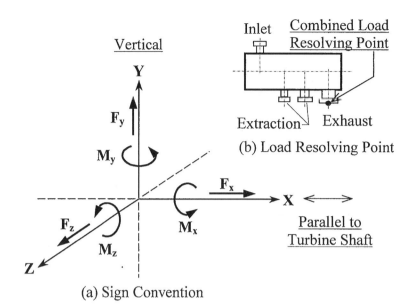

FIG. 9.9
FORCES AND MOMENTS ON A STEAM TURBINE DRIVE

the individual connection. This emphasizes the requirement for laying the piping in such a way that the loads from all individual connections more or less cancel each other.

9.4.3 Basic Piping Layout Strategy

Because of the extremely low allowable of piping load, the design of the turbine connecting piping system is extremely challenging. Engineers who stick to a routine piping design approach may get stuck in the process with endless frustrations. To get a feeling for the problem, the allowable loads for the individual connections are tabulated in Table 9.1. In this table, the force and the moment are assumed to share the allowable equally. That is, the allowable for $3F$ is $250D_e$, and the allowable for the moment is also $250D_e$. The bending stress corresponding to the allowable moment is also tabulated

TABLE 9.1
NEMA SM-23 ALLOWABLE PIPING LOADS FOR INDIVIDUAL NOZZLE CONNECTION

Pipe sizes, in.	D_e, in.	Force[1] ($83.3D_e$), lb	Moment[1] ($250D_e$), lb-ft	Bending stress,[2] psi	Bellow eff. area, in.[2]	Pressure[3] for $167D_e$, psi
2	2	167	500	10,700	–	–
4	4	333	1,000	3,740	20.87	32
6	6	500	1,500	2,120	43.28	23
8	8	666	2,000	1,430	69.74	19
10	8.67	722	2,168	880	112.60	13
12	9.33	777	2,333	640	156.10	10
14	10.00	833	2,500	560	185.00	9
16	10.67	889	2,668	460	236.30	8

[1]3FR and MR each has a half-share of the $500D_e$ allowable.
[2]Bending stresses are based on standard wall pipes.
[3]Bellow pressure force equivalent to full allowable $F = 500D_e/3$. No moment.

for standard wall pipes. To get an idea of the expansion joint application, the pressure that produces a bellow pressure thrust force equal to the allowable force is also listed.

The bending stress corresponding to the allowable moment can be considered as the allowable stress of the piping at the connection. Because the normal allowable stress on piping is in the range of 30,000 psi, an allowable stress of 3000 psi should be considered very difficult to achieve. Looking at the stress column of Table 9.1, it appears that expansion joints or some special features need to be used for pipe sizes 6 in. or greater to meet the connection load limitation. Figure 9.10 shows some basic arrangements of expansion joints.

The most direct expansion joint installation is to have an anchor to take the entire piping load and to place an expansion joint in between the anchor and the equipment connection as shown in Fig. 9.10(a). With this arrangement, the expansion joint easily absorbs all connection movements without generating unacceptable forces and moments. One problem with this arrangement, however, is the bellow pressure thrust force. From the pressure column of Table 9.1, it is clear that in order to meet the piping load limitation, the maximum internal pressure permitted is very low. Take the 8-in. pipe for instance; that maximum allowable pressure is only 19 psi for the bellow pressure thrust force to stay within the connection load limit. For a turbine exhausting to a 150-psi low-pressure steam system, the bellow pressure thrust force of the 8-in. connection is about eight times that of the allowable connection load. This is not acceptable. In fact, arrangement (a) is not acceptable for the majority of turbine installations. It is marked as "No Good" to discourage engineers from using it.

To eliminate the effect of the bellow pressure thrust force from applying to the equipment, tie-rods are added to the expansion joint as shown in Fig. 9.10(b). For the tie-rods to be effective, the pipe is supported with a sliding support. In this manner, when the system is pressurized, the pressure pushes the pipe away from the connection, but both sides of the expansion joint are held together by the tie-rods. The bellow pressure thrust force acting toward the turbine is intercepted by the tie-rods before transmitting to the equipment. This intercepted force is counterbalanced by an equal, but opposite, thrust force acting toward the elbow of the piping system. This arrangement will not absorb the axial expansion of the parts located outside the tie-rod connections. Therefore, the piping should be flexible

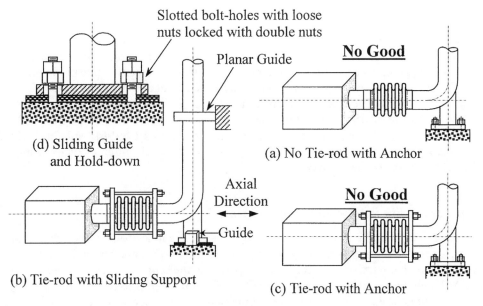

FIG. 9.10
EXPANSION JOINT ARRANGEMENTS AT TURBINE CONNECTIONS

enough to absorb the movement in the axial direction of the bellow. The resistance of piping in this direction equals the force required to push the piping the same distance as the sum of the nozzle movement and thermal expansion of the pipe assembly outside the tie-rod connections. The expansion of the expansion joint assembly inside the tie-rod connections is absorbed by the bellow. If required, the tie-rods can be extended all the way to the center of the elbow to minimize the unabsorbed expansion outside the joint assembly. The total axial direction load at the connection is the sum of this piping resistance and the friction force at the sliding support. Therefore, in addition to piping flexibility, support load and friction coefficient are also important. Support load should be kept to a minimum and a low-friction sliding pad may be needed.

The sliding support is also generally fitted with a lateral guide to take out the lateral force and prevent torsional moment at the bellow. Some elaborate designs use slotted bolt holes and double lock nuts to create a multi-directional restraint support as shown in Fig. 9.10(d). This can be constructed to function similarly to an anchor, but will allow sliding in the axial direction of the bellow. This type of restraint should be closely supervised in the installation so that the intended function is maintained. The nut should not be tight and should be locked by a double nut or equal.

For tied expansion joints installed near the equipment, one design that is often misapplied is the installation of an anchor as shown in Fig. 9.10(c), instead of using a sliding support. Less experienced engineers might think the anchor is the better design, because an anchor is generally considered a necessity in expansion joint installations. However, the anchor in this case is the cause of one of the major operating problems. The reason is simple. Once the pipe is anchored, the tie-rod is no longer functioning. The expansion of pipe and equipment will squeeze the bellow and leave the rods loose. The loose tie-rods do not transmit the bellow pressure thrust force. The anchor makes the arrangement in Fig. 9.10(c) exactly the same as the one without the tie-rod shown in Fig. 9.10(a).

Steam turbine piping is very difficult to design due to the extremely low allowable piping load on the connections. An ingenious layout, combined with a unique support and restraint system, together with special components, is generally required to come up with a satisfactory system. All parties involved, including the designer, engineer, operator, and project manager, should have this understanding to eliminate the frustration that is frequently encountered in the design of steam turbine piping system.

9.5 CENTRIFUGAL PUMPS

Pumps are the most common rotating equipment in an industrial plant. However, from a piping design standpoint, not all pumps are equal. Even two pumps that are exactly the same can have very different piping implications depending on their operational environment. Some pump piping systems are simply fit and go, whereas others require very elaborate layout and restraint systems to ensure the operability of the pumps. There are two main categories of pumps based on the standards by which they are made. Most industrial utility pumps are built with ASME B73.1M [4] and B73.2 [7] specifications and more critical pumps are generally required to meet API Std-610 standards [5]. In both categories, there are many different types of pumps available to suit different types of applications. Figure 9.11 shows some of the basic types.

9.5.1 Characteristics Related to Piping Interface

Before going to the specific allowable piping loads on pump connections, some cursory discussions on the pump characteristics related to the piping structural interface are helpful in designing the piping systems. These characteristics are the basis for planning the support scheme and making the pump selection. Although piping engineers are often left out from the pump selection process, they do have the opportunity to participate in the project meeting of a large project to voice out their opinion.

Figure 9.11(a) shows a frame-mounted overhung pump. This pump has not only the impeller overhung, but the casing is also overhung. From a piping load viewpoint, this is a very weak layout.

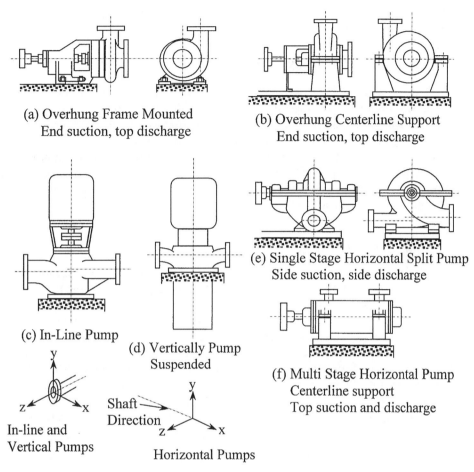

FIG. 9.11
PUMP TYPES AND NOZZLE ORIENTATIONS

Generally, this type of pump cannot even handle the weight of the discharge valve assembly. Because the pump is most likely used in low temperature applications, the weight becomes the main concern. Generally, the weight of the discharge piping has to be properly supported before the connecting piping flange is brought about to join the pump flange. Stacking of unsupported stop valve, check valve, and all fittings on the pump flange is generally unacceptable.

Figure 9.11(b) shows a centerline-supported overhung pump. The impeller is overhung, but the casing is not. The discharge flange can sustain considerable weight due to direct pedestal support. The centerline support arrangement enables it to operate at a higher temperature than that of a base support. Depending on the operating temperature, the pedestal may be air- or water-cooled to limit the vertical movement of the shaft to within the acceptable value.

Figure 9.11(c) shows an in-line pump. The driving motor and pump are stacked together in one unit. This substantially limits the potential misalignment between the two shafts. Furthermore, the pump is constructed just like a valve with few interfering parts. The in-line pump can handle almost the same amount of piping load as a control valve. It may cost more for the pump, but the cost of the overall system, including the savings in piping due to a higher allowable load, can be considerably less than the system that uses another type of pump. For tight locations, the in-line pump should be used as much as possible. In the analysis, the in-line pump is generally included as one of the piping components, with proper support and friction at the base included.

302 Chapter 9

Figure 9.11(d) shows a double-barrel suspended vertical pump. The vertical pump also has the driving motor floating over it. Therefore, the potential misalignment of the two shafts is small. However, because the pump generally involves a long overhung shaft, and the barrel casing is generally not very strong, it saves a great deal on layout spacing, but does not provide much in increasing the allowable piping loads.

Figure 9.11(e) shows a single-stage horizontal pump. It has a double suction impeller, which is located in between two supporting bearings. Both suction and discharge are in the horizontal direction, making the support somewhat easier. The base support is used for low temperature applications. A centerline support is generally used for high temperature applications.

Figure 9.11(f) shows a multi-stage horizontal pump. This is generally for high pressure applications that require better quality than usual. The pump construction is generally very robust due to the high design pressure. However, the pedestal and baseplate may not have the same proportion of strength. In this case, it makes sense to specify the double-strength baseplate if available.

9.5.2 Basic Piping Support Schemes

The most basic and important support for pump piping is the first support off the pump connection. This support, often with added horizontal restraints, controls the amount of piping load transmitted to the pump connection. The support type and the analytical modeling require careful investigation.

Figure 9.12 shows the three basic support types for end suction piping. Regardless of the temperature, the shaft centerline is generally maintained at the same elevation to minimize the misalignment with the driver. This is done with a centerline support of the casing with a water-cooled pedestal if required. The first support off the pump should also be required to maintain this shaft elevation with minimum tilting. Generally, if the temperature is lower than 200°F (100°C), a rigid support as shown in Fig. 9.12(a) can be used. The rigid adjustable support as shown in Fig. 9.12(b) can be used for temperatures up to 400°F (200°C). A spring support is generally required for temperatures higher than

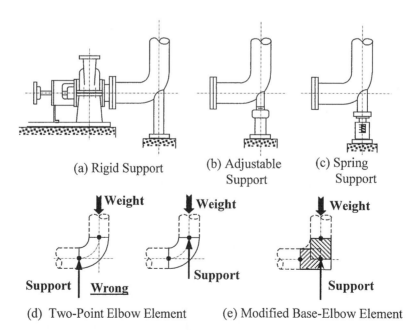

FIG. 9.12
FIRST SUPPORT AT SUCTION PIPING AND MODELING TECHNIQUE

400°F. This is just a rule-of-thumb practice considering the support point located roughly at the same elevation as the centerline of the shaft. The final decision should be made based on calculations.

The modeling of the support at an elbow, in this case and other similar cases, requires detailed consideration. Normally, the elbow is modeled as a two-point elbow element as shown in Fig. 9.12(d). Because the standard piping analysis is based on the centerline of the piping, disregarding the diametrical dimension, analytically, the support in this case can be applied at either one of the two end points located at centerline. If the support is applied at the end point not vertically in line with the actual support, an artificial bending moment will be created between the weight and the support. This is the case often wrongly applied by unsuspecting engineers in their computerized analysis. The correct point to apply the support is the one in line with the riser. At this support point, however, the expansion movement of the pipe between the fixed point of the casing and the support point has to be included as a support displacement. Otherwise, a very high but fictitious vertical force will be generated from the analysis.

The other approach in modeling the elbow is the use of a three-point modified base-elbow element as shown in Fig. 9.12(e). The base-elbow simulates the elbow with two straight elements having the length equal to the bend radius of the elbow. The elbow flexibility factor is ignored due to the reinforcing effect of the trunnion connection. Because the so-called square corner base-elbow has a longer development length than the actual elbow, it partially compensates for the ignored flexibility factor. Although the piping stress in rotating equipment piping is too low to be consequential, the elbow stress intensification factor is applied to make the analysis more conservative. In the base-elbow approach, the support is applied at the center point, joining the two straight elements. Because this point is at the shaft centerline elevation, in general no support displacement is applied.

9.5.3 Non-API Pumps

Non-API pumps include all the pumps manufactured not according to the latest edition of API Standard 610 [5] requirements. Most of these pumps are built following ASME B73.1 [4] specifications. Some of them are built using the 5th edition of API Std-610, the last edition that has a lower allowable piping load than the current edition. Some pumps are based on current API Standard 610 on construction, but fail to follow its strict quality control and quality assurance procedures. Because each type of construction has its own set of allowable piping loads, the exact type of pump used should be known before the piping system can be designed and evaluated.

ASME B73.1 pumps do not the have specific allowable piping loads. Many manufacturers use the NEMA SM-23 allowable times 1.3 as the allowable piping loads for their pumps. Because NEMA SM-23 is known for its very low allowable piping loads, the use of the 1.3-factor is considered inadequate in most cases. One of the practices used by some organizations to improve the situation is to specify heavy-duty baseplates. The so-called 2× and 3× baseplates allow two and three times, respectively, as much allowable load as that of a standard baseplate. The added costs of 2× and 3× baseplates are generally small. In the analysis, piping loads are evaluated as NEMA SM-23 with a modified factor. For 2× baseplate, a factor of $1.3 \times 2 = 2.6$ is used over the NEMA SM-23 allowable.

API Std-610 5th edition pumps do have specific allowable piping loads. However, the allowable covers only nozzles 4 in. or smaller. The allowable piping load on the pump is mostly governed by the weight of the pump. The heavier the pump, the higher the allowable piping loads. Because this version of the standard is outdated, engineers no longer perform routine evaluations based on this standard. Nowadays, manufacturers of these old edition pumps will specify the specific allowable piping load for each of their pumps. Then piping engineers would just take the allowable values given to compare with the piping analysis results.

Not fully complied API Std-610 pumps are quite popular for critical but non-mandatory applications. These pumps are built with API Std-610 requirements using heavy-duty centerline-supported steel casing, heavy-duty baseplate, steel bearing housing, and others. However, the cost is substantially smaller than the fully Std-610 complied pump by eliminating testing, documentation, software, and tractability requirements. The use of these pumps not only saves money, but also expands the scope of qualified manufacturers. More manufacturers mean more competitive pricing and availability. These pumps generally offer the same allowable piping loads as pumps built to the full API Std-610 standard. In this case, the piping loads are evaluated in the same manner as described in Section 9.5.4. Otherwise, manufacturer-specified values should be used.

9.5.4 API Standard 610 Pumps

Starting from the 7th edition, the API Std-610 has detailed allowable piping loads for each nozzle orientation. It has also extended the application to larger pipe sizes. Because the standard has outlined the step-by step evaluation of the piping loads, it leads most of the piping stress analysis computer programs to perform API Std-610 evaluation whenever pump nozzle loads are being considered. This often leads to wrong evaluations, because many analyzed pumps are non-API Std-610 pumps.

In dealing with piping loads, API Std-610 uses a two-step approach. The first step is to specify certain nominal loads that the pumps have to be designed for. These pump design loads are uniform among all the directions. The next step is to evaluate the piping loads, which are distributed, more or less, randomly — that is, the actual piping loads have widely different magnitudes in different directions.

In the discussions that follow, the traditional piping coordinate convention having the positive y axis pointing upward is used. This is consistent with the older versions of the standard and is also consistent with the piping tradition and the rest of this book. In some computer programs, this is also the only coordinate convention that is acceptable. It should be noted, however, that the current (9th)

TABLE 9.2
API STANDARD 610 NOZZLE LOADINGS [5]

	Nominal size of nozzle flange, in.								
	2	3	4	6	8	10	12	14	16
Each top nozzle (forces), lb									
F_x	160	240	320	560	850	1200	1500	1600	1900
F_y	200	300	400	700	1100	1500	1800	2000	2300
F_z	130	200	260	460	700	1000	1200	1300	1500
F_R	290	430	570	1010	1560	2200	2600	2900	3300
Each side nozzle (forces), lb									
F_x	160	240	320	560	850	1200	1500	1600	1900
F_y	130	200	260	460	700	1000	1200	1300	1500
F_z	200	300	400	700	1100	1500	1800	2000	2300
F_R	290	430	570	1010	1560	2200	2600	2900	3300
Each end nozzle (forces), lb									
F_x	200	300	400	700	1100	1500	1800	2000	2300
F_y	130	200	260	460	700	1000	1200	1300	1500
F_z	160	240	320	560	850	1200	1500	1600	1900
F_R	290	430	570	1010	1560	2200	2600	2900	3300
Each nozzle (moments), lb-ft									
M_x	340	700	980	1700	2600	3700	4500	4700	5400
M_y	260	530	740	1300	1900	2800	3400	3300	4000
M_z	170	350	500	870	1300	1800	2200	2300	2700
M_R	460	950	1330	2310	3500	5000	6100	6300	7200

Forces and moments are the ranges covering both plus and minus values.
For vertical in-line pumps, the allowable values are double of the table values.
The orientation of x, y, and z are as shown in Fig. 9.11.

edition of the standard follows the ISO-13709 [8] convention of using the z axis as the one pointing vertically upward.

(1) *External nozzle forces and moments the pump has to be designed for*

To minimize deformation of the casing and misalignment of the shaft due to external nozzle loads, a set of nominal allowable forces and moments, as given in Table 9.2, are established for the design of the pump and baseplate. This set of forces and moments should be considered as the optimal one that both manufacturers and piping engineers feel comfortable with. In other words, manufacturers think it is the maximum values that they can offer, whereas piping engineers feel it is the minimum values they can accept. The pumps are then designed to ensure that both casing deformation and shaft alignment are within the specified limits under this set of external forces and moments. There are two main criteria as follows.

The *pressure casing* shall be designed to operate without leakage or internal contact between the rotating and stationary components while being simultaneously subjected to the maximum allowable working pressure and corresponding temperature and the worst-case combination of twice the allowable nozzle loads of Table 9.2 applied through each nozzle. In addition to the deformation, the casing stress should be limited to the basic allowable stress of the material. Generally, the deformation criterion governs the design.

The *pump, baseplate, and pedestal support assembly* shall be designed and constructed with sufficient structural stiffness to limit the shaft displacement measured at the coupling on an un-grouted mock-up pump, subjected to the combined suction bending moment and discharge bending moment given in Table 9.2, to a maximum displacement not greater than that shown in Table 9.3. The combined bending moment shall be applied at either nozzle, but not at both nozzles. The combined twisting moment, M_{YC}, and the combined rocking moment, M_{ZC}, are applied separately, not at the same time.

The acceptable shaft displacement is higher for pumps with a baseplate intended for grouting. This is in recognition of the fact that grout can significantly increase the stiffness of the baseplate assembly. This grout stiffening effect is neglected in the test, making the pump more easily tested in the manufacturer's shop. After actual installation with the grout in place, the expected acceptable shaft displacement under the test loads should be in fair agreement for both types of baseplates.

(2) *Evaluation of piping loads on API Std-610 Pumps*

The pumps are designed considering all external loads given in Table 9.2 are applied simultaneously. Generally, a pump will not always be simultaneously subjected to all the forces and moments as shown in Table 9.2. The following evaluation takes into account the cases when loads in one or more directions are significantly less than those given by Table 9.2. The evaluation method is given in Appendix F of the standard. The headings and clauses given below follow the headings and clauses of the 9th edition of the standard, except that the $-y$ axis (instead of the z axis) is used for the vertical coordinate.

TABLE 9.3
STIFFNESS TEST ACCEPTANCE CRITERIA

Loading condition	Pump shaft displacement in (μm)		Direction of displacement
	Baseplate intended for grouting	Baseplate not intended for grouting	
M_{YC}	0.003 (75)	0.002 (50)	Z
M_{ZC}	0.007 (175)	0.005 (125)	Y

M_{YC} and M_{ZC} equal the sum of the allowable suction and discharge moments from Table 9.2 with y in vertical up direction. That is,

$M_{YC} = M_{Y,Suction} + M_{Y,Discharge}$

$M_{ZC} = M_{Z,Suction} + M_{Z,Discharge}$

- **F.1 Horizontal Pumps**

  <u>F.1.1</u>: Acceptable piping configuration should not cause excessive misalignment between the pump and its driver. Piping configurations that produce component nozzle loads lying within the ranges specified in Table 9.2 limit casing distortion to one-half the pump vendor's design criterion and ensure pump shaft displacement of less than 250 μm (0.010 in.).

  <u>F.1.2</u>: Piping configurations that produce loads outside the ranges specified in Table 9.2 are also acceptable without consultation with pump vendor if the conditions specified in F.1.2(a) to F.1.2(c) are satisfied. Satisfying these conditions ensure that any pump casing distortion will be within the vendor's design criteria and the shaft displacement of the pump shaft will be less than 380 μm (0.015 in). This clause is a criterion for piping design only.

  (a) The individual component forces and moments acting on each pump nozzle flange shall not exceed the range specified in Table 9.2 by a factor of more than 2.

  (b) The resultant applied forces, F_{RA}, and the resultant applied moment, M_{RA}, acting on each pump nozzle flange shall satisfy the interaction equation below.

  $$\frac{F_{RA}}{1.5F_R} + \frac{M_{RA}}{1.5M_R} \leq 2, \quad \text{for each suction and discharge nozzle} \tag{9.6}$$

  where F_R and M_R are the allowable resultant force and moment given in Table 9.2 for the nozzle evaluated.

  (c) The applied component forces and moments acting on each pump nozzle flange shall be translated and resolved to the center of the pump. The magnitude of the resultant applied force, F_{RCA}, the resultant applied moment, M_{RCA}, and the applied rocking z moment, M_{ZCA} should be limited to the relations given by

  $$\begin{aligned} F_{RCA} &< 1.5(F_{R,S} + F_{R,D}) \\ M_{ZCA} &< 2.0(M_{Z,S} + M_{Z,D}) \\ M_{RCA} &< 1.5(M_{R,S} + M_{R,D}) \end{aligned} \tag{9.7}$$

  where the combined forces and moments are calculated as outlined by Eq. (9.2). $F_{R,S}$ is the allowable resultant force given in Table 9.2 for suction nozzle, and so forth.

  <u>F.1.3</u>: Piping configurations that produce loads greater than those allowed in F.1.1 or F.1.2 shall require approval from the purchaser and the vendor before being implemented.

- **F.2 Vertical In-Line Pumps**

  Vertical in-line pumps that are supported only by the attached piping may be subjected to component piping loads that are more than double the values shown in Table 9.2, provided these loads do not cause a principal stress greater than 41 N/mm² (5950 psi) in either nozzle. For calculation purposes, the section properties of the pump nozzles shall be based on Schedule 40 pipe whose nominal size is equal to that of the appropriate pump nozzle. The principal stress can be calculated with Eq. (2.17). The internal pressure stress is not included.

  Appendix F does not give the evaluation method for vertical pumps that are not of in-line type. The common practice is to evaluate them in the same manner as horizontal pumps with the properly reoriented coordinates as shown in Fig. 9.11.

- **F.4 Evaluation Procedure**

  Most piping stress analysis software packages have a feature to automatically evaluate the pump loads at the same time the piping system stress is being analyzed. Figure 9.13 shows one of the example reports. However, regardless of the method used, the evaluation procedure is the same.

```
13 OCT 06  PENG ENGINEERING, HOUSTON   - SIMFLEX-II (RE-7.1 ) (ASME-B31.3)   PAGE  40
-------------------------------------------------------------------------------
SIMFLEX-II ROTATING EQUIPMENT EXAMPLE
                                                      DATA FILE : Rot-Equip.DA

ROTATING EQUIPMENT LOAD UNDER LOAD CASE NO. 3, LOAD ID =TH + WT
*************************************************************

CENTRIFUGAL PUMP LOAD COMPLIANCE REPORT - API STD-610, 8TH EDITION
-------------------------------------------------------------------------------
              RESOLVING PT. 50       SHAFT AXIS ORIENTATION --- = X = XX
                                     ALLOWABLE MULTIPLICATION FACTOR = 1.00

(A). FIRST NOZZLE

         POINT NO. 35           PIPE SIZE = 10.0 IN,   ORIENTATION =E

                  --------FORCES (LB)--------      --------MOMENTS (FT-LB)--------
                  FXX    FYY    FZZ    FRA         MXX     MYY     MZZ    MRA
-------------------------------------------------------------------------------
CALCULATED         335   -828   -245    927        -248   -3089    2732   4132
ALLOW-F.1.1       1500   1000   1200   2200        3700    2800    1800   5000
ALLOW-F.1.2.1     3000   2000   2400   4400        7400    5600    3600  10000
                                                           HIGH    HIGH

RESULTANTS:       FR     MR              [(FRA /1.5FR) + (MRA /1.5MR)]
-------------------------------------------------------------------------------
CALCULATED:       2200   5000           ( 927/ 3300) + ( 4132/ 7500) = .83
ALLOW-F.1.2.2:    ---                                                  2.00

(B). SECOND NOZZLE

         POINT NO. 55           PIPE SIZE = 8.0 IN,    ORIENTATION =T

                  --------FORCES (LB)--------      --------MOMENTS (FT-LB)--------
                  FXX    FYY    FZZ    FRA         MXX     MYY     MZZ    MRA
-------------------------------------------------------------------------------
CALCULATED          0    -521    663    844        3871       0       0   3871
ALLOW-F.1.1       850    1100    700   1560        2600    1900    1300   3500
ALLOW-F.1.2.1    1700    2200   1400   3120        5200    3800    2600   7000
                                                           HIGH

RESULTANTS:       FR     MR              [(FRA /1.5FR) + (MRA /1.5MR)]
-------------------------------------------------------------------------------
CALCULATED:       1560   3500           ( 844/ 2340) + ( 3871/ 5250) = 1.10
ALLOW-F.1.2.2:    ---                                                  2.00

(C). COMBINED LOAD FROM BOTH NOZZLES

CALCULATED:       FRCA (LB)= 1453,    MZZCA(FT-LB)= 3458,    MRCA, (FT-LB)=  6184
ALLOW-F.1.2.3: 1.5(FRS+FRD)= 5640,    2.0(MZS+MZD)= 6200,    1.5(MRS+MRD)= 12750

THE LOADS - COMPLY - WITH API STD-610 REQUIREMENTS
**************************************************
```

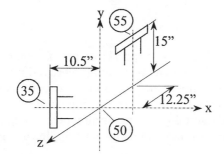

Loads at Suction Nozzle 35:
Fx=335,   Fy=-828,   Fz=-245
Mx=-248, My=-3089, Mz=2732

Loads at Discharge Nozzle 55:
Fx=0,   Fy=-521,   Fz=663
Mx=3871, My=0,   Mz=0

FIG. 9.13
COMPUTER GENERATED REPORT ON API STD-610 PUMP NOZZLE LOADS

First, piping structural analysis is performed to determine the forces and moments acting on the nozzles under specified operating conditions. The loads evaluated are based on normal operating conditions with thermal expansion and weight loads. If earthquake or other occasional loads are included, the allowable loads are generally increased. The forces and moments calculated have to be reoriented with the coordinate system used by API Std-610. In this example report, the reoriented forces and moments are as shown in the figure. The notations of the reoriented loads are given as F_{XX}, F_{YY}, etc., instead of F_X, F_Y, etc., on the analysis coordinates. To reorient the loads, the shaft orientation, XX, has to be related to the analysis coordinates. In this example, the analysis coordinates are the same as the pump orientation coordinates.

With the forces and moments acting at nozzles determined, they are evaluated by the following steps:

(a) For each nozzle, the applied forces and moments are tabulated in comparison with the allowable values given in Table 9.2. If the calculated value is higher than the allowable value, it will be flagged as "HIGH." It might still be acceptable depending on other conditions. However, if the calculated value is higher than twice of the allowable value, it is unacceptable and is flagged as "OVER." The allowable value is labeled as F.1.1, and the twice of the allowable value is labeled as F.1.2.1 (for F.1.2(a)).

(b) For each nozzle, the applied resultant force and moment are compared with 1.5 times the allowable resultant force and the allowable resultant moment as given by Eq. (9.6) required by F.1.2(b). This evaluation, labeled in the report as F.1.2.2, is required only when any of the applied force and moment components exceeds the allowable value. In a computer-generated report, it is routinely evaluated regardless of the magnitude of the applied force and moment. The allowable for this evaluation is 2.0. If the combined usage value as calculated by Eq. (9.6) exceeds 2, then the load is not acceptable and is flagged as "OVER."

(c) For combined loads from both nozzles, the forces and moments applied at individual nozzles are combined and resolved at the center of the pump. These values are evaluated with Eq. (9.7). This evaluation is required only when any of the applied force and moment components at any of the individual nozzles exceeds the allowable value. However, in general, a computer-generated report will routinely perform this evaluation regardless of the magnitude of the applied loads. This evaluation is labeled F.1.2.3 (for F.1.2(c)) in the report. If any of the three evaluations is not met, then the applied load is not acceptable and is flagged as "OVER."

(d) Compliance statement is issued when there is no "OVER" flag in all the above evaluations. Otherwise, a non-compliance statement is given.

The piping loads in this example are acceptable because no "OVER" is flagged anywhere in the report.

9.6 CENTRIFUGAL COMPRESSORS

The centrifugal compressors used in petroleum, refinery, and process services are generally designed and manufactured by API Standard 617 [3] specifications. The size and pressure rating of centrifugal compressors vary greatly. Some utility or auxiliary compressors are no more than just large blowers, but the main process compressors in some plants are very large. The piping connected to the main process compressor in an ethylene plant, liquefied natural gas plant, or ammonia plant, for instance, generally range from 24 in. to 60 in. in size. Figure 9.14 shows the schematic outline of a typical main process compressor.

API Std-617 stipulates that "compressors shall be designed to withstand external forces and moments at least equal to 1.85 times the values calculated in accordance with NEMA SM-23. Wherever

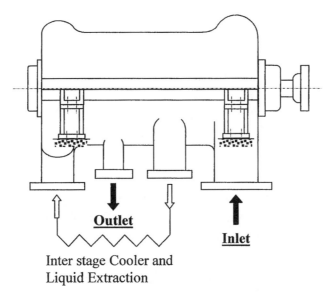

FIG. 9.14
CENTRIFUGAL PROCESS COMPRESSOR

possible, these allowable forces and moments should be increased after considering such factors as location and degree of compressor supports, nozzle strength and degree of reinforcement, and casing configuration and thickness. The allowable forces and moments shall be shown on the outline drawing." The 1.85 times NEMA SM-23 allowable means increasing the right-hand side of Eq. (9.3) from $500D_e$ to $925D_e$.

Although API Std-617 requires compressors designed for external forces and moments to be at least equal to 1.85 times the NEMA SM-23 allowable, most compressor manufacturers will give no more than 1.85 times the NEMA SM-23 allowable. Considering the notoriously low NEMA SM-23 allowable external loads for large nozzles, meeting this 1.85 times NEMA SM-23 allowable imposed by compressors is a very huge task for piping engineers to accomplish. Furthermore, NEMA SM-23 requires the combined load evaluation, which includes the loads from all the nozzles, to be resolved at the center of exhaust flange face. Not only is the allowable for combined loads smaller than the allowable for the individual nozzle, the resolving point location also creates confusion. Some engineers take the exhaust flange literally as an outlet flange, whereas some engineers consider the exhaust flange equivalent to the inlet flange in a compressor. It appears that the latter is more logical because compressors are more likely to be fixed near the larger inlet flange. Respectably, starting from the 5th edition of the standard, API Std-617 no longer requires the evaluation of the combined load. This is a big relief to piping designers, but the difficulty for the piping to meet the allowable nozzle loads remains. Because of the big line sizes and low allowable piping load, there are a few special situations pertinent to the design and analysis of compressor piping that need to be emphasized.

(a) *Radial expansion of pipe.* Thermal expansion in the diametrical direction of the pipe is generally ignored in the design and analysis of the piping system. However, when a restraint is needed near the equipment to control the piping load, expansion in the diametrical direction can be significant in large high-temperature compressor piping.

Figure 9.15(a) shows the typical arrangement where a pipe is stopped or guided. The piping is analyzed simply as being stopped or guided at the location. However, because the piping analysis is based on the centerline of the pipe, a more accurate analysis would include the radial thermal expansion of the pipe, covering the length, L, which is slightly larger than the radius of the pipe. This often-ignored radial expansion may not have significant effect on most piping

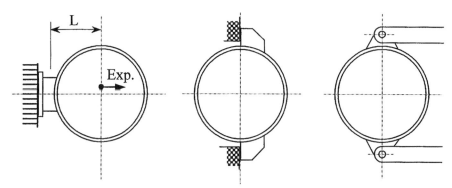

(a). The effect of the radial expansion of the pipe needs to be considered.

(b). Centerline support to reduce the effect of radial expansion of the pipe.

FIG. 9.15
THE EFFECT OF RADIAL PIPE EXPANSION

analyses, but has a very significant effect on compressor piping due to its size and low allowable load.

To get some idea of the effect of radial expansion on compressor piping, a guide is placed 20 ft away from the nozzle to investigate the general trend of the significance. The model simplifies the piping as a cantilever fixed at the compressor nozzle connection. It is further assumed that L is the same as the pipe radius, that is, $L = D/2$. The radial expansion, Δ, when stopped, generates a force as well as moment at the nozzle. For discussion purposes, only the moment is considered. From the cantilever formula given in Fig. 2.14, the moments due to stopped radial expansion at 400°F are tabulated together with the allowable piping moments at the compressor nozzle in Fig. 9.16. It is obvious that this radial displacement of the pipe produces almost the same amount of moment as allowed by the compressor nozzle for a 24-in. pipe. It produces more than twice the allowable nozzle moment for a 32-in. pipe. The moment is about 23 times (outside the chart) the allowable nozzle moment for a 60-in. pipe.

Although the moment so determined is somewhat extreme, considering 400°F is at the high end of compressor piping temperature spectrum and the flexibility of the actual guide or stop structure will reduce the moment somewhat. Nevertheless, the effect of radial expansion is still significant and should be included in the design and analysis. If radial expansion of the pipe is not acceptable, then the centerline support as shown in Fig. 9.15(b) should be used.

(b) *Support stiffness.* When the pipe size reaches 24 in. or more, there are very few structures readily available in the plant to offer effective restraining. The cross-section of the structure members in a process plant is generally smaller than the pipe cross-section. The effectiveness of the supports and restraints depends mainly on the framework to provide the stiffness. The locations that offer high supporting stiffness include the foundation of the compressor, structural steels near well-braced points, cross-braced building floors, and so forth. Therefore, qualified restraining points are not generally located at convenient locations. Most restraints may have to be placed in a skewed direction and attached to a remotely located area with a strut as shown in Fig. 9.17.

The strut assembly is included in the analysis just the same as an ordinary piping member. The pipe trunnion and the bracket connection can be treated as a branch pipe or simply as a rigid member. The main portion of the strut is generally constructed with the pipe, which is naturally treated as a pipe in the analysis. Two ball joints are treated as flexible joints with flexibility in all three rotational directions. The rotational spring rate of the flexible joint can be assigned with

Interface with Rotating Equipment   311

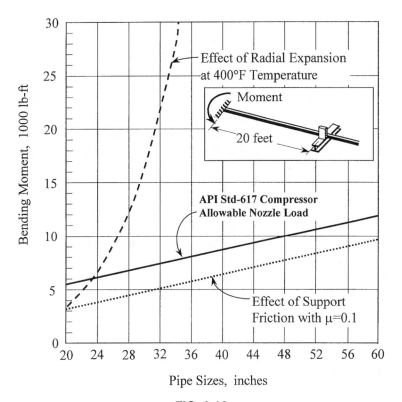

**FIG. 9.16
EFFECTS OF RADIAL EXPANSION AND FRICTION**

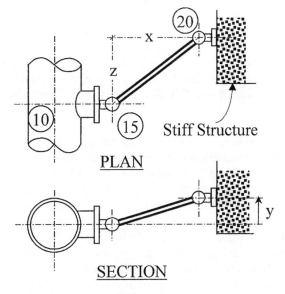

**FIG. 9.17
RESTRAINT TO STIFF STRUCTURE**

some arbitrarily small values, because these spring rates should only have very little effect on the analysis results. The stiffness of the support structure is included as the stiffness of the anchor. Normally, stiffness equal to or greater than 10^6 lb/in (1.75×10^5 N/mm) is considered perfectly rigid. Anything below this value should be included as the support stiffness in the analysis.

(c) *Nozzle flange movements.* The connecting flange of the nozzle moves due to thermal expansion of the compressor casing. Although they may be small, these nozzle movements have a significant effect on the piping loads as discussed in item (a) (pipe radial expansion). The effect of the nozzle movement is especially important when restraints are needed nearby to control the loading. Because the casing temperature distribution is quite complicated, and the compressor support system is generally not transparent to piping engineers, these nozzle movements should be provided by the manufacturer of the compressor. The movements estimated by the piping engineer, from arbitrary casing temperature distributions and assumed support functions, are generally not very accurate.

(d) *Support friction.* Again, because of the low allowable piping load on the compressor nozzle, support friction plays a very important role in the design of the compressor piping. Low friction surfaces are generally needed at supports and guides. To get an idea of the magnitude of the support friction effect, a support scheme with a 20-ft (6.1 m) span is investigated as a reference. The actual support span is generally considerably larger. With a support located 20 ft away from the nozzle, the nozzle carries a weight load equivalent to 10 ft of the pipe, and the support carries the weight of 20-ft pipe. The pipe is empty with the insulation weight ignored. When the nozzle moves, the friction force at the support produces a bending moment at the nozzle. This bending moment is compared with the nozzle allowable moment in Fig. 9.16. Even with a friction factor of 0.1, which is the design value of most common sliding plates, the effect of the friction force at one support constitutes about 50% of the allowable load for 20-in. pipes, and constitutes about 80% of the allowable load for 60-in. pipes. This simple comparison shows that low-friction sliding plates are needed at supports and guides, and friction has to be included in the design analysis.

(e) *Temporary supports for hydro test.* Compressor piping does not have any content weight during operation. Therefore, the support systems designed for the weight at operating conditions generally are not enough to carry the hydro test weight. The hangers and locked spring components can normally safely carry the hydro test weight for up to twice [9] the normal operating load. Therefore, the piping whose water weight is greater than the steel weight generally requires temporary supports for the hydro test if it is being tested hydrostatically. A 24-in. standard wall pipe, for instance, has a water weight equal to roughly twice of the steel weight, making the hydro test weight three times the normal operating weight. In this case, some temporary supports are required for the hydro test. Placement of temporary supports needs careful planning to include the arrangement of not only the support proper, but also the required supporting structure.

In addition to the piping load at the nozzle, the compressor piping should also be designed to prevent pipe shell vibration. When the impeller of the compressor rotates inside the casing, it produces a small pressure disturbance as the impeller blades pass by the vanes. This pressure disturbance has a frequency equal to the blade passing frequency, which is the impeller rotating speed times the number of blades. Because the rotor generally runs at a very high revolution, the frequency of this pressure disturbance is very high. Although this blade-passing disturbance generally does not have enough force to shake the piping system, it creates a very high pitch noise resulting from the pipe shell vibration. When the disturbance frequency coincides with the pipe shell natural frequency, it might also tear off the pipe wall. The most common preventive action taken is to place acoustic insulation on the pipe to dampen out the noise level. For certain low-pressure systems, an increase in wall thickness, above what is required for the design pressure, may also be required. These practices generally follow the experience of each company.

9.7 RECIPROCATING COMPRESSORS AND PUMPS

Reciprocating machineries are generally powered by rotating drivers, but the working elements are moving in a back-and-forth fashion. They can deliver the same volume regardless of the discharge pressure. Reciprocating compressors and pumps are used mainly to produce a high discharge pressure that is either difficult or uneconomical for the centrifugal machine to achieve.

The piping load at the nozzle is not the major problem when interfacing with reciprocating machines. The machines are very robust and can handle fairly large piping loads. The problem is switched to the pulsation of the flow due to the intermittent action of the cylinder valves. The pulsating flow causes vibration in the piping and its supporting structure. The problem associated with this pulsation is quite complicated. Detailed discussions on this problem are beyond the scope of this book. In this section, only basic phenomena are discussed so the readers have some idea about the work involved and the effects on piping and supports.

Because the pulsation problem involves the machine, dampener, piping geometry, and accessories in a multiple level of interaction, its adverse phenomena are difficult to pinpoint. The preventive study, if desired, is generally delegated to outside consultants specializing in the field. Because many installations operate quite satisfactorily without the preventive pulsation study and the preventive study does not necessarily always eliminate the problem, many plants adopt a wait-and-see policy. If the system is not operating smoothly, then all types of help are used to solve the problem. This may be the best thing one can do in normal cases. However, it should be emphasized that once the piping is shaking due to pulsation flow, the entire plant is also likely to shake. This could be a very serious situation. In some cases, an immediate shutdown of the plant is required.

In the following, only the general phenomena of pulsation produced by reciprocating machines are discussed. A few general precautionary guidelines on piping design are listed. The general dynamic analysis of the piping system will be discussed in the chapters dealing with dynamic analyses.

9.7.1 Pulsating Flow

The discharge from a reciprocating machine is not continuous. Rather, it is intermittent with an uneven rate. Figure 9.18 shows the schematic diagram of a reciprocating machine. The machine consists of a piston or plunger moving back and forth inside the cylinder. Some machines have more than one piston and cylinder. The cylinder can also be single acting or double acting. Double acting means the

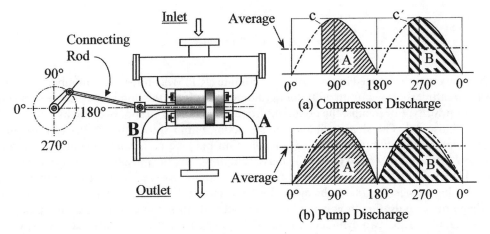

FIG. 9.18
PULSATING FLOW AT RECIPROCATING COMPRESSOR AND PUMP

piston is working both ways, having inlet and outlet ports at both ends of the cylinder. The machine shown in Fig. 9.18 is double acting.

The piston is pushed back and forth by the connecting rod connected to the crank, which is revolving at a constant speed. With a constant angular crank speed, the linear speed of the piston varies very closely to a sinusoidal shape. If the connecting rod is infinitely long, then the piston speed follows a pure sinusoidal form. With a finite-length connecting rod, the piston speed tends to be slower than given by the sine curve at the side closer to the crankshaft, and quicker toward the other side. However, for practical purposes, a sinusoidal form can be used. The discharge volume of the cylinder chamber at the B side is smaller than that of the A side due to the volume occupied by the piston rod. Again, for a general discussion purposes, the volumes at both sides of the cylinder are assumed to be similar.

The volumetric historical discharge shape of a compressor is different from the historical discharge shape of a pump. In a compressor, the gas has to be first compressed from the inlet pressure to the outlet pressure before being discharged to the outlet system. Because of its compressible nature, it takes the piston to move to point c to compress the gas to reach outlet pressure. The gas is then discharged volumetrically according to the speed of the piston. This assumes that the outlet pressure maintains a constant pressure without being influenced by the discharge of the machine. The discharge shape of the pump is quite different due to a practically incompressible nature of the liquid pumped. In a pump, the liquid starts to discharge almost instantaneously as the piston starts to move. Therefore, the volumetric shape is similar to the sinusoidal shape without the initial silent period as shown in Fig. 9.18(b). The figure shows the curve from both sides of the double-acting cylinder. They are discharged to a common header creating a combined volumetric pulsation before being discharged to the outlet piping.

The combined pulsation flow averages out by the multiple outlets spaced at strategically straddled crank phase angles. More cylinders result in a higher pulsation frequency, but lower pulsation amplitude. Figure 9.18(b) shows two humps and two valleys in one crankshaft revolution for a one cylinder (simplex) double-acting pump — that is, two cycles of pulsation per revolution of crankshaft. The shape of the pulsation has quite deviated from sinusoidal. The shape of the compressor pulsations is especially irregularly shaped. Because of its imperfect but nevertheless cyclic shape, the pulsation is regarded as the combination of many sinusoidal pulsations at crankshaft rotation frequency and its higher harmonic frequencies. The pulsation shape is matched by varying the amplitude of each of the harmonic frequencies. These harmonic frequencies, f_p (in cycles per second), are calculated as [10]

$$f_p = \frac{nN}{60}, \quad n = 1, 2, 3, \ldots \quad \text{for reciprocating compressors} \tag{9.8}$$

$$f_p = \frac{nmN}{60}, \quad n = 1, 2, 3, \ldots \quad \text{for reciprocating pumps} \tag{9.9}$$

where N is the crankshaft speed in revolution per minute (R.P.M.), and m is the number of plungers or pistons for multiplex pumps. With all these harmonic pulsations, the dominant pulsation is generally given higher priority. For pumps, the dominant frequency is easier to spot because its pulsation shape is close to sinusoidal. The single-cylinder double-acting pump shown in Fig. 9.18(b), for instance, has the dominant pulsation at two cycles of pulsation per one crankshaft revolution. Because the average flow rate is only $2/\pi = 0.637$ of the maximum flow rate, the pulsation flow has a maximum of $(1 - 0.637)/0.637 = 0.57$ of the average flow rate, and a minimum of zero flow rate. In other words, the flow is pulsating between +57% of the mean flow rate and −100% of the mean flow rate.

The shape of the pulsation flow is not very well defined. In a duplex double-acting pump, for instance, the cranks of the two cylinders are set 90 deg. apart for the best flow balance. The ideal sinusoidal flow pulsation is shown in Fig. 9.19(a), with a maximum at 17.2% above the mean average flow rate and a minimum at 17.2% below the average rate. The ideal sinusoidal shape is created by

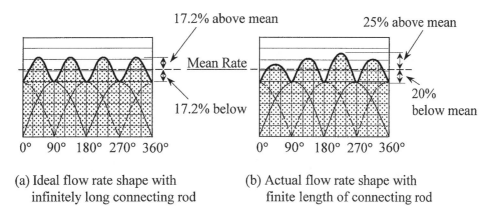

FIG. 9.19
FLOW RATE OF DUPLEX – DOUBLE ACTING PUMP

an infinite length connecting rod. For the actual finite-length connecting rod, the flow rate is slower when the piston is near the crank side. Figure 9.19(b) assumes that the piston speed, thus the flow rate, reaches the maximum as the crank rotates 10-deg. past the 90-deg. point. The combined flow rate is quite different from that of the pure sinusoidal form. In this case, the flow pulsation is quite complex. The maximum flow is 25% above the average, and the minimum flow is 20% below the average. Although there are four cycles of visible pulsation flow per crankshaft revolution, their shapes and amplitudes are all different. The shape would become even more complex once the piston rod volume is subtracted from B-side cylinders. Therefore, it is important that the manufacturer's curve be used if it is available.

The pulsation flow exiting from the compressor cylinder is less critical due to the absorbing capability of the header volume. However, it has other problems, such as acoustic resonance, that need to be addressed.

9.7.2 Pulsation Pressure

In a detailed analysis, the effect of pulsation flow to the piping is generally investigated one harmonic frequency at a time. However, for general design purposes, the overall peak-to-peak pulsation may be used as a quick estimate of the overall effect. Figure 9.20 shows the schematic layout of the pulsation flow through the machine outlet. A similar behavior also exists at the inlet. The machine discharges the average volumetric flow, Q_0, together with the volumetric pulsation, Q_1, to a header or bottle of fairly large volume, either of which is conveniently called a surge bottle. The average flow passes the entire system unchanged. However, the volumetric pulsation entering the surge bottle is partly absorbed by the volume (capacitance) of the bottle, leaving the rest, Q_2, to discharge through the piping. The pulsation flow to the piping creates the pulsation pressure, which produces the pulsation force to shake the piping system. From basic water hammer principle, the pressure change required to push a flow velocity change, Δv, during a very small time interval is determined by [11]

$$\Delta P = \rho a \Delta v = \rho a \frac{\Delta Q}{A} \tag{9.10}$$

where ΔQ is the change of volumetric flow rate, A is the flow area of the pipe, ρ is the density of the fluid, and a is the sonic speed. If the change of volumetric flow rate lasts for a certain period, the total change of pressure is

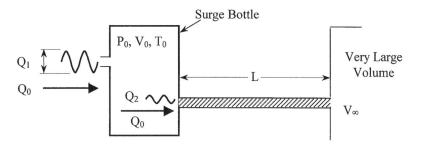

FIG. 9.20
ATTENUATION OF PULSATION FLOW BY SURGE BOTTLE

$$\sum \Delta P = \frac{\rho a}{A} \sum \Delta Q \tag{9.11}$$

So long as the time interval is less than $2L/a$. L is the pipe length. That is, for a pipe length exceeding $a/2f$, where f is the pulsation frequency, the peak-to-peak pulsation pressure is

$$P_{\text{p-p}} = \frac{\rho a}{A} Q_2 \tag{9.12}$$

This peak-to-peak pulsation pressure is needed to push the peak-to-peak Q_2 volumetric pulsation flowing through the pipe. This same pressure pulsation also compresses the fluid volume inside the surge bottle to make room for some of the incoming pulsation flow. The bigger the bottle volume and the higher the fluid compressibility, the more incoming pulsation is absorbed by the bottle, leaving less pulsation being transmitted through the pipe.

Q_2 is called the residual peak-to-peak pulsation flow, and $P_{\text{p-p}}$ given by Eq. (9.12) is called the residual peak-to-peak pulsation pressure at surge bottle. The same pressure pulsation exits the bottle to enter the piping system.

For incompressible fluids such as the ones handled by the pump, the portion of pulsation absorbed by the surge bottle is negligible. Therefore, for pumps, the residual pulsation flow is very close to the original pulsation flow. In other words, Q_2 is practically equal to Q_1 for pumps.

For compressor installations, without getting into the mathematical details, the residual pulsation is roughly determined by the ratio of the surge bottle volume and the piston displacement volume [12]. The number of single-acting (SA) cylinder displacements required to fill the surge bottle is referred to as the attenuation quotient (AQ). The magnitude of the pressure-volumetric decay is referred to as the attenuation factor (AF), which is $1/(1 + AQ)$. In other words, if the volume of the surge bottle is 9 times the SA cylinder displacement, $AQ = 9$ and $AF = 1/(1 + AQ) = 1/(1 + 9) = 0.1$. Therefore, Q_1 is attenuated to Q_2, which is equal to $0.1 Q_1$. Thus, the surge bottle achieves a tenfold reduction of the potential pulsation pressure.

9.7.3 Pulsation Dampener for Reciprocating Pumps

The surge bottle used by the reciprocating pump is somewhat more complicated than the simple volume shown in Fig. 9.20. This is mainly due to the incompressible nature of the liquid the pump handles. For liquid flow, a gas volume is generally needed to absorb the pulsation. The gas-filled surge chamber is one such example. With the compressible chamber, a large portion of the pulsation flow is absorbed by the compression or expansion of the gas volume. Thus, the residual pulsation flow through the pipe is substantially reduced. Figure 9.21 shows some of the typical pulsation dampening arrangements used in reciprocating pump installations.

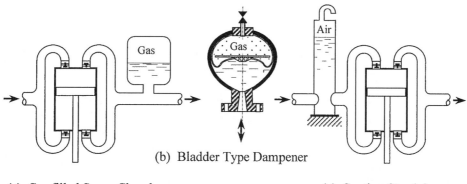

(a) Gas filled Surge Chamber (b) Bladder Type Dampener (c) Suction Standpipe

**FIG. 9.21
SURGE DAMPENER FOR RECIPROCATING PUMPS**

Chilton and Handley [13] have done extensive studies and conducted tests to publish a design chart to be used as a guideline for determining the amount of gas volume required in a surge chamber. The charts given in Fig. 9.22 can be used to determine the gas volume required for reducing the residual peak-to-peak pulsation pressure to a certain desired value.

Because pulsation flow tends to average out with multiple active cylinders, the more the active cylinders in the machine, the less gas volume is required to achieve the same proportional reduction of pulsation pressure. In addition to the type of the machine, the piston rod diameter and the connecting rod length also have a significant effect on the overall residual pulsation pressure. The charts given in

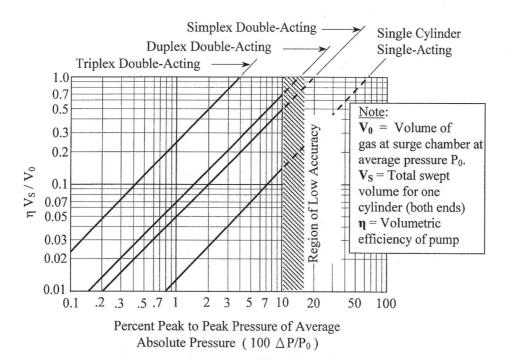

**FIG. 9.22
EFFECT OF SURGE CHAMBER GAS VOLUME ON RECIPROCATING PUMP
PRESSURE PULSATION [13]**

Fig. 9.22 are based on a connecting rod length of five times the crank arm length, which is one-half of the piston stroke. They are also based on a piston rod cross-section area of 0.2 times the piston area. Machines having different connecting rod length and/or piston rod cross-sectional area require some corrections over the charts. These correction procedures are not included in the book, because the charts only serve as general guidelines. Interested readers should consult the original paper for details.

Figure 9.22 provides the method for determining the required gas volume to reduce the pulsation pressure to a certain value. The other guideline needed is the magnitude of the acceptable, or design, pulsation pressure. Each company may have its own limit on residual pulsation pressure that can be used for design. When no specific criterion is given, the values given in Fig. 9.23 can be used. Figure 9.23 is one of the acceptable criteria widely used in the industry as reported by Scheel [12]. These values represent the ones that are achievable from the machine viewpoint and can still be managed from the piping point of view. To be consistent, both Figs. 9.22 and 9.23 are labeled for the peak-to-peak pulsation.

The gas surge chamber shown in Fig. 9.21(a) is simple and effective. However, it is very difficult to contain and maintain a constant volume of the gas during startup or shutdown. Even at a constant average pressure during operation, the gas is still constantly dissolved to the liquid, reducing its effective volume. One way to resolve this gas-escaping problem is to contain the gas inside a bladder as shown in Fig. 9.21(b). In a bladder-type dampener, the bladder is charged with inert gas to a pressure roughly equal to one-half to two-third of the operating pressure. As the system is pressurized, the gas volume is reduced to about one half to two-third of the original fully extended volume. During operation, the bottom portion of the bladder forms a diaphragm pushing up and down the gas volume to accommodate the fluctuating pulsating flow. Generally, the manufacturer of the dampener can size the required volume for a specified application. Figures 9.22 and 9.23 can also be used as references. A few notes need to be added on bladder-type dampeners. (1) Because the bottom portion of the bladder consists of a significant mass inertia, it is difficult for the dampener to respond to quick oscillations. (2) The nozzle flow area is generally very small, again making the unit unable to respond very well to quick oscillations. (3) Due to the flow restriction at the nozzle, it is generally better to use multiple units of smaller size dampeners than one big dampener in pulsation applications. A big dampener may have

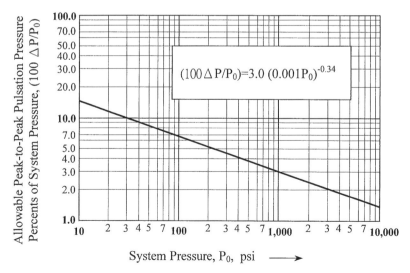

FIG. 9.23
SUGGESTED ALLOWABLE PULSATION PRESSURE [12]

the same size of nozzle as a smaller one. (4) The required volume by Fig. 9.22 is the average operating volume, which can be different from the manufacturer's specified volume.

For pumps taking fluid from an atmospheric tank or reservoir, a standpipe, as shown in Fig. 9.21(c), can be used at the suction side. The standpipe is inexpensive and much more effective than the bladder-type dampener. The function of a standpipe is equivalent to the function of a dampener with an infinitely large gas volume. The pulsation pressure equals only the elevation change of the liquid surface fluctuation required to accommodate the volumetric flow pulsation. It is important to note that pulsation at the suction side is much more critical than that of the discharge side of the pump. The suction pulsation will generally affect the performance of the discharge side, producing a disproportional large pulsation at the discharge side. A small pulsation in the suction can produce starvation of the pump, causing water hammer-type banging between the piston and the fluid when the piston has to travel a small distance of void before solidly connecting with the fluid. Therefore, it is very likely that when the pump discharge piping shakes, the cause is in the suction side rather than in the discharge side. On the other hand, the discharge pulsation at a pump has little or no effect on the suction.

9.7.4 Some Notes on Piping Connected to Reciprocating Machine

When dealing with a reciprocating compressor or pump, we are also dealing with potential vibration in the piping. With this scenario, the piping system has to be designed somewhat differently than the practices that we are normally accustomed to. The following are some of the items that need to be addressed.

(1) Independent support system. The piping vibration can propagate to the entire plant when the piping is supported from a common structure. Therefore, it is important that during the planning stage, proper spaces have been allocated so the piping can be independently supported. Furthermore, the supports offered should have sufficient stiffness to effectively control the dynamic motion of the piping. Stiff supporting members, such as concrete sleepers located at grade level, should be used if practical.

(2) Secured clamping effort. The connection between the pipe and the support structure is very critical to the effectiveness of the support. A good connection starts with good clamps, which is the first link between the pipe and support. Without a good connection, a purposely designed heavy support structure is just a waste of resources. Figure 9.24 shows some of the clamps used in vibration piping. The clamp has to be stiffened as shown in Figure 9.24(a) and 9.24(b). In Figure 9.24(a), the clamp is also lined with belting material to offer some damping effect in addition to securing a good connection. Figure 9.24(b) provides two squeezing wedges to ensure a snug fit of pipe and clamp. This is often used in large piping. Figure 9.24(c) shows hold down beams with controlled hold down force provided by coil springs or Belleville springs (washers). The spring-loaded hold down provides three-directional restraints all with limited resistance, but very high stiffness. The downward resistance is very large, but upward resistance is limited to the initial force of the spring. The lateral resistance is the friction from the weight of the piping and the initial spring force.

(3) Realistic thermal expansion design temperature. Thermal expansion of the piping needs to be accommodated without causing overstress. However, most vibration piping operates at a rather moderate temperature. If the design temperature is overly conservative, the piping designed might be too flexible to prevent vibration. Therefore, it is important to set a realistic design temperature. In addition, it is important to remember that the thermal expansion load cycles only once or twice a year, yet the vibration stress occurs hundreds of cycles per minute.

(4) Avoid acoustic and structural resonance. The pulsation flow has two potential resonance mechanisms that need to be avoided. First, the pulsation pressure wave can generate acoustic resonance if the length of any discontinuity section in the piping has an acoustic natural frequency that coincides with the pulsation frequency. This acoustic resonance has a potential of

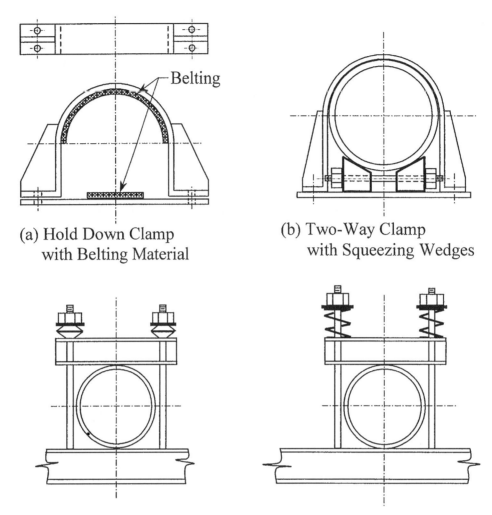

(a) Hold Down Clamp with Belting Material

(b) Two-Way Clamp with Squeezing Wedges

(c) Hold Down Beam with Controlled Hold Down Force

FIG. 9.24
HOLD DOWN CLAMPS FOR PIPING WITH POTENTIAL VIBRATION

amplifying a pulsation pressure 10 or 20 times the original pulsation pressure. The other potential resonance is the structural resonance of the piping natural flexural vibration frequency with the pulsation pressure frequency. Acoustic resonance can be largely avoided by running an analog study of the piping configuration and the equipment performance. Regarding structural resonance, one of the tactical solutions is to support the piping in such a way that no natural frequency of the piping is close to the pulsation frequency. However, this is easier said than done. As given by Eqs. (9.8) and (9.9), the pulsation has many harmonic modes. It is impossible to design the piping so that all piping natural frequencies are located away from all pulsation frequencies. One compromise policy is to make the piping so stiff that its fundamental frequency is at least 50% higher than the fundamental frequency of the pulsation. However, even this compromise approach is difficult to achieve. A more practical rule-of-thumb approach is to support the piping with support spacing reduced to one-half of the standard spacing.

(5) Avoid suspended masses. All concentrated weights, such as valves and flanges, have to be located near the support. This is required to increase the fundamental natural frequency of the piping system and to reduce potential vibration stress.
(6) Pay attention to suction side. As discussed previously in this section, discharge vibration in reciprocating pump piping often originates from problems of the suction piping. Suction piping needs to be designed with minimum flow resistance and pulsation.
(7) Supports of appendages. Appendages, such as drain valves and instrument lines, shall be properly supported. A seemingly small pipe vibration may translate to significant vibration at appendages. The drain valve, for instance, has a lower natural frequency than the main pipe due to the weight of the valve. It can resonate with the pipe shaking frequency to produce significant amplitude and stress at the root of the connection. These appendages generally cannot be supported from the surrounding structure. They are often supported with braces tied to the main pipe.
(8) Avoid using very thin pipe. Vibration stress causes high cycle fatigue. Thin pipe produces an imperfect joint that has a very high stress intensification factor against high cycle fatigue damage.

The severity of the piping vibration depends on both vibration amplitude and piping configuration. The general evaluation method is discussed in Chapter 13. Large amplitudes can still be tolerated if the piping is vibrating at a very low frequency. A low-frequency vibration also means that it is vibrating under a long span. However, psychologically, the vibration amplitude appears to be the main awareness gauge of plant personnel.

9.8 PROBLEMS ASSOCIATED WITH SOME TECHNIQUES USED IN REDUCING PIPING LOADS

With the low piping load allowed on rotating equipment nozzle connections, piping engineers have used many ingenious techniques in the attempt to make the piping meet the requirements. Some of these techniques work well in the real world, whereas other approaches only seem to work on paper. The practical solution requires practical engineering common sense without which the so-called ingenious scheme is not only a waste of resources, but is also hazardous. The following subsections discuss some of the common ideal schemes that flunked these common sense tests.

9.8.1 Excessive Flexibility

Adequate piping flexibility is required to reduce the piping load to acceptable levels. However, a good design should consider the inherent flexibility of the support structure and the use of strategically located protective restraints. Without the tactical placement of restraints, the piping system has difficulty meeting the allowable load imposed by the equipment, no matter how flexible the piping system is. Figure 9.25 shows a pump suction system, which is designed without any restraints installed. Because a restraint will increase the stiffness of the piping system, some engineers mistakenly think that the restraint can only increase the piping load. It is true that a restraint will tend to decrease the flexibility of the system as a whole and will increase the maximum stress and force in the system. However, a properly placed restraint can shift the stress from the portion of piping near the equipment to a portion remote from the equipment.

Although the above example system uses extensive loops, the piping load may still not meet the equipment allowable due to the lack of restraints. Excessive flexibility makes the system prone to vibration, because it is easily excited by a small disturbing force. In addition, the piping loops may generate excessive pressure drop that reduces the system pressure to below the saturation pressure. Once the pressure drops to below the saturation pressure, the system becomes unstable due to local vaporization inside the pipe. A system similar to the one shown in Fig. 9.25 experienced very severe

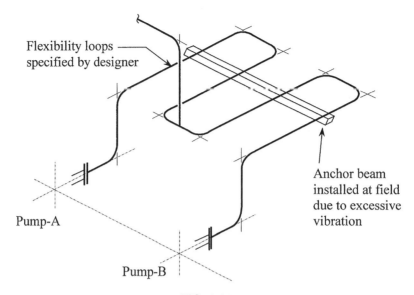

FIG. 9.25
TOO MUCH FLEXIBILITY CAUSES OPERATION PROBLEM

vibration in one petrochemical plant. The operation engineers had to put a large cross beam to anchor all the loops in the field to suppress the vibration to a seemingly comfortable level. The fact that the piping loops had to be fixed in the field shows that the loops were not needed in the first place. The cross-beam anchor may have solved the external piping vibration problem, but the internal flow disturbance remained unchanged. The voids and cavities formed by the vaporization of the fluid still happen incessantly inside the piping. The system appeared to be still, yet the banging noise inside was still audible. The banging of the fluid greatly reduces pump efficiency and the pump's service life.

9.8.2 Improper Expansion Joint Installations

Properly installed expansion joints may be the only solution to the low allowable equipment load in some piping systems. However, these expansion joints are often improperly installed under certain situations. These improper installations can escape even the sharp eyes of some experts, and thus cause severe operational problems. These situations are discussed in Section 9.4.3 and in Section 7.6, which deals with flexible connections.

9.8.3 Theoretical Restraints

In general, a piping system has some restraints to control the movement of the piping and to protect sensitive equipment from being overloaded by piping forces. However, there are also restraints that are placed in desperation by piping engineers who are trying to meet the allowable load of the equipment. These so-called computer restraints give very good computer analysis results on paper, but are often ineffective and sometimes even harmful. Figure 9.26 shows some of the situations that worked well on a computer simulation but did not work on real piping systems. These pitfalls are caused by the differences between the real system and the computer model. The following are some of the problems with theoretical restraints.

(1) Friction plays a very important role in the design of restraints placed near the equipment. Figure 9.26(a) shows a typical stop placed against a long z-direction line to protect the equipment from

Interface with Rotating Equipment   323

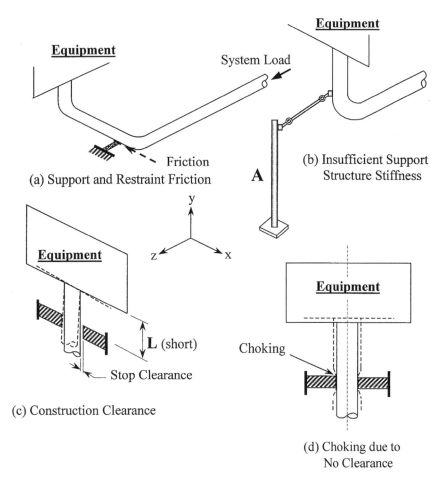

FIG. 9.26
PROBLEMS ASSOCIATED WITH THEORETICAL RESTRAINTS

being pushed by the z force. In the design calculation, if the restraint friction is ignored, the calculated equipment reaction force is generally very small. However, in reality, friction at the stop surface will prevent the pipe from expanding in the positive x direction. This friction effect can cause a high x-direction reaction force, and thus z moment to the equipment. A calculation including the friction will predict this problem beforehand. A proper type of restraint, such as a low-friction sliding plate or a strut, would then be used.

(2) An ineffective support member is another problem often encountered in protective restraints. Figure 9.26(b) shows a popular arrangement for protecting the equipment, against the z direction force in this case. The engineer's direct instinct is to always put the fix at the problem location. For instance, if the computer shows that the z-direction reaction force is too high, the natural fix is to place a z direction stop near the nozzle connection. This may be satisfactory on the computer, but in reality the stop is very ineffective. For the stop to be effective, the stiffness of the support member and its structure has to be at least one order-of-magnitude higher than the stiffness of the pipe, which is very stiff in this case due to the close proximity to the nozzle.

(3) A gap is generally required in the actual installation of a stop to ensure smooth movement of the piping. This so-called construction gap is about 1/16 in. (1.5 mm) in size. This gap can be ignored in the portion of piping away from terminal fixations. However, for a stop located close to the nozzle connection, as may be required by the computer calculation, this gap will hinder

the effectiveness of the stop due to the stiff nature of the short pipe. Figure 9.26(c) shows a typical pipe stop located at one of these close-by locations. Because of the gap, the pipe has to be bent or moved a certain distance, closing the gap before the stop becomes active. Due to the proximity of the stop to the equipment, this is almost the same as bending the equipment the same amount before the stop becomes active. This is not acceptable, because the equipment generally can only tolerate a much smaller deformation than the construction gap of the stop.

(4) Choking is another problem related to the gap at the stop. In some instances, the piping engineer may specify a no-gap installation for those stops located close to the equipment. This generally requires a special order, because it is different from the routine installation procedure. This no-gap installation may solve the ineffective stop problem mentioned above, but creates a choking problem. As the pipe expands, it expands in the diametrical direction as well as the longitudinal direction. Once the diametrical dimension of the pipe increases, the no-gap stop chokes the pipe, preventing it from moving in the axial direction. This, in turn, causes the pipe to expand toward the equipment as shown in Fig. 9.26(d), where the equipment is being pushed upward by the pipe expanding from the choked stop. This, of course, is not acceptable because it can cause severe misalignment of the shaft or even tear up the machine from its foundation.

9.9 EXAMPLE PROCEDURE FOR DESIGNING ROTATION EQUIPMENT PIPING

Rotating equipment piping is one of the most difficult systems to design due to the low allowable piping loads at equipment nozzle connections. Piping engineers have struggled for a long time to find methods to solve the problem. Yet, many piping engineers are still suffering from the fact that it is almost impossible at times to make the rotating equipment piping work, even just on paper. One of the problems is the lack of a procedure to follow. Experienced engineers emphasize that it is an art rather than craft when dealing with rotating equipment piping. In this section, we will expound on the general procedure with the hope that the discussion will help engineers grasp the essence of designing rotating equipment piping.

The problem associated with rotating equipment piping is almost invariably at the suction side. This is due to the sensitivity of the machine to the flow pressure drop on the suction and also due to the larger size of the piping normally used in the suction side. In the following, a pump suction piping will be used to demonstrate the general design procedure. The design starts with the equipment layout plot, showing the relative locations of the pieces of equipment needed for the plant. In this example, we have the locations of the pump and the storage tank, which can be a refining tower, a deaerator storage tank, and so forth. From these equipment locations, a preliminary piping plan is laid to connect the equipment as shown in Fig. 9.27. Unique to the suction, a mandatory minimum pressure is maintained at the pump entrance to ensure the proper operation of the pump. This minimum pressure, normally called the minimum required net positive suction head (NPSH), is generally specified by the pump manufacturer. Maintaining this minimum pressure is the duty of the piping engineer as well as the system engineer.

From the preliminary piping layout, shown as a solid bold line in the figure, the system engineer will figure out the system pressure drop to see if the end pressure has a sufficient margin to maintain the NPSH required by the manufacturer. This pressure drop, based on the preliminary piping layout, is the design pressure drop. Because an expansion loop or some type of offset is expected on rotating equipment piping, a margin of roughly the same as the design pressure drop is allocated on the design. To provide this margin of pressure or head, the storage tank elevation is raised as high as required. In a refining tower, for instance, this storage tank elevation affects all the connecting piping systems and other associated towers and equipment. A simple elevation change involves many other changes. Therefore, this elevation, for practical purpose, is fixed from the very beginning. Piping engineers should appreciate this basic fact and should not contemplate a change of this elevation.

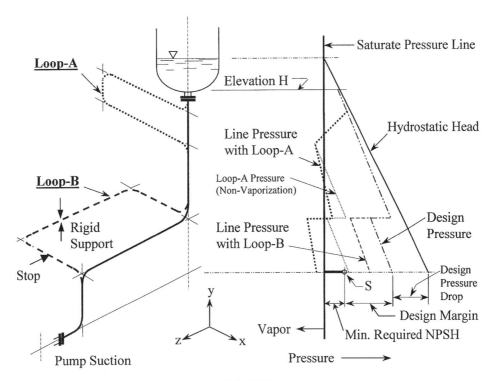

FIG. 9.27
DESIGN PROCEDURE OF ROTATING EQUIPMENT PIPING

As in many cases, the liquid handled is the condensate from the vapor. This means that the liquid at surface level is at saturation temperature or close to saturation. Because liquid temperature is maintained throughout the piping system, any pressure, including the hydrostatic head below the saturate pressure, causes vaporization of the liquid. In Fig. 9.27, the saturation pressure line is the vertical line, having the pressure equal to the pressure at liquid surface. When there is no flow, the pressure inside the piping is represented by the straight line equivalent to the saturate pressure plus hydrostatic head. As the pump operates at the design capacity, the pressure loss causes the pressure to decrease, reaching the line labeled as "Design Pressure." The design pressure at pump entrance is considerably higher than the minimum required NPSH. The difference is the margin allocated for losses due to the expansion loop and other items unaccounted for in the design.

Normally, the preliminary piping layout is not adequate to reduce the piping load at the pump connection to the allowable level. This can be easily checked by running through the analysis. In this example system, it is fairly easy to recognize that the system will produce high forces in the $-y$ and $+z$ directions. These forces will also generate high moments, mainly about the x axis. The following are procedures that may be used to solve the problem of high piping loads at the equipment connection.

(a) Include all flexibilities available. This is the time that all flexibilities should be accounted for to minimize the overall required piping length, and thus pressure drop of the system. In this particular case, the flexibility at the vessel connection is important and should be included.
(b) Use only the realistic operating temperature. Some systems may have a design flexibility temperature much higher than the actual operating temperature. These service temperatures, such as steam out or dry out temperatures, occur only during shutdown conditions. They are not critical to the operation or maintenance of the rotating equipment and should not be used in evaluating equipment connecting loads. If evaluations of those temperatures are required by the

specification, the allowable loads are generally increased by 50% of the normal allowable loads. This increase in allowable load should be given by the specification.

(c) Install strategically placed stops. In some cases, the piping connecting load can be considerably reduced by placing stops at strategic locations. In this case, it appears that a y stop may be placed at the horizontal run to reduce the y force at the pump connection. The exact location of the stop is determined by manual iterations of analyses. However, there is no stop that can really reduce the z force. Whether using the y stop only is enough depends on the temperature of the piping.

(d) Select the proper loop location. When it comes down to designing an expansion loop, the location of the loop is very important. One may be tempted to place a loop at a higher location, such as loop A, due to better space and support structure availability. However, when handling near-saturate fluids, the high location loop can be the cause of all operational problems. First, this loop uses four elbows that generate excessive pressure drop. According to the classic method of estimating pressure drop, a long radius elbow, with bend radius equal to 1½ of the pipe diameter, generates a pressure loss equivalent to a section of straight pipe 10 ft long for 6-in. pipe, and 16 ft long for 10-in. pipe. Four elbows generate a pressure loss equivalent to 40- to 60-ft-long pipe, depending on the pipe size. The loss can easily reduce the fluid pressure to below the saturate pressure at a higher location due to less hydrostatic head available at the high location. Once vaporization occurs, the fluid starts to bang and the pressure drop further increases. The situation resembles the diagram labeled "Line Pressure with Loop-A" in the figure. The expansion loop should be placed at the bottom portion as much as possible.

(e) Loop at bottom location. Although the bottom portion is generally more congested, the expansion loop should still be placed at the bottom portion in combination with original bends. The combination with original bends reduces the number of elbows required. This reduces the pressure loss of the expansion loop. Furthermore, because the bottom portion has a higher hydrostatic head, the pressure loss due to expansion loop is less likely to reduce the line pressure below saturate pressure. The diagram labeled "Line Pressure with Loop-B" in the figure represents the situation.

(f) Provide pivot points for installing stops. The loop should be laid out in such a way that pivot points are created for installing stops. Loop B, as shown in the figure, offers these pivot points. Because of the horizontal run, the pipe is pushing the pump in the $+z$ direction. This force can be reduced or even reversed by installing a stop in the z direction as shown. By fine-tuning the location of this stop, the z force at the pump connection can be reduced to a minimum. The same thing also applies to the vertical stop shown.

(g) Use expansion joint as alternatives. It is recognized that the use of expansion joint is prohibited by some design specifications. However, when there is no way to make the piping load meet the allowable limit, an expansion joint should be considered. Although the expansion joint is a specialty item requiring special engineering, it is still better than a system that seems to work on paper but is actually full of potential problems.

(h) Check the feasibility of using other types or constructions of equipment. In some cases, it is absolutely impossible to design a piping system that will meet the allowable load of the equipment. In such cases, other types of equipment or constructions should be considered. For instance, the in-line pump can take considerably more load than the separated machines/driver set and should be considered as an alternative, if feasible. Some pumps may be required to be installed on a sliding baseplate or even on a spring supported baseplate.

As previously noted, designing rotating equipment piping is more of an art than a craft, requiring considerable ingenuity and resourcefulness. One thing that piping engineers need to keep in mind is that the suction is almost always more critical than the discharge, even though the discharge piping operates at much higher pressure than the suction.

REFERENCES

[1] Rossheim, D. B., and Markl, A. R. C., 1940, "The significance of, and Suggested Limit for, the Stress in Pipelines due to the Combined Effects of Pressure and Expansion," *Transactions of the ASME*, 72, pp. 443–454.

[2] NEMA SM-23, 1991, *Steam Turbines for Mechanical Drive Service*, National Electrical Manufacturers Association, Rosslyn, VA 22209. (The original SM-20-1958 has been replaced by SM-21 and SM-22-1970 which has been replaced by SM-23-1979.)

[3] API Standard 617, *Centrifugal Compressors for General Refinery Services*, American Petroleum Institute, Washington, D. C.

[4] ASME B73.1, *Specification for Horizontal End Suction Centrifugal Pumps for Chemical Process*, ASME, New York, NY.

[5] API Standard 610, 1971, *Centrifugal Pumps for General Refinery Services*, 5th ed., American Petroleum Institute, Washington, D. C. (This edition has been replaced by later versions. Current version is 9th edition, 2003, which is renamed to Centrifugal Pumps for *Petroleum, Heavy Duty Chemical, and Gas Industry Services*.)

[6] Losey, M. D., and Trujillo, D., 1985, "Rosebudding — the Thermally Induced Bending of Pipe," ASME Publication PVP-Vol. 98-9.

[7] ASME B73.2, *Specification for Vertical In-Line Centrifugal Pumps for Chemical Process*, ASME, New York, NY.

[8] ISO-13709, 2003, "Centrifugal Pumps for Petroleum, Petrochemical and Natural Gas," International Organization for Standardization (ISO), Geneva, Switzerland.

[9] MSS SP-58, 1975, *PIPE HANGERS AND SUPPORTS – Material, Design, and Manufacture*, Manufacturers Standardization Society (MSS) of the Valve and Fitting Industry, Arlington, VA 22209.

[10] Wachel, J. C., and Bates, C. L., 1976, "Escape Piping Vibrations while Designing," *Hydrocarbon Processing*, 20, pp. 152–166.

[11] Wylie, E. B. and Streeter, V. L., 1978, *Fluid Transients*, McGraw-Hill International Book Company, New York.

[12] Scheel, L. F., 1972, *Gas Machinery*, Gulf Publishing Company, Houston, TX.

[13] Chilton, E. G., and Handley, L. R., 1955, "Pulsation Absorbers for Reciprocating Pumps," *Transactions of the ASME*, 77, pp. 225–230.

[14] The M. W. Kellogg Company, 1956, *Design of Piping Systems*, revised 2nd ed., Chapter 9, John Wiley and Sons, New York.

CHAPTER

10

TRANSPORTATION PIPELINE AND BURIED PIPING

In the modern society, products are seldom consumed in the same location where they are produced. This is true for most items, from basic farm products to very sophisticated high-technology industrial products. These products are often transported thousands of miles to reach their consumers. The difference lies in terms of the most suitable means of transporting these products into the market. For liquid and gaseous products, pipelines have proven to be the safest and most economical means of transportation onshore, and one of the most reliable methods for transporting these products across the water.

Figure 10.1 exemplifies the needs of transportation pipelines for crude oil and its refined products. The crude produced from the oil field has to be transported to the user, either a refinery or a shipping terminal, located hundreds of miles away. Because an oil refinery is generally located away from a populated area that will be consuming the refined products, the refined products will then need to be transported from the oil refinery to the consuming market. This is the situation with oil and natural gas. The main characteristic of the transportation pipeline is that it involves a large quantity of pipe, which often requires multiple suppliers for a single project. A transportation pipeline also generally runs many miles without any attachment of special components, such as elbows, tees, and other stress risers.

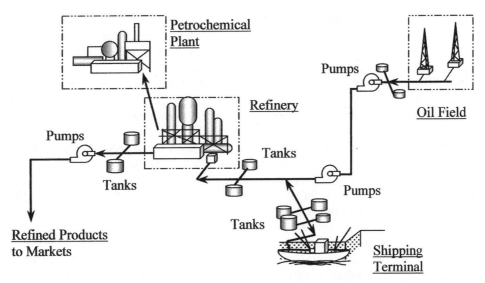

**FIG. 10.1
TRANSPORTING CRUDE OIL AND REFINED PRODUCTS**

Pipelines have been used to transport many different things. The most common pipelines are water distribution systems, which we are all familiar with. Coal, iron ore, and other solid goods, in water slurry, have also been transported through pipelines occasionally. However, this chapter deals only with gas and liquid petroleum pipelines. As pipelines are generally buried, it is necessary to combine the characteristics of the buried pipe in this discussion. However, this chapter deals only the general overall behaviors of the piping system. The local pipe shell deformation due to soil burden and road crossing is not discussed.

The design of transportation pipelines involves close cooperation among geotechnical, civil, and mechanical engineers. However, not all pipelines are equal. For instance, the design of pipelines installed in open flat wastelands in warm climate areas are essentially complete with just the calculation of the pipe wall thickness. On the other hand, the design of lines situated in frigid areas with sensitive environment and vast areas of water and mountains to cross would require the most sophisticated engineering skills and resources. This chapter covers only the basic fundamental mechanical analyses on the general behaviors of long pipelines and their interfaces with local equipment.

10.1 GOVERNING CODES AND GENERAL DESIGN REQUIREMENTS

Because many pipelines run across public spaces, they are generally regulated by government agencies. In the United States, Title 49 of the Code of Federal Regulations (49 CFR), Parts 192 [1] and 195 [2], cover the minimum safety standards for pipelines. 49 CFR emphasizes more of the administration aspects of pipelines rather than the technical matters. For technical matters, the American Society of Mechanical Engineers (ASME) Code for Pressure Piping [3] provides more detailed rules. The discussions that follow are based mainly on the ASME code, commonly referred to as ASME B31 code.

Since transportation pipelines are quite unique compared with other plant piping systems, ASME B31 code has three sections that deal specifically with them: B31.4 [4] for liquid petroleum pipeline, B31.8 [5] for gas transmission pipeline, and B31.11 [6] for slurry pipeline.

As noted above, a pipeline generally runs for miles without any fittings attached. Because of this simplicity, the stress in the majority portion of a pipeline is quite predictable. Taking advantage of this basic characteristic, the code's allowable stress for a pipeline is greatly increased, as compared to that for plant piping, to reduce the tonnage of steel required. With the large amount of piping required, the savings on steel cost translate to a high proportion of savings in the whole project. This increase in allowable stress may appear to have decreased the apparent safety factor. In reality, the geometrical and shape simplicity in a long pipeline eliminates a large chunk of uncertainty and increases the predictability of stresses. This certainty makes the pipeline's real safety factor comparable to that of other sections of the ASME B31 code.

Generally, ASME B31 code's allowable stress for a pipe is based on overall considerations of yield strength, ultimate strength, and creep strength of each specific pipe material. However, the allowable stress of a transportation pipeline is based primarily on yield strength only. In other words, the allowable stress of a pipeline is mainly to protect the pipe from gross deformation. As long as the ductility of the material meets the pipe specifications, the margin of the ultimate strength over the yield strength is considered sufficient to provide the required safety factor, assuming the pipe is not grossly deformed. As the pipeline operates at a much lower temperature than the creep temperature, no creep consideration is needed. Taking advantage of this unique characteristic, higher and higher yield strength pipes are being manufactured for pipeline use. The American Petroleum Institute (API) SPEC 5L [7] has specifications for pipes with yield strengths as high as 120 ksi (827 MPa). However, the savings on steel tonnage from using the ultra high yield pipe is offset by the price premium of the high yield pipe, higher cost in special welding and installation procedure, and so forth. The minimum thickness required for handling and bending also prevents the use of very high yield steel. For these reasons, the use of very high yield pipe is not always the best choice. Without a detailed cost analysis, most pipelines tend to use moderately high yield pipes, such as grades X42 or X46. The number following the X is the yield strength in kilo pounds per square inch (ksi) unit.

High yield steel pipes have relatively low ultimate strength to yield strength ratio. The grade X80 pipe, for instance, has a yield strength of 80 ksi (552 MPa) and an ultimate strength of 90 ksi (620 MPa). The ultimate strength to yield strength ratio is 1.125. This is quite small compared to the ratio for low yield strength steel pipes, which usually has an ultimate strength to yield strength ratio of about 2.0. The low ultimate to yield strength ratio also means a low ultimate strength to allowable stress ratio. This low ultimate strength to allowable stress ratio requires the assurance that the local stress excursion be controlled. This means that special consideration of local primary stress on certain components, such as branch connections, may be required.

The following discussions are limited to the general stress aspects of the code. Engineers should consult the full text of the code for details. Readers should refer to Chapter 4 for discussions related to the general stress aspects and requirements of other sections of the B31 code.

10.1.1 B31.4 Liquid Petroleum Pipeline

ASME B31.4 [4] covers piping systems transporting liquids, such as crude oil, condensate, natural gasoline, natural gas liquids, liquefied petroleum gas, liquid alcohol, liquid anhydrous ammonia, and liquid petroleum products, between producers' lease facilities, such as tank farms, pump stations, natural gas processing plants, refineries, ammonia plants, terminals (marine, rail, and truck), and other delivery and receiving points.

(a) *B31.4 basic allowable stress.* The basic allowable stress is used for basic pressure designs, such as calculating the required wall thickness. In the B31.4 code, it is based on the specified minimum yield strength (SMYS) and is defined as

$$S = 0.72 E_j (\text{SMYS}) \tag{10.1}$$

where 0.72 is the design factor and E_j is the weld joint factor.

In setting the above design factor, the B31.4 code considered and included the allowance for under-thickness tolerance and maximum allowable depth of imperfections provided for in the material specifications approved by the code. In other words, this design factor is based on nominal wall thickness.

The code has a specific weld joint factor for each material. These official numbers have to be used for all calculations. In general, seamless and electric resistance welded pipe has a weld joint factor equal to 1.0. The electric fusion welded pipe generally has a weld joint factor of 0.8, and the furnace butt-welded pipe has a weld joint factor of 0.6.

For pipe that has been cold worked in order to meet the SMYS and subsequently heated to 600°F (330°C) or higher (welding excepted), the basic allowable stress value shall be 75% of the value determined by Eq. (10.1).

From the basic allowable stress, the required wall thickness is determined a

$$t_n = \frac{PD}{2S} + A \tag{10.2a}$$

where t_n is the nominal wall thickness. This is a departure from the basic design of plant piping, which uses the minimum potential net thickness after subtracting the manufacturing under-tolerance.

Thickness allowance, A, is the sum of the allowances for threading, grooving, corrosion, and so forth. The wall thickness allowance for corrosion is not required if pipe and components are protected against corrosion in accordance with the requirements and procedures prescribed in Chapter VIII "Corrosion Control" of the B31.4 code.

(b) *B31.4 allowable stresses for combined loading.* Due to the very high allowable stress allocated to the pressure loading, the allowable stresses for other loadings applied to the pipeline are considerably less than that of other sections of the B31 code. Generally, the stresses due to some common load types are calculated and evaluated as follows:

(1) Internal pressure hoop stress, S_{HP}

$$S_{HP} = \frac{PD}{2(t_n - A)} \leq 1.0S \tag{10.2b}$$

where t_n is the nominal thickness of the pipe and A is the allowance including corrosion allowance and thread allowance. Corrosion allowance is not required for pipe and components that are properly corrosion-controlled according to the code requirements.

(2) External pressure. The limitation of hoop stress due to external pressure is also governed by Eq. (10.2b). In addition to the hoop stress limitation, the pipe shall also be evaluated for buckling stability protection. (See also Section 1.3.4.)

(3) Thermal expansion stress range, S_E. The thermal expansion stress range is an equivalent stress calculated based on maximum shear failure theory. It is defined as twice the value of the maximum shear stress.

For fully restrained lines:

$$\begin{aligned} S_L &= \nu S_{HP} - E\alpha(T_2 - T_1) \\ S_E &= \text{greater of } (|S_L| \text{ or } |S_L - S_{HP}|) \\ S_E &\leq 0.9(\text{SMYS}) \end{aligned} \tag{10.3}$$

For unrestrained lines:

$$S_E \leq 0.7\,(\text{SMYS}) \tag{10.4}$$

The S_E for unrestrained lines is calculated by the standard flexibility procedure applicable to all sections of the B31 code. For above-ground fully restrained lines, the weight bending stress has to be added to S_E in Eq. (10.3).

(4) Additive longitudinal stress, S_{LL}. The sum of the longitudinal stresses due to internal pressure, weight, and other sustained loadings shall not exceed 75% of the allowable stress for thermal expansion. That is, for unrestrained lines, we have

$$S_{LL} = \frac{S_{HP}}{2} + S_{LS} \leq 0.525(\text{SMYS}) \tag{10.5}$$

where S_{LS} is the longitudinal stress due to weight and other sustained loadings.

(5) Additive circumferential stress, S_H. The sum of the circumferential stresses due to internal pressure and external load in the pipe, installed under railroad or highways without use of casing, shall not exceed the basic allowable stress, S. That is,

$$S_H = S_{HP} + S_{HC} \leq 1.0S \tag{10.6}$$

where S_{HC} is the circumferential stress due to external loads on railroad or highway crossings. This stress is not automatically calculated by most pipe stress computer software packages. It needs to be calculated separately using Spangler's method [8] or others. By comparing Eqs. (10.2b) and (10.6), it is clear that the thickness of the pipe at road crossings needs to be increased over the thickness of the main pipeline.

(6) Occasional stress, S_{occ}. The sum of the longitudinal stresses produced by pressure, weight load and other sustained loads, hydraulic transient load, and wind or earthquake shall not exceed 80% of the SMYS. That is,

$$S_{OCC} = \frac{S_{HP}}{2} + S_{LS} + S_{LT} + \text{greater of }(S_{LW} \text{ or } S_{LQ}) \leq 0.8(\text{SMYS}) \tag{10.7}$$

where S_{LT}, S_{LW}, and S_{LQ} are the longitudinal stresses due to hydraulic transient, wind, and earthquake, respectively. Wind and earthquake are not considered to occur simultaneously. Moreover, some specifications do not require the combination of wind or earthquake with the hydraulic transient load.

The above stress limitations are given in subparagraphs 402.3.2 and 402.3.3 of the B31.4 code. Items (1), (3), (4), and (6) are generally evaluated automatically by the pipe stress computer program. Items (2) and (5) are generally evaluated separately either by manual calculations or with other specialized computer programs. From the above stress requirements, it is apparent that the allowable stress for pipeline thermal expansion is considerably less than that of normal plant piping. This is attributable to the high allowable stress allocated to the pressure design.

10.1.2 B31.8 Gas Transmission Pipeline

ASME B31.8 [5] covers gas transmission and distribution systems, including gas pipelines, gas compressor stations, gas metering and regulation stations, gas mains, and service lines up to the outlet of the customer's meter set assembly.

(a) *B31.8 basic allowable stress.* The basic allowable stress is used for basic pressure design of the piping. It is mainly used to determine the wall thickness of the piping under various environments. Because of the explosive nature of gas, a higher safety factor is needed at high population areas. B31.8 sets the basic allowable stress based on human activity and population density of the area. It divides the service area into four location classes as defined in Table 10.1. The allowable stress is defined as

$$S = FE_j T \text{ (SMYS)} \tag{10.8}$$

TABLE 10.1
GAS PIPELINE LOCATION CLASSES

Location class	Description	Remarks
Class 1 location	Any class location unit that has 10 or fewer buildings intended for human occupancy. It has two construction divisions: Division 1 and Division 2	Wasteland, deserts, mountains, grazing lands, farmland, offshore
Class 2 location	Any class location unit that has more than 10 but fewer than 46 buildings intended for human occupancy	Fringe areas around cities and towns, industrial areas, ranch or country estates, etc.
Class 3 location	Any class location unit that has 46 or more buildings intended for human occupancy, or an area where the pipeline lies within 100 yd (91 m) of either a building or a small, well-defined outside area that is occupied by 20 or more persons on at least 5 days a week for 10 wk in any 12-month period	Suburban housing developments, shopping centers, residential areas, industrial areas, and other populated areas not meeting Class 4 requirements
Class 4 location	Any class location unit where multi-story buildings are prevalent, and where traffic is heavy or dense, and where there may be numerous other utilities underground	Multistory means four or more floors aboveground including the first or ground floor. The depth of basement or number of basement floors is immaterial

1. A "class location unit" is an area that extends 220 yd (200 m) on either side of the centerline of any continuous 1-mile (1.6 km) length of pipeline.
2. Each separate dwelling unit in a multiple dwelling unit building is counted as a separate building intended for human occupancy.
3. Ample allowance should be made for potential future increase of buildings.
4. For pipelines shorter than 1 mile, a class location shall be assigned that is typical of the class location that would be required for 1 mile of pipeline traversing the area.
5. The length of class locations 2, 3, and 4 may be adjusted as follows: (a) a class 4 location ends 220 yd (200 m) from the nearest building with four or more stories aboveground; (b) when a cluster of buildings intended for human occupancy requires a class 2 or class 3 location, the class location ends 220 yd (200 m) from the nearest building in the cluster.

where F is the design factor as given in Table 10.2, E_j is the longitudinal weld joint factor, and T is the temperature derating factor, which is given in Table 10.3. SMYS is the specified minimum yield strngth of the pipe. The E_j factor is the same as in B31.4 discussed above in Section 10.1.1(a).

(b) *B31.8 allowable stresses for combined loading.* As in B31.4, the allowable stress for non-pressure loading in gas transmission piping is small due to the high allowable stress allocated to the pressure load. Generally, the stresses due to some common load types are calculated and evaluated as follows:

(1) Internal pressure hoop stress, S_{HP}.

$$S_{HP} = \frac{PD}{2t_n} \leq 1.0S \tag{10.9}$$

where t_n is nominal thickness of the pipe.

(2) Thermal expansion stress range, S_E.

$$S_E \leq 0.72 \, (SMYS) \tag{10.10}$$

Thermal expansion stress range, S_E, is calculated by the standard flexibility analysis procedure applicable to all sections of the B31code. The expansion stress at fully restrained portion is not mentioned in the code. Generally, the same Eq. (10.3) as given by B31.4 is used to evaluate the fully restrained portion.

(3) Occasional stress, S_{OCC}. The sum of longitudinal pressure stress, the longitudinal bending stress due to external loads, such as weight of pipe and contents, wind, etc., shall not exceed 75% of SMYS. That is,

TABLE 10.2
DESIGN FACTORS (F) FOR STEEL GAS PIPE CONSTRUCTION

Facility	Location class				
	1		2	3	4
	Div. 1	Div. 2			
Pipelines, mains, and service lines (basic factor)	0.80	0.72	0.60	0.50	0.40
Crossings of roads, railroads without casing:					
(a) Private roads	0.80	0.72	0.60	0.50	0.40
(b) Unimproved public roads	0.60	0.60	0.60	0.50	0.40
(c) Roads, highways, or public streets, with hard surface, and railroads	0.60	0.60	0.50	0.50	0.40
Crossing of roads, railroads with casing					
(a) Private roads	0.80	0.72	0.60	0.50	0.40
(b) Unimproved public roads	0.72	0.72	0.60	0.50	0.40
(c) Roads, highways, or public streets, with hard surface, and railroads	0.72	0.72	0.60	0.50	0.40
Parallel encroachment of pipelines and mains on roads and railroads					
(a) Private roads	0.80	0.72	0.60	0.50	0.40
(b) Unimproved public roads	0.80	0.72	0.60	0.50	0.40
(c) Roads, highways, or public streets, with hard surface, and railroads	0.60	0.60	0.60	0.50	0.40
Fabricated assemblies	0.60	0.60	0.60	0.50	0.40
Pipelines on bridges	0.60	0.60	0.60	0.50	0.40
Compressor station, regulating, and measuring station piping	0.50	0.50	0.50	0.50	0.40

Class 1 Division 1 requires that the pipe has been hydrostatically tested to 1.25 times the maximum operating pressure.
U.S. Code of Federal Regulation may not permit the use of design factor above 0.72.
A fabricated assembly includes the portion of pipe located within five pipe diameters in any direction from the last fitting of the assembly.

**TABLE 10.3
TEMPERATURE DERATING FACTOR, T**

Temperature, °F	Derating factor (T)
250 or less	1.000
300	0.967
350	0.933
400	0.900
450	0.867

$$S_{OCC} = \frac{S_{HP}}{2} + S_{LS} + S_{LT} + \text{greater of } (S_{LW} \text{ or } S_{LQ}) \leq 0.75(\text{SMYS}) \qquad (10.11)$$

All symbols are the same as in Eq. (10.7).
(4) Total combined stress, S_{TTL}

$$S_{TTL} = S_E + S_{OCC} \leq 1.0(\text{SMYS}) \qquad (10.12)$$

The above stresses are generally calculated and evaluated automatically by pipe stress analysis computer programs.

10.2 BEHAVIOR OF LONG PIPELINE

The stress analysis of a pipeline is quite different from that of plant piping [9]. The most fundamental difference between pipeline and plant piping is the very long length of the pipeline. A pipeline with miles in length has the potential of producing a very large amount of expansion. A reasonable estimate of the movement and its interaction with the end resistance force afforded by connecting piping and equipment are very important aspects in designing a pipeline.

10.2.1 Pressure Elongation

When estimating the potential movement of the pipeline, it is important to include pressure elongation in addition to thermal expansion. As discussed in Section 3.7, the pressure elongation per unit length of a pipe is determined as

$$e_L = \frac{S_{HP}}{E}(0.5 - v) \qquad (10.13)$$

Because the pressure hoop stress is generally very high in a pipeline, pressure elongation is very significant compared to thermal expansion. As shown in Fig. 3.10, a hoop stress of 30 ksi (207 MPa) produces a pressure elongation equivalent to a 35°F (20°C) temperature difference in thermal expansion. This is very significant in a pipeline that is generally operating at about the same amount of temperature difference. After including this pressure elongation, the total expansion per unit length of the pipe is

$$e = e_T + e_L = \alpha(T_2 - T_1) + (0.5 - v)\frac{S_{HP}}{E} \qquad (10.14)$$

10.2.2 Anchor Force

The first step in finding the potential movement of the pipeline is to determine the force required to stop the movement. This anchor force is also needed for designing the anchors placed at strategic locations to prevent the pipeline movement from causing damage to the connecting piping and equipment. From the total expansion given in Eq. (10.14), we can convert it to stress and force as

$$F = AS = AEe = A\{E\alpha(T_2 - T_1) + (0.5 - \nu)S_{HP}\} \tag{10.15}$$

where A is the cross-sectional area of the pipe material. The anchor force, F, is also the total axial driving force that generates the pipeline movement. It is also called the potential pipeline expansion force.

10.2.3 Potential Movement of Free Ends

With the potential expansion force given, the potential free end movement of the long pipeline is estimated as shown in Fig. 10.2. The expansion will be completely suppressed by an axial force or resistance force equal to the potential expansion force, which is the anchor force given in Eq. (10.15). The resistance force of the free end comes mainly from the longitudinal friction force created by the soil for a buried pipeline. As shown in the figure, assuming that the longitudinal resistance force is f, per unit length of the pipe, the length of pipe needed to generate enough friction force to reach the anchor force is

$$L = \frac{F}{f} \tag{10.16}$$

L is called the virtual anchor length. It is also called the active length. The point where the expansion is completely suppressed is called the virtual anchor point. The net expansion rate, in which the potential expansion rate subtracts the squeezing by the axial force, increases linearly from nothing at the virtual anchor point to reach the maximum full expansion rate at the end. The maximum end movement is therefore equal to the average net expansion rate multiplied by the active length. That is,

$$y_0 = \frac{e}{2}L = \left(\frac{1}{2}\frac{F}{AE}\right)\frac{F}{f} = \frac{F^2}{2AEf} \tag{10.17}$$

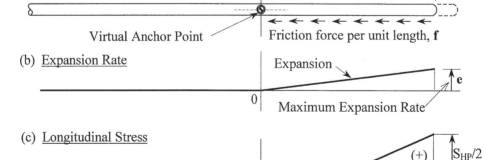

FIG. 10.2
FREE END MOVEMENT OF LONG PIPELINE

Because the potential force, F, contains the e and AE terms, y_0 is directly proportional to the square of the potential expansion rate and also to AE. It is inversely proportional to the friction resistance.

10.2.4 Movement of Restrained Ends

The pipeline end movement will be reduced by the end resistance of the connecting piping and restraints. Assuming the end resistance is Q as shown in Fig. 10.3, the potential force for expanding the pipeline will be reduced to $(F - Q)$. The exact value of Q is determined from the balance of the end movement and system resistance. Its calculation will be discussed later. For the time being, it is just an assumed number. As the potential expanding force is reduced, its maximum expansion rate and end displacement are reduced to as follows.

$$e = \frac{F - Q}{AE}, L = \frac{F - Q}{f}$$

$$y = \frac{L}{2}e = \frac{(F - Q)^2}{2AEf} \quad (10.18)$$

The end resistance reduces the movement more than just linearly. Equation (10.18) is represented by the interactive curve shown in Fig. 10.3(b). This interactive curve is a very important tool in the analysis of piping and equipment connected to a long pipeline. Because it is non-linear, the interaction is included in the piping analysis by using an iterative process. The simulation is available in some advanced computer software packages.

10.2.5 Stresses at Fully Restrained Section

The major portion of the pipeline is fully restrained either by anchor blocks or virtual anchors. For the line fixed by virtual anchors, the longitudinal pipe stress distribution is as shown in Fig. 10.2(c). At the free end, the longitudinal stress is mainly due to the pressure effect and is equal to one-half of

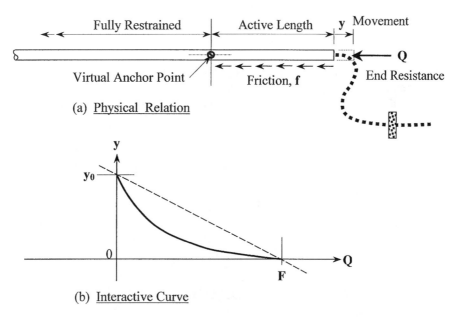

FIG. 10.3
MOVEMENT OF LONG PIPELINE WITH END RESISTANCE

the pressure hoop stress and is in tension for internal pressure. The pressure hoop stress is essentially constant throughout the system, but the longitudinal pipe stress is gradually reduced by the friction force as the pipeline moves. The sign of the stress eventually reverses to compression when enough movement and friction is produced. The longitudinal stress maintains a constant value at the fully restrained portion after reaching the virtual anchor point. Figure 10.4 shows the stress condition at the fully restrained portion.

For a buried pipeline, the weight is continuously supported; thus, no bending stress due to weight is produced. Stresses in this situation are mainly due to pressure and suppression of thermal expansion. As the longitudinal pressure stress is cut off by the anchor, the pressure in the fully restrained section does not create longitudinal elongation. Instead, it produces Poisson contraction in the longitudinal direction due to stretching in the diametrical direction. Therefore, at a fully restrained section, the longitudinal strain being suppressed is

$$e = \alpha(T_2 - T_1) - \nu \frac{S_{HP}}{E} \quad \text{for fully restrained} \tag{10.19}$$

The longitudinal stress required to suppress the above strain is

$$S_L = -Ee = \nu S_{HP} - E\alpha(T_2 - T_1) \quad \text{for fully restrained} \tag{10.20}$$

In addition to longitudinal stress, the pipe always has pressure hoop stress and pressure radial stress. The pressure radial stress is generally ignored, leaving only the hoop and longitudinal stresses. Based on the maximum shear failure theory, the combined equivalent stress is the absolute sum of the two stresses if they have opposite signs; otherwise, it is equal to the greater of the two. That is,

$$S_E = \text{greater of } (|S_L|, |S_{HP}|, \text{ or } |S_{HP} - S_L|) \tag{10.21}$$

For an internal pressure, the hoop stress is always in tension. However, the sign of the longitudinal stress depends on the magnitude of the temperature difference. As the operating temperature rises, the longitudinal stress changes gradually from tension to compression. As the longitudinal stress becomes compression, it is additive to the hoop stress to become the combined stress. Because the allowable stress for the combined equivalent stress is 0.9(SMYS) as given by Eq. (10.3), the longitudinal stress may dictate the wall thickness in installations with higher operating temperatures.

For fully restrained above-ground pipelines, the bending stress due to weight and other mechanical loads have to be added absolutely to Eq. (10.21) before comparing with the 0.9(SMYS) allowable. Bending moment has both tensile and compressive stresses.

Because longitudinal stress is a function of operating and construction temperatures, an increase in construction temperature will reduce the longitudinal stress. This may call for a special construction procedure, if practical, to backfill the soil at the time when the ambient temperature is at its highest.

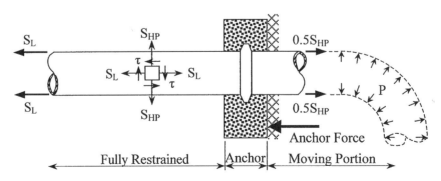

FIG. 10.4
STRESSES AT FULLY RESTRAINED SECTION

10.3 PIPELINE BENDS

Due to its potentially large movement, a pipeline has to be routed in such a manner as to avoid any potential weak link. The most common weak links of the lines are the bends. If a bend is not properly designed, the large bulging movement from the long pipeline might damage it. Also, when checking the interface with a pipeline, we normally consider the pipeline as a straight line. However, because the pipeline follows the route terrain, it cannot be straight in a mathematical sense. Therefore, it is important to find out under what condition the pipeline can be treated as a straight line mathematically. Both of these problems are related to the bend radius of the pipeline. In other words, if the bend radius is large enough for every change in direction, then the pipeline is considered functionally straight. If the pipeline is functionally straight, then no unusual bulging or overstress is expected at the bends.

The minimum bend radius is determined by the balance of the potential force (anchor force), F, and the lateral resistance, q, per unit length of the pipe. For a buried pipeline, the lateral resistance comes mainly from soil resistance. Figure 10.5 shows an arc section, or bend, located in the middle of the line. By taking the summation of the forces in the y direction, we have

$$\int_0^\varphi Rd\theta q\cos\theta - F\sin\varphi = 0 \quad \text{or} \quad Rq|\sin\theta|_0^\varphi - F\sin\varphi = 0$$

$$\text{i.e.,} \quad R = \frac{F}{q} \tag{10.22}$$

This is the minimum bend radius required to avoid excessive bulging at the bend and to enable the line to be treated as a straight line mathematically. Because the resistance, q, is generally different at different directions, the minimum bend radius required is different for side bends, sag bends, and over bends. For a buried pipeline, the resistance is greatest for sag bends and is smallest for over bends. Resisting blocks may be installed at locations where the space is not enough to accommodate the minimum bend radius required.

The minimum required bend radius is an important parameter for pipeline construction. Based on mile-to-mile geotechnical data, the minimum bend radii are specified for the construction. The field engineer knows that if the bend radius is bigger than the one specified for a particular section under construction, then no special treatment is required. Otherwise, special attention and design may be needed.

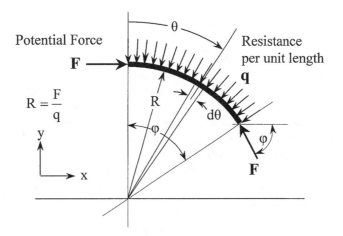

FIG. 10.5
MINIMUM BEND RADIUS

10.4 BASIC ELEMENTS OF SOIL MECHANICS

For buried pipelines, the design analyses are all related to soil-pipe interactions. This calls for an understanding of soil mechanics. This also requires data on soil properties. Soil mechanics is a very specialized field of study. Its in-depth discussion is beyond the scope of this book. Interested readers should consult appropriate books [8, 10] on the subject. However, piping engineers do need to know some of the basic elements of soil mechanics in order to use the data intelligently and to ask the right questions, if needed. The following are some of the basic items that are related to the soil-pipe interaction.

10.4.1 Types of Soils

The pipeline soil consists of many components. These components are classified by their particle sizes. There are many specifications for the classifications. Based on ASTM specification [8], we have the following components.

Gravel (76.2 mm to 2.0 mm). Among the components, gravel has the largest particle size (except for rocks). It contains a large percentage of voids for accommodating sand, water, and other smaller particles. Gravel has interlocking capabilities that can spread out the effective contact area. Specially prepared pure gravels are often used at some portions of the pipeline route to increase the effective resistance of the trench.

Coarse sand (2.0 mm to 0.42 mm) and fine sands (0.42 mm to 0.074 mm). Coarse sands are generally seen at river bottoms, whereas fine sands are seen at beaches and sand dunes. Sand has substantial voids for water and other finer soils, and is generally considered cohesionless with a very low shearing resistance.

Silt (0.074 mm to 0.005 mm) and clay (<0.005 mm). Silt and clay are cohesive soils that have substantial shear resistance. Although they have considerable water-absorbing capabilities, they hydrate and dehydrate very slowly due to the fine sizes of the particles.

The grade of soil is determined by the distribution of the above components. For pipeline engineering, only one or two of the most abundant components are generally mentioned. Based on two of the most abundant components, the soil may be referred to as gravel, gravel-sand, sand, silty sand, silt, clayed gravel, clayed sand, silty clay, clay, etc.

Gravel, gravel sand, sand, and silty sand are generally considered cohesionless, and their shear resistances are ignored in practical design calculations. Silt and clay, on the other hand, have significant shear resistance and are called cohesive soils.

10.4.2 Friction Angle

All soils possess internal friction against the relative sliding motion between two internal surfaces. For cohesive soils, friction and shear resistance are mingled to become a very complex matter. For cohesionless soils, internal friction is rather easy to picture due to the lack of shear resistance.

Figure 10.6 shows the physical depiction of soil friction. When cohesionless soils are poured to the ground from above, it will spread out like water due to gravity. However, because of friction, the area of spread is limited, creating an angle of repose at the balanced still state. At this angle of repose, a lump of soil weighing, w, will generate a sliding force, $s = w \sin\varphi$, along the surface, and a normal force $n = w \cos\varphi$, normal to the surface. For the lump of soil to stay in steady position, a friction force is needed to balance the sliding force. Because friction force is equal to the product of the normal force and the friction coefficient, we have

$$f = \mu n = \mu w \cos \varphi$$

By setting this friction force to balance the sliding force, it becomes

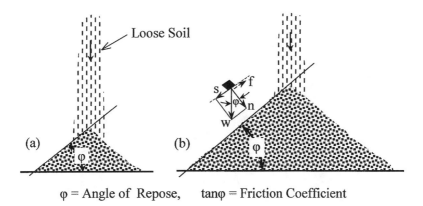

φ = Angle of Repose, tanφ = Friction Coefficient

FIG. 10.6
INTERNAL FRICTION OF COHESIONLESS SOIL

$$\mu w \cos\varphi = w \sin\varphi, \quad \text{or} \quad \mu = \frac{w \sin\varphi}{w \cos\varphi} = \tan\varphi \tag{10.23}$$

where μ is the coefficient of friction. From Eq. (10.23), the angle of repose is also called the internal soil friction angle. In soil mechanics, it is a general practice to specify the internal friction angle, φ, instead of the friction coefficient. For cohesionless soils, the internal friction angle is in the neighborhood of 30 deg.

10.4.3 Shearing Stress

Shearing stress accounts for a significant portion of the soil resistance. The shear stress at any cross-section through a mass of soil can be calculated by Coulomb's equation as

$$s_s = c + s_n \tan\varphi \tag{10.24}$$

where s_n is the compressive stress normal to the plane of the section. Tensile normal stress is not applicable. The term c is cohesion, which is equal to the shearing resistance per unit area when $s_n = 0$. For cohesionless soils, where $c = 0$, the above relation becomes

$$s_s = s_n \tan\varphi \quad \text{for cohesionless soil} \tag{10.25}$$

The values of c and φ are supplied by geotechnical engineers. This data is obtained from tests.

10.4.4 Soil Resistance Against Axial Pipe Movement

Axial soil resistance determines the potential movement of the pipeline, as shown in Eq. (10.17). An accurate estimate of the resistance is therefore very important in pipeline stress calculations. Because larger resistant forces result in smaller potential pipe movements, it is conservative to underestimate the axial soil friction resistance.

Most pipelines are buried in trenches as shown in Fig. 10.7(a). The depth of cover, H, is generally understood as the depth of soil up to the top of the pipe as shown. However, it should be noted that some specifications might consider the depth of cover to the centerline of the pipe. The minimum depth of cover as given by ASME B31.4 [4] is shown in Table 10.4. The actual depth of cover can be more than double the value given as may be needed to either protect the pipe or provide the required soil resistance.

342 Chapter 10

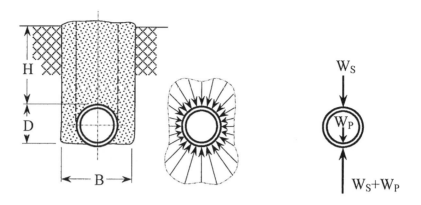

(a) Trenched Pipe   (b) Soil Pressure   (c) Idealized Soil Force

**FIG. 10.7
SOIL PRESSURE ON TRENCHED PIPE**

Axial soil resistance is the total shear resistance developed over the pipe surface as the pipe moves. As discussed previously, shear resistance comprises two parts: the cohesive force and the friction force. For practical applications, these two parts are combined into one adjusted empirical soil friction force. For steel pipes, the normal soil pressure acting on the pipe in a trench is distributed as shown in Fig. 10.7(b) [11]. Theoretically, the friction force would simply be the total normal soil force multiplied by the friction coefficient, if both the normal soil force and the friction coefficient are readily known. The fact is that both of these values are not easy to determine. A more practical approach is to idealize the soil overburden force as one concentrated force, W_s, as shown in Fig. 10.7(c). Thus, the idealized model has two forces acting on the pipe surface: the active soil force and the reaction force equal to soil force plus pipe weight. The axial soil resistance becomes

$$f = \mu(W_s + W_p + W_s) = \mu(2W_s + W_p) \tag{10.26}$$

The active soil force is the weight of the prism within the dot lines over the pipe subtracted by the shear resistance along the two dotted surfaces. It can be expressed in general terms as

$$W_s = \rho D H - \mu k H * H, \quad \text{or} \quad \frac{W_s}{\rho D H} = 1 - \frac{\mu k H^2}{\rho D H} = 1 - k'\frac{H}{D}$$

where k' is an unknown constant and ρDH is the weight of the soil prism. The term $k'H/D$ is the reduction of the active soil force by the shear force in fraction of the direct soil weight over the pipe. It shows that the effect of shear resistance in the overall effective soil weight is proportional to the depth

**TABLE 10.4
MINIMUM COVER, H, FOR BURIED PIPELINES [4]**

Location	For normal excavation		For rock excavation requiring blasting	
	in.	mm	in.	mm
Industrial, commercial, and residential areas	36	900	24	600
River and stream crossings	48	1200	18	450
Drainage ditches at roadways and railroads	36	900	24	600
Any other area	30	750	18	450

of cover, but inversely proportional to the pipe diameter. Generally, the shear resistance term can be ignored when H/D is smaller than 3. Marston's formula [8, 12] or other more detailed estimates should be used if a more accurate number is desired, or if H/D is greater than 3. That is, for H/D less than 3, we have

$$\text{For } H \leq 3D, \quad W_s = \rho DH$$

$$f = \mu(2\rho DH + W_p) \qquad (10.27)$$

Equation (10.27) overestimates the soil force by neglecting the shear resistance. However, it is used almost universally for routine pipeline designs. What the industry does to compensate for this load overestimation is to use a somewhat smaller coefficient of friction than the actual value.

Soil density, ρ, and friction coefficient, μ, are obtained from soil tests performed along the pipeline route. In cases where test data is not available, the following friction coefficients can be used [13]:

$$\text{Silt: } 0.3; \qquad \text{Sand: } 0.4; \qquad \text{Gravel: } 0.5$$

The above values are the lower bound values corresponding to the sliding friction. The static coefficient of friction can be as much as 70% higher [14].

For pipelines buried below the water table, the buoyant force for the portion of soil and pipe that are under the water table should be subtracted from the weight before it is used in Eq. (10.27).

10.4.5 Lateral Soil Force

Soil resistance in the lateral direction of the pipeline determines the stability of the line against localized bulge that might generate buckling and excessive pipe stresses. The lateral soil resistance determines the minimum bend radius required as given in Eq. (10.22). Again, the accurate lateral resistance can only be determined by geotechnical personnel based on field test data. Here, only some of the basic phenomena based mainly on cohesionless soil will be discussed. The effect of the water table is not included. Proper modification of the soil density should be made if the pipeline is located under the water table.

This section discusses only the ultimate resistance of the soil. To incorporate soil resistance into the pipeline analysis, the coefficient of elasticity or spring rate is also needed. The spring rate will be discussed in a subsequent section dealing with soil-pipe interactions.

Upward resistance. The model of upward resistance is shown in Fig. 10.8(a). The resistance force includes the weight of the soil prism, W_s, directly over the pipe and the shear resistance, S, along the two sides of the prism. In addition, pipe weight, W_p, also needs to be included. Among these three items, shear resistance is the most unpredictable. Theoretically, the shear resistance per unit length of pipeline is the shear stress given by Eq. (10.24) integrated over the full depth of cover along two verti-

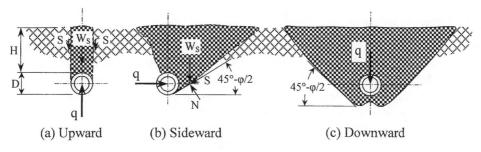

FIG. 10.8
LATERAL SOIL RESISTANCE

cal boundary lines. Cohesion stress is determined by the test, whereas friction stress can be calculated by multiplying the normal soil pressure with the friction coefficient as

$$s_f = s_n \tan\phi = \rho h K_A \tan\phi \quad \text{and} \quad K_A = \tan^2\left(45\deg. - \frac{\phi}{2}\right)$$

where K_A is the coefficient of active lateral pressure. Above is the Rankine formula for cohesionless soil. Because cohesion tends to reduce the active lateral soil pressure, a smaller K_A is used for cohesion soil. A smaller K_A results in smaller resistance. The total shear resistance is added up as

$$S = 2cH + 2\int_0^H \rho h K_A \tan\phi\, dh = 2cH + 2\left|\frac{1}{2}h^2 \rho K_A \tan\phi\right|_0^H = 2H\left[c + \frac{1}{2}\rho H K_A \tan\phi\right]$$

Summing up the total resistance, we have

$$q_u = \rho D H + 2H\left[c + \frac{1}{2}\rho H K_A \tan\phi\right] + W_p \tag{10.28}$$

where c is the cohesion, which is zero for cohesionless soil.

Horizontal resistance. When the pipe moves horizontally as shown in Fig. 10.8(b), it creates a passive soil pressure at the front surface, and at the same time receives an active soil force from the back. Because of the arch action, a void will be created behind the pipe as soon as it moves a small distance, and the active force can therefore be disregarded [15]. The only force left is the passive resistance, which can be calculated via Rankine's formula [10] for cohesionless soil as

$$q_h = \frac{1}{2}\rho(H + D)^2 \tan^2\left(45\deg. + \frac{\varphi}{2}\right) \tag{10.29}$$

Equation (10.29) is for cohesionless soil, which has a critical failure surface located at an (45 deg. − $\varphi/2$) angle from the horizontal line. For cohesive soils, the resistance is higher, with the failure surface located at a lower angle.

The wedge failure model (Fig. 10.8b) used in Rankine's formula is valid only when the depth of cover is less than the pipe diameter. It overestimates the resistance for deeper covers. However, for a three pipe diameter deep cover with dense granular soil, the overestimate is about 10% [16]. For deeper covers, the nature of failure is tunneling with pipe punching through the soil leaving the surface soil barely disturbed. The resistance in this case can be much smaller than the one given by Eq. (10.29).

Downward resistance. For cohesionless soil with shallow cover, soil flow from a downward pipe movement is as shown in Fig. 10.8(c). In pipeline construction, this involves the soil bearing capacity of undisturbed soil. Detailed geotechnical evaluation is needed for this critical work. For a general idea, the downward resistance can be roughly estimated as twice the horizontal resistance.

10.4.6 Soil-Pipe Interaction

Analysis of a pipeline requires the inclusion of all loads imposed on the piping. In addition to the common types of load, such as pressure and temperature, soil force is also very important in the design of a buried pipeline. The soil resistance force varies with the amount of movement as shown in Fig. 10.9. Rather than including the actual expected soil force in the piping stress analysis, an idealized elastic and perfect plastic resistance is generally used to reduce the number of iterative calculations required. Even with idealized elastic perfect plastic resistance, iterations are still required in the analysis to ensure that the elastic and plastic ranges are applied correctly.

Friction force. The first force to be idealized is the axial friction force. Friction force is generally considered a sudden force that comes in a snap. It comes in with full force as the pipe moves, then

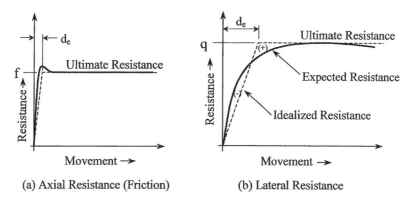

FIG. 10.9
IDEALIZED ELASTIC-PLASTIC SHAPE OF SOIL RESISTANCE

reverses to full force again when the movement reverses. This type of idealized friction is not only difficult to handle, but is also inconsistent with reality. Because the pipe and the surrounding soil all have some elasticity, it requires the pipe to move a certain amount before the full friction force is developed. Nevertheless, the movement for developing full friction is still considered small. The idealized friction is shown in Fig. 10.9(a). The pipe will move initially with elastic resistance that is proportional to the amount of movement. After reaching the elastic limit, the resistance becomes constant and is equal to the full friction force. The initial elastic behavior provides mathematical stability in the analysis of the piping. The maximum elastic movement, d_e, can be taken as 0.04 in. (1 mm) if no official data is available. The full friction force is generally taken as the dynamic friction force that exists during movement. The elastic spring constant is then calculated as

$$k_f = \frac{f}{d_{e,f}} \qquad (10.30)$$

where $d_{e,f}$ is the maximum elastic (or linear) displacement of the friction force.

Lateral resistance. The resistance curve for lateral soil resistance has a rounded shape as shown in Fig. 10.9(b). Again, it is idealized into two stages: elastic and perfect plastic. The initial elastic constant, though, can be estimated by test or calculation; it is generally very sensitive to the data gathered. Even if it can be calculated, it will still be applicable only to a very small range involving a very small displacement. A more practical and reliable approach is to estimate the amount of movement required to reach full resistance. Because the curve rounds almost the entire range, there is no clear definition of this reference movement. A linear line based on the area equalizing concept can be constructed as shown in Fig. 10.9(b). In this method, the elastic line is constructed so the area designated by (+) is roughly the same as the area designated by (−). This is a practical approach used by some engineers and appears to be a conservative one in the design of a pipeline.

Several authors have reported that the displacement required to reach the maximum resistance is about 1.5% to 2% of the pipe bottom depth [17]. Pipe bottom depth is the depth of cover plus pipe outside diameter. Once the maximum elastic displacement is determined, the elastic spring rate is obtained by dividing the maximum resistance force with the maximum linear movement. That is,

$$K_q = \frac{q}{d_{e,q}} \qquad (10.31)$$

where $d_{e,q}$ is the linear movement to reach the maximum lateral resistance. This information should be obtained from the geotechnical data sheet. If data is not available, $d_{e,q}$ can be taken as 1.0% of pipe bottom depth for vertical resistance, and 1.5% of pipe bottom depth for lateral resistance.

10.5 EXAMPLE CALCULATIONS OF BASIC PIPELINE BEHAVIORS

The general basic behaviors of a pipeline can be better understood by going through some example calculations. Assume that a crude oil pipeline with a 20-in. outside diameter operates at 1200 psi pressure and 170°F temperature, and the specific gravity of the crude is 0.85. It is decided that API-5L Grade X52 electric resistance welded pipe will be used for the main portion of the construction. The construction temperature is determined to be 50°F based on climate and the project schedule. It is being buried with 4 ft of soil cover. The soil is silty sand, having a weight density of 125 lb/ft^3 and an internal friction angle of 30 deg. (All calculations are made with U.S. customary units).

10.5.1 Basic Calculations

Basic calculations deal with the wall thickness calculation, equivalent stress at the fully restrained portion, anchor load estimates, and so forth.

(a) *Minimum wall thickness required.* The first step is to determine the basic allowable stress based on the ASME B31.4 code. X52 electric resistance welded pipe has an SMYS of 52,000 psi and a weld joint factor, E_j, of 1.00. From Eq. (10.1), the basic allowable stress is $S = 0.72 \times 1.0 \times 52,000 = 37,440$ psi. The required minimum wall thickness at the main pipeline for the design pressure is

$$t_n = \frac{PD}{2S} = \frac{1200 \times 20}{2 \times 37,440} = 0.321 \text{ in.}, \quad \text{use } 11/32 = 0.344 \text{ in.}$$

This is from Eq. (10.2a) without any allowance for corrosion. The main portion of the pipe requires a nominal wall thickness of 0.344 in. A thicker wall pipe is generally required at station piping and river and road crossings. This wall thickness is still preliminary. Depending on the thermal expansion and occasional stresses, the thickness may need to be increased.

(b) *Pressure hoop stress.* When the above commercially available wall thickness is used, the hoop stress is no longer equal to the basic allowable stress. From Eq. (10.2b), the actual pressure hoop stress is

$$S_{HP} = \frac{PD}{2(t_n - A)} = \frac{1200 \times 20}{2(0.344 - 0)} = 34,883 \text{ psi}$$

(c) *Longitudinal stress at the fully restrained portion.* Because the main portion of the pipeline is fully restrained and well supported by the soil, the longitudinal stress is produced by pressure and temperature only. From Eq. (10.20), which is applicable to both gas and liquid pipelines, the longitudinal stress is

$$\begin{aligned} S_L &= \nu S_{HP} - E\alpha(T_2 - T_1) \\ &= 0.3 \times 34,883 - 29.5 \times 10^6 \times 6.5 \times 10^{-6}(170 - 50) \\ &= -12,545 \text{ psi} \end{aligned}$$

(d) *Combined equivalent stress at the fully restrained portion.* Based on the maximum shear stress failure theory as expressed in Eq. (10.21), the combined equivalent stress is the absolute summation of the longitudinal stress and hoop stress due to opposite signs of these two stresses. That is,

$$S_E = |S_L - S_{HP}| = |-12,545 - 34,883| = 47,428 \text{ psi}; \quad \text{for } t_n = 0.344 \text{ in.}$$

Because the combined equivalent stress, S_E, is greater than 0.9(SMYS) of 46,800 psi, the pipe with 0.344-in. wall thickness does not meet the B31.4 requirements. The stress has to be reduced. Since the combined equivalent stress includes the pressure hoop stress and thermal expansion stress, the reduction can be made either on the pressure hoop stress or thermal expansion stress. A reduced operating temperature based on a more realistic estimate or an increased construction temperature based on special backfilling procedures may be sufficient to reduce the stress to within the allowable.

Assuming that both operating and construction temperatures are realistic and fixed, the reduction has to come from the pressure hoop stress. Again, assuming that the operating pressure is realistic and fixed, the pressure hoop stress reduction has to come from an increase in wall thickness.

(e) *Design wall thickness.* By doing the same exercise given above, the wall thickness is increased to 0.375 in. With a 0.375-in. wall thickness, the pressure hoop stress is reduced to 32,000 psi. The longitudinal stress is also increased due to the smaller hoop stress. That is,

$$S_L = 0.3 \times 32{,}000 - 29.5 \times 10^6 \times 6.5 \times 10^{-6}(170 - 50)$$
$$= -13{,}410 \text{ psi}; \quad t_n = 0.375 \text{ in.}$$

$$S_E = |-13{,}410 - 32{,}000| = 45{,}410 \text{ psi}; \quad t_n = 0.375 \text{ in.}$$

Since the combined equivalent stress is smaller than 0.9(SMYS), the 0.375-in. wall thickness at the fully restrained portion satisfies the B31.4 combined equivalent stress requirements.

(f) *Anchor load.* At the end of the pipeline, either an anchor is placed to stop the long pipeline movement or the potential force is estimated to analyze the effect of the long pipeline to the connecting piping and equipment. The anchor load is the same as the potential force. It is calculated based on Eq. (10.15) as

$$F = \pi D t \{E\alpha(170 - 50) + 0.2S_{HP}\}$$
$$= 23.56(23{,}010 + 0.2 \times 32{,}000)$$
$$= 692{,}900 \text{ lb}$$

A full anchor has to be designed for this anchor force. This anchor force is also the potential force that pushes the connecting piping if the pipeline is not anchored.

10.5.2 Soil-Pipe Interaction

In this example, some basic soil-pipe interactions are discussed. It includes an estimate of the longitudinal soil friction and the potential maximum movement at the free end of a long pipeline, lateral soil resistance and the minimum bend radius requirements, and so forth.

(a) *Longitudinal soil friction.* Longitudinal soil friction is very important in the analysis of pipeline movement. It determines the potential end movement of a long pipeline and also the movement of short underground piping runs. The longitudinal soil friction is calculated by Eq. (10.27) as

$$f = \mu(2\rho DH + W_p) = 0.4\left(2\frac{125}{12^3}20 \times 48 + 15.475\right) = 61.7 \text{ lb/in. pipe}$$

(b) *Potential end movement.* The free end movement of a long pipeline is one of the two major measures that the connecting piping has to be designed for. The other measure is the anchor or potential force. In other words, the connecting piping has to be either designed rigid enough to resist the anchor load or flexible enough to absorb this end movement. The potential end movement is calculated by Eq. (10.17). That is,

$$y_0 = \frac{F^2}{2AEf} = \frac{692{,}900^2}{2(\pi 20 \times 0.375) \times 29.5 \times 10^6 \times 61.7}$$
$$= \frac{480.1 \times 10^9}{85.772 \times 10^9} = 5.60 \text{ in.}$$

The virtual anchor length is calculated by Eq. (10.16) as

$$L = \frac{F}{f} = \frac{692{,}900}{61.7} = 11{,}230 \text{ in.} = 936 \text{ ft}$$

(c) *Horizontal lateral soil resistance.* The lateral soil resistance is different in each direction. This example calculates only the horizontal lateral resistance. From Eq. (10.29), the lateral soil resistance is

$$q_h = \frac{1}{2}\rho(H+D)^2 \tan^2\left(45 \text{ deg.} + \frac{\phi}{2}\right)$$
$$= 0.5\frac{125}{12^3}(48+20)^2 \tan^2\left(45 \text{ deg.} + \frac{30 \text{ deg.}}{2}\right)$$
$$= 501.7 \text{ lb/in. pipe}$$

Assuming that it takes a movement equivalent to 1.5% of the total depth to develop the full horizontal resistance, the initial spring rate of the soil resistance becomes

$$K_h = \frac{q_h}{0.015(D+H)} = \frac{501.7}{0.015(48+20)} = 492 \text{ lb/in./in. pipe}$$

The linear displacement is $0.015(48+20) = 1.02$ in.

(d) *Minimum bend radius for the horizontal bend.* To ensure that the pipeline behaves essentially like a straight pipe and to minimize the bending stress, the horizontal bends of the pipeline shall not be smaller than the minimum bend radius calculated by Eq. (10.22) as follows

$$R = \frac{F}{q} = \frac{692{,}900}{501.7} = 1381 \text{ in.} = 115 \text{ ft}$$

Some specifications might include an additional safety factor of 1.5 over the value calculated. However, as the apparent anchor force or potential force reduces rather quickly once the pipe moves, most specifications will use the bend radius as calculated above.

10.6 SIMULATION OF SOIL RESISTANCE

A pipeline is continuously supported and restrained by the soil. However, soil support and resistance are generally simulated with concentrated forces and supports at discrete points. These points are called node points in the analysis. To accurately simulate the interaction between soil and pipe, the distance between the node points has to be limited. This is because the mathematical formulation of the pipe element cannot handle multiple inflection points within an element.

The buried pipeline behaves as a beam on elastic foundation, as discussed in Chapter 5. When the pipe expands, it stabilizes into a wavy shape as shown in Fig. 10.10. Because soil force is lumped at the node points, the distance between the node points cannot exceed the half-wave length for the analysis to have any meaning at all. In general, it is recommended that at least six points should be assigned to each half-wavelength. From the deflection curve given in Fig. 5.2, the half-wavelength can be taken as π/β. Therefore, the recommended maximum node point spacing is

$$\ell = \frac{1}{6}\frac{\pi}{\beta} = \frac{\pi}{6}\sqrt[4]{\frac{4EI}{K}} \qquad (10.32)$$

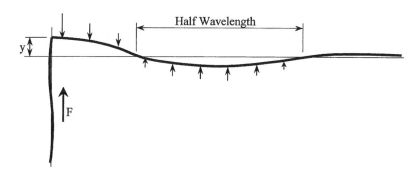

**FIG. 10.10
PIPELINE DEFLECTION CURVE**

where K is the soil spring rate and I is the moment of inertial of the pipe metal cross-section. For the example pipeline given in Section 10.5, we have

$$\ell = \frac{\pi}{6} \sqrt[4]{\frac{4 \times 29.5 \times 10^6 \times 1110}{492}}$$

$$= \frac{\pi}{6} 127.7 = 66.88 \text{ in.}$$

10.7 BEHAVIOR OF LARGE BENDS

We have previously discussed that a pipeline is constructed along the natural terrain of the route. It is not straight in the geometrical sense. However, if the bend radii of all the bends are kept larger than the minimum required bend radius given by Eq. (10.22), then the behavior of the line is considered the same as a straight line. In this section, we will investigate the behavior of a large bend using the SIM-FLEX-II [18] pipe stress analysis computer program. Figure 10.11 shows a 45-deg. horizontal bend of the pipeline discussed in Section 10.5. Two cases are investigated, the first one with both ends connected to a very long pipeline and the second case with only one end connected to the long pipeline whereas the other end is free. The first case should show if the bend meets the code stress requirement, and the second case should tell us if the buried large bend behaves like a straight line.

From Section 10.5.2(d), we know that the minimum bend radius required is 115 ft. For a 45-deg. bend the arc length is $115\pi(45/180) = 90.32$ ft. Because the node point spacing as recommended by Section 10.6 is 66.88 in., the arc is to be divided into 17 segments with a node point spacing of $90.32/17 = 5.313$ ft $= 63.76$ in. This same spacing shall also be used at the straight portion to simplify the data. Because the bending moment produced by bend movement will decrease to a negligible amount after going through one wavelength, the straight portion is modeled for one wavelength, which is roughly equal to 12 node point spaces. The final analytical model is as shown in the figure.

Soil resistance is treated as concentrated restraints acting at the node point locations. Each restraint represents the total resistance of a 5.313-ft segment. Because this is a horizontal bend, the problem is simpler since no vertical movement is involved. The first restraint to be applied is the lateral soil force, which is represented as an elastic-plastic non-linear restraint. The spring rate is given by Section 10.5.2(c) as 492 lb/in./in., which results in a total of $492 \times 5.313 \times 12 = 31,368$ lb/in. for each node point. The maximum linear displacement is also given by Section 10.5.2(c) as 1.02 in. The ultimate resistance is 31,995 lb per segment or point.

In addition to the lateral soil resistance, there is also the axial soil friction force to be considered. The axial soil friction force is given by Section 10.5.2(a) as 61.7 lb/in. pipe. The total for each segment is $61.7 \times 5.313 \times 12 = 3934$ lb. This axial soil friction can be treated as an elastic-plastic non-linear restraint applied in the tangential direction of each node point. In this example, because the pipeline

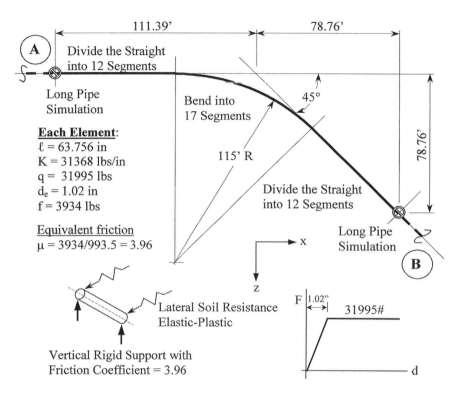

**FIG. 10.11
BEHAVIOR OF LARGE PIPELINE BENDS**

Analysis Results by SIMFLEX-II [18] Computer Program				
Cases	Maximum Lateral Displ. in	Maximum Equivalent Stress, psi	Axial Displacement at A-end, in	Axial Displacement at B-end, in
Both ends to Long Pipeline	0.70	44990	0.06	0.06
One end (A) to Long Pipeline	0.16	31520	3.20	5.40

Pipe: 20" OD, 0.375" thick; Content specific gravity = 0.85; Pressure=1200 psi
Operating temperature = 170°F; Construction Temperature = 50°F

is treated as being supported with rigid vertical supports, the axial soil friction can be treated as the friction force resulting from the support load. However, this requires an artificial coefficient of friction to match the soil friction. Since the support will see only the weight of the pipe plus its content in a computer program, the vertical load at each support is equal to $5.313 \times 185.7 = 986.6$ lb. The pipe including content weighs 185.7 lb/ft. In order for this load to generate 3934 lb friction, the friction coefficient has to be $3934/986.6 = 3.99$. The friction force, applied in this manner, acts against the direction of movement. The movement is not exactly in the axial direction at bend location, but very close.

The analysis requires the participation of the action from the connected long pipeline. This long pipe simulation follows the force-displacement relationship given by Eq. (10.18), which is shown graphically in Fig. 10.3(b). The simulation can be done by manual iteration or by automatic iteration as implemented in some computer programs [18]. To use the simulation, the connecting pipeline has to be long enough to cover the virtual anchor point. For a connection to a shorter pipeline, the whole pipeline has to be modeled with the proper soil friction and resistance included.

The analysis results are summarized in Fig. 10.11. When the bend is located at the middle of a long pipeline, both ends are connected by long pipelines, whose actions are treated with long pipe simulations. As the pipe expands, the potential force is reduced. This is the reason why the lateral displacement is only 0.70 in. with the maximum soil force equal to 69% of the ultimate soil resistance, although the bend radius is selected to balance the ultimate soil resistance. The maximum combined equivalent stress at the bend is 44,990 psi, which is smaller than the combined equivalent stress developed at the fully restrained section as given by Section 10.5.1(e). This investigation validates the general practice, which considers the bend satisfactory when its bend radius is bigger than the minimum required bend radius.

If one end of the bend is free or connected to a very short run, the lateral displacement at the bend is only 0.16 in., which is about 15% of the soil failure displacement. The maximum equivalent stress is reduced to the same value as the pressure hoop stress, because the longitudinal stress is in tension. The movement at the free end is 5.40 in., which is slightly smaller than the free end displacement of 5.60 in. as estimated in Section 10.5.2(b) for a long straight pipeline. The reduction is mainly due to lateral resistance at the bend, but is not significant. Therefore, the large bend can be considered a straight line in pipeline stress analyses.

10.8 CONSTRUCTION OF ANALYTICAL MODEL

Because a pipeline normally runs miles in length, it is impossible to model the entire line into an analysis. Even if it is possible to include the entire line in an analysis, it is still not recommended to do so due to potentially inaccurate analysis results. For a very long line, a small round-off error at one end might result in a considerable error at the other end. Because the behavior of a pipeline is rather localized, it can be divided into sections and can be analyzed section by section. Figure 10.12 shows the analytical sections of a pipeline.

The majority of the pipeline is either straight or with large bends, which are considered essentially straight as discussed in Section 10.7. The straight pipeline is considered fully restrained except near

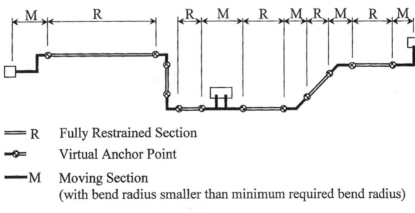

═ R   Fully Restrained Section
─○═   Virtual Anchor Point
━ M   Moving Section
       (with bend radius smaller than minimum required bend radius)

**FIG. 10.12
ANALYTICAL SECTIONS IN A PIPELINE**

the areas where sharp bends are used. The fully restrained portion, marked as "R" region, requires no analysis once the basic calculations of wall thickness, minimum bend radius, and the maximum combined equivalent stress are completed and satisfied. The areas that require analysis are the portions marked "M," signifying moving portions.

In a pipeline, the pushing force of the moving portion comes mainly from the connecting long pipeline. The movement of the long pipeline can be either limited with an anchor or included in the analysis. Most of the analyses follow the same general procedure given in the example in the previous section.

Figure 10.13 shows a scraper (pig) launching station at a pump station or compressor station. A booster station generally involves both launching and receiving stations. The scraper station is probably the first aboveground facility connected to a long underground pipeline. The launching station is also generally connected to the pump or compressor piping circuits. The analysis is performed in the same manner as any plant piping except for the interface of the long pipeline.

There are three basic methods of treating the interface of the long pipeline. The most direct way is to put an anchor at the end of the pipeline as shown in Fig. 10.13(a). Although the anchor is a sure way to handle the problem, it is very expensive due to the large load involved. The thrust load can reach several hundred tons in some cases. Even with the anchor, the expansion between the anchor and the scraper still needs to be dealt with. Without the anchor, the station has to be analyzed with the long pipeline simulation as shown in Fig. 10.13(b). Again, the simulation is done with the interaction curve given in Fig. 10.3(b). The isolation valve as well as scraper barrel are supported on a sliding plate with slotted bolt holes to allow several inches of movement. Because the resistance of the support friction and connecting piping may not significantly reduce the movement of the long pipeline, a simplified approach considering the long pipeline connection as a fixed displacement point (Fig. 10.13c) can also be used. The displacement of the fixed point is equal to the free end movement calculated by Eq. (10.17). The fixed displacement point is analytically an anchor with an applied displacement. In any case, the isolation valve and scraper barrel are allowed to move in the axial direction of the pipeline.

Figure 10.14 shows a general analytical model of handling the interface with a long pipeline. The pipeline is not anchored. Instead, it is stopped by a leg of offset between points 20 and 30. As a minimum, this offset is generally allocated with a length equal to the half-wavelength as discussed in

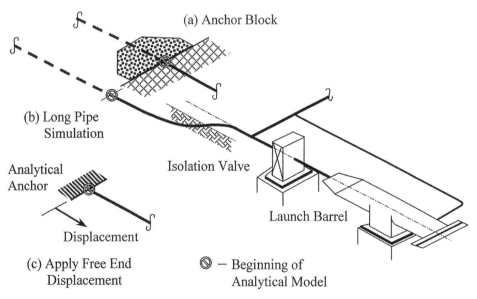

FIG. 10.13
METHODS FOR ANALYZING PIG LAUNCH STATION PIPING

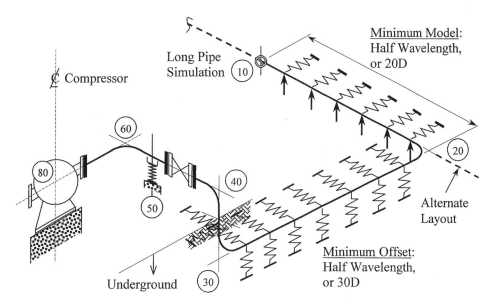

FIG. 10.14
ANALYTICAL MODEL OF COMPRESSOR STATION PIPING

Section 10.6, or 30 pipe diameters, whichever is greater. The long pipeline interface is simulated at a point a half-wavelength away from the turning corner. This distance ensures little bending moment at point 10, which simulates only the axial movement of the pipeline. The soil restraints are assigned in the same manner as in the example outlined in Section 10.7. The vertical soil resistance at the main pipeline portion between points 10 and 20 can be simulated with rigid supports to simplify the calculation process. The vertical displacements at those locations are expected to be very small.

Because the wall thickness at the station is generally thicker than at the main pipeline, the long pipe simulation should be made with the data of the main pipeline. In some cases, the main line runs straight through the station, with the station lead piping taken from a branch connection. In such cases, the long pipeline simulation for the downstream line is also required and is located again at about half a wavelength from the connection point. This downstream simulation should use the pipe and the operation data of the main downstream pipeline. Any temperature and pressure difference between upstream and downstream pipelines will cause movement at the connection point.

10.9 ANCHOR AND DRAG ANCHOR

Occasionally, the layout as given in Fig. 10.14 may not be satisfactory, mainly due to the high stress at corner 20. One solution to this problem is to extend the pipeline continuously for another 1000 ft or so, shown as the alternate layout in the figure. This will substantially, if not entirely, reduce the movement at the connection. This is not a bad idea if the extension of the line is expected in the foreseeable future.

Another approach of reducing the stress at corner 20 is to fill resilent material along the 20–30 run in the area near corner 20. The resilent material, such as spongy forms, allows the pipe to move with little resistance. The solidly compacted soil near corner 30 is needed, however, to protect the rotating equipment.

The more conventional approach to dealing with this problem is to place an anchor near the end of the main line. Generally, a pipeline anchor consists of an anchor flange encased in a fairly large concrete block formed from a fresh-cut soil bunker as shown in Fig. 10.15. Although the anchor face

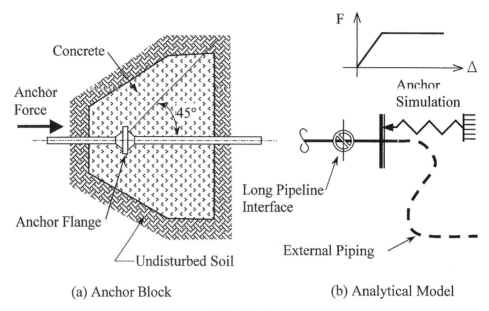

**FIG. 10.15
PIPELINE ANCHOR**

is resisted by the fresh-cut soil surface, some movement is still expected when the full anchor load is applied. The movement is mainly controlled by the anchor size. A full anchor, designed for the potential anchor force given by Eq. (10.15), has a size in the neighborhood of a 30-ft cube. Even with this mammoth size, a ¼-in. to ½-in. movement is still expected under the full operation condition.

The full anchor is expensive and difficult to construct. In actual practice, a drag anchor instead of the full anchor may be used. The drag anchor, which is ¼ to ½ the size of the full anchor, can considerably reduce the movement. This movement reduction is enough to reduce the pipe stress to within the acceptable range in many situations.

Regardless of the type of anchor used, the function of the anchor can be simulated as an elastic-plastic restraint in the analysis. Figure 10.15(b) shows the analytical model. The action of the long pipeline is simulated with the long pipe interface given in Fig. 10.3(b), and the anchor, full or drag, is simulated by an elastic-plastic restraint with the elastic spring rate and the ultimate resistance force provided by the geotechnical group. The external connecting piping system is modeled in the same manner as any other piping system.

REFERENCES

[1] 49 CFR Part 192, "Transportation of Natural and Other Gas by Pipeline: Minimum Federal Safety Standards," U.S. Code of Federal Regulations, Title 49 — Transportation.
[2] 49 CFR Part 195, "Transportation of Hazardous Liquids by Pipeline," U.S. Code of Federal Regulations, Title 49 — Transportation.
[3] ASME B31, ASME Code for Pressure Piping, An American National Standard, ASME, New York.
[4] ASME B31.4, *Pipeline Transportation Systems for Liquid Hydrocarbons and Other Liquids*, ASME, New York.
[5] ASME B31.8, *Gas Transmission and Distribution Piping Systems*, ASME, New York.
[6] ASME B31.11, *Slurry Transportation Piping Systems*, ASME, New York.
[7] API SPEC 5L, *Specifications for Line Pipe*, American Petroleum Institute, Dallas, TX.

[8] Spangler, M. G., and Handy, R. L., 1982, *Soil Engineering*, Harper & Row Publishers, New York. See Chapter 7 for soil types, Chapters 26 and 27 for soil load on and support strength of underground pipe.

[9] Peng, L. C., 1978, "Stress Analysis Methods for Underground Pipe Lines," *Part 1, Basic calculations, Pipe Line Industry*, April, 1978; *Part 2, Soil-pipe interaction, Pipe Line Industry*, May, 1978.

[10] Terzaghi, K., 1943, *Theoretical Soil Mechanics*, John Wiley and Sons, Inc., New York.

[11] Tohda, J., Yoshimura, H., Morimoto, T., and Seki, H., 1990, "Earth Pressure Acting on Buried Flexible Pipes in Centrifuged Models," *Pipeline Design and Installation*, Proceedings of the International Conference, K. K. Kienow, ed., ASCE, New York.

[12] Marston, A., 1930, "The theory of External Loads on Closed Conduits in the Light of the Latest Experiments," *Bulletin 96*, Iowa Engineering Experiment Station, Ames, IA.

[13] Ligon, J. B., and Mayer, G. R., 1975, "Friction Resistance of Buried Pipeline Coatings Studied," *Pipeline and Gas Journal*, pp. 33–36.

[14] Pacific Gas and Electric Report, 1933, "Longitudinal Movements of Underground Pipe Lines," *Western Gas*, September.

[15] Terzaghi, K., 1955, "Evaluation of Coefficient of Subgrade Reaction," *Geotechnique*, 5(4), pp. 297–326.

[16] Ovesen, N. K., 1972, "Design Methods for Vertical Anchor Slab in Sand," Vol. 1 Part 2, *Proc. ASCE JSMFD Specialty Conference on Performance of Earth and Earth-Supported Structure*, June 1972.

[17] Audibert, J., and Nyman, K., 1975, "Coefficients of Subgrade Reactions for the Design of Buried Piping," 2nd ASCE Specialty Conference on Structure Design of Nuclear Plant Facilities, New Orleans, LA, December 1975.

[18] Peng Engineering, 2007, SIMFLEX-II Pipe Stress Analysis Program, Houston, TX, www.pipstress.com.

CHAPTER

11

SPECIAL THERMAL PROBLEMS

There are a few special thermal problems that account for more than their fair share of piping failures in the field. These include thermal bowing due to uneven temperature distribution across the pipe cross-section, weather exposure of hot-cold pipe junctions, lack of insulation on flanges, and socket weld connections. These problems are seldom treated in piping or pipe stress books. Due to their importance, a separate chapter is allocated to these topics. To better visualize the situation, a number of case histories are used in the discussion.

11.1 THERMAL BOWING

General bowing behaviors and their basic formulas have been introduced in Section 4.2. This section discusses the details with some real case histories.

One of the earliest thermal bowing problems recorded was the case of cryogenic piping exposed to stratified flow [1]. The cryogenic piping systems used in transporting missile propellant, loading space shuttle fuel [2], loading liquefied natural gas (LNG) to shipping and storage places, and so forth, all have the same problem of bowing during startup operations. As the super cold liquid flows into the piping that was originally close to ambient temperature, the cold liquid settles to the bottom of the pipe with some evaporation initiated due to the heat transfer from the hot pipe wall. Because the cold vapor is heavier than the hot vapor or gas originally trapped inside the pipe, the convection that may bring the cold vapor in contact with the top portion of the pipe wall is limited. This results in a considerable temperature difference between the top and the bottom portions of the pipe wall.

The uneven temperature distribution not only has the potential to create large local thermal stress, but also has the potential to generate huge bowing displacements. The thermal stress might lead to fatigue failure of the affected area, whereas the bowing displacement can damage both the piping and connecting equipment if not properly addressed.

11.1.1 Displacement and Stress Produced by Thermal Bowing

Figure 11.1 shows the cross-section of the pipe partially filled with cryogenic fluid. The temperature distribution shape is hard to determine, but generally takes the form shown in Fig. 11.1(b). The difference between the top and bottom temperatures bows the pipe into a circular shape. To estimate the overall bowing effect, the temperature distribution is assumed to be linear as shown in Fig. 11.1(c). From Fig. 4.2 and Eq. (4.2) that with a linear temperature distribution, we have

$$R = \frac{D}{\alpha(T_1' - T_2)} \tag{11.1}$$

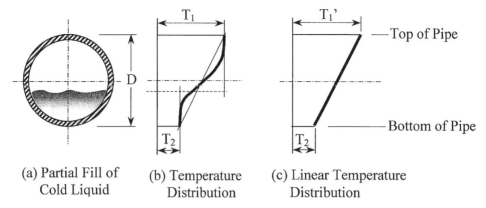

**FIG. 11.1
TEMPERATURE DISTRIBUTION DUE TO PARTIAL FILLING OF COLD LIQUID**

One way to handle the actual temperature distribution is to apply a correction factor, K, to convert the apparent top and bottom temperature difference into an equivalent linear temperature difference in terms of equal bowing effect. That is,

$$T_1' - T_2 = K(T_1 - T_2) \quad \text{with the same bowing effect}$$

Substituting the above into Eq. (11.1), we have the working formula

$$R = \frac{D}{\alpha K (T_1 - T_2)} \tag{11.2}$$

The K values for two assumed temperature distributions with various fill levels have been calculated by Flieder et al [1], and are tabulated in Fig. 11.2. Because the actual temperature distribution is difficult to determine, the use of $K = 1.0$ shall be considered not too far off from reality.

To get an idea of the magnitude of the displacements and forces generated by thermal bowing, the problem associated with cryogenic fluids will be used as an example. The reference temperature difference for cryogenic fluids is in the neighborhood of 300°F (167°C). With 12-in. (300-mm) standard wall stainless steel pipe, the bowing bend radius is

$$R = \frac{12.75}{8.16 \times 10^{-6} \times 300} = 5208 \text{ in. } (132 \text{ m})$$

With a 40-ft (12-m) free span, we have from Eqs. (4.3) and (4.4), the end rotation and mid-span deflection as

$$\theta = \sin^{-1}(240/5208) = 0.046 \text{ rad} = 2.64 \text{ deg.}$$

$$y = 5208 - \sqrt{5208^2 - 240^2} = 5.53 \text{ in. } (140 \text{ mm})$$

Suppressing this movement would require a force large enough to produce the same amount of deflection on a simply supported beam. From the beam formula given in Chapter 2, we have

$$\text{Force: } F = \frac{48EI\Delta}{L^3} = \frac{48 \times 30 \times 10^6 \times 279 \times 5.53}{480^3} = 20{,}089 \text{ lb } (89{,}360 \text{ N})$$

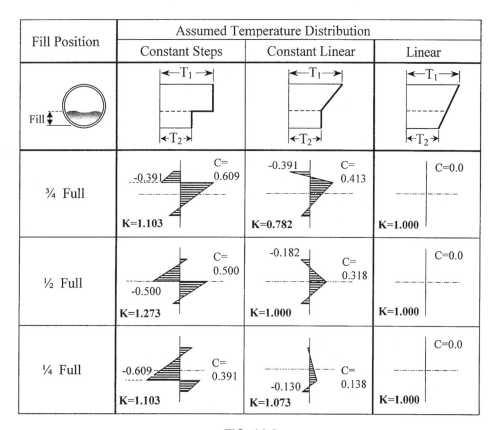

FIG. 11.2
THERMAL STRESS FACTOR, $C = S_T/\{E\alpha(T_1 - T_2)\}$ AND LINEAR CORRECTION FACTOR, K [1]

$$\text{Moment:} \quad M = \frac{FL}{4} = \frac{20{,}089 \times 480}{4} = 2{,}410{,}680 \text{ in.-lb}$$

$$\text{Stress:} \quad S = \frac{M}{Z} = \frac{2{,}410{,}680}{2 \times 279/12.75} = 55{,}100 \text{ psi}$$

The above values give only a general idea of the thermal bowing behavior. In actual piping, there are end restraints and weight suppression and so forth. A pipe stress analysis program generally has the capability of including the bowing effect in the analysis. The data of an equivalent linear bowing temperature is required to perform this analysis.

11.1.2 Internal Thermal Stresses Generated by Bowing Temperature

In addition to the large displacement and twisting created on the overall piping system, an internal local thermal stress is also generated by the uneven temperature distribution over the pipe cross-section. Local thermal stress is zero for linearly distributed temperatures [1]. However, a significant local thermal stress may be produced for non-linear temperature distributions. The thermal stress can be expressed by a generic equation as

$$S_T = CE\alpha(T_1 - T_2) \tag{11.3}$$

where C is the stress factor, whose value depends on the geometry and temperature distribution of the parts involved. E and α are the average modulus of elasticity and the average thermal expansion coefficient of the pipe cross-section, respectively. Figure 11.2 also shows the stress factors for two ideal temperature distributions. The first temperature distribution investigated is the step function, which assumes that the portion of pipe wetted by the liquid has the same temperature as the liquid while the rest of the pipe wall stays at the initial temperature. The other temperature distribution assumes that the liquid portion is uniformly at the liquid temperature, but the temperature for the rest of the pipe wall varies linearly from liquid temperature to initial temperature. In each case, three fill levels representing ¼, ½, and ¾ full are investigated. The maximum stress factors, C, in these cases range from 0.609 for a step-step distribution to 0.413 for a step-linear distribution.

To get a general idea of the stress magnitude, the αE value can be assumed as 180 psi/°F (2.25 MPa/°C) for carbon steel and 240 psi/°F (2.98 MPa/°C) for stainless steel.

At a 300°F temperature difference with a step-step temperature distribution, the local stress range can reach as high as 72,000 psi total range, with stress reversal at the same level, for stainless steel. This is the worst-case scenario with an unusual step-step temperature distribution. With the more realistic step-linear temperature distribution, the thermal stress is 30,000 psi without any reversal at the same level. Therefore, the local thermal stress is generally not a critical factor for the occasional bowing cases that are limited to certain startup conditions. The local thermal stress, however, is very important when the liquid level oscillates constantly rather than just during the startup period. A constant liquid or stratification level fluctuation can generate enough high-frequency thermal stress cycles that lead to high cycle fatigue failure generally experienced only under steady-state vibration. One of the most notorious cases is the feedwater piping cracks experienced in the nuclear power industry [3, 4]. The feedwater piping leading to the steam generator has cracked almost at the same period at more than a dozen units of nuclear power plants due to stratification of the cold and hot water. The failures were caused by a rapidly fluctuating stratification layer that generated a highly cyclic local thermal stress together with a corresponding oscillatory bowing effect.

11.1.3 Occurrences of Thermal Bowing

Thermal bowing can occur at many places, many of which are not very conspicuous to typical engineering viewpoint. Many bowing occurrences cause unexpected damage to the piping itself, or to the supporting structure, or both. Because the damage by thermal bowing mostly occurs at transient conditions, such as during startup, the thermal bowing is not usually readily recognized as such. The bowing phenomena may also have been long gone by the time the damages are discovered. For these reasons, thermal bowing damage was often thought to be caused by water hammer or something similar. One difference is that the damage by water hammer is always accompanied by a loud banging sound, which is generally difficult to miss.

Because bowing is very damaging, it is important to avoid the occurrence rather than to analyze the effect after the occurrence. The following lists some bowing phenomena that have been experienced at various piping systems. They may serve as design guides for avoiding thermal bowing.

(a) Uneven radiation. Radiant heat is absorbed only by the projected area that is facing the heat source. Un-insulated piping exposed to sunshine and heater pipes in the radiant section of a furnace are two such examples. This localized heating does not produce significant temperature variation around the pipe cross-section when enough cooling effect is provided by sufficient internal flow. The problem is the empty or low flow startup condition. For large-diameter, un-insulated empty piping, the temperature difference generated by a midday summer sun may be sufficient to damage the pipe support and connecting equipment. It is generally desirable to paint those pipes in white to reduce their radiant heat absorbing capabilities.

(b) External and internal water. Steam lines very often experience thermal bowing due to the existence of external or internal water. Figure 11.3 shows two of the typical situations. Figure 11.3(a)

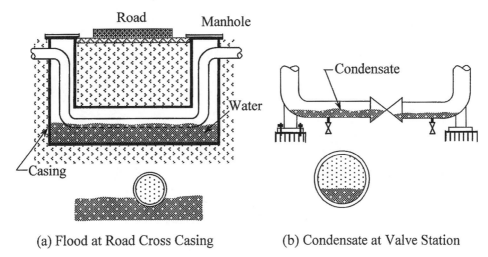

FIG. 11.3
THERMAL BOWING AT STEAM PIPING

shows the problem associated with the flooding of a road crossing casing or tunnel. When the water inside the casing rises gradually, it contacts the pipe bottom first. While the floodwater is in contact with the pipe bottom, it soaks the insulation and cools down the pipe wall. This not only creates thermal bowing, it also has the potential of condensing the steam to create water slug inside the pipe. The pipe and its supports might be damaged due to thermal bowing and the potential water hammer resulting from the water slug [5]. Prevention of flooding is therefore very important in ensuring the integrity of the piping system.

A similar situation with internal water, instead of external water, also occurs occasionally. Figure 11.3(b) shows a valve station that might have trapped condensate that is not discharged before admitting the steam. For the convenience of operation and maintenance, valve stations are generally located at ground level, the lowest point of the entire system. Although steam traps are generally provided to discharge the accumulated condensate during operation, they are generally not functional during shutdown conditions. The condensate accumulated during cooldown stays in the pipe unless discharged manually. The introduction of steam over the trapped condensate creates a large bowing temperature. Because piping and pipe supports can be instantly damaged as the bowing temperature reaches a critical point, the eventual discharge of the condensate by a steam trap or other means is too late to prevent the failure.

(c) Stratification due to stagnation. A thermal bowing temperature occurs whenever stratification exists inside the pipe. The most common stratification occurs due to stagnation of the flow that prevents enough mixing of the fluid inside the pipe. Stagnation means a very low flow rate, which is generally associated with startup, warm up, stand by, and other such conditions. Figure 11.4 shows some of the common stratifications due to a very low flow.

One of the earlier noted occurrences of thermal bowing is the startup of cryogenic piping. Because the cryogenic vapor is heavier than the hotter vapor or gas originally trapped inside the pipe, the convection heat transfer inside the pipe is very minimal. This phenomenon prolongs the heat balancing between the cold fluid and the hot pipe. Figure 11.4(a) shows stratification created by slow liquid flow. For hot fluid, the pipe wall temperature is equalized very fast due to condensation of the vapor at the top. For cryogenic fluid, the effect of the fluid temperature to the pipe cross-section is almost entirely by conduction through the pipe wall from the bottom to the top. However, because engineers are well aware of the bowing problem associated with cryogenic fluids, the startup process is well planned to

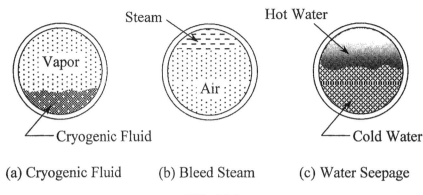

(a) Cryogenic Fluid   (b) Bleed Steam   (c) Water Seepage

FIG. 11.4
STRATIFICATION DUE TO STAGNATION

minimize the bowing effect. The commissioning of an LNG piping system, for instance, involves a very long period of well-planned dry up and cooldown [6]. A large quantity of liquid nitrogen is used to gradually cool down the system so the whole cooldown process and the final introduction of the LNG does not create more than a 40°F (22°C) top to bottom temperature difference. This 40°F difference is generally suppressed by the pipe weight without creating any problem on the support shoes.

Figure 11.4(b) shows the stratification inside steam piping due to low steam flow. In contrast to cryogenic fluids, whose bowing effect is well known, the bowing effect of steam lines is not at all well known to engineers. To reduce the thermal shock on the pipe wall, the main steam piping systems in a power plant are generally warmed up by introducing a very small amount of bleed steam at the beginning. The lines are gradually brought up to full main steam temperature little by little. Because steam is lighter than air, it tends to float over the originally trapped air and flow at the top portion of the pipe. This can create a high temperature differential between the top and bottom portions of the pipe, especially for larger pipes [7]. Purging with low-pressure steam may be needed to reduce the bowing effect.

Hot water and cold water are the same substance with a slight difference in density. Normally, they are easily mixed together without creating significant temperature gradients. However, stratification does occur when the flow rate is minute. The seepage of cold water into hotter stagnant water, or vice versa, due to valve leakage is one of these situations. The situation is more pronounced when the cold water leaks through the bottom portion of the pipe or when the hot water leaks through the top portion of the pipe as shown in Fig. 11.4(c).

As noted earlier, the stratification of hot and cold water has generated cracks in the feedwater piping at more than a dozen units of steam generators in the nuclear power industry [3]. The cracks appeared to be due to the combination of local thermal stress from the temperature gradient and bending stress created by the bowing effect. Figure 11.5 shows a schematic of the problem. During standby conditions, the steam generator is not fully functional. However, due to the consumption of auxiliary steam, small amounts of water have to be added periodically. This minor makeup is performed by an auxiliary feedwater system. Because of the low flow rate, the cold makeup flow is not well mixed with the hot water. It flows mainly through the bottom portion of the pipe, creating stratification. This stratification produces a local thermal stress as shown in Fig. 11.2. It also generates a beam bending stress due to bowing. Because the minute makeup is intermittent, the interface of the stratification fluctuates. As the local thermal stress varies greatly along the diametrical direction of the pipe wall as given in Fig. 11.2, the fluctuation of the interface position creates stress variation, or reversal, at a given point of the pipe wall. The phenomenon generates a very high number of stress cycles, resulting in high cycle fatigue experienced normally only on steady-state vibration. High cycle fatigue not only

Special Thermal Problems 363

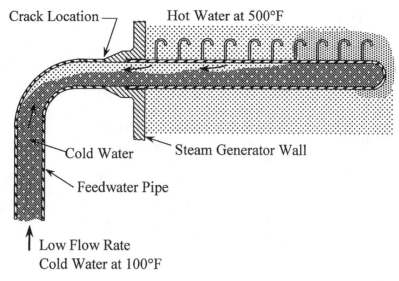

**FIG. 11.5
PIPE CRACK DUE TO THERMAL STRATIFICATION**

has a very low allowable stress amplitude that is equivalent to the endurance of the material, but also amplifies the stress intensification effect at minor discontinuity locations such as welded junctions.

(d) Common bowing at petrochemical plants. Most thermal bowing phenomena are due to transient or startup conditions. However, there are some persistent bowing situations that exist in the petrochemical industry. Figure 11.6 shows two of the situations that are common in ethylene plants. Ethylene and related hydrocarbon components are produced by pyrolysis of hydrocarbon feeds such as ethane, naphtha, and gas oil. The feed is diluted with steam and cracked in the pyrolysis furnace at about 1600°F (870°C). The heat of the high-temperature effluent gas from the furnace is used to produce high-pressure steam for process and utility use. After the steam generator, the effluent is further quenched rapidly with oil to control the components of the final product to achieve a higher proportion of the desired components.

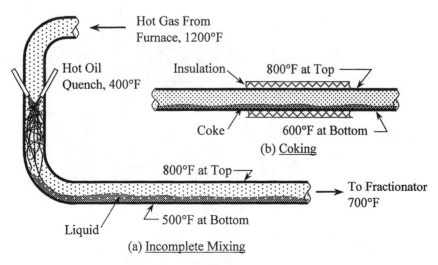

**FIG. 11.6
COMMON BOWING PHENOMENA AT PETROCHEMICAL PLANTS**

Figure 11.6 shows the places in which thermal bowing often occurs. Thermal bowing very often impairs not only the structural integrity but also the proper operation of the system. At the quick quench section, it is impossible for the quench oil and the effluent fluid to mix completely in the limited distance available. This incomplete mixing creates a large temperature difference between gas and liquid. The liquid thus settles to the bottom of the horizontal portion of the system. Although some calculation might predict annular flow for the postulated flow condition, annular flow rarely occurs. This results in quite a large temperature difference at the top and at the bottom of the pipe. Thermal bowing often twists the whole system and breaks the main line and some weak connections.

Another cause of thermal bowing is due to coking of the pipe as shown in Fig. 11.6(b). With hydrocarbon fluid operating at high temperature, coking is unavoidable. Normally, the pipe metal temperature is very close to the fluid temperature. However, when the coke accumulates at the bottom of the horizontal pipe, it functions as internal insulation, which reduces the temperature of the pipe wall behind the coke to a temperature that is considerably lower than the fluid temperature. Significant thermal bowing is, therefore, expected in all pipes that are prone to coke. Proper piping flexibility should be provided for absorbing this expected thermal bowing.

11.1.4 The Problem Created by a Tiny Line

The discussion of thermal bowing in this chapter focuses mainly on existing and potential cases to raise the awareness of engineers regarding this problem. One such case was caused by a tiny purge line that is often not included in the stress engineer's checklist. Figure 11.7 shows the schematic outline of the case. The subject line was a furnace effluent line of an ethylene plant. This was the main process line of the plant. The line functioned properly process-wise, but twisted the furnace lead piping so badly that the lead connections eventually broke. This was entirely unexpected, because the furnace leads were all provided with sufficient flexibility to absorb the estimated thermal expansion. After the

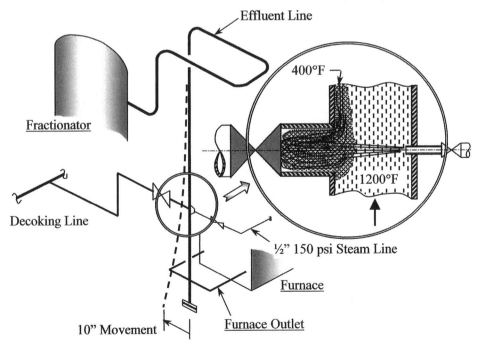

FIG. 11.7
PROBLEM CAUSED BY A TINY STEAM PURGE LINE

failure, it was estimated from support movements that the bottom of the effluent riser had moved 10 in. (254 mm) laterally. It was thought that the combination of the piping in the entire plant somehow had caused this movement. The solution naturally was to provide additional flexibility at the furnace lead pipes to absorb this 10-in. movement. Elaborate hairpin-style loops were installed for this purpose. Mathematically, it was shown to be satisfactory for absorbing this huge amount of displacement, but the problem persisted. Somehow, the hairpin loops that were expected to solve the problem had all moved in unexpected directions. As the pipe movements were completely different from those predicted by the analysis, the stress condition in the pipe and connection naturally was not the same as given by the analysis assuming a 10-in. lateral movement. The effluent riser, in addition to the 10-in. lateral movement, also rotated substantially. This rotation was not included in the redesign of the furnace lead pipes. Because the rotation would produce large vertical movements on the horizontal runs of pipe, no satisfactory support system could be easily implemented. A better solution was to determine the cause of the movement and rotation at the effluent riser.

Because the riser underwent both lateral displacement and rotation, a likely cause was believed to be thermal bowing. The question then was what caused the thermal bowing? Upon careful inspection of all the piping connected, a tiny ½-in. 150-psi steam purge line was found to be suspicious.

Because of coking, the pyrolysis furnace tubes need to be periodically de-coked with high temperature air or steam. The de-coking process goes through the furnace and exits to a discharge line by way of the effluent riser. Since the de-coking line is idle most of the time, coke may accumulate at the branch cavity. To prevent the accumulation of coke from blocking the leadoff passage, the cavity was constantly purged with steam as shown in the insert of Fig. 11.7. After purging the cavity, the returning cold steam was swept upward by the main process fluid. As the steam lost considerable momentum after purging, it stuck along the far side of the pipe wall, creating a large temperature difference between the two sides of the pipe. The entire furnace lead pipe problem was caused by this temperature difference. One solution was to purge with high-pressure superheated steam, which is readily available in an ethylene plant. From a local thermal stress consideration, the colder 150-psi, 360°F steam should never have been injected into 1200°F pipe in the first place. See Section 11.4 for further discussion on this subject.

11.2 REFRACTORY LINED PIPE

In petrochemical plants and refineries, there are situations in which the use of internally insulated pipe is desired for practical and economic reasons. Most internally insulated pipes use refractory material to resist high fluid temperatures and leave the pipe wall at near-ambient temperature. The internally insulated pipe is used in many areas. Some process fluids are simply too hot for any type of available pipe material without internal protection. Even if the fluid temperature is within the applicable temperature range of the available material, it is very often too expensive to use when the temperature is at the higher end of the applicable range. Some corrosive fluids may also require internal insulation to protect the pipe. Besides the material concern, the other major reason to use internally insulated pipe is to reduce the amount of thermal expansion by reducing the pipe wall temperature. Smaller expansion eliminates the requirement of an expansion joint or expansion loop, which may not be easily accommodated due to either process or space limitations.

11.2.1 Equivalent Modulus of Elasticity

One of the analytical concerns when using refractory insulated pipe is the simulation of the combined pipe and refractory stiffness. In contrast to external thermal insulation, whose stiffness is often ignored, the stiffness of internal refractory is generally taken into account in the analysis of the piping system. There are two practical methods of doing this: (1) increasing the analytical wall thickness of the pipe to compensate for the stiffness of refractory; (2) applying an equivalent modulus of elasticity with the actual pipe wall thickness. The latter method is generally preferred to avoid underestimating the stress due to the increase in section modulus by the increase in wall thickness.

The most common and practical approach is to use an equivalent modulus of elasticity applied on the actual pipe wall to simulate the combined effect. In other words, the weight of the system is calculated based on the actual pipe material weight, fluid weight, and refractory weight. The analysis is performed based on the actual pipe cross-section and geometry, which is applied with the total weight and an equivalent modulus of elasticity. There are several proposed methods for estimating an equivalent modulus of elasticity. The following is the one that appears to be the most practical. This method assumes that the refractory material, just like the pipe material, has similar tensile and compressive properties. This assumption is not entirely correct because refractory material is stronger in compression than in tension. However, the analysis of piping generally deals with three-dimensional systems. It is impossible to foresee which part of the piping system will have tensile stress and which part will have compressive stress. The equal tensile and compressive characteristics assumption appears to be the only one that is practical. By setting the equivalent stiffness to the sum of the stiffness of the pipe and refractory, we have

$$E_e I_p = E_p I_p + \sum (E_{fi} I_{fi}) \tag{11.4}$$

where E_e is the equivalent modulus of elasticity to be applied at the pipe material, I_{fi} is the moment of inertia of the refractory cross-section of layer i, and E_{fi} is the modulus of elasticity of the refractory material at layer i. The modulus of elasticity of the refractory material is estimated by the formula given by the American Concrete Institute as [8]

$$E_f = w^{1.5} \left(33 \sqrt{f_f} \right), \quad \text{which is in non-consistent units} \tag{11.5}$$

Equation (11.5) is expressed in non-consistent units, where E_f is modulus of elasticity in lb/in.^{2}, w is weight density in lb/ft^3, and f_f is the average rupture modulus in lb/in.^{2}. The values of w and f_f are available from refractory vendors. Generally, for low-density refractory used in the lower temperature region, the average values are $w = 60$ lb/ft^3 and $f_f = 300$ psi. The corresponding values for high-density refractory are $w = 160$ lb/ft^3 and $f_f = 3200$ psi. Generally, the equivalent modulus of elasticity, E_e, is about 25% higher than E_p, the modulus of elasticity of the pipe material.

11.2.2 Hot-Cold Pipe Junction

Most internally insulated pipes do not have internal insulation on the entire line. For instance, the valves are generally not internally insulated. Furnace leads are also generally not internally insulated. Figure 11.8 shows the insulation arrangement for an ammonia plant reformer piping. The main portion of the pipe is internally insulated to reduce the amount of expansion and also to save on pipe material cost. The leads to reforming furnaces, on the other hand, are externally insulated. This arrangement results in several hot-cold pipe junctions. The internally insulated pipe is cold, relatively speaking, and the externally insulated pipe is hot at a temperature about the same as the process fluid temperature. The detailed arrangement of the internal and external insulation at hot-cold pipe junctions is very important and is often neglected by engineers. Before going to the detailed arrangement, a failure case history may serve as a good reminder.

A case history [9]. The piping as shown in Fig. 11.8 was used to transfer a hot gas mixture from the primary reformer to the secondary reformer in an ammonia fertilizer plant. The gas mixture was operating at about 1500°F and 500 psi (815°C, 3.5 MPa). The hot pipe used alloy steel, whereas the cold pipe used carbon steel. The plant successfully operated for the first 10 years, after which maintenance was performed (as planned based on estimated creep life) to repair the refractory and to replace the hot-wall special alloy pipe. Strangely, after the revamp, the miter elbows, near the hot-cold pipe junctions as circled in the figure, developed leaking cracks about every 4 months. The revamp contractor

Special Thermal Problems 367

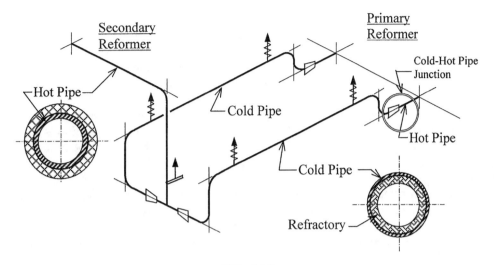

FIG. 11.8
INTERNALLY INSULATED AMMONIA PLANT REFORMER PIPING

was called in to repair and to rectify the situation following each leakage. Several changes involving re-calculation and re-support did not seem to work. Later, a large reputable constructor was contracted to handle the problem. Sophisticated analyses were performed with the spring support system completely rearranged. Yet, the cracks still developed about every 4 months. The problem was later solved by simple rearrangement of the insulation at the hot-cold pipe junctions.

Figure 11.9(a) shows a typical correct arrangement of the insulation around the hot-cold pipe junction. External insulation is required to extend over the transition cone so the pipe metal temperature is reduced gradually from internal fluid temperature to cold pipe temperature when it is exposed to the ambient environment. High temperature resistant alloy steel is used at the cone connection, and

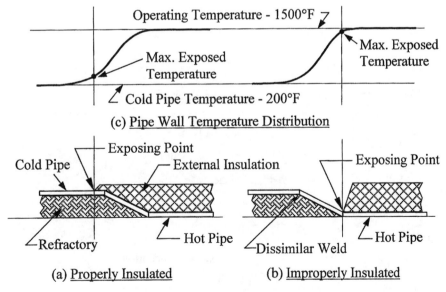

FIG. 11.9
COLD PIPE AND HOT PIPE JUNCTION

occasionally extends for a few inches beyond the cone junction to ensure that the dissimilar metal weld is located at a stabilized cold temperature region. Because the cone is surrounded by both internal and external insulation, its metal temperature reduces gradually and asymptotically to reach the cold pipe temperature, as it emerges and is exposed to the ambient environment as shown in Fig. 11.9(c). This was also the arrangement implemented by the original constructor for the piping described in the above case history.

By mistake or carelessness on the part of the draftsman, the construction drawing was erroneously marked with the insulation as given in Fig. 11.9(b). In general, the insulation notation is only casually marked on construction drawings and signifies only the existence rather than the detail of the insulation. Due to lack of experience, the revamp constructor followed the marking on the drawing and initiated the repeated cracking cycles at roughly every 4 months. With the external insulation ending at the hot pipe and cone connection, there is no refractory available for reducing the pipe metal temperature at the exposing point. The temperature of the pipe metal is roughly the same as the hot pipe temperature at the exposing point as shown in Fig. 11.9(c).

If the piping were housed indoors, then the situation given in Fig. 11.9(b) would have been less consequential. The only problem would have been the violation of the safety rule for creating burn environments. However, as most equipment and piping in a modern plant are all essentially located outdoors, the piping is constantly subject to the weather and other environmental effects. One of the detrimental environmental elements is rain shower.

Figure 11.10 shows the failure mechanism of the improperly insulated hot-cold pipe junction. With the exposed cone section at about the same temperature as the 1500°F fluid, a rain shower can quench part of the section quickly to a rain temperature of about 100°F. This shrinks the rain-impinged portion of the cone and generates a very high thermal stress to the quenched portion and its surroundings. The stress can be calculated with modern computer software if the loading and boundary conditions are known. To visualize the magnitude of the stress, the circular plate model as shown in Fig. 11.10(a) can be used. The situation can be visualized as a 1500°F uniform temperature circular plate having its center portion cooled to 100°F uniformly. Because of this center cold spot, the following stresses are generated [10, 11].

$$s_r = s_t = \frac{1}{2}E\alpha(T_2 - T_1) \cong 126{,}000 \text{ psi} \quad \text{Within the cold spot}$$

$$s_r = -s_t = \frac{1}{2}E\alpha(T_2 - T_1)\frac{a^2}{r^2} \cong 126{,}000\frac{a^2}{r^2} \quad \text{Outside the cold spot}$$

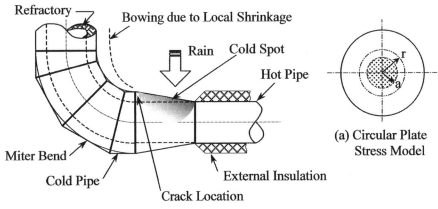

FIG. 11.10
FAILURE MECHANISM OF IMPROPERLY ARRANGED INSULATION

Special Thermal Problems 369

A large tensile stress is created at the crotch area of the miter bend junction. The above rough estimate shows that the stress generated from the quench of the rain shower is enough to cause the failure.

In addition to the thermal stress created by a temperature gradient, the cold spot also produces a bowing effect, which has a tendency to bend the pipe into the shape shown by the dotted line. The free bowing shape, however, is suppressed by the stiffness of the system generating a large beam bending stress, again with tensile stress at the crotch area of the miter bend junction. The combination of this beam bending and thermal gradient stress, together with a dissimilar material weld, makes this type of failure almost a certainty. The calculation of the stress is superficial and is not required. The situation has to be avoided by all means.

11.3 UN-INSULATED FLANGE CONNECTIONS

The process piping code ASME B31.3 [12] paragraph 301.3.2 stipulates that the design temperature of un-insulated flanges, including those on fittings and valves, shall not exceed 90% of the fluid temperature. In other words, the design temperature of the un-insulated flange can be taken as 10% less than that of an insulated flange. This is a reasonable rule considering the large exposed surface of the flange. The rule provides some relief for high-temperature flanges operating at the higher end of the allowable stress table. A 100°F degree difference in temperature at that range may result in a substantial increase in the allowable stress. This rule has tempted engineers to adopt un-insulated designs for some high-temperature flanges.

The un-insulated flange construction is fine when the piping is properly sheltered from environmental effects. However, because most piping systems are installed outdoors in modern plants, the lack of insulation creates serious problems. The rain shower effect as discussed in the previous section is also applicable to un-insulated flanges. Nagging leakages have been experienced in many of the un-insulated high-temperature flange connections. It is preferred that the flanges be designed with insulation or at least with a rain hood installed.

Lately, due to energy conservation issues, some plants have covered up bare flanges with insulation. This is good, but may have violated code rules and jeopardized structural integrity if the original un-insulated flanges were designed with the reduced design temperature allowed by un-insulated flanges.

11.4 UNMATCHED SMALL BRANCH CONNECTIONS

In a process plant, there are occasions when a small line carrying secondary fluid is tied or dumped to a main process line. The temperature of the fluid from the small inserted line can be quite different from that of the main process fluid. This temperature difference can create problems at the connection and the area nearby. Because the small line is not the main process line, its importance is often overlooked.

The problems associated with small off-temperature connections occur in two areas. One is the connection itself that might generate a high thermal stress, and the other is the cold spot created by the wetting or the impingement of the colder fluid on the main process pipe wall. The cold spot, as discussed in previous sections, can create both high thermal stress and thermal bowing. The discussion assumes that the branch flow is colder than the main process flow.

Figure 11.11 shows some of the construction details used in small branch connections. Normally, the connection is constructed as shown in Fig. 11.11(a), if no special instruction is given. This simple direct connection will generate a high thermal stress at the junction due to the temperature gradient. This type of construction is not suitable for 300°F (170°C) or higher temperature differences between the branch flow and main flow. When the temperature difference is greater, a nozzle sleeve is generally used as shown in Fig. 11.11(b) to reduce the thermal stress. However, the cold fluid can wet the top portion of the pipe wall nearby if the branch flow rate is small, and it will impinge the bottom portion

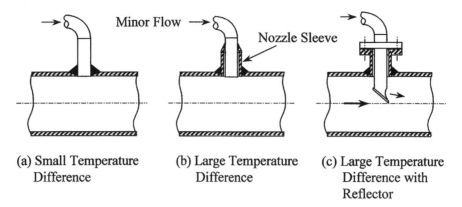

**FIG. 11.11
SMALL INSERTS TO HOT PROCESS LINE**

of the main pipe if the branch flow rate is high. To reduce this possibility, the nozzle can be extended to the center of the main pipe with the nozzle end cambered. The cambered nozzle end reflects the flow toward and along the centerline of the main pipe, as shown in Fig. 11.11(c).

11.5 SOCKET-WELDED CONNECTIONS

Smaller piping, 2 in. or smaller, are generally connected with socket-welded joints. This simplifies the alignment and expedites the construction. For thinner pipes, the automatic alignment of the socket also helps to maintain a certain level of joint quality. However, socket welds also create frequent failures. A single failure of the small socket weld can sometimes require the shutdown of certain systems or even an entire plant. This is especially so in a nuclear plant where even a slight compromise of safety is not tolerated.

The failures of socket welds can be divided into two categories: those caused by a very low cycle of operation and those caused by high cycle fatigue. The low cycle failures, due to fatigue or otherwise, were well known as soon as the socket weld was introduced, but the high cycle fatigue failures drew attention only rather recently. High cycle fatigues are caused by vibration, very often amplified from the minute vibration of the main pipe to which the small socket welded branch is connected. The small branch very often has heavy appendages attached. The heavy appendages, such as values, produce high vibration bending on the joint due to large inertia resistance.

Originally, most of the low cycle failures were attributed to the bottoming out of the socket as shown in Fig. 11.12(b). It was believed, and is still believed, that the failure is due to shrinkage of the weld and the joining components during the welding process. With bottomed-out construction, the joining parts do not have the freedom to move around during welding. A slight mismatch of temperature on the joining parts is enough to plant the seed for failure. However, tests [13, 14] have shown that, although it may cause micro-cracks at the root of the fillet welds in worst-case situations, the welding on bottomed-out sockets does not create serious problems. Nevertheless, the gap at the socket does alleviate crack problems during operation. This is reason enough for the code to set the gap requirement on socket welds as shown in Fig. 11.12(a).

In addition to the concern of welding, the bottomed-out socket also has a thermal stress problem due to potential jack up actions, which can be much more serious than the welding problem in some cases. Typically, socket welded small piping is used in secondary functions, such as for blow-offs and drains. These functions are intermittent with sudden discharge when called for. This sudden flow pro-

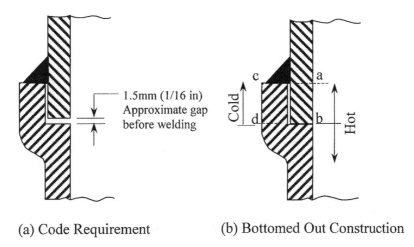

**FIG. 11.12
SOCKET WELD CONNECTION**

duces large temperature differences between the pipe and the socket. As shown in Fig. 11.12(b), when the temperature for pipe section a-b is much higher than the temperature at socket section c-d, the differential expansion of the entire a-b piece is absorbed and concentrated at the fillet weld when the joint is bottomed out. This jack up action generates very high stress at the weld, promoting premature failure. The gap provided at the joint relieves the expansion of a-b without creating any jack up action to the weld. It is apparent that, for high temperature discharge, the gap is essential for ensuring the integrity of the joint.

Lately, from tests on joints at ambient temperature, it has been found that the bottomed-out socket welded joint has considerably higher fatigue strength [13, 14] than that of the gapped joint. This is mainly due to friction resistance between the pipe end and the socket bed. However, this higher fatigue strength is possible only at ambient or low temperature situations when there is no jack up action from thermal expansion. Based on common sense and the above-mentioned test data, occasionally, vibration experts will suggest the use of the bottomed-out socket weld connection for vibration lines when they discover another crack has developed on a socket welded joint. The suggestion may be considered for cold vibration lines, such as the ones connected to reciprocating compressor circuits. However, because the bottomed-out socket welded construction is in violation of the piping code, it has to be clearly identified in the design specifications as an exception to the code rule. The general feeling is to always provide the gap as required by the code.

Because the gap is important in socket welded connection, it requires a reliable procedure for its implementation. Since it is very difficult to see the gap in the finished joint, its implementation relies mainly on the honesty of the pipe fitter. Most pipe fitters will insert the pipe all the way to the bottom of the socket and make a mark on the pipe. They would then pull out the mark for about 1/16 in. (1.5 mm) before welding. There are also gapped spring rings available in the market for this purpose. With a gap ring, the pipe fitter only has to insert the ring into the socket first, and then insert the pipe all the way into the socket until resisted by the ring.

REFERENCES

[1] Flieder, W. R., Loria, J. G., and Smith, W. J., 1961, "Bowing of Cryogenic Pipelines," *Transactions of the ASME, Journal of Applied Mechanics*, vol.83, September, no.3, pp. 409–416.

[2] Howard, F. S., 1991, *Cryogenic Transfer System Mechanical Design Handbook*, Report, NASA Special Project Branch, DM-MED-1.

[3] U.S. NRC IE Bulleting No. 79-13, 1979, "Cracking in Feedwater System Piping," October 16, 1079.
[4] Bush, S. H., 1992, "Failure Mechanisms in Nuclear Power Plant Piping Systems," *Transactions of the ASME, Journal of Pressure Vessel Technology*, **114**, pp. 389–395.
[5] Kirsner, W., 2002, "Flooded Manholes & Submerged Steam Lines — Understanding the danger of nucleate boiling," *HPAC Engineering*, May.
[6] Venendaal, B., 1979, "Dryout, Cooldown Keyed Cove Point Commissioning," *Pipeline and Gas Journal*, **206**(7), pp. 28–38.
[7] Machacek, S., and Zelenka, T., 1978, "Design and Operation of a Large Diameter (1.67 m) Steamline," ASME paper 78-PVP-88, presented at ASME/CSME PVP Conference, Montreal, Canada, June 1978.
[8] Getz, R. C., 1977, "Analytical Method Check Stresses in Hot, Large-OD Lined Pipe," *The Oil and Gas Journal*, June 20, pp. 82-84.
[9] Peng, L. C., and Peng, T. L., 1998, "Thermal Insulation and Pipe Stress," *Hydrocarbon Processing*, May.
[10] Goodier, J. N., 1937, "Thermal Stress," *ASME Journal of Applied Mechanics*, vol.59, March, no.1. (Reprinted in *Pressure Vessel and Piping Design – Collected Papers, 1927-1959*, pp. 528–531, ASME).
[11] Young, W. C., 1989, *Roark's Formulas for Stress & Strain*, 6th ed., p. 722, McGraw-Hill, New York.
[12] ASME B31.3, "Process Piping," *ASME Code for Pressure Piping*, An American National Standard, ASME, New York, NY
[13] Higuchi, M., Hayashi, M., Yamauchi, T., Ilda, K., and Sato, M., 1995, "Fatigue Strength of Socket Welded Pipe Joint," *ASME PVP. Vol. 313-1, International Pressure Vessels and Piping Codes and Standards: Vol. 1 — Current Applications*, ASME, New York
[14] Higuchi, M., Nakagawa, A., Ilda, K., Hayashi, M., Yamauchi, T., Saito, M., and Sato, M., 1998, "Experimental Study on Fatigue Strength of Small-Diameter Socket-Welded Pipe Joints," *Transactions of the ASME, Journal of Pressure Vessel Technology*, **120**, pp. 149–156.

CHAPTER

12

DYNAMIC ANALYSIS — PART 1: SDOF SYSTEMS AND BASICS

A static load is applied slowly and gradually to the piping or other structural system. It does not involve any time factor, nor the inertia of the mass. On the other hand, a dynamic load is applied rapidly to the piping or other structural systems. The manner in which the dynamic load is applied, in addition to the magnitude of the load, substantially affects the system response. In dealing with dynamic loads, not only is the time factor important; the mass that represents the inertia of the structure is also important.

This chapter deals with some fundamental concepts of dynamic loads and the common types of loads that are related to piping engineering. Due to space limitations and to allow piping design engineers to get into the subject quickly, only the practical basics are discussed. Readers who are interested in more theoretical treatments should consult specialized books on structural dynamics [1–3] for further details.

Besides earthquakes, the dynamic load in piping comes mainly from fluid transient phenomena. In this book, only the basics of fluid transient loads are discussed. Detailed discussions of acoustic and thermal fluid transient loads are beyond the scope of this book. This book deals mainly with the treatment of dynamic forces in the design analysis of a piping system.

The scope of dynamic analysis is divided into two chapters in this book. This chapter discusses single degree of freedom (SDOF) systems and some basic fundamentals. The multi-degree of freedom (MDOF) systems and some more elaborate applications are discussed in the next chapter.

12.1 IMPACT AND DYNAMIC LOAD FACTOR

When a load is not applied gradually to a structural system, it will impact the system just like other sudden changes we are familiar with. If the application of a load is very fast, it is regarded as a shock. Figure 12.1 shows a simple structural system that will be used to explain the effect of impact. The simple structure system is a coil spring, having negligible mass, but with one end anchored to the floor.

With a static loading, as shown in Fig. 12.1(a), the spring deforms slowly and gradually as the load is applied and gradually increased. When the weight load reaches its preset magnitude, W, the deformation reaches, Δ_{st}, which is determined by

$$\Delta_{st} = \frac{W}{k} \tag{12.1}$$

where k is the spring constant of the spring or structure and Δ_{st} is the static deformation of the spring. Because the load is applied very gradually, the velocity of the weight is negligible. The deformation stops as soon as the load stops increasing. Therefore, Δ_{st} is the final deformation corresponding to the weight load applied.

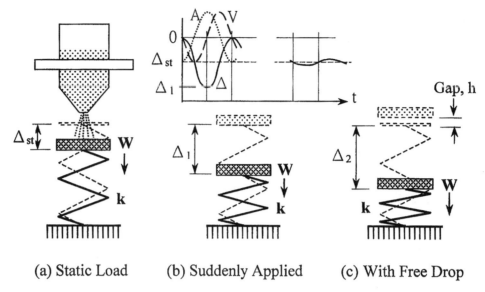

(a) Static Load (b) Suddenly Applied (c) With Free Drop

FIG. 12.1
DYNAMIC LOAD FACTORS

One impact type that is often used as a benchmark comparison is the suddenly applied load. As shown in Fig. 12.1(b), the whole weight load, W, is held at the top of the spring before being released suddenly. In this case, the spring force balances the weight load when it deforms to Δ_{st}. The weight at this balanced position, however, is not still, but is instead at its maximum velocity, as shown in the graph. This velocity pushes the weight and the spring further downward, generating an additional push up spring force. This unbalanced push up spring force causes the weight to decelerate. The weight keeps moving downward until its velocity decelerates to zero. The spring force at this final deformed position is larger than the weight, thus pulling the weight back up. The motion of the weight oscillates back and forth in a cyclic form. The oscillating amplitude will be gradually reduced to zero by the damping of the system. The weight will eventually settle at the static balanced position, Δ_{st}. The largest deformation can be calculated by energy balance. Because the kinetic energy of the weight at its maximum displacement location is zero, due to zero velocity, the participating energy at this point includes only potential energy and stored internal structural energy. They are balanced as follows

$$W \cdot \Delta_1 = \frac{1}{2}k\Delta_1 \cdot \Delta_1$$

or

$$\Delta_1 = 2\frac{W}{k} = 2\Delta_{st} \quad \text{for suddenly applied load}$$

where $\frac{1}{2}k\Delta_1$ is the average spring force. By comparing the above with Eq. (12.1), it is clear that the deformation caused by the suddenly applied load is 2.0 times the deformation of the static load of the same magnitude. This 2.0 factor is called the dynamic load factor (DLF). That is,

$$\Delta = (\text{DLF})\Delta_{st} \quad \text{or} \quad \text{DLF} = \frac{\Delta}{\Delta_{st}} \tag{12.2}$$

where Δ is the maximum deformation produced by the load and Δ_{st} is the deformation due to the static load of the same magnitude. From above, the DLF of a suddenly applied load is 2.0. That is, the suddenly applied load can be considered as the static load with twice the magnitude —which means that a suddenly applied load can be treated in a static analysis as a static load having a magnitude twice the actual load.

When a load reaches the structural system with an initial velocity like the weight load with a free drop, as shown in Fig. 12.1(c), it is called an impact load. The DLF of the weight load with a free drop can be calculated, again by energy balance, as follows.

$$W \cdot (h + \Delta_2) = \frac{1}{2} k \Delta_2 \cdot \Delta_2 \quad \text{or} \quad 2\Delta_{st} h + 2\Delta_{st} \Delta_2 = \Delta_2^2$$

or

$$\Delta_2 = \frac{2\Delta_{st} + \sqrt{4\Delta_{st}^2 + 8\Delta_{st} h}}{2} = \Delta_{st} \left(1 + \sqrt{1 + 2\frac{h}{\Delta_{st}}}\right)$$

$$\text{i.e.,} \quad \text{DLF} = \frac{\Delta_2}{\Delta_{st}} = 1 + \sqrt{1 + 2\frac{h}{\Delta_{st}}}$$

Because Δ_{st} is independent of the free drop gap, h, the DLF increases as the gap increases. When the free drop gap is zero, the DLF is 2.0 as derived previously. With a free drop gap, the DLF is always greater than 2.0.

The above examples are applicable only to a sustained constant load with the magnitude of the load maintaining the same throughout the deformation. In such cases, the DLF is always greater than 1.0. However, depending on the duration of the load and the mass of the structure, the DLF can also be smaller than 1.0, and sometimes even approaches zero.

12.2 SDOF STRUCTURES

The dynamics of piping and structures is very complicated. However, because we are mainly dealing with linear systems, most of the structural behaviors are linear combinations of the behaviors of simple systems. The simplest system is the Single Degree of Freedom (SDOF) system. The fundamental characteristics of the SDOF structure are the basics of structural dynamics.

The beam shown in Fig. 12.2(a) can be considered as a segment of pipe simply supported. The actual pipe mass is uniformly distributed. However, in practical piping analysis, the weight is lumped as a concentrated mass at certain point or points. This is because it is very difficult to handle distributed mass in a dynamic analysis. In this example, the uniform pipe mass is lumped as a concentrated mass at the mid-span point. Generally, only one-half of the total mass of the segment is considered the effective mass at the point. The rest of the mass is distributed to the end points. The mass, or the point,

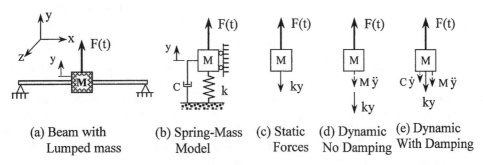

FIG. 12.2
SINGLE DEGREE OF FREEDOM SYSTEM

generally has six degrees of freedom: three in translation and three in rotation, along and about three orthogonal coordinate axes. Depending on the shape of the system and the type of load, generally not all six degrees of freedom are participating in the activity. For a force applied in the y direction at the center of the beam, only the y-translation degree of freedom is participating in the action. Therefore, it is categorized as an SDOF system.

12.2.1 Working Formula for SDOF Systems

Mathematically, the SDOF system is represented by the spring-mass system as shown in Fig. 12.2(b). The spring constant is determined by the beam stiffness — that is, the force required to deflect the beam for a unit distance. The beam formula given in Chapter 2 can be used for this calculation. In addition to the spring constant, the spring-mass system also includes a damping apparatus representing the resistance from internal and external frictions.

Taking the mass as a free body, we have forces acting on the free body as shown in Fig. 12.2(c), (d), and (e) depending on the presence of motion and damping. It should be noted that the mass also generates a static beam deflection due to gravity. However, this static deflection is calculated separately and its response is combined with the dynamic response afterward if required. The dynamic analysis calculates only the response due to dynamic loads.

From the general free body diagram given in Fig. 12.2(e) with positive y in the vertical up direction, the following SDOF working equation is derived from the equilibrium of forces.

$$M\ddot{y} + c\dot{y} + ky = F(t) \tag{12.3}$$

Where, $\ddot{y} = \dfrac{d^2y}{dt^2}$ is acceleration of the mass and $\dot{y} = \dfrac{dy}{dt}$ is the velocity of the mass. $F(t)$ is the applied force that is a function of the time. The applied force generally changes with time. The damping c applied on velocity term is regarded as viscous damping, which has a resistance force against the velocity and has a magnitude proportional to the velocity.

12.2.2 Un-Damped SDOF Systems

The fundamental characteristics of SDOF systems can be better visualized with an un-damped system. The damping generally contributes in the decaying of the motion. It does not substantially affect the characteristics of the system, but it does add considerable complexity in the solution of the motion.

(a) Natural frequency. The first step in dealing with structural dynamics is the determination of the natural frequency of the system. Natural frequency is a fundamental property of free vibration without an applied force. One way to find out the natural frequency is to pull the system for a small displacement then release it to count the number of oscillations per unit time. Mathematically, the natural frequency is determined from Eq. (12.3) by setting c and $F(t)$ to zero. That is, for free vibration we have

$$M\ddot{y} + ky = 0 \tag{12.4}$$

Equation (12.3) is a linear differential equation with constant coefficients. M, c, and k are constant coefficients. The equations without a forcing function, such as the one given by Eq. (12.4), are called homogeneous equations. In general, these types of equations can be solved by assuming that $y = e^{mt}$, with constant m to be determined from actual substitution to the equation. However, an equation with only even order terms like Eq. (12.4), can be solved by assuming $y = \sin \omega t$ or $y = \cos \omega t$. By substituting $y = \sin \omega t$ and $y = \cos \omega t$ separately into the equation, we find that both $y = \sin \omega t$ and $y = \cos \omega t$ are the solutions of the equation with

$$\omega_n = \omega = \sqrt{\dfrac{k}{M}} \tag{12.5}$$

Because this ω parameter is for natural vibration, we will denote it with the n subscript as ω_n. The general solution of Eq. (12.4) is

$$y = C_1 \sin \omega_n t + C_2 \cos \omega_n t \qquad (12.6)$$

Constants C_1 and C_2 are determined by initial conditions. Assume we have $y = y_0$ and $\dot{y} = \dot{y}_0$ initially at $t = 0$. Substituting $y = y_0$ and $t = 0$ in Eq. (12.6), we find $C_2 = y_0$. We then differentiate Eq. (12.6) with respect to t and substitute $\dot{y} = \dot{y}_0$ and $t = 0$, we have $C_1 = \dot{y}_0/\omega_n$. Substituting C_1 and C_2 to the above equation, we have the solution for a system without any forcing function as

$$y = \frac{\dot{y}_0}{\omega_n} \sin \omega_n t + y_0 \cos \omega_n t \qquad (12.7)$$

For determining the natural frequency, we can pull the mass for a distance y_0, and then release it to measure the frequency. In this case, the initial velocity is zero, so we have $y = y_0 \cos \omega_n t$. This solution is plotted in Fig. 12.3. The mass oscillates according to sinusoidal form and repeats itself for every $2\pi/\omega_n$ time interval. This time interval is called the natural period of the system. That is,

$$\text{Natural period} = T = \frac{2\pi}{\omega_n} = 2\pi \sqrt{\frac{M}{k}} \qquad (12.8)$$

Natural frequency is defined as the number of vibration cycles per unit time. It is the inverse of the natural period as follows:

$$\text{Natural frequency} = f_n = \frac{1}{T} = \frac{1}{2\pi} \sqrt{\frac{k}{M}} \qquad (12.9)$$

In contrast to the natural frequency f_n, ω_n is called the natural circular frequency. The unit for natural period is generally in seconds and the unit for frequency is in cycles per second, which is also occasionally referred to as Hertz (Hz). Thus, 100 Hz means 100 cycles/sec, and so forth. Natural period and, thus, natural frequency are inherent characteristics of the structural system. The loading that is applied to the system does not alter them. They will be somewhat affected by system damping. However, as will be shown later, the effect of damping on the natural period and frequency is very small and can be ignored in practical analyses.

(b) Un-damped system with constant force. After solving the natural period and natural frequency, we are ready to investigate some loading conditions. The first case, which is the simplest, is the one with a constant force — that is, $F(t) = F_0$. By setting $c = 0$ and $F(t) = F_0$ in Eq. (12.3) for un-damped forced vibration, we have

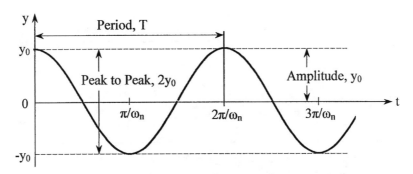

FIG. 12.3
FREE VIBRATION OF SDOF SYSTEM WITH INITIAL DISPLACEMENT

$$M\ddot{y} + ky = F_0 \tag{12.10}$$

The particular solution for this type of forcing function can be found by letting $y = AF_0$, where A is an undetermined constant. By substituting $y = AF_0$ in the equation, we find $A = 1/k$ or $y = F_0/k$. The general solution of the equation is the combination of this particular solution with the solution of the homogenous equation. The general solution in this case is

$$y = C_1 \sin \omega_n t + C_2 \cos \omega_n t + \frac{F_0}{k} \tag{12.11}$$

C_1 and C_2 are again determined from initial conditions. For an initially stationary system with both displacement and velocity zero at $t = 0$, we find $C_2 = -F_0/k$ and $C_1 = 0$. The solution for the initially inactive system becomes

$$y = -\frac{F_0}{k} \cos \omega_n t + \frac{F_0}{k} = \frac{F_0}{k}(1 - \cos \omega_n t) \tag{12.12}$$

The displacement follows the sinusoidal shape as shown in Fig. 12.4. The maximum values are reached when $\cos \omega_n t = -1$, which is equivalent to $\omega_n t = \pi, 3\pi, 5\pi$, etc. The maximum displacement is $2F_0/k$, which is twice the value of the static displacement F_0/k. This again shows that the DLF of a suddenly applied load is 2.0.

(c) Un-damped system with harmonic load. Another basic load to be investigated is the harmonic load because of its importance and simplicity. Assuming the force is a sinusoidal force $F(t) = F_0 \sin \omega t$, we have

$$M\ddot{y} + ky = F_0 \sin \omega t \tag{12.13}$$

where ω is the forcing circular frequency, which is generally different from the natural circular frequency ω_n.

Since differentiating $\sin \omega t$ with respect to time twice comes back to $\sin \omega t$ again, the particular solution for Eq. (12.13) can be found by assuming that $y = A \sin \omega t$, where A is an undetermined constant. By substituting $y = A \sin \omega t$ into Eq. (12.13), we find

$$A = \frac{F_0}{k - M\omega^2} = \frac{F_0/k}{1 - \omega^2(M/k)} = \frac{F_0/k}{1 - (\omega/\omega_n)^2}$$

or

$$y = \frac{F_0/k}{1 - (\omega/\omega_n)^2} \sin \omega t \tag{12.14}$$

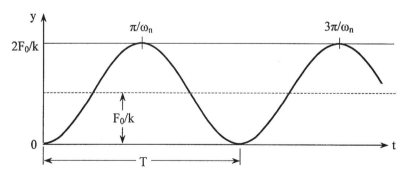

FIG. 12.4
RESPONSE OF UN-DAMPED SDOF SYSTEM TO SUDDENLY APPLIED FORCE

Because $F_0/k = y_{st}$ is the static displacement due to F_0 force, we can write

$$y = \frac{y_{st}}{1-(\omega/\omega_n)^2} \sin\omega t \tag{12.15}$$

The above is a particular solution of the force equation. The general solution is the combination of this particular solution with the solution of the homogeneous equation. That is, the general solution for this case is

$$y = C_1 \sin\omega_n t + C_2 \cos\omega_n t + \frac{y_{st}}{1-(\omega/\omega_n)^2} \sin\omega t \tag{12.16}$$

The first two terms are un-damped free vibration, whereas the last term is un-damped forced vibration. Again, C_1 and C_2 are to be determined from the initial condition. Assume that the system starts from a resting condition with both displacement and velocity zero at $t = 0$. After substituting the zero initial displacement condition into Eq. (12.16), we find $C_2 = 0$. Differentiating Eq. (12.16) and substituting the zero initial velocity condition, we find

$$C_1 = -\frac{y_{st}}{1-(\omega/\omega_n)^2} \frac{\omega}{\omega_n}$$

The solution for the system starting from a resting condition is

$$y = \frac{y_{st}}{1-(\omega/\omega_n)^2} \left[\sin\omega t - \frac{\omega}{\omega_n} \sin\omega_n t\right] \tag{12.17}$$

or

$$\text{DLF} = \frac{y}{y_{st}} = \frac{1}{1-(\omega/\omega_n)^2} \left[\sin\omega t - \frac{\omega}{\omega_n} \sin\omega_n t\right] \tag{12.18}$$

Equation (12.18) consists of two parts. The first term inside the brackets is due to forced vibration, and the second term inside the brackets is free natural vibration. The maximum value of DLF occurs when $\sin\omega t$ and $\sin\omega_n t$ are additive and at their maximum level. This occurs at some point of time when $\sin\omega t = 1$ and $\sin\omega_n t = -1$, or vice versa. After substituting these values into Eq. (12.18), we have the maximum DLF as

$$(\text{DLF})_{max} = \pm\frac{1+\omega/\omega_n}{1-(\omega/\omega_n)^2} = \pm\frac{1}{1-\omega/\omega_n} \tag{12.19}$$

For general applications, Eq. (12.19) often overestimates the response as the free natural vibration part, represented by the second term of Eq. (12.18), which is damped out rather quickly in actual structural systems.

Therefore, for practical applications dealing with steady-state vibration, the free natural vibration term is often ignored. After omitting the free vibration term, the maximum DLF in this case becomes

$$(\text{DLF})_{max} = \frac{1}{1-(\omega/\omega_n)^2} \tag{12.20}$$

In any case, when the applied forcing frequency equals the natural frequency, the DLF becomes infinite. This situation is called resonance, and the forcing frequency is called the resonant frequency. Theoretically, when there is no damping, the system will vibrate at an infinitely large amplitude when the applied forcing frequency is in resonance with the natural frequency. This is one of the conditions to be avoided in a vibratory system. The absolute value of DLF becomes less than 1.0 when the applied frequency is greater than $\sqrt{2}$ times of the natural frequency.

(d) Un-damped system subjected to impulse loads. In practical applications, we frequently deal with impulse loads with a relatively short duration. The commonly encountered impulse loads include rectangular pulse, suddenly applied triangular pulse, symmetrical triangular pulse, sinusoidal impulse, ramped constant force, and so forth. An actual load may have a shape different from any of these idealized shapes. However, it can often be approximated to one of these idealized shapes. The response of the impulse load is generally calculated in two steps. For instance, take the rectangular pulse as shown in Fig. 12.5; the first step is to use Eq. (12.12) to calculate the forced response due to the suddenly applied constant load, up to the end of the load duration. That is, for a rectangular pulse, we have

$$\text{For } 0 \leq t \leq t_d: \quad y = \frac{F_0}{k}(1 - \cos \omega_n t) \quad \text{-- first step}$$

This calculates the response of the system while the force exists. Substituting $t = t_d$ into the equation, the displacement and velocity at the end of the impulse force are calculated. These two quantities are needed, as initial conditions, for calculating the response after the applied force terminates. The second step is a free vibration having initial displacement and velocity as calculated above. Equation (12.7) applies to this condition. That is,

$$\text{For } t > t_d: \quad y = \frac{\dot{y}_{td}}{\omega_n} \sin \omega_n(t - t_d) + y_{td} \cos \omega_n(t - t_d) \quad \text{-- second step}$$

Within the first step, the maximum response, as shown in Fig. 12.4, is reached when $\omega_n t = \pi$, or $t = \pi/\omega_n = T/2$. This means that if the time duration is longer than one-half of the natural period of the system, the maximum response falls in the first step. However, if the duration of the force is shorter than $T/2$, the maximum response occurs at the free vibration portion after the force ends. This requires the calculation of the second step response beyond the force duration. Therefore, for a rectangular pulse, the DLF is

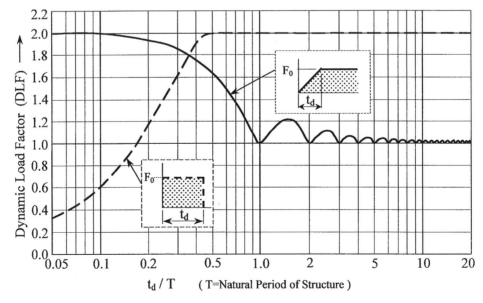

FIG. 12.5
DYNAMIC LOAD FACTORS FOR IMPULSE LOADS RECTANGULAR SHAPED AND RAMPED CONSTANT LOADS

$$\text{DLF} = \frac{y_{\max}}{F_0/k} = 2 \quad \text{when } t_d \geq 0.5T$$

$$\text{DLF} = \sqrt{(\dot{y}_{td}/\omega_n)^2 + (y_{td})^2} \quad \text{when } t_d < 0.5T$$

The second expression will be further explained later in Fig. 12.9.

The above procedure is generally applicable to all idealized impulse loads mentioned. The response time-history can be calculated at both the forced and free vibration stages. However, the often-needed practical information for engineering design is the DLF corresponding to the maximum displacement produced. Because the elastic structural force, moment, and stress are all proportional to the displacement, we can then apply the maximum DLF on the applied force to convert it into a static equivalent force for performing static analysis. The values of DLF for some common shapes of impulse loads are plotted in Figs. 12.5 and 12.6. Some of the detailed calculations are outlined in a number of books on structural dynamics [1, 3].

With these figures, one can get a feeling of the load's effect on a system by estimating the natural frequency of the system. From the natural frequency and the duration of the impulse, the maximum DLF is determined from the appropriate figure. By applying the DLF on the maximum load, a static equivalent load is arrived. This static equivalent load should give us an idea of the effect of the impulse on the system. With non-reversing loading, the maximum DLF is less than or equal to 2.0 for all loads. However, with reversing of the load, such as in the case of full sine wave impulse, the DLF can be as

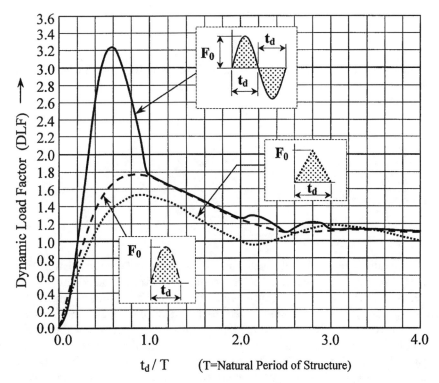

FIG. 12.6
DYNAMIC LOAD FACTORS FOR IMPULSE LOADS HALF AND FULL SINUSOIDAL, SYMMETRICAL TRIANGULAR

high as 3.25 without damping. The DLF can be near zero when the time duration is very short. For short durations, as compared to the natural period of the system, the impulse force terminates before the system has any chance to respond.

12.2.3 Damped SDOF Systems

Damping is a resistance force against structural motion. It reduces the response of the system. Therefore, it is conservative to ignore the damping in dynamic analysis. However, omission of the damping can lead to an uneconomical design of the structure. Damping can be roughly divided into two categories. One is viscous damping that is proportional to the relative velocity of the structural parts involved. Viscous damping comes mainly from the viscosity of internal and surrounding matters that resists the change of shapes. The other category of damping is Coulomb damping, which has a value proportional to the contact force on a sliding surface but acts against the moving direction. Support friction, for instance, belongs to Coulomb damping. We will limit our discussion to only viscous damping. The Coulomb damping, if required, may be approximately lumped into the viscous damping for consideration.

The working formula for the SDOF system with viscous damping is given in Eq. (12.3), which is duplicated as follows

$$M\ddot{y} + c\dot{y} + ky = F(t) \tag{12.3}$$

(a) Free vibration on viscously damped system. In an un-damped system, once disturbed by pulling a small displacement then releasing it, for instance, the system will vibrate forever as shown in Fig. (12.3) with the same amplitude. The amplitude of free vibration in a damped system, on the other hand, diminishes very quickly. To investigate the nature of free vibration on a damped system, the forcing function, $F(t)$, is set to zero. That is, for free vibration of a damped SDOF system, we have

$$M\ddot{y} + c\dot{y} + ky = 0 \tag{12.21}$$

This type of linear differential equation can be solved by assuming $y = e^{st}$, where s values are determined from actual substitution to the equation. By substituting $y = e^{st}$ to the equation, we have

$$Ms^2 + cs + k = 0$$

This a quadratic equation whose solution is

$$s_i = -\frac{c}{2M} \pm \sqrt{\left(\frac{c}{2M}\right)^2 - \frac{k}{M}}, \quad i = 1, 2 \tag{12.22}$$

The solution of Eq. (12.21) becomes

$$y = C_1 e^{s_1 t} + C_2 e^{s_2 t} \tag{12.23}$$

Depending on the values of s_1 and s_2, this solution has two different characteristics. When $(c/2M)^2$ is greater than (k/M), the expression inside the square root is positive, so both s_1 and s_2 are real numbers, but are negative. With both s_1 and s_2 as negative real numbers, the solution is constantly decaying. It is not an oscillatory vibration. This is called an over-damped system. Over-damping seldom occurs in a practical structural system and shall be skipped in this discussion. However, the damping that initiates this non-oscillatory condition is called critical damping, c_c, which is a very important benchmark parameter. Critical damping is the smallest damping that prevents the system from moving oscillatory. At critical damping, the expression inside the square root is zero. Therefore, we have

$$\left(\frac{c_c}{2M}\right)^2 = \frac{k}{M}, \quad \text{or} \quad c_c = 2M\sqrt{\frac{k}{M}} = 2M\omega_n \tag{12.24}$$

In most practical applications, the expression inside the square root in Eq. (12.22) is less than zero. In this case, s_1 and s_2 are both imaginary numbers. After substituting imaginary s_1 and s_2 into Eq. (12.23) and performing some mathematical manipulations, the general solution of the under-damped system becomes

$$y = e^{-\frac{c}{2M}t}(C_1 \sin \omega_d t + C_2 \cos \omega_d t) \tag{12.25}$$

where ω_d is the natural circular frequency of the damped system, and is given in Eq. (12.26). Because $2M = c_c/\omega_n$, we have

$$c/2M = c/(c_c/\omega_n) = (c/c_c)\omega_n = \zeta\omega_n$$

where $\zeta = c/c_c$ is the damping ratio. From above mathematical manipulations, the natural circular frequency of the damped system is expressed as

$$\omega_d = \sqrt{\frac{k}{M} - \left(\frac{c}{2M}\right)^2} = \sqrt{\omega_n^2 - \zeta^2\omega_n^2} = \omega_n\sqrt{1-\zeta^2} \tag{12.26}$$

This shows the relationship between damped and un-damped natural circular frequencies. With a 10% damping ratio, which is on the high end for practical structure systems, the damped natural frequency is $\omega_d = 0.995\omega_n$. The difference between damped and un-damped natural frequencies is so small that it is often ignored in actual applications. In other words, the frequencies are generally calculated based on an un-damped system. For simplicity, the term ω_n will be used later on, after the discussion of damped free vibration, for both un-damped and damped natural circular frequencies.

The constants C_1 and C_2 of Eq. (12.25) can be determined from the initial condition having $y = y_0$ and $\dot{y} = \dot{y}_0$ at $t = 0$, just as in the case of the un-damped system. Therefore, we have the solution for damped free vibration with initial condition as

$$y = e^{-\zeta\omega_n t}\left[\frac{\dot{y}_0 + y_0\zeta\omega_n}{\omega_d}\sin \omega_d t + y_0 \cos \omega_d t\right] \tag{12.27}$$

Comparing the above equation to Eq. (12.7) of the un-damped system, the free vibration of the damped system is continuously decayed by the factor $e^{-\zeta\omega_n t}$. Furthermore, the constant of the velocity term (first term) also includes the initial displacement.

(b) Free vibration on viscously damped system with initial displacement. Investigation of free vibration with an initially displaced system adds to our understanding of the basic dynamic characteristics of the structural system, either damped or un-damped. We have discussed the un-damped system previously. In this section, we will explore the nature of the damped system, specifically the under-damped system. By setting $\dot{y} = 0$ in Eq. (12.27), we have the free vibration of initially displaced system as

$$y = e^{-\zeta\omega_n t}\left(\frac{y_0\zeta\omega_n}{\omega_d}\sin \omega_d t + y_0 \cos \omega_d t\right) \tag{12.28}$$

The time-history displacement curve is as shown in Fig. 12.7. The expression inside the brackets can be treated as a cosine curve with the circular frequency, ω_d. Because of the sine component, the peaks will shift a little from the pure cosine curve — that is, the phase angle is somewhat different than the original cosine curve. However, because the sine component is relatively small, this shift is very small and can be ignored for practical purposes. In other words, for practical purposes, the first term inside the bracket can be ignored. That is, for the initially displaced system, we have

$$y = y_0 e^{-\zeta\omega_n t}\cos\omega_d t \tag{12.29}$$

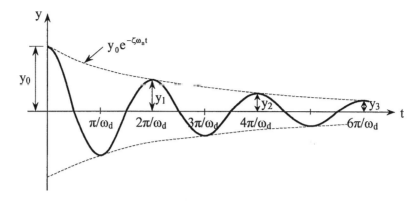

FIG. 12.7
FREE VIBRATION OF SDOF DAMPED SYSTEM WITH INITIAL DISPLACEMENT

The displacement is decayed with time by the time function $e^{-\zeta\omega_n t}$. Because the peaks occur at every $2\pi/\omega_d$ periods, the decay from a peak to the next peak can be expressed by the ratio of amplitudes at $t_n = 2n\pi/\omega_d$ and $t_{n+1} = 2(n+1)\pi/\omega_d$ as

$$\frac{y_n}{y_{n+1}} = \frac{e^{-\zeta(2n\pi)(\omega_n/\omega_d)}}{e^{-\zeta[2(n+1)\pi](\omega_n/\omega_d)}} = e^{2\pi\zeta(\omega_n/\omega_d)} \tag{12.30}$$

Taking the natural logarithm of both sides of Eq. (12.30), we have

$$\ln\left(\frac{y_n}{y_{n+1}}\right) = 2\pi\zeta\frac{\omega_n}{\omega_d} \tag{12.31}$$

Substituting the relation from Eq. (12.26), the preceding equation becomes

$$\ln\left[\frac{y_n}{y_{n+1}}\right] = \frac{2\pi\zeta}{\sqrt{1-\zeta^2}} \tag{12.32}$$

From Eq. (12.32), it is clear that the damping ratio ζ of a structural system can be determined if the amplitudes of the consecutive peaks can be measured. This is the main reason why in structural dynamics, the damping ratio instead of damping is given or specified. In general, the difference between two consecutive peaks is too small to measure accurately. In practice, the values of two peaks separated by a few cycles are measured for better accuracy. Assuming the two peaks are separated by j cycles, then we have

$$\ln\left[\frac{y_n}{y_{n+j}}\right] = \frac{2j\pi\zeta}{\sqrt{1-\zeta^2}} \cong 2j\pi\zeta \quad \text{for low damping systems}$$

(c) Alternative equation for SDOF damped vibration. The equation of the SDOF damped system can be written in an alternative form, which expresses the damping in a more familiar term of damping ratio. Dividing both sides of the equation by M and substituting $c/M = 2\zeta\omega_n$ and $k/M = \omega_n^2$, Eq. (12.3) becomes

$$\ddot{y} + 2\zeta\omega_n\dot{y} + \omega_n^2 y = \frac{F(t)}{M} \tag{12.33}$$

(d) Forced vibration on damped system. Damping has less effect on a short-duration impulse load to which the structure reaches the maximum response before going through cyclic oscillation. On the other hand, it has considerable effect on a harmonic load, which produces steady-state vibration.

For damped SDOF vibration, only the systems subject to harmonic loading will be discussed here to obtain a general idea of its characteristics. For a system with a harmonic forcing function, the working equation is

$$M\ddot{y} + c\dot{y} + ky = F_0 \sin \omega t \quad (12.34)$$

The solution of the homogeneous equation of the above equation is given in Eq. (12.25). With a forcing function, we need to find the particular solution that satisfies the whole equation. The particular solution to this type of force can be found by assuming $y = A \sin \omega t + B \cos \omega t$, with A and B to be determined by actual substitution to the equation. After substitution and reduction, the particular solution is found as [3]

$$y = \frac{F_0}{k} \frac{1}{(1-\beta^2)^2 + (2\zeta\beta)^2} \left[(1-\beta^2) \sin \omega t - 2\zeta\beta \cos \omega t \right] \quad (12.35)$$

where $\beta = \omega/\omega_n$.

The complete solution of Eq. (12.34) is the complementary function given by Eq. (12.27) plus the above particular solution for the harmonic forcing function. However, the complementary function represents system free vibration, which is damped out very quickly. Therefore, the free vibration portion is generally ignored when dealing with a harmonic forcing function, which produces high cycle steady-state vibration.

When dealing with a harmonic forcing function, only the particular solution as given by Eq. (12.35) is counted. To picture the meaning of the solution, we will combine the sine and cosine terms inside the brackets into one sinusoidal form with a shift of phase angle, that is, $\rho \sin(\omega t + \theta)$. This can be done via the following relations:

$$A \sin \omega t + B \cos \omega t = \rho \sin(\omega t + \theta) = \rho [\sin \omega t \cos \theta + \cos \omega t \sin \theta]$$

that is,

$$\rho \cos \theta = A, \quad \rho \sin \theta = B, \quad \text{or} \quad \rho^2 \cos^2 \theta = A^2, \quad \rho^2 \sin^2 \theta = B^2$$

or

$$\rho^2 (\cos^2 \theta + \sin^2 \theta) = \rho^2 = A^2 + B^2, \quad \text{and} \quad \frac{\rho^2 \sin^2 \theta}{\rho^2 \cos^2 \theta} = \tan^2 \theta = \frac{B^2}{A^2}$$

With $\rho = \sqrt{(1-\beta^2)^2 + (2\zeta\beta)^2}$, Eq. (12.35) can be rewritten as follows:

$$y = \frac{F_0}{k} \frac{1}{\sqrt{(1-\beta^2)^2 + (2\zeta\beta)^2}} \sin(\omega t + \theta) \quad (12.36)$$

$$\theta = \tan^{-1}\left(\frac{-2\zeta\beta}{1-\beta^2}\right) \quad \text{or} \quad \theta = -\tan^{-1}\left(\frac{2\zeta\beta}{1-\beta^2}\right) \quad (12.37)$$

The displacement has the maximum values when $\sin(\omega t + \theta) = \pm 1.0$. Because F_0/k is the maximum static displacement, we have the maximum magnification on the displacement, or the maximum DLF, as

$$(\text{DLF})_{\max} = \frac{y_{\max}}{F_0/k} = \frac{1}{\sqrt{(1-\beta^2)^2 + (2\zeta\beta)^2}} \quad (12.38)$$

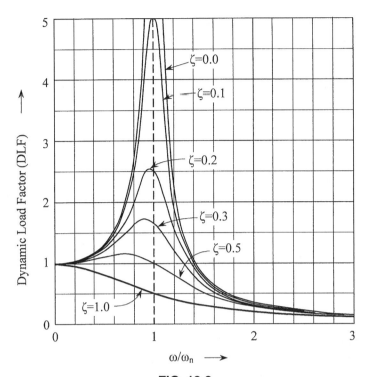

FIG. 12.8
MAGNIFICATION FACTORS OF HARMONIC LOAD ON SDOF SYSTEMS

The maximum DLF depends on the damping ratio and the frequency ratio. The values for some damping ratios are plotted in Fig. 12.8. When $\beta = \omega/\omega_n = 1$, the force resonates with the system. The situation is called resonance, which produces a very high magnification on the response. At resonant state, $\beta = 1$, so the DLF $= 1/(2\zeta)$. Without damping, the DLF at resonance state is infinite as previously found in Eq. (12.20). For piping systems, the damping ratio varies from 2% to 8% depending on the pipe size, insulation, and frequency. With 5% damping, the DLF is 10 at resonance. A small oscillating force can produce a very large displacement at this resonance state.

12.2.4 Summary of the Characteristics of SDOF Vibration

The characteristics of SDOF vibration are also applicable to general structural systems with many degrees of freedom. These basic characteristics of structural dynamics are summarized in the following:

(1) The most fundamental characteristic of a structural system is its natural frequency, which is proportional to the square root of the stiffness and is inversely proportional to the square root of the mass. A stiff system has a higher natural frequency, but a large mass reduces the natural frequency.
(2) Damping in a practical structural system has very little effect on natural frequency. Therefore, damping is generally ignored in calculating natural frequency.
(3) The response to an impulse load depends on the ratio of the load duration and the natural period of the system. With a short duration, the system does not have time to fully respond to the load before the load disappears. The DLF for the suddenly applied constant load is 2, but can be close to zero with a very short duration impulse.

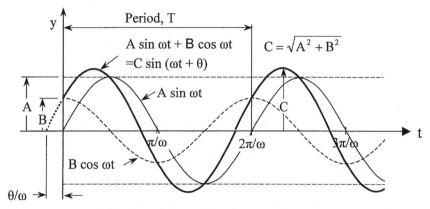

(a) Combination of sine and cosine waves with same frequency

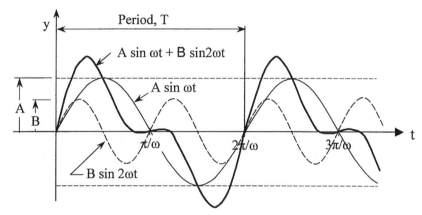

(b) Combination of two sine waves with different frequency

**FIG. 12.9
COMBINATION OF SINE AND COSINE WAVES**

(4) The structural response to an external dynamic load consists of two parts: forced vibration and natural free vibration. Both parts are significant for short impulse loads, but only the forced vibration portion survives in a long-lasting harmonic or periodic load that leads to a steady-state vibration.

(5) For harmonic or periodic loading, the structural response can be very large when the forcing frequency is the same or near the natural frequency of the structural system. The state when the forcing frequency is the same as the natural frequency is called resonance.

(6) The combination of a sine and a cosine wave with identical frequency is a pure sine or cosine wave with bigger amplitude and a shifted phase angle. That is, $A \sin \omega t + B \sin \omega t = C \sin(\omega t + \theta)$ as shown in Fig. 12.9(a). On the other hand, the combination of sine and cosine waves with non-identical frequency leads to distorted waves. Figure 12.9(b) shows that the combination of two sine waves having one frequency twice the other frequency does not result in a pure sine wave. This property affects the procedure of harmonic analysis to be discussed later.

12.3 DAMPING

All motions encounter some type of resistance; otherwise, the world would have no order. The resistance to piping movement is lumped together and called damping. Damping is roughly classified into two categories: viscous damping and Coulomb damping. Viscous damping has a resistance force proportional to the relative velocity of the affected parts, whereas Coulomb damping has a resistance force proportional to the contact force between affected parts. The damping included in the customary dynamic equation, such as Eq. (12.3), is viscous damping. However, the damping value used in actual practice does include a portion of Coulomb damping. The design damping value is taken based on experimental data, which is more of the combined effect than just the pure viscous or Coulomb damping.

The design damping value is generally specified as a percentage of critical damping, which is defined in Eq. (12.24). It is clear from Fig. 12.8 that the response of the structure to a dynamic force varies considerably with the damping value. Therefore, proper selection of the design damping value is very critical to the design of structural systems including piping systems.

Damping in piping depends on pipe material, size, insulation, fluid content, support type, and so forth. It also varies with the stress level of the material and deflection of the piping. In other words, the accurate damping value is very difficult to obtain. Conservative values should be used if no accurate data is available. However, excessive conservatism may also result in very impractical designs.

Routine dynamic analysis of piping systems started during the boom time of nuclear power plant construction in the early 1970s. At that time, the damping for piping systems was mainly based on expected stress levels as proposed by Newmark [4]. The recommendations were officially adopted by the Nuclear Regulatory Commission (NRC) Regulatory Guide 1.61 [5] in its initial issue in 1973. At that time, the damping for piping was given as shown in Table 12.1. Many nuclear power plants have been built with this criterion. At that time, roughly 1000 snubbers [6] were required for each unit of the nuclear power plant to ensure that the piping was safe under the specified earthquake design criteria including the damping. These 1000 snubbers not only took up substantial construction costs, but also prolonged the scheduled maintenance downtime. To ensure the functionality of these snubbers, inspection and testing were required at each scheduled downtime. Because a lot of snubbers were located in very tight places with limited access, the inspection and testing could take longer than the time required for refueling and other maintenance. Therefore, it was important that the use of snubbers be limited to the minimum number required.

There are several avenues that can be used to reduce the number of snubbers. One is to use more accurate calculations such as time-history analyses and the analyses using multiple support response spectra instead of envelope spectra. (See the next chapter for response spectra earthquake analysis). The other is to optimize the system by using as many rigid struts, instead of snubbers, as possible. Another option is to reduce the design criteria. Where the design criteria are concerned, however, the stress limit is pretty much set and has little room for relaxation. The damping, on the other hand, is still not a very well defined quantity and shows a great prospect for fine adjustment.

TABLE 12.1
DAMPING IN PIPING (AS PERCENT OF CRITICAL DAMPING)

Newmark Recommendation [4]		U.S. NRC Regulatory Guide 1.61, 1973 [5]		
Stress category	Damping	Load category		Damping
Stress below ¼ yield point	0.5			
Stress no more than ½ yield point	1.0	Operation Basis Earthquake	$D > 12$ in. $D \leq 12$ in.	2.0 1.0
Stress just below yield point	2.0	Safe Shutdown Earthquake	$D > 12$ in. $D \leq 12$ in.	3.0 2.0
U.S. NRC Regulatory Guide 1.61, Revision 1, 2007 [5]				
Operation Basis Earthquake	Damping 3.0	Safe Shutdown Earthquake		Damping 4.0

To find out the possibility of increasing the design damping, direct experiments seem to be the most straightforward approach. Because the damping value affects millions of dollars in construction and maintenance costs, there is a very good incentive for the utilities to be involved. For example, Japanese utilities were all very enthusiastic about the experiments because they have used very low damping of 0.5% in the design of their piping systems [7]. Considerable amount of tests have been performed both in the United States and overseas. Based on available test data, WRC Pressure Vessel Research Committee published a summary [8] in 1984, supplemented with additional information in 1986 [9]. Based on these WRC summaries and other considerations, U.S. NRC revised the damping values in Regulatory Guide 1.61 [5] Revision 1, which was published in 2007. The revised values, also included in Table 12.1, are 4% for safe shutdown earthquake (SSE) and 3% for operation basis earthquake (OBE) conditions. Higher damping may be allowed subject to restrictions given by the Regulatory Guide 1.61. For instance, the frequency-dependent damping as shown in Fig. 12.10 can be used for response spectra analysis using the envelope spectra method, which is considered the most conservative method in analyzing earthquake response. Figure 12.10, if used, should also be used in its entirety throughout the frequencies and analyses.

The above damping values are mainly for earthquake analysis. For other types of loads, some adjustment may be required. For steady-state vibration analysis, for instance, a lower value should be used because of the small stress and displacement amplitude involved.

12.4 SONIC VELOCITY VERSUS FLOW VELOCITY

Piping design also involves loads from fluid transient phenomena. Pipeline surge load, safety relief valve discharge load, and steam turbine trip load are some of the often-encountered fluid transient loads. To effectively handle these loads, some basic understanding of fluid dynamics is required. One of the most fundamental characteristics of fluid dynamics related to piping analysis is the sonic velocity. Because this book permits only a brief introduction of applicable formulas, interested engineers should consult specialized textbooks [10–12] for detailed discussions and derivations of these formulas.

There are two distinct velocities in fluid that are of interest to piping engineers: flow velocity and pressure wave or sonic velocity. Flow velocity is the velocity of bulk fluid substance. This is the one related to the flow rate that we are all familiar with. On the other hand, sonic velocity is somewhat

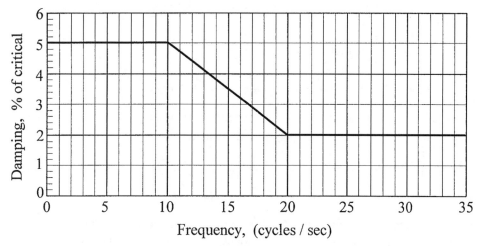

FIG. 12.10
FREQUENCY-DEPENDENT DAMPING FOR PIPING [5]

abstract, but visible. The two velocities can be explained with domino tiles, which many of us have played with before. The physical movement of the domino tile is the flow velocity, but the transmission of the movement is the wave velocity. Normally, the tiles just fall down but do not move forward; therefore, the flow velocity is zero. However, the fall down motion, if transmitted forward very quickly, creates a high wave velocity. Generally, flow velocity and sonic velocity are two independent phenomena. They can be at different magnitudes and also in opposite directions.

12.4.1 Sonic Velocity

To provide some idea about sonic velocity, we shall start with the fundamentals of gas properties. We will mainly deal with ideal gases. Some adjustments are then made for real gases. In an ideal gas, there is the basic relationship of the states consisting of pressure, volume, and temperature. For a unit mass of gas, we have

$$\frac{p_1 v_1}{T_1} = \frac{p_2 v_2}{T_2} = R$$

or

$$pv = \frac{p}{\rho} = RT \qquad (12.39)$$

This basically says that, for a given quantity (measured by its mass) of gas, if the temperature is maintained at a constant level, then squeezing the volume will increase the pressure. If the pressure is maintained at a constant value, then the volume increases as the temperature increases. This is just the thermal expansion phenomenon experienced in most substances. The interesting thing is that they are all simply related without any odd exponential. The constant, R, is called the gas constant, which is inversely proportional to molecular weight. That is,

$$R = \frac{\mathcal{R}}{M_W} \qquad (12.40)$$

where $\mathcal{R}$ is the universal gas constant, and M_W is molecular weight. The universal gas constant has a value of

$$\mathcal{R} = 1545.4 \frac{\text{ft.lbf}}{°\text{R(lbm} - \text{mole)}} = (1545.4)(32.2) \frac{\text{ft.lbf}}{°\text{R(slug} - \text{mole)}} \qquad \text{(English units)}$$

$$\mathcal{R} = 8312 \frac{\text{m.N}}{°\text{K(kg} - \text{mole)}} \qquad \text{(S.I. units)}$$

Equation (12.39) is valid only with absolute temperatures, defined as

$$T = 459.67 + °F = °R \qquad \text{(English units, degree Rankine)}$$
$$T = 273.15 + °C = °K \qquad \text{(SI units, degree Kelvin)}$$

Sonic velocity, relative to the fluid through which it is passing, can be calculated by

$$a = \sqrt{dp/d\rho} = \sqrt{kp/\rho} = \sqrt{kRT} = \sqrt{k\mathcal{R}T/M_W} \qquad (12.41)$$

which assumes that the wave propagation front goes through an isentropic process with $p\rho^{-k} =$ constant. k is the ratio of the constant pressure specific heat to constant volume specific heat. As shown in Fig. 12.11, when heating a mass of gas under constant volume, the heat needed is just that for

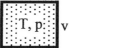 

(a) Constant Volume   (b) Constant Pressure

FIG. 12.11
HEATING THE GAS

increasing the internal energy. However, if the heating is under constant pressure as shown in Fig. 12.11(b), the heat needed is that for increasing the internal energy, as in constant volume heating, plus the equivalent of heat required for pushing the expanding volume through the pressure resistance. The latter portion is the heat equivalent of the work $p(\Delta v)$. Therefore, the amount of heat required to increase the gas by 1° under constant pressure conditions is higher than the heat required to increase the same amount of gas by 1° under constant volume conditions. The ratio of the two heats required is called the specific heat ratio, k, which can be theoretically expressed as

$$k = \frac{n+2}{n} \tag{12.42}$$

where n is the number of degrees of freedom of the gas molecule. For monatomic gases, such as argon, helium, and neon, $n = 3$ with $k = 5/3$. For diatomic gases, such as hydrogen, oxygen, and nitrogen, $n = 5$ with $k = 7/5$. Air is mainly composed of oxygen and nitrogen, and has a k value of 1.4. Superheated steam has a k vale of about 1.3, whereas saturated steam has a k value of about 1.1.

From Eq. (12.41), it is clear that sonic velocity is proportional to the square root of the absolute temperature. The gas has a higher sonic velocity at a higher temperature. However, the velocity change from the temperature is not very drastic because the absolute temperature scale changes more gradually in proportion to a conventional temperature scale. The equation also shows that the velocity of sound is inversely proportional to the square root of the molecular weight. This effect of molecular weight is more drastic as the molecular weight of the gas varies considerably among different types of gases. For instance, the molecular weight of hydrogen is 2, and for oxygen it is 32. Therefore at the same temperature, the speed of sound in hydrogen gas is 4 times the speed of sound in oxygen gas.

The speed of sound is calculated with universal gas constant $\mathcal{R}$. In English units, because the conventional force/mass units are not consistent, we have two values that may be used. The one to use depends on the subject of discussion. For static thermodynamics, we generally use the pound mass mole, so the specific volume is based on the volume per unit weight of gas. In fluid mechanics, such as calculating the velocity of sound, we have to use the slug-mole constant, because lbf = slug · ft/sec². To get some idea of the magnitude, the velocity of sound in air (molecular weight = 29) at ambient temperature 70°F (21°C) is

$$a = \sqrt{1.4(1545.4)32.2(70+459.67)/29} = 1128 \text{ ft/sec} \qquad \text{(English units)}$$

$$a = \sqrt{1.4(8312)(21+273.15)/29} = 344 \text{ m/sec} \qquad \text{(SI Units)}$$

For hydrogen gas having a molecular weight of 2, the velocity of sound is 4295 ft/sec (1310 m/sec) at ambient temperature. This is comparable to the velocity of sound in liquid as given in the following.

Equation (12.41) is used mainly for gasses, but is equally applicable to liquids, which do not have a universal liquid constant. Without using the gas constant, the original sonic velocity for any fluid as given by Eq. (12.41) can be re-written as

$$a = \sqrt{\frac{dp}{d\rho}} = \sqrt{\frac{1}{\rho}\left(\rho\frac{dp}{d\rho}\right)} = \sqrt{\frac{1}{\rho}\left(\frac{dp}{vd(1/v)}\right)} = \sqrt{\frac{1}{\rho}\left(\frac{dp}{(-v/v^2)dv}\right)} = \sqrt{\frac{K}{\rho}} \quad (12.43)$$

where $K = -dp/(dv/v)$ is the bulk modulus of the liquid. The bulk modulus at ambient temperature for freshwater is 319,000 psi; for seawater, 344,000 psi with a specific gravity of 1.03; for kerosene, 209,000 psi with a specific gravity of 0.81; for average crude oil, 219,000 psi with a specific gravity of 0.91. The velocity of sound in freshwater at ambient temperature is

$$a = \sqrt{\frac{319,000 \times 144}{62.4/32.2}} = 4868 \text{ ft/sec } (1484 \text{ m/sec})$$

Similarly, the sonic velocity of seawater is 4990 ft/sec (1521 m/sec) and that of crude oil is 4240 ft/sec (1293 m/sec). It is clear that the sonic velocity of common liquid fluids changes only slightly, and is roughly the same as the sonic velocity of hydrogen gas. The flexibility of the pipe wall tends to reduce the sonic velocity of liquid inside a pipe. The reduction depends mainly on diameter to thickness ratio. For steel pipe with $d/t = 50$, the reduction is about 20% [10] and the reduction would be 30% if $d/t = 100$.

12.4.2 Flow Velocity

In a piping system, whether the flow velocity reaches the sonic velocity or even surpasses the sonic velocity depends mainly on the fluid medium it carries. In liquid applications, it is safe to say that flow velocity never reaches the value of sonic velocity because the sonic velocity is very high in liquid and the high density of liquid requires very a high energy to drive it to sonic velocity. On the other hand, the flow of gas reaches sonic velocity rather easily, but only surpasses sonic velocity by coincidence. To reach supersonic velocity, the flow has to reach sonic velocity first, and then be guided by a properly contoured expanding nozzle to further increase the velocity. As supersonic flow creates shock waves when the velocity is reduced, supersonic flow should be avoided by all means in piping applications. Therefore, the favorable contour of the passage is never provided. The only chance of achieving supersonic velocity is by coincidence. This section deals mainly with the flow velocity of gases and vapors.

Figure 12.12 shows a simple model of a gas flow system. The system is assumed to be adiabatic with no heat transfer. Pressure and temperature at the upstream reservoir are maintained at a constant value, and the fluid is stagnant with zero velocity at the reservoir. The flow is controlled by the downstream backpressure, P_B, regulated by the gate valve. The gas flows through the nozzle when backpressure P_B is lower than the upstream pressure P_0. Initially the flow velocity varies with some function of the pressure ratio P_0/P_B. As backpressure reduces to a certain value, the velocity reaches sonic velocity at the throat. Once it reaches sonic velocity, the velocity will not increase any further with further reduction in backpressure. This is known as the choking phenomenon. The least pressure difference that produces a sonic flow velocity depends on the magnitude of the upstream pressure. The ratio of downstream and upstream pressure at this transition point is called the critical pressure ratio. The critical pressure ratio and the critical temperature ratio depend only on the specific heat ratio, k. Using the * (superscript) to denote the sonic critical condition, we have [11]

$$\frac{P^*}{P_0} = \left(\frac{2}{k+1}\right)^{k/(k-1)} = 0.5283, \quad \text{for } k = 1.4 \quad (12.44)$$

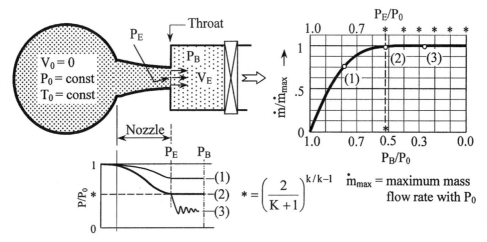

FIG. 12.12
GAS FLOWS THROUGH CONVERGENT NOZZLE [11]

$$\frac{T^*}{T_0} = \frac{2}{k+1} = 0.8333, \quad \text{for } k = 1.4 \tag{12.45}$$

These are the values for diatomic gases such as hydrogen and oxygen with $k = 1.4$. For superheated steam with $k = 1.3$, the critical pressure ratio is 0.546, whereas saturated steam with $k = 1.1$ has a critical pressure ratio of 0.585. For monatomic gases such as argon and helium with $k = 1.67$, the critical pressure is 0.4867. As a rule of thumb, 0.5 can be used as a general common sense value. In other words, for gas or vapor, if the upstream pressure is more than twice of the downstream pressure, then sonic velocity flow is expected. The steam flow through a safety valve to the atmosphere is generally discharging at sonic velocity. The flow at a pressure reducing station is also generally sonic unless multi-stage reduction and energy dissipation devices are provided.

When P_B/P_0 is lower than the critical pressure ratio, the flow is sonic and at the same time has an exit pressure, P_E, which is higher than the backpressure, P_B. With further lowering in backpressure, P_B, P_E/P_0 maintains the same value as the critical pressure ratio regardless of the magnitude of the backpressure. Therefore, it is important that this exit pressure is included in the design of the piping subject to safety valves and other exit forces. The exit pressure eventually expands away to equalize with the backpressure.

The flow has two distinctly different stages. When the pressure ratio is greater than the critical ratio, that is, the backpressure is close to the upstream pressure, the flow is subsonic, and the flow rate is calculated by [12]

$$\left(\frac{\dot{m}}{A}\right) = \frac{P_0}{\sqrt{T_0}} \sqrt{\frac{M_W}{R}\left(\frac{2k}{k-1}\right)\left(\frac{P_B}{P_0}\right)^{\frac{2}{k}}\left[1 - \left(\frac{P_B}{P_0}\right)^{\frac{k-1}{k}}\right]} \tag{12.46}$$

The flow rate reaches the maximum when the pressures reach the critical pressure ratio. This maximum flow rate per unit throat area is independent of the backpressure, and is calculated by $\rho^* a$ as

$$\left(\frac{\dot{m}}{A}\right)_{max} = \frac{P_0}{\sqrt{T_0}} \sqrt{\frac{M_W k}{R}\left(\frac{2}{k+1}\right)^{\frac{k+1}{k-1}}} = CP_0\sqrt{\frac{M_W}{T_0}} \tag{12.47a}$$

$$\text{with} \quad C = \frac{1}{\sqrt{R}} \sqrt{k \left(\frac{2}{k+1} \right)^{\frac{k+1}{k-1}}} \qquad (12.47b)$$

where $\dot{m}$ is the mass flow rate and A is the nozzle throat area. All variables are in consistent units. Equation (12.47a) is used for ideal gases. For real gases or vapors on a non-perfect nozzle, a compressibility factor, z, and nozzle efficiency, η, have to be added. That is, for real gases or vapors on a real nozzle, we have

$$\left(\frac{\dot{m}}{A} \right)_{max} = \eta C P_0 \sqrt{\frac{M_W}{T_0 z}} \qquad (12.48)$$

where the compressibility factor, z, is related to Eq. (12.39) as $pv = zRT$. The z value is close to 1.0 for low-pressure dry gases, and in the range of 0.8 for high-pressure superheated steam applications. We shall only recognize the existence of this z value, which is available in most engineering handbooks.

It should be noted that some industrial standards [13, 14] use conventional units requiring an adjustment factor, J_A, on Eq. (12.47b). That is, for non-consistent units, we use

$$\left(\frac{\dot{m}}{A} \right)_{max} = \eta C' P_0 \sqrt{\frac{M_W}{T_0 z}} \qquad (12.49)$$

$$\text{with} \quad C' = J_A C = \frac{J_A}{\sqrt{R}} \sqrt{k \left(\frac{2}{k+1} \right)^{\frac{k+1}{k-1}}}$$

When U.S. conventional units are used with $\dot{m}$ in lbm/hr, A in in.^{2}, P_0 in psi; the J_A required is the proper one that converts these units to slug/sec, ft^2, and lbf/ft^2 consistent units, respectively. That is, $J_A = (32.2 \times 3600) \times (1/144) \times (144) = 115{,}920$, or

$$\frac{J_A}{\sqrt{R}} = \frac{115{,}920}{\sqrt{1545.4 \times 32.2}} = 520 \quad \text{(for U.S. conventional units)}$$

For maximum flow rate of steam, there is a simple formula frequently used. Based on experimental data, in the 19th century Napier had discovered that the maximum steam flow rate is proportional to the product of the upstream pressure and cross-sectional flow area. That is,

$$\left(\frac{\dot{m}}{A} \right)_{max} = K_N P_0 \quad \text{(Napier's law)}$$

where K_N is the proportional constant, or Napier constant. Based on Eq. (12.48), Napier's law is justified nowadays by the fact that in a certain range of steam temperatures and pressures, the product of absolute temperature and compressibility factor, $(T_0)z$, remains fairly constant. In other words, the compressibility factor reduces at about the same rate as the increases in absolute temperature. In this case, Eq. (12.48) can be simplified as $(\dot{m}/A) = K_N P_0$. However, because the value of $(T_0)z$ is only approximately constant through a small range of temperatures and pressures, Napier's law is good only for quick approximate estimates. Furthermore, the K_N constant is not a pure constant, but has a dimension of the inverse of velocity. Therefore, it has different values for different units used. With U.S. conventional units of lbm/hr, in.^{2}, and psi for flow rate, flow area, and pressure, respectively, $K_N = 51.45$ [15]. If lbm/sec is used for the flow rate, then $K_N = 0.01429 = 1/70$.

The flow rate in a pipe can reach supersonic velocities when there is a diverging passage following a converging nozzle or orifice. Figure 12.13 shows the operating characteristics of a converging-diverging nozzle. It also shows the pressure distribution along the nozzle and flow rate against the backpressure P_B. The flow rate is shown as varying from backpressure conditions (1) through (5). Condition (1) has a relatively high backpressure as compared to the upstream reservoir pressure. In this case, the

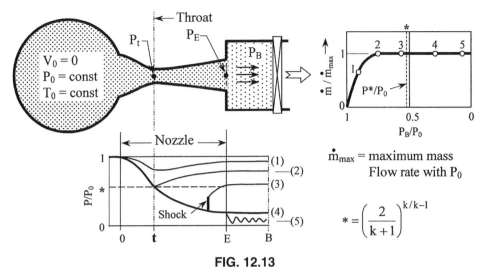

FIG. 12.13
GAS FLOW THROUGH CONVERGENT-DIVERGENT NOZZLE [11]

flow rate increases as the backpressure reduces. The minimum pressure occurs at the throat area. The flow is subsonic throughout the nozzle. As the backpressure reduces to condition (2), the pressure at the throat reaches critical pressure, and the flow is sonic at the throat, but the flow is still subsonic elsewhere. The flow rate reaches the maximum at this condition for the given reservoir pressure, P_0. The backpressure is higher than the critical pressure due to the recompression effect of the divergent section. With further lowering of the backpressure, the pressure at the throat maintains at critical pressure but the fluid expands further downstream, reducing the pressure below the backpressure. This creates supersonic flow in the divergent section before a shock occurs to reduce the flow into subsonic and compresses the fluid up to backpressure. Condition (4) is the design case for a smooth supersonic flow.

We will not get into the details of the shock wave and supersonic flow. What we are interested in is some of the characteristics of convergent-divergent nozzles. Supersonic flow is achieved only through a convergent-divergent nozzle. The divergent section of the nozzle does not increase the flow rate, which is determined by the reservoir pressure and temperature as given by Eq. (12.47a). However, the divergent section does allow higher backpressure for which the maximum flow rate is attained.

12.5 SHAKING FORCES DUE TO FLUID FLOW

In addition to the static loads, a piping system also needs to be designed to resist the shaking forces generated by upset fluid flow conditions. Here, we emphasize the shaking force instead of just the force, because only the unbalanced shaking force generates sensible movement of the system. The forces that do not have the shaking effect are called static or balanced forces.

As shown in Fig. 12.14, each leg of a piping system is subject to a potential shaking force. A shaking force due to fluid flow is in the axial direction of the piping leg. The forces are acting at places where a change of direction occurs. In this system, the potential damaging axial forces are acting on legs 2-3 and 3-4, because they are acting in the most flexible directions and locations of the system. Since the fluid flow in the pipe is generally considered one-dimensional along the centerline of the piping, the fluid force at each piping leg is also considered one-dimensional. In a piping leg lying in the x direction as shown in Fig. 12.15, the equilibrium of all pertinent x direction forces results in the following:

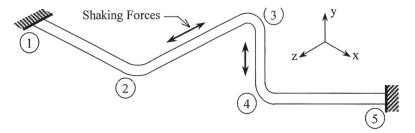

FIG. 12.14
FORCES THAT CAUSE THE SYSTEM TO SHAKE

$$M_\mathrm{P}\ddot{x} + \frac{d}{dt}\int_v V_x \rho\, dv + C\dot{x} + Kx = 0$$

where M_p is the mass of the pipe material of the pipe leg, and C and K are damping and stiffness of the leg, respectively. The quantity inside the integral is the total x momentum of the fluid inside the pipe leg. Its time rate of change is the main shaking force of the leg. The shaking force is generated by any change in the velocity, density, or the total mass contained. It is clear that in a steady-state flow, there is no shaking force generated because velocity, density, and total mass are not changed.

Because fluid velocity, V_x, is influenced by the displacement, velocity, and also the acceleration of the pipe, there are interactions between fluid dynamics and structural dynamics. However, for general piping applications, the fluid flow is calculated assuming that the pipe is fixed. That is, that the thermo-hydraulics of the fluid inside the pipe is calculated disregarding the movement of the piping. After the flow condition is so calculated, the fluid forces acting at the pipe leg are determined as shown in Fig. 12.16. As this 2-3 leg is located in the z direction, only the z-direction fluid forces exist. This is based on the one-dimensional flow concept adopted by piping engineers. Because the thermo-hydraulic fluid flow analysis generally provides only the profile of pressure, temperature, velocity, and density, along the entire system, the shaking force of each leg is estimated from these quantities. Here, we have two main forces acting at each end of the pipe leg, and the overall friction force acting on the inside surface of the pipe. At each end of the pipe leg, the acting force consists of the pressure force and momentum force, two parts calculated as follows:

Pressure force: $F_\mathrm{P} = PA$

Momentum Force: $F_\mathrm{M} = \dot{m}V = \rho A V^2$

The friction force is given as

Friction force: $F_\mathrm{F} = \tau \pi D L$

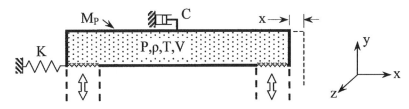

FIG. 12.15
FORCES ON PIPING ELEMENT

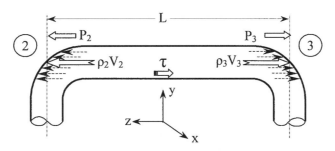

**FIG. 12.16
AXIAL FORCES ACTING ON A PIPING LEG**

where τ is the friction shear stress per unit inside surface area of the pipe and L is the length of pipe segment in consideration. The momentum force is acting toward the elbow at both ends. The sum of F_P and F_M can be considered as the total end force.

In the leg shown in Fig. 12.16, the shaking force at a certain instant is the difference of the two end forces minus the friction force. Assuming the fluid flows from 2 to 3, the net shaking force of the leg is calculated by

$$F_x = P_2 A + \rho_2 V_2^2 A - P_3 A - \rho_3 V_3^2 A - \tau L \sqrt{4\pi A} \qquad (12.50)$$

Since the shaking force results only from transient flow, the thermo-hydraulic analysis often calculates only the fluctuation quantities off from the mean flow. In practical applications, not all terms in the above equation are considered. The friction term is often ignored because it is produced by the total main flow rate. The momentum terms are often ignored when dealing with traveling surge pressure waves.

12.6 SAFETY VALVE RELIEVING FORCES

One of the most common dynamic fluid forces often encountered in piping is the relieving force from a safety relief valve. Safety valve relieving systems are generally divided into two categories: open discharge and close discharge. In an open discharge system, the fluid is simply discharged into the atmosphere. The closed discharge system collects the discharged fluid in a drum or header for proper recycling or disposal. The fluid force is treated differently in each type of discharge system.

12.6.1 Open Discharge System

For non-toxic, non-hazardous fluids, the over-pressured fluid may be discharged to the atmosphere either directly or through a separate vent pipe or silencer. Figure 12.17 shows the most basic installation of the open discharge safety valve. The over-pressured fluid is simply discharged into the atmosphere. No bent pipe or extension piping is connected. In this case, the most apparent dynamic force is the reaction of the discharge fluid momentum. Because the operating pressure is generally much higher than twice the atmospheric pressure, the flow is sonic at the valve orifice location, and is most likely either sonic or supersonic at the valve elbow exit location. This is a critical flow condition. As shown in Fig. 12.12, the exit pressure is higher than the atmospheric pressure with a critical flow condition. Therefore, a pressure force also exists at the end of the safety valve elbow. The total force at the end of the discharge elbow becomes

$$F_1 = \dot{m} V_1 + (P_1 - P_a) A_1 = \rho_1 A_1 V_1^2 + (P_1 - P_a) A_1 \qquad (12.51)$$

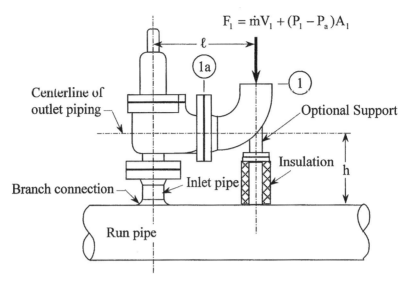

FIG. 12.17
SAFETY VALVE RELIEVING FORCES

where P_1 is the exit pressure and P_a is the atmospheric pressure. Subscript 1 corresponds to the end of the safety valve elbow location. The combination of $(\rho A V^2 + PA)$ is also called the impulse function, or the total end force, as often referred to by piping engineers.

The preceding equation requires the flow rate, flow velocity, and safety valve elbow exit pressure. The valve vendor generally supplies the flow rate, so the main task here is to find the elbow exit pressure and the exit velocity. As the flow inside the valve chamber is a very complicated phenomenon involving sonic, shock, supersonic, and re-compression phenomena, the calculation of this exit pressure and velocity is complex and uncertain. The inconsistent units used by some of the calculation formulas may also add to the confusion. Therefore, it is useful to find a means to estimate these quantities quickly and conservatively, or, as an alternative, to avoid calculating them at all. Because the valve elbow is short, the friction can be assumed to be small. In this case, the flow inside the valve chamber and elbow is isentropic, which preserves the impulse function based on the momentum equation. That is,

$$F_1 = \dot{m}V_1 + (P_1 - P_a)A_1 = \dot{m}V^* + P^*A_T$$

where superscript * denotes sonic condition at throat or valve orifice, and A_T is the valve orifice flow area. The mass rate is the same at both the orifice and the elbow exit. $V^* = a$ is the sonic velocity at the throat. Substituting flow rate, sonic velocity, and critical pressure from Eqs. (12.47a), (12.41), and (12.44), respectively, to the above relation, we have the force as calculated at the orifice location as

$$F_1 = \frac{P_0 A_T}{\sqrt{T_0}} \sqrt{\frac{M_W k}{R}\left(\frac{2}{k+1}\right)^{\frac{k+1}{k-1}}} \left(\sqrt{\frac{kRT^*}{M_W}}\right) + \left(\frac{2}{k+1}\right)^{k/(k-1)} P_0 A_T$$

$$= P_0 A_T \left(\sqrt{\frac{k^2 T^*}{T_0}\left(\frac{2}{k+1}\right)^{\frac{k+1}{k-1}}} + \left(\frac{2}{k+1}\right)^{k/(k-1)}\right)$$

Substituting $T^* = \dfrac{2}{k+1} T_0$ from Eq. (12.45), the preceding equation becomes

$$F_1 = P_0 A_T (1+k)\left(\frac{2}{k+1}\right)^{\frac{k}{k-1}} = P_0 A_T C_F \qquad (12.52)$$

where C_F is the reaction force coefficient, which is dependent only on the specific heat ratio, k. $C_F = 1.268$ for $k = 1.4$; $C_F = 1.255$ for $k = 1.3$; and $C_F = 1.228$ for $k = 1.1$. C_F increases with an increase in k. At the maximum k value of 1.67 for monatomic gases, the maximum $C_F = 1.30$. That is, the force coefficient, C_F, varies in a narrow range of between 1.20 and 1.30.

Equation (12.52) depends on the upstream stagnation pressure and the valve orifice flow area. The orifice flow area and flow rate are the items generally supplied by the valve vendor. When using the vendor supplied flow rate to calculate the reaction force, it is important to note that the vendor flow rate is generally taken as 90% of the maximum rate. Therefore, the vendor's flow rate or the specified flow rate has to be increased by 1.11 times before being used in the calculation of the reaction force. Some designers would increase this value by another 10% for the variation of the test results. If the orifice flow area is not known, then it can be calculated from the flow rate using Eq. (12.47a).

For water steam applications, ASME B31.1 [16] has given a fairly comprehensive cookbook method for the design and analysis of the related piping. B31.1 uses semi-empirical formulas and the steam table to calculate P_1 and V_1 from the flow rate and stagnation enthalpy of the inlet steam. The current (2008) edition uses U.S. conventional units for these calculations. It is recommended that engineers working with other measurement units calculate the force with the code-specified units, rather than converting the formulas into appropriate units. If required, the calculated pound force could then be converted into the desirable unit of force. B31.1 also includes the procedure to design a vent stack, if one needs to be provided. Because the code is constantly revised, engineers should consult the latest edition of the code for details.

The general shape of the reliving force time history is shown in Fig. 12.18. Due to the inertia of the fluid column before the valve, the flow starts out proportionally more slowly than the actual valve opening, but overshoots somewhat when the valve is fully opened. This is typical for all natural phenomena. The force calculated from Eq. (12.52) or the B31.1 method is the sustained steady-state force. In many designs, the force history is idealized as a ramp-sustained force without going through detailed thermo-hydraulic analysis. The actual force or the idealized ramp force can be directly applied in a time-history analysis (see next chapter for time history analysis). To apply the idealized ramp force statically, a DLF as given in Fig. 12.5 has to be applied. The DLF depends on the ratio of the ramp time and system natural period. The natural period of the system is given by Eq. (12.8), and can be approximately calculated as

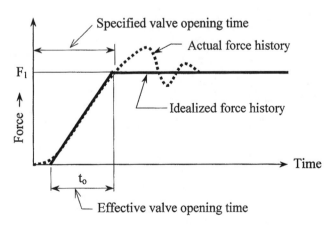

FIG. 12.18
TIME HISTORY OF SAFETY VALVE RELIEVING FORCE

$$T = 2\pi\sqrt{\frac{M}{k}} = 2\pi\sqrt{\frac{Mh^3}{3EI}} \qquad (12.53)$$

which assumes a mass-less uniform pipe cantilever model with a concentrated mass located at the end. M is the total mass, in consistent unit, of the safety valve assembly including the valve, piping, flanges, and attachments. If pound-mass, inch, second, and pound-force units are used as in B31.1, the equation becomes

$$T = 2\pi\sqrt{\frac{Wh^3}{32.2 \times 12 \times 3EI}} = 0.1845\sqrt{\frac{Wh^3}{EI}} \quad \text{(pound-mass, sec, inch, pound-force units)}$$

where W is the weight in pound-mass unit.

The period given by Eq. (12.53) is for the mode of vibration that is in the horizontal direction perpendicular to the force. Although it is not directly correlated with the vertical force, it is considered conservative, because the lateral vibration period is mostly longer than the axial period. A longer natural period corresponds to a higher DLF. Because the calculated period is only approximate and is not directly applicable, some compensation is needed for proper application of the force. B31.1 has enveloped Fig. 12.5 to construct a design curve as shown in Fig. 12.19.

Another parameter for determining the DLF is the ramp time, which can be taken as 75% of the full valve opening time as shown in Fig. 12.18. The full valve opening time can be obtained from most vendors. However, it should be noted that the often-assumed 0.040-second opening time might require some adjustment. Although the use of 0.040-second was based on previous test data, some later tests have shown shorter opening times. Based on some tests conducted in 1982, Auble [17] found that the typical opening time for Crosby [18] valves was 0.010 second, and for Dresser [19] valves was 0.015 second. The tests conducted by Wheeler and Siegel [20], also in 1982, showed an opening time of about 0.030 second for Crosby valves. Because a shorter opening time results in a higher DLF, the specific vendor has to be consulted for the most recently established opening time.

The valve discharge force produces a bending moment on the piping and especially on the branch connection, which is a high stress concentration point. To reduce this bending moment, the moment arm ℓ has to be kept as short as possible. Furthermore, because the branch connection can take a considerably higher bending moment in the longitudinal in-plane direction than in the circumferential out-plane direction, the exit elbow is better placed in plane, or close to in-plane, with the header pipe

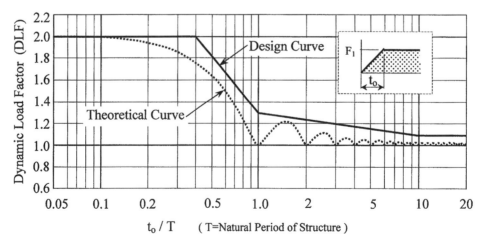

FIG. 12.19
DESIGN DYNAMIC LOAD FACTORS FOR RAMP IMPULSE LOAD

as shown in Fig. 12.17. For critical systems, a support column directly underneath the exit elbow may be needed. Thermal insulation on the support column is recommended to reduce the relative expansion of the valve nozzle and the support column. It is obvious that in this case the natural period in the forcing direction becomes very short due to the high stiffness. Because a shorter natural period increases the ramp-time to period ratio, the support column also tends to reduce the DLF.

12.6.2 Closed Discharge System

When dealing with hazardous or toxic fluids such as radioactive steam and most hydrocarbons, the over-pressured fluid is relieved to a close system for recycling, treatment, or proper disposal. In a closed system, the maximum flow is generally the same as the open discharge system, unless it is choked by the friction of excessive piping length. Therefore, the maximum reaction force produced by the fluid leaving an elbow, and the impulse force produced by the fluid entering an elbow can be considered the same as the F_1 force calculated for the open discharge system. If the friction force is ignored, then the force can be considered the same throughout the system.

For a piping leg between points n and $n+1$ as shown in Fig. 12.20, there is an F_1-shape force acting on end n and another F_1-shape force acting at end $n+1$. These two forces have the same maximum magnitudes, but are in opposite directions. Under the steady-state condition, there is no net shaking force because the two forces balance out each other. This balanced situation is maintained even if the friction is included. The situation is different during a transient condition, such as the initial phase of the safety valve relieving.

When the safety valve starts to pop open, the flow starts from nothing to the maximum as shown in Fig. 12.18. The flow compresses the fluid inside the discharge pipe and transmits the pushing effect to the downstream either by wave motion or by actual flow velocity. In any case, for a safety valve discharge, the wave speed and the flow speed are considered the same. The force will have the same time history shape throughout the piping, but the arriving time is different at each point. This is called a traveling wave, which occurs in many types of piping hydraulic transients. Because of this arriving time difference, each pipe leg experiences a net shaking force, whose magnitude depends mainly on the length of the pipe leg.

To estimate the net force on a piping leg, we consider both ends of the pipe leg to receive the same force time history but at different starting time points. As shown in Fig. 12.20(a), the force arrives at end n at time t_n, but the same force arrives at end $n+1$ Δt time later at t_{n+1}. Since the forces at both ends have different signs, the net force at any given time instant is the difference of the two end forces. Figure 12.20(a) shows the actual net force time history, whereas Fig. 12.20(b) shows the idealized net force time history. The idealized net force is constructed with the idealized ramp relieving force. The net shaking force time history has the maximum force of F_{max}.

Because the force wave travels at sonic velocity, the force arriving time at both ends differ by

$$\Delta t = L/a$$

where a is the sonic velocity with respect to the fluid inside the pipe. If the leg is sufficiently long, the time difference becomes greater than the valve opening time. In this case, the force at n-end reaches the maximum before $(n+1)$-end has any force to counterbalance it. The maximum net shaking force of the leg, in this case, is the same as the maximum relieving force, F_1. When Δt is smaller than the effective valve opening time t_o, the maximum net force can be calculated by the idealized forcing function as

$$F_{max} = F_1 \frac{\Delta t}{t_o} = F_1 \frac{L}{at_o} \qquad (12.54)$$

which shows the longer the leg, the greater the net force.

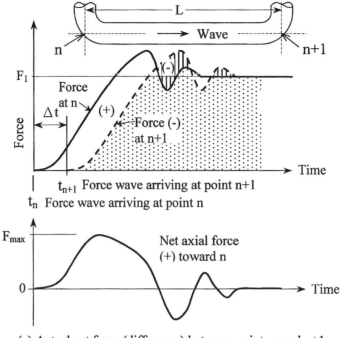

(a) Actual net force (difference) between points n and n+1

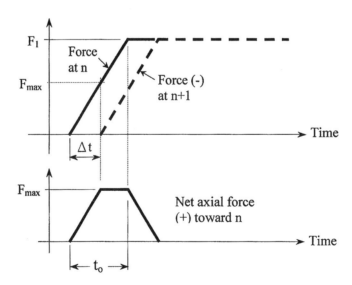

(b) Idealized net force (difference between points n and n+1

FIG. 12.20
NET SAFETY VALVE FORCE ACTING ON A PIPE LEG

If the pipe leg is very long with a Δt greater than the effective valve opening time, the net force starts when the flow or wave front reaches the first elbow. It eventually reaches the maximum then maintains at a constant force the same as F_1. The net force starts to reduce when the initial flow or wave front reaches the second elbow. It eventually reduces to zero, and the flow becomes steady when the peak of the force reaches the second elbow.

The discussion given above is based on the opening of the valve. A similar net shaking force, in a reverse direction, also results during the closing of the valve. The opening force and closing force are generally well separated by a few seconds, so they are considered as two independent events in the structural response. The two forces can overlap each other during short pops, but with considerably smaller forces. The above discusses only the concept of idealized discharge forces. For general safety/relief valve discharging forces, the approaches developed by Moody [21] and other authors can be used.

The net shaking force at each pipe leg, applied with a proper DLF, can be used in performing an equivalent static analysis of the piping system. The DLF can be taken from the appropriate shape or combination of shapes given in Figs. 12.5 and 12.6. Although the forces are all acting at different times, it is necessary to apply the forces all at the same time in a static analysis. The time-history analysis, on the other hand, can consider the actual force shapes and arriving times at different locations.

12.7 STEAM TURBINE TRIP LOAD

The steam turbine that drives an electric generator has one function that is somewhat beyond the control of the operator — that is, the trip of the electrical output load. During normal startup and shutdown, the load is added or reduced gradually to avoid any serious transient in either steam flow or electric circuit. However, when the electric circuit is tripped, the load is lost completely and suddenly. In this case, the turbine inlet valve has to be closed quickly or the turbine faces running at a dangerously high speed. Even the residual steam contained in the steam chest and inlet leads has enough power to drive the load-free turbine into an undesirable speed.

When the flow of steam is stopped or slowed, a pressure wave results. This is the basic water hammer principle we are all aware of. The pressure wave transmitting throughout the piping system creates shaking forces in the piping, in addition to a pressure rise in the piping and piping components. Because of the high compressibility, the shaking force due to the steam valve closure is not as severe as that in a liquid line. This is why the steam hammer effect had not drawn any attention before some support damages were discovered in the 1960s. Theoretical treatment on the subject was first published by Coccio [22] in 1966. Since then, the steam hammer phenomenon has been included in the routine design practice.

From the basic water hammer theory [10], we know that a sudden reduction in fluid flow velocity in the pipe produces an increase in pressure. The increase in pressure forms a pressure wave traveling upstream along the pipe at sonic velocity. For a small reduction of velocity, $(\Delta V)_i$, the increase of pressure, $(\Delta P)_i$, is

$$(\Delta P)_i = \rho a (\Delta V)_i \tag{12.55}$$

where ρ is the density of the fluid and a is the sonic velocity through the fluid.

When the fluid velocity reduction is large, but not all of a sudden, as shown in Fig. 12.21, the increase in pressure can be constructed with a series of small sudden changes. ΔV_1 generates pressure change ΔP_1, and ΔV_2 produces ΔP_2, and so forth. The total pressure rise occurs at the time when the valve is fully closed, and is equal to

$$(\Delta P)_T = \rho a V \tag{12.56}$$

The typical total pressure rise for a turbine trip is in the 10% to 15% range of the operating pressure. The lower percentage is for high-pressure main steam and the higher percentage is for low-pressure reheat steam. Although the system has a tendency to generate higher pressure-change as shown by the dotted line, this extra pressure change is often ignored in practical calculations. This is mainly due to the fact that its value is not easy to estimate without going through a detailed thermo-hydraulic analysis. The produced pressure rise forms a wave traveling upstream with sonic velocity. As the pressure rises, the density and temperature also increase. This will then increase the sonic velocity in the

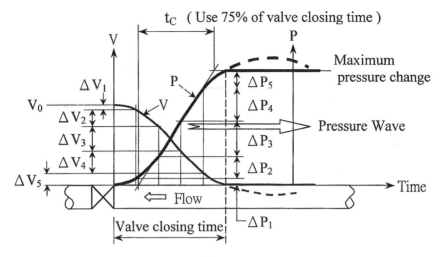

**FIG. 12.21
PRESSURE WAVE DUE TO VALVE CLOSING**

fluid trailing the wave, causing the wave front to pile up a little. However, because the change is not very great, the changes in density and temperature are ignored. Sonic velocity is considered constant throughout the process. In steam hammer analysis, we also often ignore the momentum flux change due to flow velocity change. The piping shaking force is estimated purely on the pressure change. Therefore, the total surge force is calculated by multiplying the surge pressure with the flow area. That is,

$$F_T = \rho a A V \tag{12.57}$$

where density ρ can be found from the steam table for the known pressure and temperature. Flow velocity V is generally determined from the mass flow rate and density. Sonic velocity a can be found via Eq. (12.41) using $k = 1.28$ for superheated steam.

The time-history shape of the force is the same as the shape of the pressure rise given in Fig. 12.21. This steam hammer force time history follows the general shape of the safety valve relief force time history shown in Fig. 12.18, except that here the force is F_T and the valve opening time is replaced by the valve closing time. Again, the effective close time, t_C, can be taken as 75% of the specified valve closing time. Since the valve closing time dictates the net shaking force of each piping leg, an accurate closing time is essential for an accurate estimate of the force. Because all turbine manufacturers are sophisticated technical organizations, they should be able to provide the valve closing time when requested. It appears that larger units tend to have shorter closing times due to the higher potential energy of the steam. The published closing times range from 0.22 second [22] (early fossil plants), to 0.126 second [23] (modern fossil plants) and 0.06 second [24] (nuclear plants).

The method of obtaining the shaking force at each leg is by finding the difference of the forces acting at both ends of the leg. This is exactly the same approach used in the closed discharge safety valve force given in Fig. 12.20. However, because the steam hammer force comes mainly from the pressure wave, rather than the actual flow in the case of a safety valve, its pressure wave reflects at the ends and travels back and forth many times with diminishing magnitude. In practical calculations, only the first complete reflection is considered. The reflections after the first one are ignored because they are much smaller forces.

Using an idealized ramp shape of the pressure force, the force history at each point along the pipe is constructed as shown in Fig. 12.22. At point 1, for instance, the pressure rises linearly as the valve stops. It reaches the maximum at t_C, then remains constant until the reflection wave front reaches

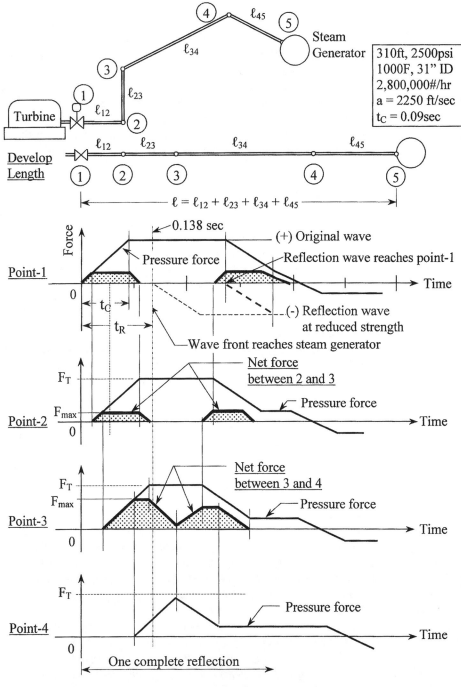

**FIG. 12.22
PIPING FORCE DUE TO STEAM TURBINE TRIP**

the point. The wave reaches the steam generator at t_R, which is the total pipe length divided by the sonic velocity. The reflection wave is then generated and pushes upstream at a negative pressure of somewhat smaller magnitude than the original wave due to friction and exit loss. The reflection wave reaches point 1 after another t_R of time has passed. The pressure starts to reduce from that time instant and gradually to the magnitude that is the difference of the strength between the original wave and the reflection wave. The same procedure is used to construct the pressure history of all the points along the pipe. The points located near the end of the pipe will see the incoming pressure wave and the reflection wave separated by a much shorter time span than that at points located near the valve. Point 4, for instance, never sees the full total pressure surge.

The net shaking force at each piping leg is taken as the difference of the pressure existing at both ends of the pipe leg under consideration. The maximum shaking force, F_{max}, is determined by the length of the leg. If the length of the leg is greater than the sonic velocity times the effective valve close time, the maximum shaking force is the same as the maximum surge force, F_T. If the leg length is shorter than the sonic velocity multiplied by the valve closing time, the maximum shaking force is determined by direct proportion as

$$F_{max} = F_T \frac{\ell_{ij}}{t_c a} \quad \text{(for } \ell_{ij} \leq t_c a\text{)} \tag{12.58}$$

where ℓ_{ij} is the length of the pipe leg located between point i and point j. This is essentially the same equation as Eq. (12.54), because the shaking forces are calculated the same manner in both cases. Moreover, in both cases, the magnitude of the shaking force is proportional to the length of the pipe leg. The shaking forces, therefore, may be ignored at very short legs.

The net shaking force for each leg has two separated humps for a complete reflection cycle. The humps are widely separated for the legs near the turbine, but are close together for the legs located near the steam generator. The time-history of the shaking force is marked as the shaded area in Fig. 12.22. The piping is then analyzed either dynamically using the time-history method, or statically by applying proper DLFs. The static method requires the application of all forces at the same time. The time history analysis of this example is given in the next chapter, which deals with general dynamic analyses.

REFERENCES

[1] Biggs, J. M., 1964, *Introduction to Structural Dynamics*, McGraw-Hill, New York, NY.
[2] Hurty, W. C., and Rubinstein, M. F., 1964, *Dynamics of Structures*, Prentice-Hall, Inc., Englewood Cliffs, NJ.
[3] Clough, R. W., and Penzien, J., 1975, *Dynamics of Structures*, McGraw-Hill, New York, NY.
[4] Newmark, N. M., 1971, "Seismic Response of Reactor Facility Components." *Symposium on Seismic Analysis of Pressure Vessel and Piping Components*, Presented in ASME 1st National Congress on Pressure Vessels and Piping, San Francisco, CA, May, 1971, ASME.
[5] U.S. NRC, 1973, Revision 1, 2007, Regulatory Guide 1.61, "Damping Values for Seismic Design of Nuclear Power Plants," U.S. Nuclear Regulatory Commission, Office of Standards Development, Washington, D. C.
[6] Fox, J. T., Szumski, D. R., Gibraiel, S. A., and Panucci, K. R., 1988, "Snubber-Reduction Program Cuts Costs, Boosts Safety," *Power*, April, vol.132, no.4, pp. 103–105.
[7] Shibata, H., et al., "A Study on Damping Characteristics of Piping Systems in Nuclear Power Plants," *Seismic Analysis of Power Plant Systems and Components*, PVP Vol. 73, pp. 151–178.
[8] WRC-300, 1984, "Technical Position on Damping Values for Piping — Interim Summary Reports," *Welding Research Council (WRC) Bulletin* 300, December.
[9] WRC-316, 1986, "Technical Position on Damping Values for Insulated Pipe — Summary Report," *WRC Bulletin* 316, July.
[10] Parmakian, J., 1955, *Waterhammer Analysis*, Dover Publications, Inc., New York, NY.
[11] Shapiro, B., 1953, *The Dynamics and Thermodynamics of Compressible Fluid Flow*, The Ronald Press Company, New York, NY.

[12] Streeter, V. L., and Wylie, E. B., 1975, *Fluid Mechanics*, 6th ed., McGraw-Hill Book Company, New York, NY.
[13] ASME, "Capacity Conversions for Safety Valves," B&PV Codes, Section VIII, *Pressure Vessels*, Division 1, Appendix 11, ASME, New York, NY.
[14] API RP-520, "Recommended Practice for the Design and Installation of Pressure-Relieving Systems in Refineries, Part 1 — Design," American Petroleum Institute, Division of Refining, Washington, D.C.
[15] ASME, B&PV Codes, Section I, "Power Boilers," ASME, New York, NY.
[16] ASME B31.1, "Nonmandatory Appendix II — Rules for the Design of Safety Valve Installations," *Power Piping*, ASME, New York, NY.
[17] Auble, T. E., 1982, "Full Scale Pressurized Water Reactor Safety Valve Test Results." ASME paper 82-WA/NE-11.
[18] The Crosby Valve and Gage Company, Wrentham, MA 02093.
[19] Dresser Industries, Valve and Control Division, Alexandria, LA 71301.
[20] Wheeler, A. J., and Siegel, E. A., 1982, "Measurements of Piping Forces in A Safety Valve Discharge Line." ASME paper 82-WA/NE-8.
[21] Moody, F. J., 1985, "Pipe Forces Caused by a Fluid Density Change Through a Discharge Safety/Relief Valve," ASME Special Publication PVP-Vol.98-7.
[22] Coccio, C. L., 1966, "Steam Hammer in Turbine Piping System," ASME Paper 66-WA-FE32.
[23] Rooney, J. W., et al, 1990, "Typical Turbine Trip Loads in a Fossil Power Plant," *ASME* PVP-Vol.-188.
[24] Bostrom, T. E., 1979, "Comparison of Steam Hammer Dynamic Testing with Analysis for Main Steam Piping," ASME Paper 79-PVP-22.

CHAPTER 13

DYNAMIC ANALYSIS — PART 2: MDOF SYSTEMS AND APPLICATIONS

This chapter, the second one on dynamic analysis, focuses on general applications of dynamic analysis to piping systems. It includes the response spectra method for earthquake analysis, harmonic analysis of steady-state vibration, and the time-history analysis for general fluid transient loads. We will start with the general formulations of the approaches of the analysis [1, 2].

13.1 LUMPED-MASS MULTI-DEGREE OF FREEDOM SYSTEMS

A piping system is a continuous system that can deform into infinitely different ways and patterns. However, the patterns of deformation requiring sharp reflection between two closely located points are not likely to occur in the practical world. This allows the system to be divided into a finite number of discrete points without losing practical accuracy. The deformation between any two adjacent points is assumed to follow a certain pattern that requires the least energy or the pattern that is easiest to achieve. This approach is called the finite element method. Practically all dynamic analyses of piping system are done by this method and are performed via computers. A detailed discussion of the finite element method is beyond the scope of this book. Interested engineers should read relevant books [3, 4] for details. This chapter will cover only several basic concepts that are needed to accurately and effectively use the computerized tools.

The finite element method divides the continuous system into a finite number of elements. The behavior of each element is determined by the movements of the boundary points of the element. One set of boundary point conditions always generates the same pattern of behavior within the element. Therefore, it is important to note that to evaluate certain behavior of the system, the element should be small enough to preclude multiple inflections inside the element. The points at the boundaries of the elements are called the node points. Once the desired element length is determined and the system is subdivided properly, the equation of motion is constructed by force equilibrium as follows.

$$[M]\{\ddot{X}\} + [C]\{\dot{X}\} + [K]\{X\} = \{F\} \tag{13.1}$$

This is equivalent to Eq. (12.3) for single degree of freedom (SDOF) systems, except that Eq. (13.1) represents N simultaneous linear differential equations. N generally represents the total degrees of freedom of the system and is equal to six times the total number of node points, that is, $N = 6p$, where p is the number of node points. The expanded form of the equation is given as follows

$$\begin{bmatrix} m_{11} & m_{12} & \cdots & m_{1n} \\ m_{21} & m_{22} & \cdots & m_{2n} \\ \cdots & \cdots & \cdots & \cdots \\ m_{n1} & m_{n2} & \cdots & m_{nn} \end{bmatrix} \begin{bmatrix} \ddot{x}_1 \\ \ddot{x}_2 \\ \vdots \\ \ddot{x}_n \end{bmatrix} + \begin{bmatrix} c_{11} & c_{12} & \cdots & c_{1n} \\ c_{21} & c_{22} & \cdots & c_{2n} \\ \cdots & \cdots & \cdots & \cdots \\ c_{n1} & c_{n2} & \cdots & c_{nn} \end{bmatrix} \begin{bmatrix} \dot{x}_1 \\ \dot{x}_2 \\ \vdots \\ \dot{x}_n \end{bmatrix} +$$
$$\begin{bmatrix} k_{11} & k_{12} & \cdots & k_{1n} \\ k_{21} & k_{22} & \cdots & k_{2n} \\ \cdots & \cdots & \cdots & \cdots \\ k_{n1} & k_{n2} & \cdots & k_{nn} \end{bmatrix} \begin{bmatrix} x_1 \\ x_2 \\ \vdots \\ x_n \end{bmatrix} = \begin{bmatrix} f_1 \\ f_2 \\ \vdots \\ f_n \end{bmatrix} \qquad (13.2)$$

{X} is the displacement vector consisting of all displacements and rotations that correspond to all degrees of freedom of the system. Mathematically, written in transposed form, {X} represents

$$\{X\} = [x_1\, x_2\, x_3\, x_4\, x_5 \ldots x_{n-3}\, x_{n-2}\, x_{n-1}\, x_n]^T$$

However, in engineering terms, again written in transposed form, {X} can be expressed as

$$\{X\} = [d_{x1}\, d_{y1}\, d_{z1}\, r_{x1}\, r_{y1}\, r_{z1}\, d_{x2}\, d_{y2} \ldots d_{yp}\, d_{zp}\, r_{xp}\, r_{yp}\, r_{zp}]^T$$

where d_{x1}, d_{y1}, and d_{z1} are the displacements of point 1 in x, y, and z directions, whereas r_{x1}, r_{y1}, and r_{z1} are the rotations of point 1 in x, y, and z directions, respectively. Similarly, the {F} vector also consists of the forces and moments applicable to all degrees of freedom.

Because elastic force is determined by the relative displacement, whereas inertia force is based on the absolute acceleration, Eq. (13.1) is valid only when there is no support or anchor displacement. When there is a motion at the anchor or support, the acceleration vector should be derived from the absolute displacement that includes both the relative displacement and the anchor or support displacement.

In the field of piping engineering, Eq. (13.1) is solved differently for each type of load involved. The most commonly used solution methods include the response spectra method for handling earthquake problems, modal or direct integration for solving impulse and other well-defined time-history forces, harmonic analysis for dealing with pulsation and other steady-state periodic loadings, and the static equivalent method for other poorly defined loads. For all solution methods, proper mass lumping is the first task in ensuring an accurate analysis.

13.1.1 Mass Lumping

A piping system consists of many continuous beams of pipe with mass uniformly distributed. It also has some concentrated masses representing valves, fittings, and other components located at strategic points. The concentrated mass points are automatically handled by static analysis models, which are routinely constructed properly by most piping engineers. However, the continuous beams, straight or curved, require additional attention when performing a dynamic analysis. One often-committed mistake in dynamic analysis is to construct the dynamic analytical model in the same manner that the static analytical model was constructed. This often results in inaccurate and meaningless analyses.

As shown in Fig. 13.1, unlike in the static case when a beam generally deforms into one predictable form, a beam can dynamically deform into many different forms. These deformation forms are called the natural mode shapes. Each mode shape is associated with a natural frequency. Because the mode with the lowest frequency is most easily excited by an external force, only the modes at lower frequencies are of interest in practical piping analysis. It is very difficult for modes at very high frequencies to get excited, and thus they are often ignored. The three mode shapes shown represent the six modes with the lowest frequencies. In this cantilever beam, each mode shape actually represents two modes: one mode in one direction and the other in an orthogonal direction. In other words, if the beam has

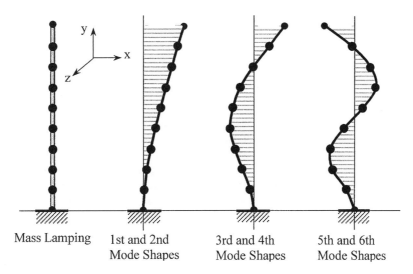

**FIG. 13.1
MASS LAMPING SCHEME**

a mode vibrating in x direction, there is also another mode vibrating in the z direction with the same natural frequency.

In a static analysis model, the beam is sufficiently defined by two end points. With this static model, the first mode and its second mode twin can be produced due to the availability of the mass at end of the beam. Although the frequency determined will not be accurate, at least the tendency of the vibrating mode is clear. The static model will completely fail to predict the third and higher modes because there is no mass available to deflect the mid portion of the beam opposite to the top end point. Therefore, with an otherwise proper static analysis model, many important dynamic responses could be omitted if used in performing a dynamic analysis.

To achieve reasonable accuracy in dynamic analysis, many more node points than are required for the static analysis are needed. Depending on the highest frequency of the vibrating mode required, the general rule is that at least two mass points are needed for each half-wavelength [5]. The maximum element length can be calculated by dividing the half-wavelength with three equal spans. That is,

$$\ell_{max} = \frac{1}{3}\sqrt{\frac{1.57}{f}}\sqrt[4]{\frac{EI}{m}} \tag{13.3}$$

where m is the mass of the pipe, including insulation and content, per unit length, and f is the cutoff frequency above which the modes are ignored. Some computer software packages, such as SIM-FLEX-II [6], can automatically subdivide both the straight and curved piping segments into a series of elements based on the criterion given by Eq. (13.3).

When the analysis uses a lump mass approach, the mass matrix needed for Eq. (13.1) is a diagonal matrix. This diagonal matrix, instead of a full matrix, greatly simplifies the solution process. The lump mass method also generally ignores the rotational inertia of the mass.

Besides the lumped mass approach, there is also a consistent mass matrix approach available. The consistent mass matrix approach considers that the element mass follows the predetermined deflection pattern as in the case of the stiffness matrix. The consistent mass matrix is therefore filled with the same amount of entries as in the stiffness matrix. Theoretically, the consistent mass matrix method requires fewer node points than the lump mass matrix method to achieve the same degree of accuracy in the analysis. However, because an element is unable to deform into more than one inflection in any case, there is also a limit on the maximum element length in the consistent mass matrix method.

Generally, the saving on the number of nodes in a consistent mass matrix is of less of an incentive than the simplicity offered by the lump mass method.

13.1.2 Free Vibration and Modal Superposition

The dynamic behavior of a piping system depends greatly on the free or natural vibration of the system. For an SDOF system, we only have to deal with one natural frequency, and the system can only move in one direction. However, in a multi-degree of freedom (MDOF) structural system, such as a piping system, there are many natural frequencies, each with its vibration shape or mode. To investigate the free vibration, just like in the SDOF case, we set the external force to zero. Since we have already established in the SDOF case that damping has a very small effect on natural frequency, we can also assume zero damping for the free vibration analysis. By removing the force and damping terms from Eq. (13.1), the equation for free vibration becomes

$$[M]\{\ddot{X}\} + [K]\{X\} = 0 \qquad (13.4)$$

Because differentiating a sine or cosine function twice will revert to the original function, a harmonic oscillation should satisfy the equation. By considering a sinusoidal oscillation, the $\{X\}$ vector is assumed to be

$$\{X\} = \{\Phi\}\sin\omega_n t \qquad (13.5)$$

Substituting Eq. (13.5) to Eq. (13.4), we have

$$-\omega_n^2[M]\{\Phi\} + [K]\{\Phi\} = 0 \quad \text{or} \quad \left([K] - \omega_n^2[M]\right)\{\Phi\} = 0 \qquad (13.6)$$

To have a non-trivial solution of $\{\Phi\}$, $([K] - \omega_n^2[M])$ has to be zero. Equation (13.6) is the standard eigenvalue problem, which can be solved by using standard solution schemes [4, 7]. Theoretically, the equation can produce N eigenpairs as follows:

$$(\omega_{n1}^2, \{\Phi_1\}), \left(\omega_{n2}^2, \{\Phi_2\}\right), \ldots \left(\omega_{nN}^2, \{\Phi_N\}\right)$$

where ω_{ni}^2 is the eigenvalue, $\{\Phi_i\}$ is the eigenvector, and ω_{ni} is the circular frequency of the ith natural vibration mode. The eigenvectors are natural vibration shapes. Each vector shows only the relative values and can be multiplied with any scale factor. The three deformation shapes shown in Fig. 13.1 are eigenvectors. They are also called mode shapes. It is important to note that they are not actual displacements, but are just the proportions of the displacements.

One of the important characteristics of the eigenvector is the orthogonality of the natural modes. Because of the orthogonality, the natural modes are also called normal modes. The orthogonality is expressed as

$$\{\Phi_i\}^T[M]\{\Phi_j\} = 0 \quad \text{for all } i \neq j \qquad (13.7)$$

Because the eigenvectors may be multiplied with any scale factor, it is convenient to scale them into the following relations

$$\{\Phi_i\}^T[M]\{\Phi_i\} = 1 \quad \text{and} \quad [\Phi]^T[M][\Phi] = [I]_D \qquad (13.8)$$

where $[I]_D$ is the unit diagonal matrix or identity matrix. The vectors so scaled are called normalized. For convenience, *all $\{\Phi\}$ vectors are assumed to be normalized in this book*. With normalized vectors, Eq. (13.6) yields the following relation

$$[\Phi]^T[K][\Phi] = \left[\omega^2\right]_D \qquad (13.9)$$

where $[\omega^2]_D$ is a diagonal matrix.

Due to the orthogonality, the N normal modes in a sense constitute an N-dimensional space with N-independent coordinates. They can be used, in the same manner as in ordinary three-dimensional space coordinates, to define any vector. The displacement vector $\{X\}$, therefore, can be defined by the N normal vectors as

$$\{X\} = \xi_1\{\Phi_1\} + \xi_2\{\Phi_2\} + \cdots = \sum \xi_i\{\Phi_i\} = [\Phi]\{\xi\} \tag{13.10}$$

where $[\Phi]$ is the matrix consisting of $\{\Phi_i\}$ as columns, and $\{\xi\}$ is the vector containing ξ_i, which are coordinates or scale factors in Φ space. Substituting Eq. (13.10) to Eq. (13.1), we have

$$[M][\Phi]\{\ddot{\xi}\} + [C][\Phi]\{\dot{\xi}\} + [K][\Phi]\{\xi\} = \{F\} \tag{13.11}$$

Pre-multiplying the above with the transpose of $[\Phi]$, the equation becomes

$$[\Phi]^T[M][\Phi]\{\ddot{\xi}\} + [\Phi]^T[C][\Phi]\{\dot{\xi}\} + [\Phi]^T[K][\Phi]\{\xi\} = [\Phi]^T\{F\} \tag{13.12}$$

Substituting Eqs. (13.8) and (13.9), the above equation becomes

$$[I]_D\{\ddot{\xi}\} + [\Phi]^T[C][\Phi]\{\dot{\xi}\} + \left[\omega^2\right]_D\{\xi\} = [\Phi]^T\{F\} \tag{13.13}$$

Because both $[I]_D$ and $[\omega^2]_D$ are diagonal matrices, the above simultaneous equations are decoupled into N independent equations representing N SDOF systems, if the damping matrix $[C]$ is orthogonal. However, because the real damping is difficult to assess and apply, we shall be satisfied to regard $[C]$ as orthogonal and assign the damping for each mode using the damping ratio. Taking the form as expressed in Eq. (12.33), Eq. (13.13) becomes N SDOF systems as

$$\ddot{\xi}_i + 2\zeta_i\omega_i\dot{\xi}_i + \omega_i^2\xi_i = \{\Phi_i\}^T\{F\} \quad i = 1,\ldots,N \tag{13.14}$$

which also implies that

$$[\Phi]^T[C][\Phi] = [2\zeta\omega]_D \tag{13.15}$$

where $[2\zeta\omega]_D$ is diagonal matrix

The modal superposition converts the N simultaneous differential equations of the MDOF system into N independent SDOF systems as expressed in Eq. (13.14). These N independent SDOF systems are solved one by one using SDOF techniques. Depending on the type of the force $\{F\}$, the solution approach is different for different types of forcing functions. Two often-used approaches are the response spectra method and the modal integration time-history method. The response spectra method converts the forcing function into response spectra, and the modal integration method integrates the equation with the actual forcing function a small time step by a small time step.

The natural frequency and mode shape are generally calculated one by one starting from the mode with the lowest frequency. The mode with the lowest frequency is called the fundamental mode. In theory, there are N natural modes. However, because only the modes with lower frequencies get significant response to the excitation, only certain n modes with lower frequencies are actually calculated for the analysis. Therefore, Eq. (13.14) is generally calculated for the lower n modes only. The cutoff point is generally determined by the frequency limit given by the design specification.

13.2 PIPING SUBJECT TO GROUND MOTION

A piping system has to be designed for any support or anchor motion, if applicable. The most often encountered support and anchor motion comes from an earthquake that produces the ground motion. For most non-critical piping systems, the earthquake is treated as a static load, which is proportional to the weight of the piping and components. The so-called g factor is applied statically at the vertical and two horizontal directions. The magnitude of the load is generally determined according to the American Society of Civil Engineers [8] or Uniform Building Code [9]. For critical piping systems, such as nuclear piping systems, a dynamic analysis that produces more realistic and accurate results is generally preferred.

As noted previously, the displacement vector {X} in Eq. (13.1) has different meanings at different terms. The elastic force and damping force are based on relative displacements, but the acceleration force is based on absolute displacement. There is no difference between them if no support or anchor motion exists. When there is support or anchor motion, the acceleration vector has to be revised to cover the support and anchor motion as

$$\{X\}_{abs} = \{X\} + \{X\}_g = \{X\} + [Q]\{S_g\} \tag{13.16}$$

where $\{S_g\}$ is the vector containing all independent support motions, and $[Q]$ is the influence or relation matrix relating each ground motion to each degree of freedom of the system.

In a system with uniform support motion at each direction, we have three independent motions in three coordinate directions. Therefore, with uniform support motion, we have $[Q]$ and $\{S_g\}$ as expressed in Eq. (13.17).

$$[Q] = \begin{vmatrix} 1 & 0 & 0 \\ 0 & 1 & 0 \\ 0 & 0 & 1 \\ 0 & 0 & 0 \\ 0 & 0 & 0 \\ 0 & 0 & 0 \\ 1 & 0 & 0 \\ 0 & 1 & 0 \\ . & . & . \\ . & . & . \end{vmatrix} = |Q_1 Q_2 Q_3| \quad \{S_g\} = \begin{vmatrix} s_{g1} \\ s_{g2} \\ s_{g3} \end{vmatrix} = \begin{vmatrix} s_{gx} \\ s_{gy} \\ s_{gz} \end{vmatrix} \tag{13.17}$$

For a non-uniform support motion, $[Q]$ is determined as the static displacement of each degree of freedom responding to a unit displacement applied at all support locations of the group with the same set of motion. For instance, if there are two groups of independent motions, generally $[Q]$ will have $2 \times 3 = 6$ relationship vectors, namely, $\{Q_1\}$ through $\{Q_6\}$.

After substituting the absolute displacement $\{X\}_{abs}$ to Eq. (13.1), and assuming the external force, $\{F\}$, is non-existent, we have the equation for ground motion as

$$[M](\{\ddot{X}\} + [Q]\{\ddot{S}_g\}) + [C]\{\dot{X}\} + [K]\{X\} = 0$$

or

$$[M]\{\ddot{X}\} + [C]\{\dot{X}\} + [K]\{X\} = -[M][Q]\{\ddot{S}_g\} \tag{13.18}$$

The above equation can be solved by the normal mode superposition approach. By applying the same transformation that resulted in Eq. (13.14), we have

$$\ddot{\xi}_i + 2\zeta_i\omega_i\dot{\xi}_i + \omega_i^2\xi = -\{\Phi_i\}^T[M][Q]\{\ddot{S}_g\}$$

or

$$\ddot{\xi}_i + 2\zeta_i\omega_i\dot{\xi}_i + \omega_i^2\xi = -\{\Phi_i\}^T[M](\{Q_1\}\ddot{s}_{g1} + \{Q_2\}\ddot{s}_{g2} + \{Q_3\}\ddot{s}_{g3} + \cdots) \tag{13.19}$$

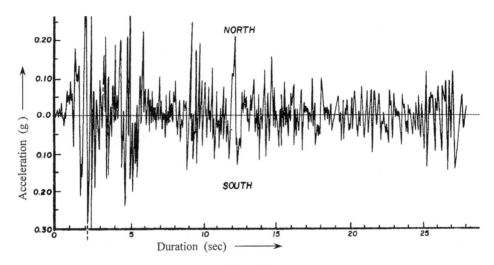

FIG. 13.2
ACCELEROGRAPH RECORD OF EL CENTRO, CALIFORNIA EARTHQUAKE MAY 18, 1940, NORTH – SOUTH DIRECTION

where $\{\Phi_i\}^T [M]\{Q_j\}$ is the mode participation factor of the ith mode with respect to the jth ground motion. It should be remembered that Φ vectors are normalized so that $\{\Phi_i\}^T [M]\{\Phi_i\} = 1$.

Equation (13.19) can always be solved by integration if the $\ddot{S}_g(t)$ is well defined. For ground motion due to earthquake, $\ddot{S}_g(t)$ is more of a probability nature, and is generally solved by the response spectra method.

13.2.1 Response Spectra Method

The earthquake ground motion as shown in Fig. 13.2 [10] is fairly complicated for routine design analysis of the structure system. In other words, the design analysis generally does not go through integrating the motion a small time step by a small time step. Instead, the motion is used to generate a response spectra chart, such as shown in Fig. 13.3 [10] for routine design analysis. The curve is generally averaged and smoothed from the actual calculated data.

The response spectrum is determined by running the time-history motion, either by computation or instrumentation, through a series of SDOF harmonic oscillators. Mathematical computation is generally used for economic reasons and also for ease of inputting the motion and adjusting the frequency and damping ratio. The maximum response of each oscillator throughout the entire duration of the motion is the response spectrum of the ground motion with respect to the frequency of the SDOF harmonic oscillator. After running through the computations for many harmonic oscillators with various natural frequencies, a curve or chart is constructed for design use. It is important to note that the response spectra curve already includes the damping effect. The curve is always labeled with the damping ratio with which it was constructed. Different damping values have different response spectra curves.

The quantity given by the response spectra can be in acceleration, velocity, or displacement. The one given in Fig. 13.3 is acceleration in g unit, a very popular unit but not a consistent one. As the modal displacement coefficient, ξ, in Eq. (13.19) is a relative displacement; the displacement spectra are for relative displacements. Relative displacements are the ones used for calculating internal elastic forces, moments, and stresses of the piping system. However, because the response spectra are generated with respect to SDOF harmonic motion, the acceleration spectra and velocity spectra can be defined as follows

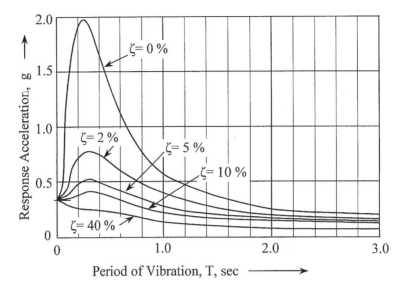

FIG. 13.3
AVERAGE ACCELERATION SPECTRUM CURVE [10] - EL CENTRO, MAY 18, 1940, NORTH – SOUTH DIRECTION

$$R_a = \omega^2 R_d \text{ and } R_v = \omega R_d \quad (13.20)$$

where R_a, R_v, and R_d are the response spectra of acceleration, velocity, and displacement, respectively. R_a is expressed in consistent units, but not in g units. The acceleration and velocity so defined are called pseudo-acceleration and pseudo-velocity, respectively. Pseudo-acceleration is very close to absolute acceleration and is the same as absolute acceleration when there is no damping. This can be verified by the balancing of elastic force and inertia force. The elastic force is KR_d, whereas the acceleration force is Ma. By equating elastic force and acceleration force, that is, $KR_d = Ma$, we have

$$\text{Absolute acceleration} = a = \frac{K}{M}R_d = \omega^2 R_d$$

when there is no damping.

Absolute acceleration is often required for qualifying components, such as valves and controllers, to withstand the design earthquake. The actual meaning of pseudo-velocity is not very important. It can be considered as just a convenient working parameter. From now on, the relative and pseudo prefixes will be dropped. They will be simply called displacement, velocity, and acceleration.

The design response spectra used in designing piping systems generally do not come directly from the ground motion response spectra. Ground motion is generally modified, most often amplified, by the main building or tower structure to which the piping is attached or supported. The spectra used in piping design are the amplified ground motion spectra. Because a piping system is generally attached or supported at many points, it is affected by different response spectra at different points. However, these support points can be grouped based on their location similarity, such as the same floor of the same building, and so forth. The piping system can then be considered as subject to a few independent support or spectra groups.

For simplicity purposes, the analysis can be analyzed by assuming that all the support points have the same set of response spectra. A set generally includes one each for the three orthogonal directions. In this case, the spectra to be used are the envelope spectra that encompass all the peaks of all the applicable groups of spectra. Assuming that the piping is supported at three main locations with

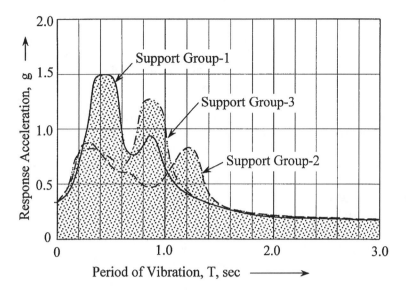

FIG. 13.4
ENVELOPE RESPONSE SPECTRA (SHADED) - DAMPING RATIO ς=0.02, N-S DIRECTION

three groups of applicable response spectra, then in each direction, an envelope of these three groups of spectra for the corresponding direction is constructed either on paper as shown in Fig. 13.4, or by other means. The envelope method is a fairly conservative approach.

Once the applicable response spectra curve is available, the maximum modal displacement coefficients are calculated mode by mode using Eq. (13.19), that is,

$$\xi_i = \frac{\ddot{\xi}_1}{\omega_i^2} = \frac{1}{\omega_i^2} \{\Phi_i\}^T [M](\{Q_1\} R_{a1} \& \{Q_2\} R_{a2} \& \{Q_3\} R_{a3} \& \ldots)$$

or

$$\xi_i = \xi_{i,1} \& \xi_{i,2} \& \xi_{i,3} \& \ldots, \text{ etc. } \quad \xi_{i,j} = \frac{1}{\omega_i^2} \{\Phi_i\}^T [M]\{Q_j\} R_{aj} \qquad (13.21)$$

where the symbol "&" represents the yet-to-be-determined combination method; ξ_{ij} is the displacement coefficient of the ith mode due to jth independent motion; and R_{aj} is the acceleration response spectra, in consistent unit, of the jth independent support motion. A set of support motions in x, y, and z directions are considered as three independent motions.

13.2.2 Combination of Response Spectra Analysis Results

In an analysis of a structural system subject to ground or support motion, the modal response is determined by Eq. (13.21), whereas the total response is determined by the superposition expressed in Eq. (13.10). Both equations assume that the signs of the responses are properly maintained, so the individual components of the response are summed directly. However, in the response spectra method, only the maximum value without consideration of the sign is available. The loss of the proper sign requires special treatment on the combination of the results. The combination method for earthquake analysis generally follows the U.S. Nuclear Regulatory Commission (NRC) Regulatory Guide 1.92 [11].

(a) Spatial Components Combination

Because the proper signs of the nodal displacements are required for calculating the forces and stresses of the piping system, the mode shape has to be maintained as long as possible during the combination. The combination that does not alter the mode shape shall be performed first. This prompts the combination of the spatial components as given in Eq. (13.21) to be performed before the inter-mode combination. The modal forces and moments of the piping can still be properly calculated after the spatial combination is completed, but not after the inter-mode combination.

The combination of the spatial components is done statistically by the square root of the sum of the squares (SRSS) method. This is justified for independent motions, which will not likely have their maximum values occur at the same time. Furthermore, most independent motions are in mutually orthogonal directions, such as in x, y, and z motions; the combination by SRSS is the only method logical in this case. However, if the motions are not independent, such as when the piping at the same floor is separated into two groups, the combination has to be done absolutely. This is the reason why the supports should not be arbitrarily divided into groups. Due to the probability nature, the equation is solved for one independent motion at a time. For independent motions, the total result for each mode is calculated by combining the spatial and independent modal results as

$$\xi_i = \sqrt{\xi_{i,1}^2 + \xi_{i,2}^2 + \xi_{i,3}^2 + \cdots} = \sqrt{\sum (\xi_{i,j})^2} \qquad (13.22)$$

where ξ_{ij} is defined in Eq. (13.21). Once the modal displacement coefficient, or coordinate, is determined, the internal modal forces and moments of the piping system are calculated using the modal displacement vector $X_i = \xi_i \Phi_i$ and the stiffness matrix $[K]$. The element forces and moments are calculated using element stiffness matrices.

(b) Modal Combination

The total response of the system is determined by combining the responses from all modes. This combination is termed as modal combination.

Modal combination includes internal modal forces and internal modal moments, as well as modal displacements. The internal forces and moments have to be calculated before the modal results are combined. Engineers may be incorrectly tempted to postpone the calculation of the internal forces and moments until after the final combined displacements are calculated. This might save some computations, but it is not correct, because the final combined displacements have lost the proper signs for correctly determining the internal forces and moments.

Again, because the maximum values of different modes are not likely to occur at the same time, the statistical SRSS method can generally be applied for modal combination too. However, to take into account the possibility of the maximum values occurring at the same time for modes vibrating at similar frequency, a modification is needed for closely spaced modes. The modes with frequencies spaced less than 10% apart (10% rule) are called close modes. More specifically, all modes that satisfy the following relation are called closely spaced modes:

$$\frac{\omega_j - \omega_i}{\omega_i} \leq 0.1 \quad \text{where } \omega_i \leq \omega_j \qquad (13.23)$$

where ω_i and ω_j are the natural circular frequencies of ith and jth modes, respectively. The grouping starts from the lowest frequency mode.

The SRSS rule for modal combination applies to the modal groups instead of individual modes. That is, the modal results are combined absolutely within the closely spaced modes to become modal group results. The modal group results are then combined by the SRSS method, that is,

$$R = \sqrt{\sum_1^{n_g} \left(\sum_1^{n_m} |R_i| \right)^2} \qquad (13.24)$$

where R is the combined response for each component of force, moment, displacement, and so forth. R_i is the modal response of the ith mode. The notation n_g represents the number of closely spaced modal groups, whereas n_m represents the number of modes in each closely spaced group. For widely spaced modes, each group contains only one mode. If all modes are widely spaced, then $n_m = 1$ and $n_g = n$, the number of modes.

Equation (13.24) is the so-called on-off interface approach, which combines the modal results either by SRSS or by absolute (ABS) method. This method is generally too conservative without considering the degree of closeness and other factors. A more gradually interfaced combination, called complete quadratic combination (CQC), as shown in Eqs. (13.25) and (13.26), can also be used [11].

$$R = \sqrt{\sum_{i=1}^{n}\sum_{j=1}^{n} \varepsilon_{ij}(R_i R_j)} \tag{13.25}$$

$$\varepsilon_{ij} = \frac{\sqrt{\zeta_i \zeta_j}}{(\zeta_i + \zeta_j)/2}\left[1 + \left(\frac{f_i - f_j}{\zeta_i f_i + \zeta_j f_j}\right)^2\right]^{-1} \tag{13.26}$$

where ζ_i and f_i are damping ratio and frequency, respectively, of the ith mode and so forth. It should be emphasized that the signs of R_i and R_j in Eq. (13.25) are important and shall be preserved. Equation (13.26) is called the Rosenblueth [12] correlation coefficient. Another equally popular correlation is called the Der Kiureghian [13] correlation coefficient as given by Eq. (13.27).

$$\varepsilon_{ij} = \frac{8\sqrt{\zeta_i \zeta_j}(\zeta_i + r\zeta_j)r^{3/2}}{(1 - r^2)^2 + 4\zeta_i \zeta_j r(1 + r^2) + 4(\zeta_i^2 + \zeta_j^2)r^2} \tag{13.27}$$

where $r = \omega_i/\omega_j = f_i/f_j$.

The interaction between any two modes depends on the damping difference as well as the frequency difference. For any two modes with the same damping ratio and frequency, the cross modal coefficient, ε_{ij}, becomes 1.0, and the two modes are combined algebraically. Again, it should be emphasized that the two modes with identical frequency are combined algebraically, not absolutely, in CQC method.

The combinations are operated on all components of force, moment, displacement, rotation, acceleration, support load, and so forth. Occasionally, there are attempts to just combine the displacement components and calculate all other components by multiplying the combined displacements with the stiffness matrix. This is incorrect because the signs of the displacements will be lost after the combination. Without the proper signs of the displacement, calculation of the element force and moment is impossible.

13.2.3 Comparison of Modal Combination Methods

The methods of combining modal results present some confusion to many piping engineers. We have the SRSS, ABS, CQC, and algebraic methods; they all have been used in one situation or another. However, we do not really have a clear picture of as to when and why a certain method is used.

For earthquake response, the response spectra approach and the probabilistic nature call for SRSS combination in most occasions. The question is: When there are two modes that are vibrating at almost the same pace, would not their peaks coincide with each other most of the time? The question is valid, but also opens a can of worms, as they say in the field. For closely spaced modes, the simplest solution is to add them absolutely, which is the most direct and reliable, but it is also considered too conservative. To trim all the fat, the CQC method is recommended. It appears that the CQC method offers a more realistic combination than the ABS method in most situations. To get some picture of the problem, two sets of modal responses with identical frequency are combined with the above-mentioned methods for comparison.

420 Chapter 13

In piping, as in many other common structures, the situations that most likely to get two modes with identical natural frequency are as follows: (1) two modes lie on two mutually perpendicular planes of a straight section of pipe or structure; (2) at two similar configurations remotely connected and separated by considerable restraints. The latter case has little significance on their combination, because the two modes are not geographically related. In other words, they do not interact with each other. Therefore, only cases with two modes occupying two perpendicular planes of the same section of pipe are of interest. The cantilever as given in Fig. 13.1 is a good candidate to illustrate this discussion. The comparison table shown in Table 13.1 is based on further simplification of using only one mass located at the top of the beam. The same comparison applies also to general beams with multiple mass points.

For a straight beam or column structure, there are two natural modes vibrating in mutually perpendicular planes with the same frequency. Take the cantilever shown in Fig. 13.1, for instance, where every natural vibrating mode in the x-y plane brings another similar mode in the z-y plane. Due to the symmetric nature, both modes have the same shape and natural frequency. However, in actual applications, the orientations of the modes do not always lie on principal coordinate planes. Depending

TABLE 13.1
COMPARISON OF DIFFERENT METHODS IN COMBINING CLOSELY SPACED MODES

Mode Orientations (Two modes with same frequency)		A 0°		B 45°		C 30°	
Components		X -	Z -	X -	Z -	X -	Z -
Mode Shapes (Same Scales)	Mode-1	1.0	0.0	0.7071	-0.7071	0.866	-0.500
	Mode-2	0.0	1.0	0.7071	0.7071	0.500	0.866
Mode Participation Coefficients	X-Quake Mode-1	1.0		0.7071		0.866	
	X-Quake Mode-2	0.0		0.7071		0.500	
	Z-Quake Mode-1	0.0		-0.7071		-0.500	
	Z-Quake Mode-2	1.0		0.7071		0.866	
Responses on unit Response Spectrum	X-Quake Mode-1	1.0	0.0	0.5	-0.5	0.750	-0.433
	X-Quake Mode-2	0.0	0.0	0.5	0.5	0.250	0.433
	Z-Quake Mode-1	0.0	0.0	-0.5	0.5	-0.433	0.250
	Z-Quake Mode-2	0.0	1.0	0.5	0.5	0.433	0.750
SRSS Mode Combinations	X-Quake	1.0	0.0	0.707	0.707	0.791	0.612
	Z-Quake	0.0	1.0	0.707	0.707	0.612	0.791
ABS Mode Combinations	X-Quake	1.0	0.0	1.0	1.0	1.0	0.866
	Z-Quake	0.0	1.0	1.0	1.0	0.866	1.0
CQC Mode Combinations	X-Quake	1.0	0.0	1.0	0.0	1.0	0.0
	Z-Quake	0.0	1.0	0.0	1.0	0.0	1.0
Total Response SRSS Combined X- and Z- Quakes	SRSS	1.0	1.0	1.0	1.0	1.0	1.0
	ABS	1.0	1.0	1.414	1.414	1.323	1.323
	CQC	1.0	1.0	1.0	1.0	1.0	1.0

on minute mathematical, as well as structural, impurity associated with the computation, the modes may lie at an angle from the principal coordinate planes. In Table 13.1, three modal orientations are investigated. In case A, the modes lie on the principal planes, in case B the modes are located on planes bisecting the principal planes, and in case C the modes found on the planes 30 deg. from the principal planes.

For case A, everything appears good no matter which combination method is used. However, in case B and case C, the results differ considerably depending on the combination method used. Case B is similar to the case investigated by Wilson's team [13] that resulted in their alarming concerns. Wilson's conclusions were based on a single direction of earthquake, which tends to amplify the irregularities on a three-dimensional analytical model. Take the X-quake in case B, for instance; the SRSS method under-predicts the x component response by about 30%, but overestimates the z component by about 70%. (The use of "infinite times," i.e., 0.707/0.0, is spared for comparing on equal footing [14].) On the other hand, the ABS method gives the correct x component, but over-predicts the z component by 100%. The CQC method, in this case, is equivalent to the algebraic summation, which is generally used in time-history analysis.

In piping analysis, three directions of earthquake response spectra are generally applied, with two horizontal spectra that are either identical or at similar magnitudes. In this case, the combined results, as tabulated in the combined X- and Z-quakes block, are the same for both SRSS and CQC combination, whereas the results by ABS method are always on the conservative side. Whether this is enough to show that SRSS is adequate for general piping analysis is not clear, but it appears that the old reliable SRSS method is actually better than originally thought. The ABS combination of closely spaced modes (which is implemented in most computer programs), although conservative, is quite inconsistent. This is also true for the old CQC method adopted by Revision 1 of Regulatory Guide 1.92 [11], which used the absolute value of all the cross mode product terms.

13.2.4 Puzzles of Absolute Closely Spaced Modal Combination

In performing earthquake analysis, especially on simple structures such as the stick or stack model, we often came across very puzzling results from very reliable computer software packages. Because the stick model is often the first analytical model an engineer uses to verify the program, these puzzling results are very disturbing. Engineers are simply confused by the fact that different computer programs can give very different results using the same model under the same response spectra. Even with the same program, we might one day get one set of results while another day, with a very slight modification, get a set of completely different results. This, of course, raises questions regarding the validity of the programs. In fact, all these are caused by imperfect procedure and have nothing to do with the computer programs.

In piping engineering, the piping is generally considered as three-dimensional, that is, even a simple system is handled with a three-dimensional approach. A piping analysis program generally cannot even handle a simple system in a two-dimensional fashion. Take the stick model shown in Fig. 13.1, for instance; it was handled as a three-dimensional object even though it could be solved with a two-dimensional formulation. That is the reason why each frequency has two vibration modes: one in a certain plane, and the other at a plane perpendicular to the first one. These two modes are, of course, orthogonal to each other, a basic property of eigenvectors. This is the most common situation, if not the only situation — a structural system has two natural modes with identical or very close frequencies.

Using the stick model as an example, the calculated mode shape orientations, most of the time, will follow the direction of principal coordinate axes. However, this is much less than a certainty, because the orientation is controlled by a slight mathematical, as well as structural, impurity. The same system can have a different calculated mode shape orientation just by changing the number of modes calculated and so forth. The modes may orient in one way when only two modes are calculated, and orient in completely different way when four modes are calculated. Therefore, for a stick model, we might have calculated mode shapes of the first two modes oriented in many different ways as shown in Table 13.1.

The response summary in Table 13.1 assumes that the system is subjected to the same x- and z-direction earthquake response spectra. The table assumes three situations of mode orientations. In actual applications, all these orientations are probable, and something in between is also equally probable. From the combination results tabulated, it is obvious the ABS combination for closely spaced mode can generate fairly unpredictable, but nevertheless conservative, results.

Table 13.1 comparison is based on assumed mode shape orientations. The actual mode shape orientation of the stick model is beyond the control of the analysts. The orientation, however, can be controlled by adding a small branch or appendage, which always exists in a real system. Figure 13.5 shows a system having a vertical tower with a short horizontal branch. The system is subjected to the same response spectra at both horizontal directions. By common sense, we will expect the system to generate the same force and stress regardless of how the branch is oriented. Therefore, system (a) should have the same force and stress as in system (b). However, with the ABS combination for closely spaced modes, system (b) produces 41% more force and stress than system (a). This is not a puzzle, but just the side effect of the ABS combination on closely spaced modes.

13.2.5 Compensation for the Higher Modes Truncated

With continuously distributed mass, a piping system has infinite vibration modes. Since it is impossible to calculate all the modes in the real world, only limited modes at the lower frequency region are included in a practical analysis. The modes not included are cut off from the analysis. The contributions of those cutoff modes, therefore, are not included in the analysis resulting in a non-conservative design. The contribution of the cutoff mode somehow needs to be included in the analysis. Because these cutoff modes generally vibrate coinciding with the rigid motion of the support, the compensation is called rigid body compensation or residual mode compensation.

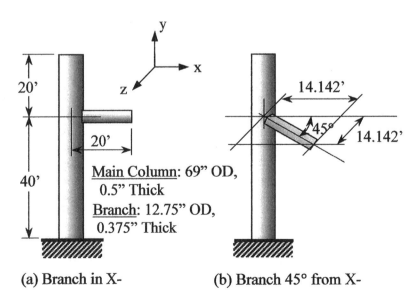

(a) Branch in X-   (b) Branch 45° from X-

FIG. 13.5
EFFECT OF BRANCH ORIENTATION ON ABSOLUTE MODAL COMBINATION

From a typical response spectra curve, such as the one shown in Fig. 13.3, we know that if the system is sufficiently rigid, it will essentially experience a constant acceleration called rigid body acceleration or zero period acceleration (ZPA). The total contribution of the cutoff modes is considered as part of this ZPA. The response of the calculated modes also includes part of this ZPA. Therefore, the first step in determining the contribution due to cutoff modes is to figure out the portion of ZPA included in the calculated modes. Several approaches [5, 15–17] have been suggested for this application. A standard approach is being developed by the U.S. NRC Regulatory Guide 1.92 [11] for use in nuclear facility design. The following follows the approach developed by Biswas and Duff [15].

Following the same concept in Eq. (13.16), the system acceleration for rigid body response can be written as

$$\{\ddot{X}_R\} = [Q]\{A_R\} \tag{13.28}$$

where $\{A_R\}$ is the ZPA vector consisting of all components from all support groups. From Eqs. (13.10) and (13.21), the rigid body acceleration can also be expressed by modal superposition as

$$\{\ddot{X}_R\} = \sum_{i=1}^{N} \left(\{\Phi_i\}^T [M][Q]\{A_R\}\{\Phi_i\} \right) \tag{13.29}$$

where N is the total number of degrees of freedom. Because rigid acceleration is a constant value, the algebraic modal summation is used due to certainty. When an analysis takes n modes into consideration, the portion of the rigid body motion already accounted for is

$$\{\ddot{X}_{Rn}\} = \sum_{i=1}^{n} \left(\{\Phi_i\}^T [M][Q]\{A_R\}\{\Phi_i\} \right) \tag{13.30}$$

where n is the number of modes included in the analysis. The residual rigid acceleration needs to be added to the analysis is the difference between the values given by Eqs. (13.28) and (13.30), that is,

$$\{\ddot{X}_{Rr}\} = \{\ddot{X}_R\} - \{\ddot{X}_{Rn}\}, \quad \{\ddot{X}_{Rr}\} \geq \{0\} \tag{13.31}$$

The residual acceleration is converted to system inertia force, which is then used to calculate the system response statically. The displacements, forces, and moments so calculated are then added absolutely to the corresponding components calculated by using only n modes. The preceding method appears to be the most natural and kink-free. Engineers might also be tempted to calculate residual displacement directly following the same procedure used for calculating residual acceleration, and then use the residual displacement to calculate system response. However, this residual displacement approach often introduces a mathematical kink on the calculation of the forces and moments. This is due to the fact that the calculation of forces and moments require perfect balance of displacement and rotation distribution, which is often unachievable due to summation and subtraction of so many imperfect mode shapes.

The compensation process can also be viewed as a way to make the system acceleration be at least equal to ZPA. This rule-of-thumb approach may give engineers a better feeling about the whole procedure. Using this common sense approach, the residual acceleration is calculated by subtracting the rigid acceleration with the calculated system acceleration using limited modes.

13.2.6 Design Response Spectra

The response spectra method depends on the availability and reliability of the design response spectra. Although seismic ground motions are vastly different at different locations and at different times, a statistical study of a large number of ground motions has concluded that design response spectra can be constructed once the maximum acceleration is known. Based on the work done by Newmark et al [18], the U.S. NRC has published two seismic design response spectra in Regulatory Guide 1.60 [19] for designing nuclear power plants. Two charts, one for horizontal direction and the other for

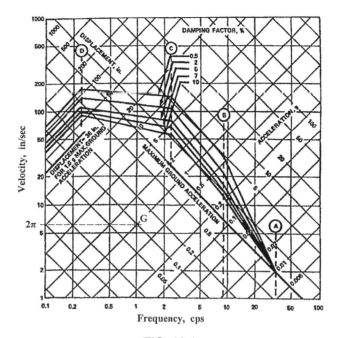

FIG. 13.6
HORIZONTAL DESIGN RESPONSE SPECTRA [19] - SCALED TO 1.0 G HORIZONTAL GROUND ACCELERATION

vertical direction (Figs. 13.6 and 13.7), are provided. Both charts are scaled to 1.0 g horizontal ground acceleration. The design horizontal acceleration differs from site to site. For a site to be designed for 0.5 g horizontal acceleration, for instance, the design response spectra would be one-half of the value given by the figures.

The figures are given with four logarithm scales. These so-called tripartite log-log graphs are a curiosity to piping engineers, but are often used by seismic engineers for easy assembly of the vast data recorded for many different earthquakes. From these figures, one can read response displacement, velocity, and acceleration of a given frequency directly. One may also use only the two main coordinates, frequency and velocity, in which case it should be remembered that velocity is a pseudo-velocity. It can be converted to displacement and acceleration using the relations given by Eq. (13.20).

For readers who are curious about the way the coordinates are constructed, the layout of the graph is briefly explained as follows. Using frequency and velocity as main coordinates, take the logarithm of Eq. (13.20) and remember that $\omega = 2\pi f$; thus, we have

$$\ln(R_a) = \ln(2\pi) + \ln(f) + \ln(R_v); \quad \ln(R_d) = \ln(R_v) - \ln(f) - \ln(2\pi)$$

Therefore, if we use the same logarithm scale for all the quantities involved, we have the displacement coordinate located with same increment of velocity, but also the same decrement of frequency, that is, 45 deg. toward increasing velocity and decreasing frequency. The acceleration coordinate, on the other hand, is located 45 deg. toward increasing velocity and increasing frequency. It is interesting to note that by using cycles/sec (cps), in., in./sec, and g as units for frequency, displacement, velocity, and acceleration, respectively, the main coordinate axes of frequency, displacement, and acceleration, by coincidence, intersect at the same point. Consider the point G in Fig. 13.6; this is the point for the 1.0-in. displacement at 1 cps. At this same point, the acceleration $a = d\omega^2 = 1.0 \times (2\pi*1)^2 = 39.478$ in./sec^2 = 0.102 g. At this point, the velocity $v = d\omega = 1.0 \times 2\pi = 2\pi$ in./sec, which anchors the whole coordinate system. The 1.0 cps, 1.0 in, and 0.1 g coordinates intersect at the same point, and so do the 10 cps, 0.1 in., and 1.0 g coordinates.

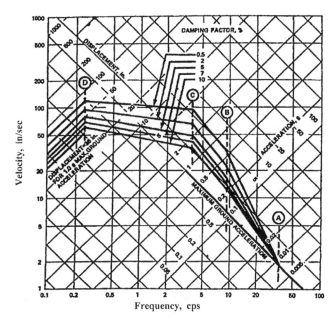

FIG. 13.7
VERTICAL DESIGN RESPONSE SPECTRA [19] - SCALED TO 1.0 G HORIZONTAL GROUND ACCELERATION

At a given site, the vertical design response spectra is roughly the same as the horizontal response spectra at the high frequency range, but is only about 2/3 as big in the low frequency range. The design response spectra given by Figs. 13.6 and 13.7 are for ground motion. In general, they are not used directly for piping design. The response spectra used in piping design are called building or floor response spectra. They are derived by a series of quite complex procedures. First, the design ground motion response spectra curve of the proper damping is used to generate the artificial earthquake time-history movement. This artificial time-history motion shall produce a set of response spectra that closely matches the design spectra throughout the whole frequency range. The artificial time-history motion is then used in time-history analysis of the building to which the piping is attached. The weights of the piping and other equipment are estimated and added to the building analytical model. The

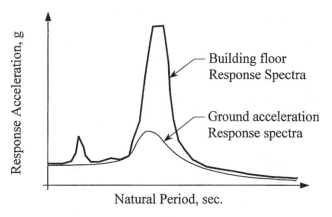

FIG. 13.8
BUILDING AMPLIFICATION ON RESPONSE SPECTRA

time-history analysis is required to produce time-history response of the building structure. Finally, the time-history response of the building is used to generate the building response spectra, with various damping values, at various points of the building. In many cases, each floor of the building is considered a rigid plane of the building and has its unique response spectra. The building response spectra to be used in piping design are, therefore, highly influenced by the soil-structure interaction and building mass-stiffness characteristics. They are generally amplified at a certain range of the frequency as shown in Fig. 13.8. The peak response at higher building elevation can be amplified by as much as several hundred percent. The response spectra at different floors of the same building are generally amplified at the same frequency range. They are, therefore, not entirely independent. In this case, the envelope spectra should be used if the piping runs through multiple floors. Considering them as independent response spectra groups might lead to an un-conservative analysis.

13.3 ACCOUNT FOR UNCERTAINTIES

A piping analysis involves many uncertainties such as material properties, support stiffness, damping values, approximation in modeling, mathematical round off errors, and so forth. In static analysis, a 10% variation in these uncertainties generally also means a 10% variation in analysis results. These predictable variations of analysis results are, in a way, taken care of by the design safety factor included in setting the allowable stresses.

In dynamic analysis, though a 10% variation in these material and other uncertainties generally also causes a roughly10% variation in calculated natural frequency, but could produce a much greater variation in analysis results due to potential resonance effect. At near-resonance frequency range, a slight change of frequency may mean a substantial change in response. This larger-than-usual change in response is generally beyond the amount implied by the design safety factor. Proper additional account is needed for the effect of the uncertainties in dynamic analyses.

The large variations of dynamic response mainly originated from the variation of the natural frequency calculated. In other words, it is due to the uncertainty of the calculated natural frequency. Depending on the type of application, there are several ways of accounting for the effect of natural frequency uncertainties.

For building response spectra, there are spikes corresponding to the building natural frequencies. The spectrum at these spikes changes sharply against the frequency. To account for the uncertainties, the spectra curve is broadened before being used in the design. Figure 13.9 shows the 10% broadening

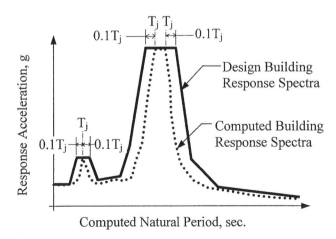

FIG. 13.9
10% BROADENED RESPONSE SPECTRA

scheme. The original calculated response spectra curve is given as a dotted line. The peaks are broadened by 10% in each direction of every peak natural period. The curve is then maintained parallel to the contour of the spike, down to the base. The base portion is then smoothed out. This broadened response spectra curve ensures that the peak response is not missed even if the calculated frequency is 10% different than the actual frequency.

For harmonic analysis, which is to be discussed in the next section, a slightly different approach is used. In harmonic analysis, the response spectra can be calculated by Eq. (12.38), which is re-plotted in Fig. 13.10 for $\zeta = 0.1$ and $\zeta = 1.0$. For a given calculated frequency, the actual frequency lies inside the probable frequency range as shown in the figure due to uncertainties involved in the calculation. This probable frequency range produces a probable response range, which is considerably bigger than the frequency range. Assuming a 10% probable frequency uncertainty, a calculated frequency ratio of 0.75 actually represents a probable natural frequency ratio of between 0.675 and 0.825. In a system with 0.1 damping ratio (10%) as shown in the figure, the calculated dynamic load factor (DLF) is 2.162, whereas the DLF corresponding to the probable frequency range varies from 1.783 to 2.782. The upper range deviation is (2.782-2.162)/2.162 = 0.287, that is, a 10% frequency variation produces a 29% response variation in this case. The response variation is larger for piping systems, which generally have a damping ratio of 5% or less.

In harmonic analysis, to account for the uncertainties, the response calculation is conducted in four steps. First, the response of the calculated frequency is calculated. Then, the probable frequency range is set by the shift criteria, which is generally set to 10%. If the 10% shift criterion is adopted,

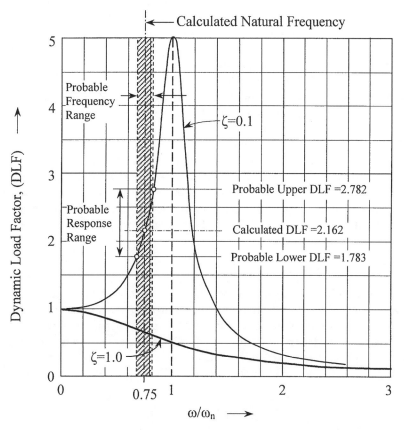

FIG. 13.10
FREQUENCY SHIFT TECHNIQUE ON HARMONIC ANALYSIS

the probable frequency is located between 10% below the calculated frequency and 10% above the calculated frequency. The responses are then calculated for the upper and lower probable frequencies. In addition, it is also checked to ensure that the resonance does not occur inside the probable frequency range. If it does, then the resonant response is calculated. The maximum probable response is the maximum value calculated from all the steps.

For impulse loading, the method of accounting for the uncertainties is similar to the broadening of the spectra. Take the ramp loading as shown in Fig. 12.19, for instance, where the theoretical DLF is the one shown in dotted line. The theoretical DLF consists of many half-waves. To account for the uncertainty of the natural period and ramp duration, the curve has to be smoothed and broadened for design use. The solid line is the curve used by the American Society of Mechanical engineers (ASME) B31.1 [20] for designing piping systems subject to a safety relief valve force.

13.4 STEADY-STATE VIBRATION AND HARMONIC ANALYSIS

It is generally accurate to say that all piping systems experience some type of steady-state vibration. This is because of the large amount of potential causes and sources of vibration present in the surrounding. For instance, the equipment to which the piping is connected generally has some type of vibration due to internal turbulence of fluid or rotating parts. Distillation columns, chemical reactors, hydrocarbon fractionators, and offshore platforms, for example, all vibrate noticeably due to internal or external sloshing of fluid. Even the well-controlled microscopic vibration at rotating equipment can excite significant piping vibration under favorable conditions. In addition to these anchor or support vibrations, a piping system also faces many vibration sources on the piping itself. The most notorious ones are the pulsation flow from reciprocating pumps and compressors, blade passing turbulence of centrifugal pumps and compressors, vortex shedding turbulence from both external and internal flows, and general turbulence of internal flow. The water hammer type loads are discussed separately, because they do not generally produce steady-state vibration.

Piping vibration is very unpredictable. It is very difficult to take into account during the design phase. Therefore, besides the systems handling predictable pulsation flow that are studied beforehand and treated with special support systems, most of the vibration problems are handled when problems arise. However, because all piping systems vibrate more or less, it is very difficult to judge which system is acceptable and which is not. The acceptability involves both stress and psychological considerations. Plant engineers are often alarmed by perceptible vibrations that are actually harmless. It is within human nature to feel uncomfortable about any piping that is vibrating noticeably. Kellogg [21] has observed that an amplitude as small as 1/16 in. (1.5 mm) in large-size piping is sufficient to cause alarm if the piping is in an enclosed building, and ¼ in. (6 mm) if it is located in an open structure. This is the psychological aspect of vibration tolerance. In the following, we will discuss the stress level aspects.

13.4.1 Basic Vibration Patterns

To correlate field vibration, we have to understand the basic vibration patterns. We shall use the SDOF formulas presented in previous chapter to find out what are the basic vibration patterns. For a general piping system with many degrees of freedom, it is just a matter of doing the superposition of many SDOF systems. The force equilibrium of an SDOF system subjected to harmonic motion is given by Eq. (12.34), whose general solution is the combination of the complementary function given by Eq. (12.27) and the particular solution given by Eq. (12.35). Therefore, the general solution is

$$y = e^{-\zeta \omega_n t} \left[\frac{\dot{y}_0 + y_0 \zeta \omega_n}{\omega_n} \sin \omega_n t + y_0 \cos \omega_n t \right] \\ + \frac{F_0}{k} \frac{1}{(1-\beta^2)^2 + (2\zeta\beta)^2} \left[(1-\beta^2) \sin \omega t - 2\zeta\beta \cos \omega t \right]$$

(13.32)

The symbol ω_d in Eq. (12.27) is replaced with ω_n for simplicity.

The motion of the system consists of two parts. The first part of the motion is due to free vibration, which oscillates with a frequency the same as natural frequency. This part of the motion would damp out quickly if the disturbance discontinued. The second part of the motion is due to harmonic force F_0. If the force persists, the motion oscillates continuously at a frequency that is the same as the forcing frequency.

From the preceding phenomena, we can conclude that the piping vibration has two general basic patterns. When the harmonic force dominates, the system vibrates at the same frequency as the forcing frequency. In other words, the observed frequency is the same as the forcing frequency. This phenomenon would help the engineers in their search for the source of the vibration. On the other hand, if the harmonic force is insignificant, then the piping vibrates in one of its natural frequencies. This type of vibration is generally created by random impulses inside the piping. It is also generally accompanied by banging noises, which occasionally can be very loud.

13.4.2 Allowable Vibration Displacement and Velocity

The main concern in steady-state vibration is the significance of a given vibration. The most direct and accurate approach in assessing the vibration is to calculate or measure the stress. Direct calculation of stress, discussed in the next section, is possible if the geometry and forcing function are well defined. However, with a given vibration, the direct measurement of the stress related characteristic is possible and preferred.

The most directly measurable quantity of the vibration is the displacement, which can be measured not only by electronic equipment, but also by simple ruler or even by eyesight. ASME OM, Part-3 [22] has outlined the procedure to determine both allowable displacement and velocity. Engineers should follow the latest edition of ASME OM-3 to evaluate the displacement and velocity. The following discusses the general background for setting the allowable displacement and velocity.

Evaluation of vibration displacement and velocity involves numerous uncertainties. Therefore, it is necessary to be very conservative. This is even more so when dealing with displacement, because the same amount of displacement does not necessarily produce the same stress in a given system. A quicker displacement produces higher stress and vice versa. As shown in Fig. 13.1 for cantilever beam and Fig. 13.11(a) for simple supported beam, a pipe can vibrate in many different shapes. A same

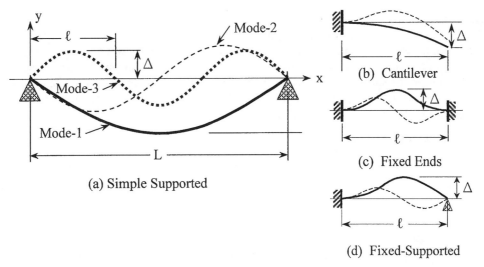

FIG. 13.11
VIBRATION OF BEAMS

amount of displacement produces much higher stress vibrating at mode 3 than vibrating at mode 1. This shows the requirement of evaluating the vibration displacement in conjunction with system geometry and vibration mode.

Figure 13.11(a) shows the first three vibrating modes for a simple supported pipe beam. It is clear that the third mode vibrates in the same proportion or pattern as the first mode, except that the beam length, ℓ, is reduced to one third of the pipe span, L. Therefore, the natural frequency of the nth mode is calculated in the same manner as the first mode, by properly adjusting the beam length. The natural frequency of the nth mode of the simple supported beam is calculated as [1]

$$\omega_n = \frac{n^2\pi^2}{L^2}\sqrt{\frac{EI}{m}} = \frac{\pi^2}{(L/n)^2}\sqrt{\frac{EI}{m}} = \frac{\pi^2}{L_n^2}\sqrt{\frac{EI}{m}} \tag{13.33}$$

where L_n is the half-wavelength of mode n, and m is the mass per unit length of the beam. Even for straight beams with other end conditions (Fig. 13.11b, c, and d) that do not have the exact proportional shapes for all modes, they also have the natural frequencies given by an equation similar to the above. The second mode shape of one type of the beams is often the same as the fundamental (first) mode shape of one of the four basic beams shown. As shown in Fig. 13.11(c), the second mode shape of the fixed-fixed beam is the same as the first mode shape of the fixed-supported beam. The second mode shape of the fixed-supported beam is the same as the first mode shape of the same beam, coupled with the first mode shape of the simple supported beam.

The vibration stress of the pipe beam can be estimated from the shape of the deflection. The beams can be interpreted, in a broader sense, as the ones either fixed by actual anchors and support, or defined by the inflection points of the mode shapes. The beam representing mode 3 in Fig. 13.11(a) belongs to the latter interpretation. The vibration stress is mainly from the bending moment, which is proportional to beam curvature as

$$M = -EI\frac{d^2y}{dx^2} \quad \text{and} \quad S_b = \frac{M}{Z} = \frac{MD}{2I} \tag{13.34}$$

Equation (13.34) was given previously as Eq. (5.4). Because only the absolute value of the stress is of interest, the (−) will be omitted in this discussion. Once the deflection shape is known, the bending stress can be readily calculated.

(a) Deflection shapes of common beams

Although the pipe mass is uniformly distributed, its inertial force at a given point is proportional to the vibration acceleration, thus displacement, at that point. A vibration beam is therefore subject to forces distributed proportional to the displacement rather than uniformly. The vibration deflection shape is in between the shape for a concentrated load and the shape for a uniform load. From Biggs [1], the first mode vibration deflection curves together with the natural frequencies of the uniform beams given in Fig. 13.11 are listed as follows:

- Simple supported beam (Fig. 13.11a)

$$y = \Delta \sin a_1 x$$

$$a_1 = \frac{\pi}{\ell}, \quad \omega = a_1^2\sqrt{\frac{EI}{m}} = \frac{\pi^2}{\ell^2}\sqrt{\frac{EI}{m}} \tag{13.35}$$

- Cantilever beam (Fig. 13.11b)

$$y = 0.5\Delta[-0.7341(\sinh a_1 x - \sin a_1 x) + \cosh a_1 x - \cos a_1 x]$$

Dynamic Analysis — Part 2: MDOF Systems and Applications 431

$$a_1 = \frac{0.597\pi}{\ell}, \quad \omega = \frac{(0.597\pi)^2}{\ell^2}\sqrt{\frac{EI}{m}} \tag{13.36}$$

- Fixed ends beam (Fig. 13.11c)

$$y = 0.63\Delta[-0.9825(\sinh a_1 x - \sin a_1 x) + \cosh a_1 x - \cos a_1 x]$$

$$a_1 = \frac{3}{2}\frac{\pi}{\ell}, \quad \omega = \left(\frac{3\pi}{2\ell}\right)^2\sqrt{\frac{EI}{m}} \tag{13.37}$$

- Fixed-supported beam (Fig. 13.11d)

$$y = 0.6624\Delta[-1.0007(\sinh a_1 x - \sin a_1 x) + \cosh a_1 x - \cos a_1 x]$$

$$a_1 = \frac{5}{4}\frac{\pi}{\ell}, \quad \omega = \left(\frac{5\pi}{4\ell}\right)^2\sqrt{\frac{EI}{m}} \tag{13.38}$$

(b) Maximum bending moments

The stress before applying any stress intensification is called un-intensified stress. The un-intensified stress is calculated by dividing the bending moment by the section modulus. The maximum un-intensified stress does not occur at the same point that produces the maximum displacement, Δ. The maximum stress locations, however, are well known for these beams. By differentiating the displacement curve with respect to x twice, we will have the equation for the curvature, which is proportional to bending moment as given by Eq. (13.34). By simply taking the curvature value at the maximum bending stress location, multiplied with EI, we have the maximum bending moments of the first vibration mode for the beams as

- Simple supported beam, at center span point

$$M = 1.0 EI\left(\frac{\pi}{\ell}\right)^2 \Delta \tag{13.39}$$

- Cantilever beam, at anchor point

$$M = 1.0 EI\left(\frac{0.597\pi}{\ell}\right)^2 \Delta = 0.356 EI\left(\frac{\pi}{\ell}\right)^2 \Delta \tag{13.40}$$

- Fixed ends beam, at fixed points

$$M = 1.26 EI\left(\frac{3\pi}{2\ell}\right)^2 \Delta = 2.835 EI\left(\frac{\pi}{\ell}\right)^2 \Delta \tag{13.41}$$

- Fixed-supported beam, at fixed point

$$M = 1.325 EI\left(\frac{5\pi}{4\ell}\right)^2 \Delta = 2.07 EI\left(\frac{\pi}{\ell}\right)^2 \Delta \tag{13.42}$$

The first mode moment for a fixed-supported beam is also applicable to the second mode moment of the fixed-ends beam. This will be used later as the basic moment in developing allowable velocity criterion.

The maximum bending moment can be expressed in term of the maximum vibration displacement in a general form as

$$M = C_{ED} EI\left(\frac{\pi}{\ell}\right)^2 \Delta \tag{13.43}$$

where C_{ED} is the end coefficient with respect to displacement. C_{ED} ranges from 0.356 for a cantilever beam to 2.835 for a fixed-ends beam.

(c) Stress due to displacement and allowable vibration displacement

The nominal un-intensified vibration stress can be calculated from Eq. (13.43) by dividing the maximum moment with the section modulus as

$$S_{bn} = \frac{M}{Z} = C_{ED}\frac{EI}{Z}\left(\frac{\pi}{\ell}\right)^2 \Delta = C_{ED}\frac{ED\pi^2}{2\ell^2}\Delta \tag{13.44}$$

The above stress does not yet include any stress intensification of fitting and discontinuity. When evaluating vibration stress, the stress used is based on the theoretical stress intensification given by Class-1 nuclear code [23] as C_2K_2, which is roughly equal to twice the stress intensification factor, i, used in B31 [20] codes, that is, $C_2K_2 = 2i$. In setting the allowable vibration displacement, the potential maximum stress intensification throughout the system has to be used. After multiplying the stress intensification, the intensified bending stress becomes

$$S_b = C_2K_2C_{ED}\frac{E\pi^2}{2}\frac{D}{\ell^2}\Delta \tag{13.45}$$

The allowable vibration displacement is the one that generates a bending stress equal to the allowable stress, which is the endurance strength of the material divided by the allowable stress reduction factor. That is,

$$\Delta_{allow} = \frac{S_{allow}}{C_2K_2C_{ED}}\frac{2}{E\pi^2}\left(\frac{\ell^2}{D}\right) \tag{13.46}$$

where allowable stress $S_{allow} = S_{el}/\alpha$. S_{el} is the endurance strength, which equals 1.0 times of the allowable alternating stress at 10^{11} cycles, and 0.8 times of the allowable alternating stress at 10^6 cycles. The allowable stress reduction factor, α, is 1.3 for carbon and low alloy steels, and 1.0 for austenitic stainless and high alloy steels.

Equation (13.46) can be simplified by substituting some known data for routine applications. Using U.S. conventional units of in.-lb system and substituting $E = 29.3 \times 10^6$ psi, the equation becomes

$$\Delta_{allow} = \frac{S_{allow}}{C_2K_2C_{ED}}\left(\frac{1}{145 \times 10^6}\right)\left(\frac{\ell^2}{D}\right) = \frac{S_{el}}{10,000C_2K_2\alpha}\left(\frac{1}{14,500C_{ED}}\right)\left(\frac{\ell^2}{D}\right)$$

By using feet as the unit for the span length as in ASME OM-3, the allowable displacement, in inch, becomes

$$\Delta_{allow} = \frac{S_{el}}{10,000C_2K_2\alpha}\left(\frac{1}{101C_{ED}}\right)\left(\frac{\ell_f^2}{D}\right), \quad (in., lb, \ell_f \text{ in ft}) \tag{13.47}$$

The expression $1/(101C_{ED})$ is equivalent to the value of K in ASME OM-3. From Eqs. (13.39) to (13.42), K values are calculated as:

$K = 0.010$, for simple supported beam
$K = 0.027$, for cantilever beam
$K = 0.003$, for fixed or continuous beams

Equation (13.47) is not very easy to apply due to the difficulties encountered in obtaining the end coefficient, C_{ED}, and the span length, ℓ, in a general piping system. The effective span length is determined by the vibration mode as well as the support and piping layout. Because the allowable displacement is proportional to the square of the effective span length, the uncertainty of the span length is a huge roadblock in applying this equation.

The allowable vibration displacement is based on simple amplitude, which is zero-to-peak value. In contrast, the measurements made in the field often are in double amplitude, which are peak-to-peak values. If the measurement is in peak to peak, the measured value has to be halved before comparing with the allowable vibration displacement.

(d) Stress due to velocity and allowable vibration velocity

The maximum bending moments given by Eqs. (13.39) to (13.42) can be rewritten in terms of vibration velocity, V. The vibration velocity V is equal to $\Delta\omega$, with the circular frequency, ω, as expressed by Eqs. (13.35) to (13.38). Substituting the corresponding expression of ω in Eqs. (13.39) to (13.42), we have

- Simple supported beam, at center span point

$$M = 1.0\text{EI}\left(\frac{\pi}{\ell}\right)^2 \Delta = 1.0\text{EI}\Delta\omega\sqrt{\frac{m}{\text{EI}}} = 1.0V\sqrt{\text{EI}m} \tag{13.48}$$

- Cantilever beam, at anchor point

$$M = 1.0\text{EI}\left(\frac{0.597\pi}{\ell}\right)^2 \Delta = 1.0V\sqrt{\text{EI}m} \tag{13.49}$$

- Fixed-ends beam, at fixed points

$$M = 1.26\left(\frac{3\pi}{2\ell}\right)^2 \Delta = 1.26V\sqrt{\text{EI}m} \tag{13.50}$$

- Fixed-supported beam, at fixed point (also for second mode of fixed ends)

$$M = 1.325\left(\frac{5\pi}{4\ell}\right)^2 \Delta = 1.325V\sqrt{\text{EI}m} \tag{13.51}$$

Because it is desirable to cover the situation for as many modes as possible, the evaluation of the allowable vibration velocity will include the first two modes of each beam type. The preceding equations derived for the first mode are, in fact, also applicable to the second mode, except for the fixed-ends beam whose second mode maximum bending moment is the same as the first mode moment for the fixed-supported beam. By close investigation of the mode shapes given by all straight beams as shown in Fig. 13.11, it is clear that by including the first two mode shapes of each type of the beams actually covers all mode shapes of the straight beam regardless of the end conditions.

The maximum bending moment can be expressed in term of the maximum vibration velocity in a general from, using second mode moment of the fixed-ends beam as basis, as

$$M = C_{\text{EV}} 1.325 \sqrt{\text{EI}m}\, V \tag{13.52}$$

where C_{EV} is the end coefficient with respect to velocity. C_{EV} is 1.0 for fixed-ends and fixed-supported beams, and 0.755 for both simple supported and cantilever beams.

The nominal un-intensified bending stress is calculated by dividing the bending moment with section modulus. By using common approximate formulas for section modulus and moment of inertia, we have

$$\begin{aligned}S_{\text{bn}} &= \frac{M}{Z} = 1.325 C_{\text{EV}} \sqrt{\frac{\text{EI}m}{Z^2}} V = 1.325 C_{\text{EV}} \sqrt{\frac{E(\pi r^3 t)(2\pi r t)\rho C_{\text{W}}}{(\pi r^2 t)^2}} V \\ &= 1.874 C_{\text{EV}} \sqrt{E\rho C_{\text{W}}}\, V \end{aligned} \tag{13.53}$$

where ρ is the density of the pipe metal and C_W is the ratio of the total pipe weight, including content and insulation, to the pipe metal weight.

The stress given by Eq. (13.53) is an un-intensified stress. To compare the allowable stress, the potential maximum stress intensification, C_2K_2, throughout the system has to be included. The equation also shows that the stress is proportional to the square root of the mass involved. This means that the effects of all masses, in addition to the pipe mass, have to be considered. The additional uniform mass, such as the one due to insulation and fluid content, is already taken care of by the C_w factor; the only thing left is concentrated masses. Assuming the concentrated weight increases the stress by a factor of C_X, we have the intensified vibration stress as

$$S_b = 1.874 C_2 K_2 C_X C_{EV} \sqrt{C_W} \sqrt{E\rho} V \qquad (13.54)$$

The allowable vibration velocity is the one that generates a bending stress equal to the allowable stress, which is the endurance strength of the material divided by the allowable stress reduction factor. That is,

$$V_{\text{allow}} = \frac{S_{\text{allow}}}{1.874 C_2 K_2 C_X C_{EV} \sqrt{C_W} \sqrt{E\rho}} \qquad (13.55)$$

The factors involved are as follows: C_2K_2 (roughly the same as twice the value of the stress intensification factor, i, given by ASME B31 code) are defined by the ASME nuclear piping code [23]; C_{EV}, the end coefficient with respect to velocity, is defined by the relationship given by Eq. (13.52); C_W equals the sum of pipe weight and content weight divided by the pipe weight; C_X is the concentrated weight factor yet to be determined. The effect of concentrated mass can be determined from the basic beam formulas for the simple supported beam, as well as the fixed beam, in the same manner. Because ASME OM-3 uses the fixed beam as the basis, the effect of concentrated weight on the fixed beam is calculated in the following.

- For fixed-ends beam with concentrated mass

To have a direct comparison of uniformly distributed mass and concentrated mass, we will first investigate the situation when all pipe mass is concentrated at the mid-span of the beam. When the pipe vibrates, the concentrated mass generates a dynamic force, F, acting at the mid-span point. From fixed beam formulas, we have

$$\Delta = \frac{1}{192} \frac{F\ell^3}{EI}, \quad k = \frac{F}{\Delta} = \frac{192 EI}{\ell^3}$$

$$\omega = \sqrt{\frac{k}{M_{\text{mass}}}} = \sqrt{\frac{192 EI}{\ell^3 (m\ell)}} = \frac{13.856}{\ell^2} \sqrt{\frac{EI}{m}}$$

M_{mass} is the concentrated mass, which is equal to the total pipe mass. The maximum bending moment, occurring at mid-span as well as at both ends, is

$$M = \frac{1}{8} F\ell = \frac{1}{8}(m\ell)\Delta\omega^2 \ell = \frac{m\ell^2}{8} V\omega = 1.732 V \sqrt{EIm} \qquad (13.56)$$

Comparing Eq. (13.56) with Eq. (13.50), it is clear that with the same vibration velocity, the stress generated by concentrated mass is $1.732/1.26 = 1.375$ times the stress generated by the uniformly distributed beam having the same total mass. Therefore, the C_X factor is determined by the ratio of concentrated weight to total pipe weight of the span. That is,

$$C_X = 1 + 1.375 \sqrt{\frac{M_{\text{mass}}}{(m\ell)}} \qquad (13.57)$$

ASME OM has the allowable velocity formula set up in a different format, using allowable factors instead of stress coefficients, for two parameters. It uses $C_1 = 1/C_X$ and $C_4 = 1/C_{EV}$ allowable factors to compensate for the concentrated mass and end condition, respectively. It also uses $C_3 = (C_W)^{1/2}$ to account for the content and insulation weights. A C_5 correction factor is also used for piping vibrating at higher than the first natural frequency of the span. Writing in ASME OM-3 format, Eq. (13.55) becomes

$$V_{\text{allow}} = \frac{C_1 C_4}{C_3 C_5} \left(\frac{S_{\text{el}}}{\alpha}\right) \frac{1}{1.874 C_2 K_2 \sqrt{E\rho}} \tag{13.58}$$

where C_4 is the correction factor for the end condition that is different from the fixed ends. It is also used to compensate for non-straight configurations. The C_4 values to be used are 1.0 for fixed ends and fixed supported beams; 1.325 for cantilever and simple supported beams; 0.74 for equal leg Z bend; and 0.83 for equal leg U bend. U bend and Z bend are defined in Fig. 13.12. $C_2 K_2$ is the stress index defined by the ASME code. $C_2 K_2 < 4.0$ for most piping systems. C_5 is the correction factor used when the measured frequency differs from the first natural frequency of the piping span. The C_5 correction factor has less effect on allowable velocity than on allowable displacement. As shown in Fig. 13.11(a), for the simple supported beam, for instance, the same amount of stress is generated with the same amount of velocity regardless of the mode that the piping is vibrating in. Furthermore, Eq. (13.58) already represents the higher stress condition of the first two modes of all the common straight beams.

In U.S. conventional units with the in-lb-sec system, Eq. (13.58) can be simplified for steel using $E = 29.3 \times 10^6$ psi and $\rho = 0.283/(32.2 \times 12) = 732.4 \times 10^{-6}$ lb-sec^2/in.^{4}. That is, for steel, we have

$$V_{\text{allow}} = \frac{C_1 C_4}{C_3 C_5} \left(\frac{S_{\text{el}}}{\alpha}\right) \frac{0.00364}{C_2 K_2} \quad (\text{in./sec}) \tag{13.59}$$

This is the ASME OM-3 equation for determining the allowable vibration velocity for carbon steel piping system. Theoretically, a higher velocity shall be allowed for an austenitic stainless steel piping system due to lower modulus of elasticity, but this additional allowable is ignored due to considerable uncertainties involved in the evaluation.

By taking the conservative values of all the factors involved, an absolute benchmark vibration velocity can be determined for screening use. Substituting Eq. (13.59) with the following conservative values:

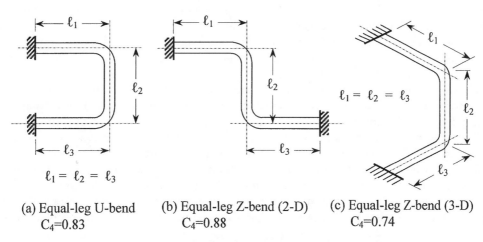

(a) Equal-leg U-bend $C_4=0.83$

(b) Equal-leg Z-bend (2-D) $C_4=0.88$

(c) Equal-leg Z-bend (3-D) $C_4=0.74$

FIG. 13.12
CORRECTION FACTOR FOR U-BEND AND Z-BEND

C_1 = $1/C_X$ = 0.15 (for $M_{\text{mass}}/(m\ell)$ = 17)
C_2K_2 = 4.0
C_3 = 1.5 (for 16 in. std pipe with water)
C_4 = 0.7 (for equal leg Z bend)
C_5 = 1.0
S_{el}/α = 7690 psi (see Section 13.4.4)

We have

$$V_{\text{allow}} = \frac{0.15 \times 0.7}{1.5 \times 1.0}(7690)\frac{0.00364}{4.0} = 0.5 \text{ in./sec} \quad (12.7 \text{ mm/sec})$$

This conservative allowable velocity of 0.5 in/sec (12.7 mm/sec) can be used as a screening tool. In other words, the vibration is acceptable if the vibration velocity is less than 0.5 in/sec. This velocity screening criteria is easier to use than the displacement criteria. Vibration velocity can be directly measured with an electronic tool. If an electronic or other direct measuring tool is not available, the velocity can be estimated from the displacement measured by a ruler, together with the vibration frequency counted with a stopwatch. The velocity equals the displacement divided by circular frequency, which is 2π times the frequency in cycles per second. It is important to note that the displacement used is the single amplitude value from zero to peak. However, the displacement measured by a ruler generally is the peak-to-peak value, which is twice of the single amplitude. For piping vibrating at 2 cps, for example, the 0.5 in/sec velocity is equivalent to $0.5/(2\pi f) = 0.5/(4\pi) = 0.04$ in. displacement amplitude, which is equivalent to 0.08 in. peak-to-peak displacement. A piping vibrating at 0.5 cycle per second would have an allowable peak-to-peak displacement of 0.32 in.

(e) Summary of allowable displacement and velocity

ASME OM-3 provides methods and procedures for evaluating both vibration displacement and vibration velocity. Wachel [24] also presented a rather comprehensive summary in his 1995 paper on allowable vibration displacement. These procedures, however, are not clear-cut and generally require some field experience for effective use. The following are brief summaries of allowable vibration displacement and velocity.

The allowable vibration displacement is given by Eq. (13.46). The allowable displacement is proportional to the square of the effective span length, ℓ, and inversely proportional to the end condition coefficient, C_{ED}, with respect to displacement. Because a piping system is generally three-dimensional and the effectiveness of support against vibration is questionable, the estimate of these two important parameters is very difficult. For instance, most standard piping supports do not have any moment resisting capability, which makes the determination of end condition and span length very difficult, if not impossible. The C_{ED} value can differ by as much as 10 times on a simple straight pipe span, just because of different end conditions. Judging from the above, the allowable vibration displacement criterion does not appear to be very reliable.

The allowable vibration velocity is given by Eqs. (13.55) and (13.58). The allowable velocity is inversely proportional to the square root of the additional weight due to content, insulation, and concentrated mass. It is also inversely proportional to the end coefficient, C_{EV}, with respect to velocity. The additional weight is easy to estimate and the end coefficient with respect to velocity changes only modestly among different configurations. From the simple supported beam, cantilever bean, fixed beam, all the way up to U-shape and Z-shape configurations, the C_{EV} values vary only from 0.80 to 1.35. In other words, the allowable vibration velocity is fairly insensitive to the piping geometry and end condition. The allowable vibration velocity criterion appears to be much more reliable than the vibration displacement criterion. The average pipe stress engineer should have no problem in applying it properly.

ASME OM-3 gives a screening velocity of 0.5 in/sec for quick checking if the vibration is acceptable. This screening velocity can be converted to screening displacement if the vibration frequency is known. Once again, it should be noted that the displacement so obtained is the simple amplitude value, which is half of the peak-to-peak displacement the field engineers are generally talking about. Knowing the background of this 0.5 in/sec value, some adjustment may be made if the condition is different than the one used in deriving this allowable value. For instance, if the pipe has no content or insulation such as the case of gas piping, the C_3 factor becomes 1.0 and the allowable velocity can be increased by 50%. The same thing also applies to the concentrated weight. If the weight is less than 17 times of the pipe span weight, then the allowable velocity can be increased accordingly. However, this type of adjustment should only be carried out with caution.

For systems with small appendages such as vent valves and drain valves, the vibration at appendages has to be evaluated separately using local geometry and velocity.

13.4.3 Formulation of Harmonic Analysis

The analysis of steady-state vibration is generally referred to as harmonic dynamic analysis because of the sinusoidal functions involved. For practical reasons, the analysis is done with one forcing frequency at a time, even if the piping system is subject to more than one frequency of excitation.

(a) *Static equivalent solution of un-damped systems*

The simplest harmonic forcing function acting in a structural system is the one that has all the applied forces in the same phase angle. In this case, the general force equilibrium equation of structural dynamics as given by Eq. (13.1) becomes

$$[M]\{\ddot{X}\} + [C]\{\dot{X}\} + [K]\{X\} = \{F\} \sin \omega t \tag{13.60}$$

where $\{F\}$ is the forcing vector and $\sin \omega t$ is a scalar function.

Even with this type of simple harmonic loading, the solution is not very simple. Just like the SDOF system discussed in Section 12.2.3 (see "Forced vibration on damped system"), the solution of Eq. (13.60) can be expressed as $\{X\} = \{A\}\sin \omega t + \{B\}\cos \omega t$, with constant vectors $\{A\}$ and $\{B\}$ to be determined by actual substitution. This is not simple when dealing with complex structures. However, if damping is ignored, the solution becomes simple as it is suffice to assume $\{X\} = \{A\}\sin \omega t$. Substituting $\{X\} = \{A\}\sin \omega t$ and $[C] = [0]$. For an un-damped system, we have

$$(-\omega^2[M] + [K])\{A\}\sin \omega t = \{F\}\sin \omega t$$

or

$$\{A\} = (-\omega^2[M] + [K])^{-1}\{F\} \tag{13.61}$$

This is the static equivalent approach of an un-damped system, which involves just one inversion or, more likely, one static solution of the combined matrix. However, because damping is very important and so is the adjustment for uncertainties, Eq. (13.61) generally is not used in actual design analysis.

(b) *General form of harmonic forcing function*

Harmonic motion is a circular motion that speeds up gradually and slows down gradually. This type of natural and smooth movement is the normality of all natural motions. Mathematically, it is expressed with a sinusoidal function, which consists of two independent parts: amplitude and circular frequency. A harmonic forcing function is either the result or the cause of a harmonic motion. The following are some of the harmonic forcing functions that are often encountered in a piping system:

(1) Vortex shedding forces

When water flows pass an object, visible eddies or vortices form at the wake behind the object. At certain ranges of flow speed, the vortices become regular and shed alternately from each side of the object. The shed vortices resemble a street downstream of the object, thus called the vortex street. The shedding of the vortex generates an impact force on the object. The alternating shedding of the vortices generates oscillatory forces acting on the object in the direction perpendicular to the flow direction. This oscillating force is considered a harmonic force having a frequency given by the following relation

$$\frac{fD}{V} = N_S \tag{13.62}$$

where N_S is Strouhal number, which can be taken the value of 0.22 for cylindrical objects within the velocity range generally encountered in piping applications.

The maximum force generated by the shedding of the vortex is proportional to the stagnation pressure of the flow. The resulting harmonic force per unit length of pipe can be expressed as

$$F_K = C_K \left(\frac{1}{2}\rho V^2\right) D \sin \omega t \tag{13.63}$$

where F_K is the so-called Kármán force and C_K the Kármán force coefficient. In his book on mechanical vibration, Den Hartog [25] reported that the C_K value is not known with great accuracy, but $C_K = 1.0$ is satisfactory for a range of Reynolds number from 10^2 to 10^7. The circular frequency is defined by $\omega = 2\pi f$.

When the piping or a portion of piping is expected to be subjected to a steady stream of fluid flow, the potential vibration from the vortex shedding force has to be examined. Piping located inside a furnace, distillation column, reactor, or high in the open air has the potential of experiencing vibration due to vortex shedding from the surrounding fluid flow. Figure 13.13(a) shows a part of piping which is located in such potential vibration area. An analysis of this type of load requires that the portion of piping subject to the harmonic force be divided into many sections with each section applied with a proper harmonic force calculated from Eq. (13.63) multiplied with the proper section length.

In the design phase, the vortex shedding phenomenon is handled based mainly on experience. Special attention is paid only to the areas known to have a potential vortex problem. The areas prone to vortex vibration are provided either with a vortex breaker or with a stiff piping system to avoid resonance. The latter approach may not be very practical at the open-air environment where the wind blows at different velocities at different times. Nevertheless, it is important that piping engineers are aware of the phenomena, so proper judgment can be made when the situation arises.

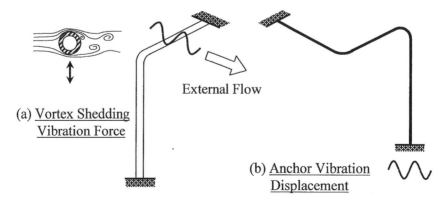

FIG. 13.13
EXTERNAL VIBRATION SOURCES

Vibration of open-air piping from vortex shedding often occurs at such a low wind velocity, that the association of the wind effect is often disregarded. A very gentle ocean breeze during sunset is enough to cause considerable vibration on piping if resonance occurs. Take a 6-in. standard wall liquid pipeline rigidly supported at every 20 feet, for instance, the fundamental natural frequency of the piping is about 10 cps. The resonance wind velocity in this case is about 25 ft/sec. A pipe with longer support span can be resonant at a much lower wind velocity.

(2) Anchor vibration displacement

The equipment and support vibration imparted to the piping is measured by the vibration displacement and is treated as anchor or support harmonic displacement in the analysis.

Equipment, such as the fractionator and chemical reactor, all have inherent vibration due to sloshing of fluid inside. The vibration displacement in this case applies to the piping end or ends that are attached to the equipment as shown in Fig. 13.13(b). On the other hand, for piping installed on an offshore platform or onboard a ship, the platform or ship vibration applies to all anchors and supports of the piping.

The anchor and support vibration displacement in piping is handled differently than the force. There are different methods that can be used to analyze the effect of the anchor and support vibration displacement in piping. One popular method is to convert the vibration displacement into vibration force, then treat it in the manner as an ordinary vibration force. In this method, the anchor and support stiffness is realistically assumed to be very stiff, but not perfectly rigid as in the theoretical case. To simulate the vibration displacement, a very large vibration force is applied at the anchor or support point forcing the pipe to move the same amount as the given vibration displacement. For ($d \sin \omega t$) vibration displacement, the magnitude of the equivalent vibration force is

$$F \sin \omega t = Kd \sin \omega t \qquad (13.64)$$

where K is the stiffness or spring constant of the anchor or support in the forcing direction and d is the vibration displacement amplitude. After the conversion, the vibration displacement is handled in the same manner as an ordinary vibration force.

(3) Pulsation force

Pulsation flow is the main source of harmonic force in piping systems. All piping connected to a reciprocating pump or compressor is subject to some type of pulsation flow. A centrifugal pump or compressor may also deliver pulsation flow when operating in the surge range.

When dealing with a periodic forcing function, one thing all designers want to avoid is resonance. In pulsation flow, there are two potential aspects of resonance to deal with. One is the acoustic resonance between pulsation fluid and the acoustic chamber or tube formed by the piping and connecting volumes. Similar to all resonance, the piping system will generate very large acoustic pulsation pressure when the natural frequency of the piping acoustic is the same or is in the neighborhood of the pulsation flow frequency. The other aspect of the resonance is, of course, the elastic resonance of the piping structure and the pulsation pressure force.

Figure 13.14 shows some basic cases of natural acoustic frequencies of pipe sections. Three types of pipe section are given: both ends closed, both ends open, and one end closed. Three natural modes, from the lowest frequency, are shown with the frequency formulas given. The mode shapes are just proportional values, which vibrate between plus and minus with the same proportion. The curve is for pressure distribution. In this case, the zero pressure points are called nodes, and the maximum pressure points loops. For a frictionless system, the pressure and the velocity are 90 deg. out of phase. Therefore, the maximum pressure point is also the zero velocity point, and a zero pressure point is located at the maximum velocity point.

The magnitude of pulsation pressure has to be taken into account in the calculation of wall thickness and other static stress calculations. In addition to the magnitude of the pressure, the shaking

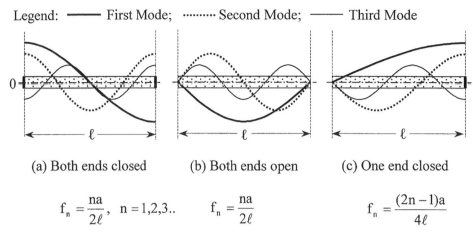

**FIG. 13.14
NATURAL ACOUSTIC MODES OF PIPE SECTION**

effect of the pressure wave phenomena also needs to be considered in the design of the piping system. This procedure applies to all hydraulic transient related phenomena.

Pulsation flow generates two categories of harmonic pressure wave forces: one corresponds to the force required for pushing the oscillatory fluid mass through the piping, and the other is the acoustic resonance pressure of a volume chamber in response to oscillatory fluid motion. The former is a harmonic pressure wave traveling through the piping and is called the traveling harmonic pressure wave. The latter vibrates at the same frequency as the pulsation flow frequency, and has the magnitude of pressure determined from the relation between natural acoustic frequency of the system and the pulsation flow frequency. This latter one is a pressure profile confined by the dimension of the system volumetric chamber and is called the standing wave. Because most pulsation flow consists of several frequencies, the highest pulsation pressure does not necessarily correspond to the fundamental pulsation frequency.

Although it is recognized that the pulsation flow generates both traveling and standing pressure waves, it is generally considered that the standing wave has a much higher potential of creating damaging pressure pulsation due to potential acoustic resonance. Traveling pressure waves are considered tolerable when the pulsation flow is reduced to the acceptable value by bottles, dampeners, and other pulsation reducing apparatus. Both traveling wave and standing wave produce shaking forces to the piping based on the differential pressure between both ends of a straight section.

Figure 13.15 shows a simple Z bend in a piping system. In fluid flow sense, a piping system is considered as one-dimensional with axial forces acting at every end or corner point. In this system, the main critical forces are the y-direction forces acting at leg 2-3, because they are acting in the most flexible direction of the piping. During normal steady-state non-pulsation flow, F_{y2} is equal but opposite to F_{y3}, thus producing no net shaking force. For simplicity, friction is ignored in this case. With either or both traveling and standing pressure waves, F_{y2} and F_{y3} may have different magnitudes as well as different phase angles at any given instant. The relation between F_{y2} and F_{y3} for both types of wave can be explained with the development length given in Fig. 13.15(b).

For a traveling wave, the amplitude is the same throughout the system but the phase angles are different at different points. Assuming a traveling pressure wave with amplitude P_T, the pressure wave function at reference point 1 is

$$P_1 = P_T \sin \omega t \quad \text{at point 1 (the reference point)}$$

Dynamic Analysis — Part 2: MDOF Systems and Applications 441

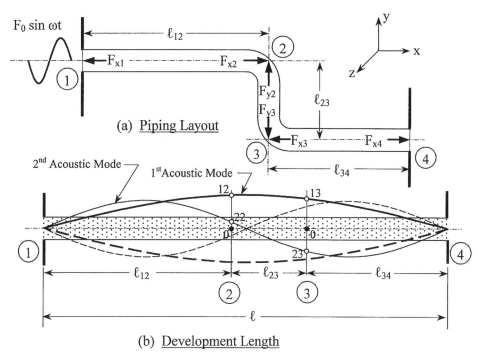

FIG. 13.15
HARMONIC LOADING INSIDE THE PIPE (OPEN – OPEN)

As it travels downstream with sonic velocity a, the corresponding wave at point 2 arrives ℓ_{12}/a of time later. Therefore, the wave at point 2 becomes

$$P_2 = P_T \sin \omega \left(t - \frac{\ell_{12}}{a} \right) = P_T \sin \left(\omega t - \frac{\omega \ell_{12}}{a} \right) \quad \text{at point 2}$$

where ℓ_{12}/a is the time lag and $\omega \ell_{12}/a$ is the phase angle of point 2 with respect to point 1. The wave at point 3 can be expressed in a similar manner, using ℓ_{13} instead of ℓ_{12}, where $\ell_{13} = \ell_{12} + \ell_{23}$, etc. The pressure force at point n is obtained by multiplying the pressure with the pipe inside cross-sectional area as

$$F_{nT} = AP_n = AP_T \sin \left(\omega t - \frac{\omega \ell_{1n}}{a} \right), \quad \ell_{1n} = \ell_{12} + \cdots \ell_{(n-1)n} \quad (13.65)$$

F_{nT} is the traveling wave force at point n. A similar force exists at every corner and end point of the piping system. Each point generally has two components of force. Take point 2, for instance; it has F_{x2} and F_{y2} — two components each with the same F_{2T} magnitude. The analysis procedure is discussed in the next section.

For a standing wave, the wave shape is generally sinusoidal, the same as the structural vibration mode shapes of a straight beam. The open-open pipe volume has a pressure wave shape similar to the structural mode shape of a simple supported beam. The pressure at points located within each half-wave oscillates at the same frequency and phase, but with different amplitude. Assuming the pressure at loop point (maximum pressure point) oscillates at

$$P_0 = P_s \sin \omega t \quad \text{at loop point}$$

where P_s is the maximum pulsation pressure amplitude for a standing wave. The pulsation pressures at other points are scaled with a sinusoidal relationship. Let ℓ_h be the half-wavelength of the standing wave, then the pressure pulsation at point located ℓ_n distance away from the loop point is

$$P_n = \cos\left(\frac{\pi \ell_n}{\ell_h}\right) P_0 \sin \omega t \quad \text{at a point } \ell_n \text{ away from loop point}$$

where P_n is the pressure at point n located ℓ_n distance away from the loop point. The pressure force at point n is the pressure multiplied with the inside cross-sectional area of the pipe. That is,

$$F_{ns} = A\cos\left(\frac{\pi \ell_n}{\ell_h}\right) P_s \sin \omega t \tag{13.66}$$

F_{nS} is the standing wave pressure force at point n. A similar force exits at every corner and end point of the piping system. Each point again generally has two components of force. At point 2, it has F_{x2} and F_{y2} — two components each with the same F_{2s} magnitude. Because a section of pipe can generate multiple modes of standing wave, it is important to know which mode is being excited to generate the wave. Take the pipe shown in Fig. 13.15(b), for instance; if the wave is corresponding to the first mode, the pressures at points 2 and 3 are represented by 0-12 and 0-13 amplitudes, respectively. It is important to note that 0-12 and 0-13 are in the same sign, meaning that the net force acting on pipe section 2-3 is the difference of the two. However, if the wave is corresponding to the second mode, then the pressures at points 2 and 3 are represented by 0-22 and 0-23, respectively. These two pressures have opposite signs, in which case the net shaking force working on section 2-3 is the sum of the two forces.

The pressure amplitude of the standing wave is normally obtained from a system acoustic simulation conducted via either digital or analog method. This type of simulation is generally performed by specialized vibration specialists. To obtain the traveling wave pressure amplitude, non-linear terms of the mathematical relation have to be included. This is not normally done in the simulation. Without a precise value, the conservative traveling pressure wave assuming non-reflective long pipeline, as given by Eq. (9.12), can be used.

(c) Harmonic Analysis Procedure

Since all harmonic forces acting in a piping system may not have a definite phase relationship among them, the harmonic loads are separated into groups. Each harmonic analysis case is performed for the group of loads having a fixed phase relationship. Therefore, the traveling wave and standing wave forces are analyzed separately. Furthermore, to simplify the analytical procedure, each analysis handles only one forcing frequency. If the excitation forces have more than one frequency, they are separated into groups with the same frequency. Each analysis case calculates pipe forces and moments as well as displacements and rotations. The final results are obtained by combining absolutely the sum of the individual case results. Harmonic analysis is generally performed using the modal superposition method discussed in Section 13.1.2.

In each harmonic analysis, the force vector is expressed in the form as given by either Eq. (13.65) or Eq. (13.66). The combined form encompassing both is used here as follows:

$$\{F(t)\} = \{F \sin(\omega t - \theta)\} = [\sin(\omega t - \theta)]_D \{F\}$$

where $\{F\}$ is the amplitude vector representing the force amplitude at each degree of freedom and θ is the phase angle with respect to the wave at a common reference point. θ is either zero or constant for standing waves. $[\sin(\omega t - \theta)]_D$ is a diagonal matrix. The solution for the above format of applied force cannot be readily decoupled with the modal superposition method due to the matrix of phase angles. Instead, each element of the force vector is separated into two components as

$$f_i = f_i \sin(\omega t - \theta_i) = f_i[\sin \omega t \cos \theta_i - \cos \omega t \sin \theta_i] = f_{iS} \sin \omega t + f_{iC} \cos \omega t$$

where $f_{iS} = f_i \cos\theta_i$ and $f_{iC} = -f_i \sin\theta_i$ are components for the sine function and cosine function, respectively. The load vector becomes two vectors: one with the $\sin\omega t$ function and the other with the $\cos\omega t$ function as follows:

$$\{F(t)\} = \{F_S\} \sin\omega t + \{F_C\} \cos\omega t \tag{13.67}$$

The above equation has two scalar functions, $\sin\omega t$ and $\cos\omega t$, instead of the matrix function $[\sin(\omega t - \theta)]_D$. $\{F_S\}$ and $\{F_C\}$ are force amplitude vectors for sine and cosine functions, respectively.

The analysis is then performed by the superposition of two forcing vectors. Each forcing vector is analyzed separately, then combined. From Eq. (13.14) we have the relation for the ith mode as

$$\ddot{\xi}_i + 2\zeta_i\omega_{ni}\dot{\xi}_i + \omega_{ni}^2\xi_i = \{\Phi_i\}^T\{F_S\} \sin\omega t + \{\Phi_i\}^T\{F_C\} \cos\omega t$$

or

$$\ddot{\xi}_i + 2\zeta_i\omega_{ni}\dot{\xi}_i + \omega_{ni}^2\xi_i = F_{Si}\sin\omega t + F_{Ci}\cos\omega t, \quad i = 1, N \tag{13.68}$$

where ω_{ni} is the natural circular frequency of the ith natural mode. The above has decoupled the system into N-SDOF systems, each in the form of Eqs. (12.33) and (12.34), with normalized unit effective mass. The solution for the steady-state forced vibration of Eq. (13.68) is given by Eqs. (12.36) and (12.37), with $k/M = k = \omega^2$ for a unit effective mass. That is,

$$\xi_i = \frac{1}{\omega_{ni}^2\sqrt{(1-\beta_i^2)^2 + (2\zeta_i\beta_i)^2}} [F_{Si}\sin(\omega t - \alpha_i) + F_{Ci}\cos(\omega t - \alpha_i)]$$

and

$$\alpha_i = \tan^{-1}\left(\frac{2\zeta_i\beta_i}{1-\beta_i^2}\right) \tag{13.69}$$

where $\beta_i = \omega/\omega_{ni}$. Expanding $\sin(\omega t - \alpha_i)$ and $\cos(\omega t - \alpha_i)$ and collecting terms, we have

$$\xi_i = [\xi_{Si} \sin\omega t + \xi_{Ci} \cos\omega t] \tag{13.70}$$

where

$$\xi_{Si} = \frac{F_{Si}\cos\alpha_i + F_{Ci}\sin\alpha_i}{\omega_{ni}^2\sqrt{(1-\beta_i^2)^2 + (2\zeta_i\beta_i)^2}}, \quad \xi_{Ci} = \frac{F_{Ci}\cos\alpha_i - F_{Si}\sin\alpha_i}{\omega_{ni}^2\sqrt{(1-\beta_i^2)^2 + (2\zeta_i\beta_i)^2}}$$

Equation (13.70) represents a sinusoidal curve as shown in Fig. 12.9(a). That is,

$$\xi_i = \xi_{i0} \sin(\omega t + \theta_{i0})$$

where ξ_{i0} and θ_{i0} are the resultant coordinate amplitude and phase angle for the ith mode, respectively. They are calculated by

$$\xi_{i0} = \sqrt{(\xi_{Si})^2 + (\xi_{Ci})^2}, \quad \theta_{i0} = \tan^{-1}\left(\frac{\xi_{Ci}}{\xi_{Si}}\right) \tag{13.71}$$

Engineers are often tempted to directly use the resultant coordinate given above to obtain the modal displacement amplitude. However, this will lead to an unpredictable analysis because the resultant given in the preceding equation has lost the sign, which is needed to properly combine with other modes. The proper approach is to maintain two sets of results, one with $\sin\omega t$ and the other with $\cos\omega t$. The two sets of individual modal displacements are combined by modal superposition to obtain two combined response displacement vectors as

$$\{X\} = [\Phi]\{\xi\} = [\Phi]\{\xi_S\}\sin\omega t + [\Phi]\{\xi_C\}\cos\omega t = \{X_S\}\sin\omega t + \{X_C\}\cos\omega t$$

$$\{X_S\} = [\Phi]\{\xi_S\}, \quad \{X_C\} = [\Phi]\{\xi_C\} \tag{13.72}$$

where $\{\xi_S\} = \{\zeta_{S1}, \zeta_{S2}, \ldots\}^T$ and $\{\zeta_C\} = \{\zeta_{C1}, \zeta_{C2}, \ldots\}^T$.

$\{X_S\}$ and $\{X_C\}$ are then used to obtained two sets of results including pipe forces, moments, and accelerations. The two sets of results are then combined with relations given by Eq. (13.71) to obtain the final results.

13.4.4 Evaluation of Vibration Stress

The calculated vibration stress needs to be evaluated for its acceptance. The acceptance criterion is high cycle fatigue (HCF) involving almost infinite number of operating cycles. Therefore, the stress is limited to the allowable endurance limit of the piping material.

The evaluation of vibration stress can be confusing even to experienced pipe stress engineers. The confusion is, in part, due to different codes and standards adopted by different industries. In B31 and Class-2/Class-3 nuclear piping codes, the stress calculated is not the real stress when the stress intensification factor is involved. As discussed in Chapter 3, that for the convenience of not identifying common girth weld locations, the stress intensification factor used in non-nuclear piping sets 1.0 for girth weld as basis. Because the real stress intensification factor for a girth weld is close to 2.0, the stress intensification factor using girth weld as the basis is actually only about one-half of the theoretical intensification factor. This results in a calculated stress of only one-half of the theoretical stress.

Vibration stress evaluation is based on fatigue, specifically HCF. ASME Design fatigue curves [23, 26], as shown in Fig. 13.16, are the main tool for routine evaluations. The nature of HCF fatigue is somewhat different than low cycle fatigue (LCF) with 10^4 or less operating cycles. The material defects, notches, and especially weld residual mean stresses are more tolerable in LCF due to local yielding at higher allowable stress levels, but they have significant effect in HCF at the elastic state. The design curves have A, B, and C, three curves depending mainly on residual stress and the mean stress

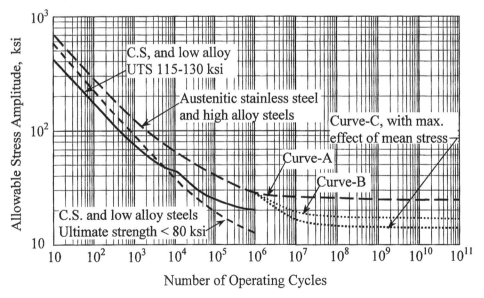

FIG. 13.16
TYPICAL ASME DESIGN FATIGUE CURVES [23, 26]

level. In piping using elastic analysis, only curves B and C are applicable, and curve C is generally used without going through detailed weld stress analyses. For vibration stress, we are mainly concerned with allowable endurance strength. ASME OM-3 defines the allowable endurance strength as

$$S_{elA} = \frac{S_{el}}{\alpha} \tag{13.73}$$

where S_{el} = 0.8 of the allowable alternative stress at 10^6 cycles when the curve terminates at 10^6 cycles. S_{el} = 1.0 of the allowable alternative stress at 10^{11} cycles when the data is available. α is the allowable stress reduction factor, which is 1.3 for carbon and low alloy steels with the fatigue curve terminated at 10^6 cycles and 1.0 for austenitic and high alloy steels with fatigue curve extends to 10^{11} cycles. This is another source of confusion on the evaluation of vibration stress, as two factors of the same nature are involved in carbon and low alloy steel with the fatigue curve terminated at 10^6 operating cycles. According to Wachel [24], it appears that the first factor of 0.8 was first introduced to convert the 10^6-cycle stress to 10^7-cycle stress, which was originally believed to be the endurance limit, but was later proved to be incorrect. The α factor of 1.3 was later introduced to convert the 10^7-cycle stress to the real endurance limit reached at about 10^{11} cycles. Therefore, with a fatigue curve ending at 10^6 cycles, the endurance limit is taken as the 10^6-cycle stress multiplied by 0.8/1.3 = 0.615 factor.

Based on the above definition, the allowable endurance strength can be taken as 7.69 ksi (=12.5 × 0.615) for carbon and low alloy steels with an ultimate tensile strength less than 80 ksi, and as 12.3 ksi (=20 × 0.615) for high strength steels with an ultimate tensile strength between 115 and 130 ksi. For austenitic stainless steels and high alloy steels, the curve C value of 13.6 ksi shall be used without going through detailed weld stress analysis. The 0.5 in/sec vibration screen velocity given in Section 13.4.2 (d) ("Stress due to velocity and allowable vibration velocity") is based on an allowable stress of 7.69 ksi, which is equivalent to the allowable for the most common carbon and low alloy steels with ultimate strength of less than 80 ksi. Therefore, the screen velocity can be increased proportionally for the pipe with higher allowable endurance strength.

The vibration stress is calculated mainly from bending moment as

$$S_{alt} = \frac{C_2 K_2}{Z} M \leq S_{elA} \tag{13.74}$$

where C_2 and K_2 are stress indices defined and used in Class-1 nuclear piping code [23], and Z is the section modulus of the pipe cross-section.

Because most of the piping systems are non-nuclear, C_2 and K_2 are not available in the codes that most of us are using. ASME OM-3 has related C_2 and K_2 with non-nuclear stress intensification factor as

$$C_2 K_2 = 2i \tag{13.75}$$

where i is the stress intensification factor given by ASME B31 and Class-2/Class-3 nuclear piping codes. As noted previously in this section, the B31 and Class-2/Class-3 stress intensification factor is only one-half of the theoretical stress intensification. By doubling the i value, we have increased the stress to the same level of theoretical stress. Equation (13.75) is straightforward. We can simply double the i values and perform the analysis as usual. However, since the i values of most common piping components are automatically calculated by most computer programs, we may use the original i and just double the calculated stress. The i method used by B31 and Class-2/Class-3 codes assumes that every inch of the piping consists of a girth butt weld without identifying the actual girth weld locations. This simplified approach works just fine for thermal expansion problems, but may need rethinking when dealing with vibration problems.

Another thing to remember is that the allowable vibration stress is based on the single amplitude of zero to peak value. This is different than the familiar allowable thermal expansion stress, which is based on the stress range due to yielding and shakedown involved in the LCF process.

13.5 TIME-HISTORY ANALYSIS

A structural dynamic problem with a defined time-history forcing function can be solved using the one small time step at a time approach. This step-by-step approach is called time-history analysis. Taking the forcing function shown in Fig. 13.17, as an example, the force can be simulated by a series of rectangular forces with very short, but equal time duration. This short time duration is called the time step. Although the series of rectangular forces will never be exactly the same as the original forcing function, they are close enough for practical purpose if the time step used is short enough. Using this stepwise approach, any structural dynamic problem can be reduced to a series of structural dynamic problems, performed one time-step at a time, with simple forces such as the above constant rectangular forces. In each time-step, an analysis is performed, using the results of the previous step as initial values, to find the responses at the end of the time-step. Because the analysis of one time step is roughly the same as performing a static analysis of the system, a time-history analysis of 100 time steps is equivalent of performing 100 static analyses.

The step-by-step time-history analysis is generally performed via one of two methods: modal integration method and direct integration method. The modal integration method uses the modal superposition approach discussed in Section 13.1.2. It requires that the natural frequencies and mode shapes of the system be calculated first. On the other hand, the direct integration method integrates the original equations as given by Eq. (13.1) directly without knowing the natural frequency and mode shape of the system. The modal integration method was favored in the past due to the familiar classic modal superposition theory and clear physical pictures of the mode shape and natural frequency. It also needs less computing power, which was a critical factor in the past. However, the modal integration method does have a problem in terms of the difficulty in calculating and including enough natural modes in the analysis. Nowadays, the direct integration approach has gained ground due to the availability of huge computing power. Furthermore, because modal superposition is applicable only to linear structures, the direct integration method has to be used for non-linear structures resulting from special phenomena such as pipe rupture and piping whip.

13.5.1 Treatment of Damping

The damping of the piping system is treated differently for different methods of integration. For modal integration with the controlling equation given by Eq. (13.14), the value of damping, expressed

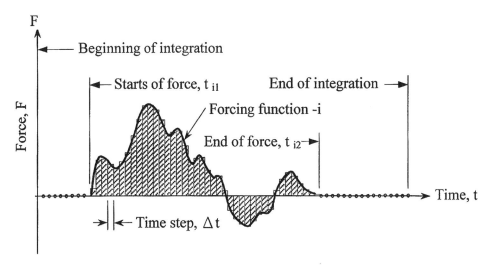

FIG. 13.17
DIVIDING THE FORCING FUNCTION INTO STEPS

as a damping ratio, can be directly included in the calculation. Furthermore, each natural mode can have any specific damping included. The inclusion of the damping is straightforward in modal integration. However, the inclusion of damping in direct integration is somewhat more complicated.

The direct integration works directly on Eq. (13.1). One major problem with this equation is the definition of the damping matrix $[C]$, because the damping of a structural system is not well defined. The application generally resorts to experimental data given as the damping ratio. That is, in the design specification, we would specify what damping ratio to use for a given type of structure subject to a certain type of loading. To relate this specified design damping ratio to the damping matrix, the following damping matrix is used [27, 28].

$$[C] = \alpha[M] + \beta[K] \tag{13.76}$$

where α and β are the coefficients determined from the damping ratio specified. It is obvious that α and β are not pure constants, but are rather with proper consistent units. By multiplying the mode shape matrix and pre-multiplying the transpose of the mode shape matrix on the above equation, we have

$$[\Phi]^T [C][\Phi] = [\Phi]^T \{\alpha[M] + \beta[K]\}[\Phi]$$

From the relations of Eqs. (13.8), (13.9), and (13.15), the damping ratio of the mode having a natural circular frequency of ω_i is

$$\zeta_i = \frac{\alpha}{2\omega_i} + \frac{\beta \omega_i}{2} \tag{13.77}$$

which shows that by using this method, the damping value varies with the circular frequency of a given natural mode of the piping. On the other hand, the desired damping at certain frequency can be set by proper selection of the α and β values. For each pair of (ζ_i, ω_i), an α and β relation is established by Eq. (13.77). For two variables, only two independent relational equations are needed to fully determine the values, that is, α and β can be selected in such a way that any two circular frequencies can have their desired damping values.

In design calculations, we often have only one damping value to deal with. In this case, we may consider the specified damping to occur at the two ends of the expected frequency spectrum for determining the α and β values, and have the damping elsewhere determined by Eq. (13.77).

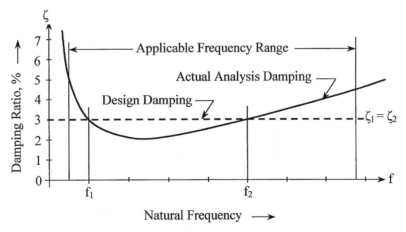

FIG. 13.18
DAMPING IN DIRECT INTEGRATION WITH 3% SPECIFIED DAMPING

Assuming that a piping system is being designed with a damping ratio of 0.03 (3%), and the expected analysis natural frequency spectrum ranges from 0.5 cps to 8 cps, we can make the 3% design damping ratio to apply at two benchmark frequencies, 1 cps and 5 cps. By substituting ($\zeta_1 = 0.03$, $\omega_1 = 2\pi$) and ($\zeta_2 = 0.03$, $\omega_2 = 10\pi$) into Eq. (13.77), we find $\alpha = 0.314$ and $\beta = 0.0016$. The damping ratios at other frequencies are as shown in Fig. 13.18. The figure also shows that the damping ratios between two benchmark frequencies are smaller than the specified value, and the damping ratios outside the benchmark frequencies are higher than the specified value.

13.5.2 Integration Schemes

Both modal and direct integration methods use the same integration schemes. The fundamental scheme of step-by-step time-history integration is to assume certain behavior within a small time step, then based on this assumption and the data at the start of the time step, to calculate the response at end of the time step. Naturally, the assumption of the behavior within the time step greatly affects the response at end of the time step. Guided by accuracy and stability, several integration schemes are proposed and used. Currently, the most popular schemes are: linear acceleration, Newmark's [29] mid-point constant acceleration, and Wilson's [30] θ method. These schemes are schematically outlined in Fig. 13.19.

From the assumed acceleration distribution within the time step, the velocity and displacement at end of the time step can be expressed in terms of the acceleration at end of the time step. The step-by-step integration is performed by substituting the acceleration, velocity, and displacement at end of the time step into Eq. (13.1) to obtain the acceleration at end of the time step. From the newly calculated acceleration, the velocity and displacement at end of the time step are calculated. Alternatively, the acceleration and velocity at end of the time step may also be expressed in terms of the displacement at end of the time step to solve the displacement at end of the time step first, then to calculate the acceleration and velocity from this displacement. Here, we will formulate the relation using the acceleration as the basis. For convenience, equilibrium Equation (13.1) can be rewritten at the end of the time step as

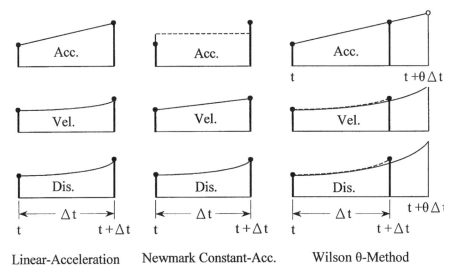

FIG. 13.19
INTEGRATION SCHEMES

$$[M]\{\ddot{X}\}_{t+\Delta t} + [C]\{\dot{X}\}_{t+\Delta t} + [K]\{X\}_{t+\Delta t} = \{F\}_{t+\Delta t} \tag{13.78}$$

which is then solved by one of the following integration schemes step by step.

For linear acceleration within the time step, we can have the velocity and displacement at end of the time step expressed in term of acceleration at end of time step as

$$\{\dot{X}\}_{t+\Delta t} = \{\dot{X}\}_t + \frac{\Delta t}{2}\left[\{\ddot{X}\}_t + \{\ddot{X}\}_{t+\Delta t}\right] \tag{13.79a}$$

$$\{X\}_{t+\Delta t} = \{X\}_t + \Delta t\{\dot{X}\}_t + \frac{(\Delta t)^2}{6}\left[2\{\ddot{X}\}_t + \{\ddot{X}\}_{t+\Delta t}\right] \tag{13.79b}$$

For Newmark mid-point constant acceleration, the velocity and displacement at end of time step are

$$\{\dot{X}\}_{t+\Delta t} = \{\dot{X}\}_t + \frac{\Delta t}{2}\left[\{\ddot{X}\}_t + \{\ddot{X}\}_{t+\Delta t}\right] \tag{13.80a}$$

$$\{X\}_{t+\Delta t} = \{X\}_t + \Delta t\{\dot{X}\}_t + \frac{(\Delta t)^2}{4}\left[\{\ddot{X}\}_t + \{\ddot{X}\}_{t+\Delta t}\right] \tag{13.80b}$$

Substituting above acceleration, velocity, and displacement at end of the time step into Eq. (13.78), we can find the acceleration at end of the time step. Once the acceleration is found, it is substituted into the above equations to calculate the velocity and displacement at the end of the time step. The integration continues until the time reaches the time duration specified.

For Wilson θ method, the relationships between acceleration and velocity, and between acceleration and displacement are the same as in the linear acceleration scheme, except that the time step is extended to $(\theta \Delta t)$ as

$$\{\dot{X}\}_{t+\theta\Delta t} = \{\dot{X}\}_t + \frac{\theta\Delta t}{2}\left[\{\ddot{X}\}_{t+\theta\Delta t} + \{\ddot{X}\}_t\right] \tag{13.81a}$$

$$\{X\}_{t+\theta\Delta t} = \{X\}_t + \theta\Delta t\{\dot{X}\}_t + \frac{(\theta\Delta t)^2}{6}\left[2\{\ddot{X}\}_t + \{\ddot{X}\}_{t+\theta\Delta t}\right] \tag{13.81b}$$

The Wilson θ method operates in the same manner as the linear acceleration scheme. However, the integration calculates the acceleration at end of $(t + \theta\Delta t)$ time, with a calculation $(\theta\Delta t)$ time step a fraction of time more than the intended (Δt) time step. The θ value is generally taken as 1.40 to ensure unconditional stability. After the acceleration at $(t + \theta\Delta t)$ is calculated, it is used to calculate the acceleration, velocity, and displacement at intended $(t + \Delta t)$ time by

$$\{\ddot{X}\}_{t+\Delta t} = \{\ddot{X}\}_t + \frac{1}{\theta}\left[\{\ddot{X}\}_{t+\theta\Delta t} - \{\ddot{X}\}_t\right] \tag{13.82a}$$

$$\{\dot{X}\}_{t+\Delta t} = \{\dot{X}\}_t + \Delta t\{\ddot{X}\}_t + \frac{\Delta t}{2\theta}\left[\{\ddot{X}\}_{t+\theta\Delta t} - \{\ddot{X}\}_t\right] \tag{13.82b}$$

$$\{X\}_{t+\Delta t} = \{X\}_t + \Delta t\{\dot{X}\}_t + \frac{(\Delta t)^2}{2}\{\ddot{X}\}_t + \frac{(\Delta t)^2}{6\theta}\left[\{\ddot{X}\}_{t+\theta\Delta t} - \{\ddot{X}\}_t\right] \tag{13.82c}$$

There are many other integration schemes used by various industries. However, the preceding three schemes are the most common ones used in the analysis of piping systems. These schemes differ only in the assumption of the displacement in relation to acceleration. Generally, linear acceleration appears to be the most natural, but the Newmark scheme is considered as the most accurate and the Wilson θ method offers the most stable solution.

Because time-history analysis in piping can be performed practically by using only a computer, we will not be too concerned with the details of the analysis formulation. As we will most likely use one of the commercially available software packages to perform the analysis, what we need to know is the effective approach and the correct manner of using the software we have chosen.

13.5.3 Time Step, Stability, and Accuracy

In time-history analysis, the size of the time step is very important. Accurate simulation of the force and structural response requires the time step to be very small; however, a small time step might produce excessive round-off error due to more cycles of calculation required. Therefore, appropriate time steps should be used. It generally takes about five equal time steps to adequately simulate a half-wave of forcing function or mode shape. The time increment of each step can be set as 1/10 of the forcing period or the natural period of the highest natural frequency taken into account in the analysis. That is, $\Delta t = T/10$, where T is the lesser of the forcing period or the shortest natural period of the piping system included in the analysis.

In the forcing function, we generally do not have a definitive forcing period. What we have is a function consisting of irregular shapes of pulse or spike. In this case, we will try to make at least five time steps within a pulse and at least two time steps within a spike. The situation for the natural period is more complicated because a piping system theoretically has modes with infinite small natural periods. In a continuous structural system, there is a definite lowest natural frequency, but with an indefinite highest natural frequency. In modal integration, we can manually limit the highest mode to be included in the analysis, but in direct integration all the modes, regardless of frequency, are automatically included. In this case, the choice of time step can only be based on the effective modes to be included and leaves the stability to the integration scheme.

Because of the mathematical approximation of the finite elements and the round-off error introduced in each time step of calculation, the response analyzed differs from the actual response by a small amount per time step. When enough errors are introduced, the calculated response can reach a point where it is completely irrelevant to the actual response. It might also be in the opposite direction or even grow exponentially into a senseless region. The instability can also be created by artificial amplification of the high frequency modes, which are actually not expected to respond to the applied load under consideration. Therefore, to ensure the stability, there are two things that can be done. One is to make both the time step and finite element size very small, and the other is to ensure that high frequency modes are not excited artificially.

When a system is divided into a finite number of elements, the calculated natural period tends to be longer than the actual period. A large time step will also tend to yield longer natural period. Therefore, the accuracy of the calculation may be measured by the calculated period versus the actual period. Figure 13.20 shows the behavior of an un-damped system subject to an initial displacement. The system is expected to vibrate at its natural period and with the same amplitude forever. Even without instability, the calculated response will have the period elongated and the amplitude shrunken. All integration schemes produce period elongation due to finite size of the element and time step. As for the amplitude decay, some schemes such as Newmark's method do not produce amplitude decay by careful balance of integration parameters, and some schemes produce amplitude decay to achieve unconditional stability. The Wilson θ method, for instance, generates some artificial damping to suppress response from high frequency modes. This artificial damping naturally creates amplitude decay.

When rigid supports or restraints are involved, special attention is needed to ensure the correctness of the analysis. In many computer software packages, a rigid support or restraint is automatically assigned with a very high, if not infinite, stiffness. High stiffness means high frequency of interaction. Therefore, the modes at very high frequency range have to be included. This also means a very small

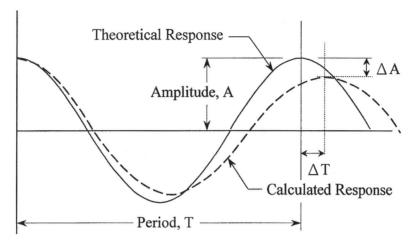

**FIG. 13.20
PERIOD ELONGATION AND AMPLITUDE DECAY**

time step, 1/10 of the shortest period, has to be used. Very often, because the number of modes and the time-step we use are not enough to obtain a response, the result that we come up with is not correct at all. Therefore, it is important that the realistic stiffness for the support and restraint has to be used in dynamic analysis.

Time-history analysis is still not considered routine for the time being. Analysis results can be quite different using different models and calculation schemes. It is very difficult even for experienced engineers to discover the instability that might have resulted. One means to quick check the stability and accuracy is to cut the time step by half and perform a confirmation analysis. If convenient, we may also cut the element size by half to perform another confirmation analysis. In piping, this element size reduction can be done easily by asking the program to subdivide the piping system for a higher cutoff frequency. If the results from the confirming analysis are essentially the same as the original results, we are pretty much assured that the analysis is stable.

13.5.4 Example Time-History Analysis

As a summary, we shall use an example analysis to demonstrate the procedure, benefit, and results of a time-history analysis. The steam piping system shown in Fig. 13.21 is the same system as given in Fig. 12.22 of the last chapter when the static equivalent approach using the SDOF concept was discussed. This is just a conceptual system. The real system is considerably more complex.

When the static equivalent approached is used, the net force of each leg has to be constructed. The natural frequencies of the system also have to be estimated. Then, from the shape of the net force and the main participating natural frequency a DLF is estimated for each force. The static equivalent force is taken as the maximum net force times the DLF. The system is then applied with all the static equivalent forces at the same time to find out the system response. This approach sounds simple enough for most engineers. However, this process has a few shortcomings. It could be tedious to obtain the net force, which is the difference of the two forces acting at both ends of a given piping leg. Calculation of the natural frequencies is always required. After the natural frequencies are calculated, the main participating frequency or frequencies have to be determined. This may not be very obvious for complicated systems. The application of all forces at the same time is too conservative most of the time, and can be also non-conservative in other times, due to sign confusion. The static equivalent approach is a good alternative when the time-history approach is not available.

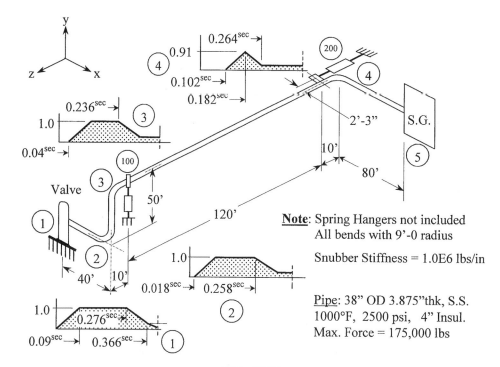

FIG. 13.21
TIME-HISTORY ANALYSIS OF STEAM HAMMER LOADING

Using the time-history approach, the above procedures, described for the static equivalent method, are mostly spared. The forces occurring at each end of the piping leg are applied directly as they are. The method will take the difference automatically through the calculation. The forces are all applied at the same time, but with proper sign and time-lag relation specified. This makes the method more accurate than the static equivalent method. Frequency calculation is required only for the modal integration method, but is not mandatory for the direct integration method. Nevertheless, having some idea about frequency distribution is also helpful for the direct integration method.

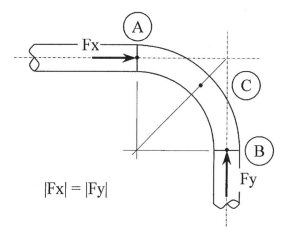

FIG. 13.22
FORCE APPLICATION

As with all dynamic analyses, the stiffness of the supports and restraints has very great effect on the results of the analysis. Realistic restraint stiffness has to be used at all times. For snubbers and struts, the stiffness naturally depends on the size of the pipe where it is attached. Larger pipe supports have higher support stiffness. For pipe sizes 30 in. or above, the stiffness of restraint is in the neighborhood of 10^6 lb/in, and for small pipes, it is in the range of about 10^4 lb/in. In this example analysis, we use 10^6 lb/in as a reference.

The first step in a time-history analysis is to make the decision on the size of the time step. We generally use about 1/10 of the shortest period of the main participating modes, or 1/5 of the shortest duration of the forcing pulse. The piping displacement and stress normally are excited easily at longer periods due to the relatively flexible piping proper. The support, which is generally oriented in the same direction of the applied force, is much more difficult to become excited. Without going through system natural mode analysis, we can use the benchmark frequency of 33 cps as cutoff, as recommended in earthquake design. Using 33 cps as the highest frequency, the required time step is $\Delta t = 0.1/33 = 0.003$ second. In this example, we know from natural frequency analysis that the modes participating with the snubber-200, for instance, are the sixth and eighth modes with frequencies at 5.714 cps and 6.861 cps, respectively. Therefore, we may use a time step of $\Delta t = 0.1/6.861 = 0.0145$ second. In addition to the natural piping frequency, we also have to consider the shape of the forcing function. In our example, we have ramped-constant force with the shortest duration of the ramp at 0.08 second. To simulate the force properly, the time step should be smaller than $\Delta t = 0.08/5 = 0.016$ second. This example analysis will be integrated with a time step of 0.005 second, with the confirming analysis performed at 0.002 second. The total duration of the integration has to be longer than the termination points of all the forces plus one-half of the longest period of the system. The additional time after the forces ended is required to check the maximum response in the free vibration region. In this example, since the longest period, from frequency analysis, is 1.1 seconds, and the force ends at 0.4 second, the total integration time has to be longer than $0.4 + 1.1/2 = 0.95$ second. To investigate the general behavior at the free vibration region, a longer 2.0-second duration is used in this example analysis.

The fluid end forces, shown in the shaded graphs, are applied directly as they are, without pre-calculating the net forces of the legs. The same force is applied at inlet and outlet points of the bend, because both points are experiencing the same fluid force. It is temping to apply the force at the center point, C, of the bend. However, this center point application would have created unwanted bending moments due to offsets between A and C, and between C and B. The correct approach is to apply the forces at the ends of the elbow as shown in Fig. 13.22. F_x and F_y, in this case, have the same time-history force, but act at different locations and in different directions. Actually, the magnitudes of both forces correspond to the theoretical force occurring at point C.

The system is analyzed using SIMFLEX [6] pipe stress computer software. Both modal integration and direct integration are investigated. The modal integration uses the linear acceleration scheme, and the direct integration uses both the Newmark and Wilson schemes. All three integration schemes are investigated. The results for the snubber-200 loads are shown in Figs. 13.23 and 13.24. It is clear that all three integration schemes yield very consistent results. Snubber-200 is installed mainly to resist the unbalanced load of leg 3-4. The unbalanced force is the difference between force 3 and force 4. It resembles a ramp force with the maximum magnitude equal to about 69% of the maximum steam hammer force. That is, the ramp force has a maximum value of $0.69 \times 175,000 = 120,000$ lb. The time duration of the ramp is 0.062 second. Because the two participating natural modes have natural periods of 0.175 second and 0.146 second, respectively, the t_d/T ratio is about 0.386. Using the curve from Fig. 12.5, we find a DLF of 1.78. Therefore, we will expect a snubber load of about $1.78 \times 120,000 = 213,000$ lb. The time-history results show a maximum load of 185,000 lb, which is very close to the above rough estimate. The deviation may come from many areas. Two of the likely areas of deviation are as follows: the snubber does not absorb the full load and the unbalanced force is not a real ramp force.

454 Chapter 13

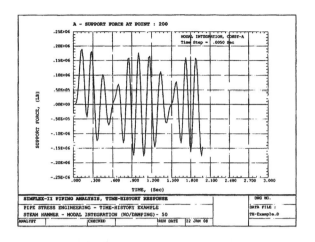

(a) Modal Integration

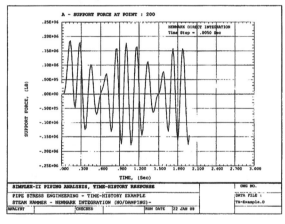

(b) Newmark Method
Direct Integration

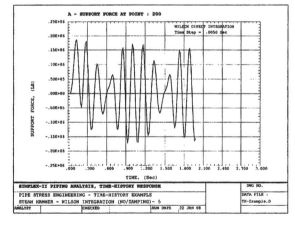

(c) Wilson θ-Method
Direct Integration

FIG. 13.23
SNUBBER-200 LOAD, WITHOUT DAMPING

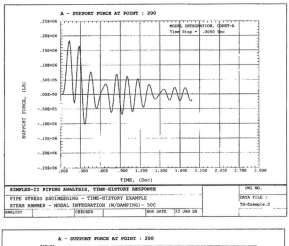

(a) Modal Integration
(3% Damping)

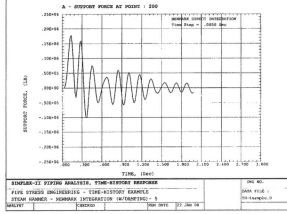

(b) Newmark Method
Direct Integration
(3.5% Damping)

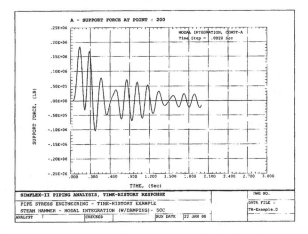

(c) Modal Integration
With 1/2 Time-Step
(3% Damping)

**FIG. 13.24
SNUBBER-200 LOAD, WITH DAMPING**

Without considering damping, the results clearly show the existence of beats due to two closely spaced main participating modes. When 3% of damping is included, the free vibration damped very quickly after the applying forces terminated. For direct integration, the damping is approximate. Using the scheme developed in Fig. 13.18, the damping for frequency between 5.714 cps and 6.861 cps is about 3.5%, which results in a somewhat smaller response than the modal integration that uses the exact 3% of damping. Results from a confirming analysis, using a 0.002-second time step, show exact agreement with the original analysis. The smaller time-step, however, does yield a somewhat smoother response curve.

REFERENCES

[1] Biggs, J. M., 1964, *Introduction to Structural Dynamics*, McGraw-Hill, New York, NY.
[2] Clough, r. W., and Penzien, J., 1975, *Dynamics of Structures*, McGraw-Hill, New York, NY.
[3] Przemieniecki, J. S., 1968, *Theory of Matrix Structural Analysis*, McGraw-Hill Company, New York, NY.
[4] Bathe, K.-J., and Wilson, E. L., 1976, *Numerical Methods in Finite Element Analysis*, Prentice-Hall, Inc., Englewood Cliffs, NJ.
[5] Munson, D. P., Keever, R. E., Peng, L. C., and Broman, R., 1974, "Computer Application to the Piping Analysis Requirements of ASME Section III, Subarticle NB-3600," *Pressure Vessels and Piping, Analysis and Computers*.
[6] SIMFLEX-II Pipe Stress Analysis Program, Peng Engineering, Houston, TX, www.pipestress.com.
[7] Wilkinson, J. H., and Reinsch, C., 1971, *Linear Algebra, Handbook for Automatic Computation*, Volume II, F. I. Bauer, A. S. Householder, F. W. J. Olver, H. Rutishauser, K., Samelson, and E. Stiefel, eds., Springer-Verlag, New York.
[8] ASCE 7-02, 2002, "Minimum Design Loads for Building and Other Structures," American Society of Civil Engineers, Washington, D. C.
[9] ICBO, 1997, "Uniform Building Code," International Conference of Building Officials, Whittier, CA.
[10] U.S. AEC (U.S. NRC) TID-7024, 1963, *Nuclear Reactors and Earthquakes*, U.S. Atomic Energy Commission, Division of Technical Information, Washington, D. C.
[11] U.S. NRC, 1976, Regulatory Guide 1.92, Rev.1, "Combining Modal Responses and Spatial Components in Seismic Response Analysis," Rev. 2, July 2006, U.S. Nuclear Regulatory Commission, Office of Standards Development, Washington, D. C.
[12] Rosenblueth, E., and Elorduy, J., 1969, "Responses of Linear Systems to Certain Transient Disturbances," *Proceeding of the Fourth World Conference on Earthquake Engineering*, Santiago, Chile, January 13–18, 1969, Vol. 1, pp. 185–196.
[13] Wilson, E. L., Der Kiureghian, A., and Bayo, E. P., 1981, "A Replacement for the SRSS Method in Seismic Analysis," *Earthquake Engineering and Structural Dynamics*, 9(2), pp. 187–192.
[14] Peng, L. C., 1989, "The Art of Checking Pipe Stress Computer Programs," PVP-Vol.169, *Design and Analysis of Piping and Components*, ASME Book No. H00484-1989, ASME, New York.
[15] Biswas, J. K., and Duff, C. G., 1978, "Response Spectra Method with Residual Terms," ASME Paper 78-PVP-79, presented at the Joint ASME/CSME Pressure Vessels & Piping Conference, Montreal, Canada, June, 1978.
[16] Lindley, D. W., and Yow, T. R., 1980, "Model Response Summation for Seismic Qualification," *Proceeding of the 2nd ASCE Conference on Civil Engineering and Nuclear Power*, Vol. VI, Paper 8-2, Knoxville, TN, September 15–17, 1980.
[17] Vashi, K. M., 1981, "Computation of Seismic Response from Higher Frequency Modes," *Transactions of the ASME, Journal of Pressure Vessel Technology*, **103**, pp. 16–19.

[18] Newmark, N. M., Blume, J. A., and Kapur, K. K., 1973, "Design Response Spectra for Nuclear Power Plants," ASCE Structural Engineering Meeting, San Francisco, April 1973.

[19] U.S. NRC, 1973, Regulatory Guide 1.60, Rev.1, "Design Response Spectra for Seismic Design of Nuclear Power Plants," U.S. Nuclear Regulatory Commission, Office of Standards Development, Washington, D. C.

[20] ASME B31.1, "Power Piping," ASME, New York, NY.

[21] The M. W. Kellogg Company, 1956, *Design of Piping Systems*, revised 2nd ed., John Wiley & Sons, Inc., New York, NY.

[22] ASME OM — 1987, "Requirements for Preoperational and Initial Start-Up Vibration Testing of Nuclear Power Plant Piping Systems," *Operation and Maintenance of Nuclear Power Plants*, Part-3 ASME, New York, NY.

[23] ASME B&PV Code-III, "Rules for Construction of Nuclear Facility Components," Boiler & Pressure Vessel Code, Section-III, ASME, New York, NY.

[24] Wachel, J. C., 1995, "Displacement Method for Determining Acceptable Piping Vibration Amplitudes," a paper presented at the 1995 Joint ASME/JSME Pressure Vessels and Piping Conference, PVP-Vol.313-2, ASME, New York.

[25] Den Hartog, J. P., 1956, *Mechanical Vibrations*, 4th ed., McGraw-Hill Book Company, New York, p. 305.

[26] Jaske, C. E., 2000, "Fatigue-Strenght-Reduction Factors for Welds in Pressure Vessels and Piping," *Transactions of the ASME, Journal of Pressure Vessel Technology*, **122**, pp. 297–304.

[27] Wilson, E. L., and Clough, R. W., 1962, "Dynamic response by Step-by-Step Matrix Analysis," Paper No. 45, *Symposium on the Use of Computers in Civil Engineering*, October 1962, Laboratorio Nacional de Engenharia Civil, Lisbon, Portugal.

[28] Berg, G. V., "The Analysis of Structural Response to Earthquake Forces," University of Michigan Industry Program of the College of Engineering, Report No. IP-291.

[29] Newmark, N. M., 1959, "A Method of Computation for Structural Dynamics," *Journal of the Engineering Mechanics Division, Proceedings of ASCE*, **85**(EM3), pp. 67–94.

[30] Wilson, E. L., Farhoomand, I., and Bathe, K. J., 1973, "Non-linear Dynamic Analysis of Complex Structures." *International Journal of Earthquake Engineering and Structural Dynamics*, **1**, pp. 241–252.

APPENDIX A

STANDARD NOMINAL PIPE WALL THICKNESS (INCHES; 1 IN. = 25.4 MM)

Size (in.)	OD (in.)	Schedule No. 10	20	30	40	60	80	100	120	140	160	5S	10S	STD 40S	XS 80S	XXS
0.50	0.840	0.083	−1.00	−1.00	0.109	−1.00	0.147	−1.00	−1.00	−1.00	0.187	0.065	0.083	0.109	0.147	0.294
0.75	1.050	0.083	−1.00	−1.00	0.113	−1.00	0.154	−1.00	−1.00	−1.00	0.218	0.065	0.083	0.113	0.154	0.308
1.00	1.315	0.109	−1.00	−1.00	0.133	−1.00	0.179	−1.00	−1.00	−1.00	0.250	0.065	0.109	0.133	0.179	0.358
1.25	1.660	0.109	−1.00	−1.00	0.140	−1.00	0.191	−1.00	−1.00	−1.00	0.250	0.065	0.109	0.140	0.191	0.382
1.50	1.900	0.109	−1.00	−1.00	0.145	−1.00	0.200	−1.00	−1.00	−1.00	0.281	0.065	0.109	0.145	0.200	0.400
2.00	2.375	0.109	−1.00	−1.00	0.154	−1.00	0.218	−1.00	−1.00	−1.00	0.343	0.065	0.109	0.154	0.218	0.436
2.50	2.875	0.120	−1.00	−1.00	0.203	−1.00	0.276	−1.00	−1.00	−1.00	0.375	0.083	0.120	0.203	0.276	0.552
3.00	3.500	0.120	−1.00	−1.00	0.216	−1.00	0.300	−1.00	−1.00	−1.00	0.437	0.083	0.120	0.216	0.300	0.600
3.50	4.000	0.120	−1.00	−1.00	0.226	−1.00	0.318	−1.00	−1.00	−1.00	−1.00	0.083	0.120	0.226	0.318	0.636
4.00	4.500	0.120	−1.00	−1.00	0.237	−1.00	0.337	−1.00	0.437	−1.00	0.531	0.083	0.120	0.237	0.337	0.674
5.00	5.563	0.131	−1.00	−1.00	0.258	−1.00	0.375	−1.00	0.500	−1.00	0.625	0.109	0.134	0.258	0.375	0.750
6.00	6.625	0.134	−1.00	−1.00	0.280	−1.00	0.432	−1.00	0.562	−1.00	0.718	0.109	0.134	0.280	0.432	0.864
8.00	8.625	0.148	0.250	0.277	0.322	0.406	0.500	0.593	0.718	0.812	0.906	0.109	0.148	0.322	0.500	0.875
10.00	10.75	0.165	0.250	0.307	0.365	0.500	0.593	0.718	0.843	1.000	1.125	0.134	0.165	0.365	0.500	−1.00
12.00	12.75	0.180	0.250	0.330	0.406	0.562	0.687	0.843	1.000	1.125	1.312	0.156	0.180	0.375	0.500	−1.00
14.00	14.00	0.250	0.312	0.375	0.437	0.593	0.750	0.937	1.093	1.250	1.406	0.156	0.188	0.375	0.500	−1.00
16.00	16.00	0.250	0.312	0.375	0.500	0.656	0.843	1.031	1.218	1.437	1.593	0.165	0.188	0.375	0.500	−1.00
18.00	18.00	0.250	0.312	0.437	0.562	0.750	0.937	1.156	1.375	1.562	1.781	0.165	0.188	0.375	0.500	−1.00
20.00	20.00	0.250	0.375	0.500	0.593	0.812	1.031	1.281	1.500	1.750	1.968	0.188	0.218	0.375	0.500	−1.00
22.00	22.00	0.250	0.375	0.500	0.656	0.875	1.125	1.375	1.625	1.875	2.125	0.188	0.218	0.375	0.500	−1.00
24.00	24.00	0.250	0.375	0.562	0.688	0.969	1.218	1.531	1.812	2.062	2.344	0.218	0.250	0.375	0.500	−1.00
26.00	26.00	0.312	0.500	−1.00	−1.00	−1.00	−1.00	−1.00	−1.00	−1.00	−1.00	−1.00	−1.00	0.375	0.500	−1.00
28.00	28.00	0.312	0.500	0.625	−1.00	−1.00	−1.00	−1.00	−1.00	−1.00	−1.00	−1.00	−1.00	0.375	0.500	−1.00
30.00	30.00	0.312	0.500	0.625	−1.00	−1.00	−1.00	−1.00	−1.00	−1.00	−1.00	0.250	0.312	0.375	0.500	−1.00
32.00	32.00	0.312	0.500	0.625	0.688	−1.00	−1.00	−1.00	−1.00	−1.00	−1.00	−1.00	−1.00	0.375	0.500	−1.00
34.00	34.00	0.312	0.500	0.625	0.688	−1.00	−1.00	−1.00	−1.00	−1.00	−1.00	−1.00	−1.00	0.375	0.500	−1.00
36.00	36.00	0.312	0.500	0.625	0.750	−1.00	−1.00	−1.00	−1.00	−1.00	−1.00	−1.00	−1.00	0.375	0.500	−1.00
42.00	42.00	−1.00	−1.00	−1.00	−1.00	−1.00	−1.00	−1.00	−1.00	−1.00	−1.00	−1.00	−1.00	0.375	0.500	−1.00

Note: −1.00 means not available.

APPENDIX B

DIMENSION OF BUTT-WELDING FITTINGS

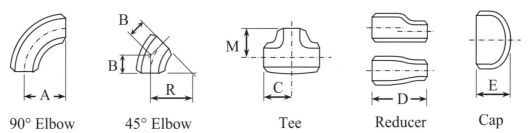

90° Elbow   45° Elbow   Tee   Reducer   Cap

All dimensions in inches (1 in. = 25.4 mm)

Pipe size	Pipe O.D.	90° Elbow Long R. A	90° Elbow Short R. A	45° Elbow B	Tee, full size C	Tee, full size M	Reducer D	Cap E
1/2	0.840	1 1/2		5/8	1	1		1
3/4	1.050	1 1/8		7/16	1 1/8	1 1/8	1 1/2	1 1/4
1	1.315	1 1/2	1	7/8	1 1/2	1 1/2	2	1 1/2
1 1/4	1.660	1 7/8	1 1/4	1	1 7/8	1 7/8	2	1 1/2
1 1/2	1.900	2 1/4	1 1/2	1 1/8	2 1/4	2 1/4	2 1/2	1 1/2
2	2.375	3	2	1 3/8	2 1/2	2 1/2	3	1 1/2
2 1/2	2.875	3 3/4	2 1/2	1 3/4	3	3	3 1/2	1 1/2
3	3.500	4 1/2	3	2	3 3/8	3 3/8	3 1/2	2
3 1/2	4.000	5 1/4	3 1/2	2 1/4	3 3/4	3 3/4	4	2 1/2
4	4.500	6	4	2 1/2	4 1/8	4 1/8	4	2 1/2
5	5.563	7 1/2	5	3 1/8	4 7/8	4 7/8	5	3
6	6.625	9	6	3 3/4	5 5/8	5 5/8	5 1/2	3 1/2
8	8.625	12	8	5	7	7	6	4
10	10.750	15	10	6 1/4	8 1/2	8 1/2	7	5
12	12.750	18	12	7 1/2	10	10	8	6
14	14	21	14	8 3/4	11	11	13	6 1/2
16	16	24	16	10	12	12	14	7
18	18	27	18	11 1/4	13 1/2	13 1/2	15	8
20	20	30	20	12 1/2	15	15	20	9
22	22	33	22	13 1/2	16 1/2	16 1/2	20	10
24	24	36	24	15	17	17	20	10 1/2
26	26	39	26	16	19 1/2	19 1/2	24	10 1/2
28	28	42	28	17 1/2	20 1/2	20 1/2	24	10 1/2
30	30	45	30	18 1/2	22	22	24	10 1/2
32	32	48	32	20	23 1/2	23 1/2	24	10 1/2
34	34	51	34	21	25	25	24	10 1/2
36	36	54	36	22 1/4	26 1/2	26 1/2	24	10 1/2
42	42	63	48	26			24	12

Appendix C

Appendix C 463

THERMAL EXPANSION RATE – FROM AMBIENT (70°F) TO TEMPERATURE, (IN./100 FT) (1 IN./100 FT = 0.833 MM/M)

MATL	-325	-200	-100	70	200	300	400	500	600	700	800	900	1000	1100	1200	1300	1400	1500
CS	-2.37	-1.71	-1.15	0.00	0.99	1.82	2.70	3.62	4.60	5.63	6.70	7.81	8.89	10.04	****	****	****	****
HS	-2.37	-1.71	-1.15	0.00	0.99	1.82	2.70	3.62	4.60	5.63	6.70	7.81	8.89	10.04	11.10	****	****	****
CM	-2.37	-1.71	-1.15	0.00	0.99	1.82	2.70	3.62	4.60	5.63	6.70	7.81	8.89	10.04	11.10	****	****	****
LC	-2.37	-1.71	-1.15	0.00	0.99	1.82	2.70	3.62	4.60	5.63	6.70	7.81	8.89	10.04	11.10	12.22	13.34	****
IC	-2.22	-1.62	-1.08	0.00	0.94	1.71	2.50	3.35	4.24	5.14	6.10	7.07	8.06	9.05	10.00	11.06	12.05	****
HC	-2.04	-1.46	-0.98	0.00	0.86	1.56	2.30	3.08	3.90	4.73	5.60	6.49	7.40	8.31	9.20	****	****	****
SS	-3.85	-2.73	-1.75	0.00	1.46	2.61	3.80	5.01	6.12	7.50	8.80	10.12	11.48	12.84	14.20	15.56	16.92	18.47
3T	-3.61	-2.56	-1.64	0.00	1.37	2.45	3.53	4.61	5.69	6.77	7.85	9.05	10.25	11.45	12.77	14.09	15.29	16.49
MO	-2.62	-2.02	-1.39	0.00	1.22	2.21	3.25	4.33	5.46	6.64	7.85	9.12	10.42	11.77	13.15	****	****	****
NS	-2.25	-1.76	-1.17	0.00	1.01	1.84	2.69	3.58	4.50	5.46	6.43	7.43	8.41	****	****	****	****	****
AL	-4.68	-3.44	-2.27	0.00	2.00	3.67	5.39	7.17	****	****	****	****	****	****	****	****	****	****
CI	****	****	****	0.00	0.90	1.64	2.42	3.24	4.11	5.03	5.98	****	****	****	****	****	****	****
BZ	-3.98	-2.78	-1.81	0.00	1.56	2.79	4.05	5.33	6.64	7.95	****	****	****	****	****	****	****	****
BS	-3.88	-2.70	-1.76	0.00	1.52	2.76	4.05	5.40	6.80	8.26	****	****	****	****	****	****	****	****
CN	-3.15	-2.19	-1.53	0.00	1.33	2.40	3.52	****	****	****	****	****	****	****	****	****	****	****
NC	-3.25	-2.30	-1.48	0.00	1.23	2.30	3.48	4.59	5.72	6.88	8.06	9.26	10.49	11.74	13.02	14.39	15.80	17.25

Note: **** means data not available. (Check the code book for the most updated values.)
MATL: CS, low carbon steel; HS, high carbon steel; CM, carbon-moly steel; LC, low chrome steel
IC, intermediate chrome; HC, high chrome steel; SS, austenitic stainless steel; 3T, 310 SS steel
MO, monel; NS, 3.5% nickel steel; AL, aluminum; CI, cast iron
BZ, bronze; BS, brass; CN, copper nickel; NC, nickel iron chrome

APPENDIX D

MODULUS OF ELASTICITY (1.E6 PSI) NOTE: 1 PSI = 6.895 KPA

MATL	-325	-200	-100	70	200	300	400	500	600	700	800	900	1000	1100	1200	1300	1400	1500
CS	31.4	30.8	30.2	29.5	28.8	28.3	27.7	27.3	26.7	25.5	24.2	22.4	20.4	18.0	****	****	****	****
HS	31.2	30.6	30.0	29.3	28.6	28.1	27.5	27.1	26.5	25.3	24.0	22.3	20.2	17.9	15.4	****	****	****
CM	31.1	30.5	29.9	29.2	28.5	28.0	27.4	27.0	26.4	25.3	23.9	22.2	20.1	17.8	15.3	****	****	****
LC	31.6	31.0	30.4	29.7	29.0	28.5	27.9	27.5	26.9	26.3	25.5	24.8	23.9	23.0	21.8	20.5	18.9	****
IC	32.9	32.3	31.7	30.9	30.1	29.7	29.0	28.6	28.0	27.3	26.1	24.7	22.7	20.4	18.2	15.5	12.7	****
HC	31.2	30.7	30.1	29.2	28.5	27.9	27.3	26.7	26.1	25.6	24.7	22.2	21.5	19.1	16.6	****	****	****
SS	30.3	29.7	29.1	28.3	27.6	27.0	26.5	25.8	25.3	24.8	24.1	23.5	22.8	22.1	21.2	20.2	19.2	18.1
3T	30.3	29.7	29.1	28.3	27.6	27.0	26.5	25.8	25.3	24.8	24.1	23.5	22.8	22.1	21.2	20.2	19.2	18.1
MO	27.8	27.3	26.8	26.0	25.4	25.0	24.7	24.3	24.1	23.7	23.1	22.6	22.1	21.7	21.2	****	****	****
NS	29.6	29.1	28.5	27.8	27.1	26.7	26.1	25.7	25.2	24.6	23.9	23.2	22.4	****	****	****	****	****
AL	11.3	10.9	10.7	10.2	9.7	9.4	8.9	8.3	****	****	****	****	****	****	****	****	****	****
CI	****	****	****	13.4	13.2	12.9	12.6	12.2	11.7	11.0	10.2	****	****	****	****	****	****	****
BZ	14.8	14.6	14.4	14.0	13.7	13.4	13.2	12.9	12.5	12.0	****	****	****	****	****	****	****	****
BS	15.9	15.6	15.4	15.0	14.6	14.4	14.1	13.8	13.4	12.8	****	****	****	****	****	****	****	****
CN	23.3	22.9	22.7	22.0	21.5	21.1	20.7	****	****	****	****	****	****	****	****	****	****	****
NC	30.5	29.9	29.4	28.5	27.8	27.4	27.1	26.6	26.4	25.9	25.4	24.8	24.2	23.8	23.2	22.6	22.0	21.4

Temperature (°F)

Note: **** means data not available. (Check the code book for the most updated values.)

MATL:
CS – low carbon steel;
IC – intermediate chrome;
MO – monel;
BZ – bronze;
HS – high carbon steel;
HC – high chrome steel;
NS – 3.5% nickel steel;
BS – brass;
CM – carbon-moly steel;
SS – austentic stainless steel;
AL – aluminum;
CN – copper nickel;
LC – low chrome steel
3T – 310 SS steel
CI – cast iron
NC – nickel iron chrome

APPENDIX E

VALVE AND FLANGE DATA

CLASS — 150

(A) WEIGHT (lb) - *Warning: Valve weight varies greatly from vendor to vendor*

SIZE (in.)	OD (in.)	BLIN	WNF	SLPN	LAPJ	FGTV	GTV	FGLV	GLV	FCKV	CKV
0.50	0.840	1	2	1	1	5	3	5	3	5	3
0.75	1.050	2	2	2	2	8	6	8	6	7	5
1.00	1.315	3	2	3	3	14	14	14	14	16	16
1.25	1.660	3	3	3	3	18	18	18	18	20	20
1.50	1.900	4	4	4	4	27	27	25	25	30	30
2.00	2.375	5	6	6	6	45	42	49	37	32	25
2.50	2.875	9	9	9	9	64	54	66	53	53	40
3.00	3.500	10	11	9	9	74	63	88	74	61	45
3.50	4.000	15	14	13	13	90	75	110	95	90	75
4.00	4.500	19	17	15	15	109	93	138	121	124	106
5.00	5.563	23	22	18	18	155	140	190	155	140	120
6.00	6.625	29	27	22	22	185	171	248	221	223	188
8.00	8.625	48	42	33	33	310	267	425	386	360	323
10.00	10.750	78	60	51	51	473	430	650	580	500	429
12.00	12.750	118	88	72	72	675	613	1,120	1,010	748	637
14.00	14.000	142	113	96	110	1,030	960	1,525	1,360	1,155	964
16.00	16.000	185	142	108	143	1,368	1,348	2,025	1,910	1,600	1,400
18.00	18.000	229	160	140	166	1,775	1,788	−1	−1	2,080	1,800
20.00	20.000	298	196	181	211	2,138	2,108	−1	−1	2,460	2,100
22.00	22.000	372	246	213	253	2,664	2,551	−1	−1	3,000	2,600
24.00	24.000	446	295	245	295	3,190	2,993	−1	−1	3,500	3,000

(B) LENGTH (in.)

SIZE (in.)	OD (in.)	BLIN	WNF	SLPN	LAPJ	FGTV	GTV	FGLV	GLV	FCKV	CKV
0.50	0.840	0.44	1.88	0.62	0.62	4.25	4.25	4.25	4.25	4.25	4.25
0.75	1.050	0.50	2.06	0.62	0.62	4.63	4.63	4.63	4.63	4.63	4.63
1.00	1.315	0.56	2.19	0.69	0.69	5.00	5.00	5.00	5.00	5.00	5.00
1.25	1.660	0.62	2.25	0.81	0.81	5.50	5.50	5.50	5.50	5.50	5.50
1.50	1.900	0.69	2.44	0.88	0.88	6.50	6.50	6.50	6.50	6.50	6.50
2.00	2.375	0.75	2.50	1.00	1.00	7.00	8.50	8.00	8.00	8.00	8.00
2.50	2.875	0.88	2.75	1.12	1.12	7.50	9.50	8.50	8.50	8.50	8.50
3.00	3.500	0.94	2.75	1.19	1.19	8.00	11.13	9.50	9.50	9.50	9.50
3.50	4.000	0.94	2.81	1.25	1.25	8.50	11.50	10.00	10.00	10.00	10.00
4.00	4.500	0.94	3.00	1.31	1.31	9.00	12.00	11.50	11.50	11.50	11.50
5.00	5.563	0.94	3.50	1.44	1.44	10.00	15.00	14.00	14.00	13.00	13.00
6.00	6.625	1.00	3.50	1.56	1.56	10.50	15.90	16.00	16.00	14.00	14.00
8.00	8.625	1.13	4.00	1.75	1.75	11.50	16.50	19.50	19.50	19.50	19.50
10.00	10.750	1.19	4.00	1.94	1.94	13.00	18.00	24.50	24.50	24.50	24.50
12.00	12.750	1.25	4.50	2.19	2.19	14.00	19.80	27.50	27.50	27.50	27.50
14.00	14.000	1.38	5.00	2.25	3.12	15.00	22.50	31.00	31.00	31.00	31.00
16.00	16.000	1.44	5.00	2.50	3.44	16.00	24.00	36.00	36.00	34.00	34.00
18.00	18.000	1.56	5.50	2.69	3.81	17.00	26.00	−1.00	−1.00	38.50	38.50
20.00	20.000	1.69	5.69	2.88	4.06	18.00	28.00	−1.00	−1.00	38.50	38.50
22.00	22.000	1.81	5.75	3.13	4.22	19.00	30.00	−1.00	−1.00	42.00	42.00
24.00	24.000	1.88	6.00	3.25	4.38	20.00	32.00	−1.00	−1.00	51.00	51.00

VALVE AND FLANGE DATA (*CONTINUED*)

CLASS — 300

(A) WEIGHT (lb) - *Warning: Valve weight varies greatly from vendor to vendor*

SIZE (in.)	OD (in.)	BLIN	WNF	SLPN	LAPJ	FGTV	GTV	FGLV	GLV	FCKV	CKV
0.50	0.840	3	3	3	3	9	6	7	5	6	4
0.75	1.050	4	4	4	4	14	7	11	6	10	5
1.00	1.315	5	5	4	4	19	19	25	19	20	19
1.25	1.660	7	7	5	5	27	27	35	24	29	26
1.50	1.900	9	9	8	8	34	34	45	28	38	33
2.00	2.375	10	10	9	9	74	49	79	67	56	46
2.50	2.875	16	14	14	14	95	80	105	87	85	61
3.00	3.500	20	19	17	17	108	85	137	108	109	84
3.50	4.000	25	22	21	21	140	105	175	145	145	110
4.00	4.500	31	29	26	26	165	120	217	177	175	134
5.00	5.563	39	36	32	32	285	225	300	260	250	195
6.00	6.625	56	48	45	45	407	328	395	346	314	255
8.00	8.625	90	76	67	67	648	410	690	605	561	452
10.00	10.750	146	110	100	110	760	780	1,031	910	704	600
12.00	12.750	209	163	140	164	1,020	890	1,730	1,403	1,125	843
14.00	14.000	267	217	195	220	1,898	1,595	−1	−1	1,700	1,440
16.00	16.000	349	288	262	282	2,583	2,108	−1	−1	2,320	1,900
18.00	18.000	440	355	331	335	3,393	2,868	−1	−1	3,050	2,580
20.00	20.000	545	431	378	428	4,060	3,535	−1	−1	3,650	3,180
22.00	22.000	693	532	478	523	5,232	4,058	−1	−1	4,700	3,650
24.00	24.000	841	632	577	617	6,405	4,580	−1	−1	5,760	4,120

(B) LENGTH (in.)

SIZE (in.)	OD (in.)	BLIN	WNF	SLPN	LAPJ	FGTV	GTV	FGLV	GLV	FCKV	CKV
0.50	0.840	0.56	2.06	0.88	0.88	5.50	5.50	6.00	6.00	6.00	6.00
0.75	1.050	0.62	2.25	1.00	1.00	6.00	6.00	7.00	7.00	7.00	7.00
1.00	1.315	0.69	2.44	1.06	1.06	6.50	6.50	8.00	8.00	8.50	8.50
1.25	1.660	0.75	2.56	1.06	1.06	7.00	7.00	8.50	8.50	9.00	9.00
1.50	1.900	0.81	2.69	1.19	1.19	7.50	7.50	9.00	9.00	9.50	9.50
2.00	2.375	0.88	2.75	1.31	1.31	8.50	8.50	10.50	10.50	10.50	10.50
2.50	2.875	1.00	3.00	1.50	1.50	9.50	9.50	11.50	11.50	11.50	11.50
3.50	4.000	1.19	3.19	1.75	1.75	11.50	11.50	13.25	13.25	13.25	13.25
4.00	4.500	1.25	3.38	1.88	1.88	12.00	12.00	14.00	14.00	14.00	14.00
5.00	5.563	1.38	3.88	2.00	2.00	15.00	15.00	14.75	14.75	15.75	15.75
6.00	6.625	1.44	3.88	2.06	2.06	15.90	15.90	17.50	17.50	17.50	17.50
8.00	8.625	1.63	4.38	2.44	2.44	16.50	16.50	22.00	22.00	21.00	21.00
10.00	10.750	1.88	4.62	2.62	3.75	18.00	18.00	24.50	24.50	24.50	24.50
12.00	12.750	2.00	5.12	2.88	4.00	19.80	19.80	28.00	28.00	28.00	28.00
14.00	14.000	2.12	5.62	3.00	4.38	30.00	30.00	−1.00	−1.00	33.00	33.00
16.00	16.000	2.25	5.75	3.25	4.75	33.00	33.00	−1.00	−1.00	34.00	34.00
18.00	18.000	2.38	6.25	3.50	5.12	36.00	36.00	−1.00	−1.00	38.50	38.50
20.00	20.000	2.50	6.38	3.75	5.50	39.00	39.00	−1.00	−1.00	40.00	40.00
22.00	22.000	2.63	6.50	3.97	5.75	43.00	43.00	−1.00	−1.00	44.00	44.00
24.00	24.000	2.75	6.62	4.19	6.00	45.00	45.00	−1.00	−1.00	53.00	53.00

VALVE AND FLANGE DATA (*CONTINUED*)

CLASS — 400

(A) WEIGHT (lb) - *Warning: Valve weight varies greatly from vendor to vendor*

SIZE (in.)	OD (in.)	BLIN	WNF	SLPN	LAPJ	FGTV	GTV	FGLV	GLV	FCKV	CKV
0.50	0.840	4	4	4	4	12	6	12	6	10	4
0.75	1.050	5	6	5	5	16	8	16	8	13	5
1.00	1.315	5	7	5	5	38	30	38	30	27	19
1.25	1.660	7	8	7	7	56	45	56	45	40	29
1.50	1.900	10	12	9	9	74	60	74	60	52	39
2.00	2.375	12	13	11	11	92	75	92	75	65	49
2.50	2.875	19	20	17	17	152	126	152	126	107	82
3.00	3.500	24	27	20	19	180	150	180	150	127	97
3.50	4.000	35	32	27	27	226	185	226	185	161	120
4.00	4.500	39	41	32	31	263	215	263	215	188	140
5.00	5.563	50	49	37	37	386	330	386	330	270	214
6.00	6.625	71	67	54	52	541	460	541	460	380	299
8.00	8.625	115	104	82	79	923	800	923	800	643	520
10.00	10.750	181	152	117	138	1,236	1,060	1,236	1,060	864	689
12.00	12.750	261	212	164	187	1,696	1,450	1,696	1,450	1,188	942
14.00	14.000	354	277	235	254	2,853	2,500	2,853	2,500	1,977	1,625
16.00	16.000	455	351	310	337	3,695	3,230	3,695	3,230	2,564	2,099
18.00	18.000	572	430	380	415	4,470	3,900	4,470	3,900	3,105	2,535
20.00	20.000	711	535	468	510	6,602	5,900	6,602	5,900	4,537	3,835
22.00	22.000	892	632	573	631	7,200	6,340	7,200	6,340	4,980	4,121
24.00	24.000	1,073	777	676	752	8,014	7,000	8,014	7,000	5,564	4,550

(B) LENGTH (in.)

SIZE (in.)	OD (in.)	BLIN	WNF	SLPN	LAPJ	FGTV	GTV	FGLV	GLV	FCKV	CKV
0.50	0.840	0.56	2.06	0.88	0.88	6.50	6.50	6.50	6.50	6.50	6.50
0.75	1.050	0.62	2.25	1.00	1.00	7.50	7.50	7.50	7.50	7.50	7.50
1.00	1.315	0.69	2.44	1.06	1.06	8.50	8.50	8.50	8.50	8.50	8.50
1.25	1.660	0.81	2.62	1.12	1.12	9.00	9.00	9.00	9.00	9.00	9.00
1.50	1.900	0.88	2.75	1.25	1.25	9.50	9.50	9.50	9.50	9.50	9.50
2.00	2.375	1.00	2.88	1.44	1.44	11.50	11.50	11.50	11.50	11.50	11.50
2.50	2.875	1.12	3.12	1.62	1.62	13.00	13.00	13.00	13.00	13.00	13.00
3.00	3.500	1.25	3.25	1.81	1.81	14.00	14.00	14.00	14.00	14.00	14.00
3.50	4.000	1.38	3.38	1.94	1.94	15.00	15.00	15.00	15.00	15.00	15.00
4.00	4.500	1.38	3.50	2.00	2.00	16.00	16.00	16.00	16.00	16.00	16.00
5.00	5.563	1.50	4.00	2.12	2.12	18.00	18.00	18.00	18.00	18.00	18.00
6.00	6.625	1.62	4.06	2.25	2.25	19.50	19.50	19.50	19.50	19.50	19.50
8.00	8.625	1.88	4.62	2.69	2.69	23.50	23.50	23.50	23.50	23.50	23.50
10.00	10.750	2.12	4.88	2.88	4.00	26.50	26.50	26.50	26.50	26.50	26.50
12.00	12.750	2.25	5.38	3.12	4.25	30.00	30.00	30.00	30.00	30.00	30.00
14.00	14.000	2.38	5.88	3.31	4.62	32.50	32.50	32.50	32.50	32.50	32.50
16.00	16.000	2.50	6.00	3.69	5.00	35.50	35.50	35.50	35.50	35.50	35.50
18.00	18.000	2.62	6.50	3.88	5.38	38.50	38.50	38.50	38.50	38.50	38.50
20.00	20.000	2.75	6.62	4.00	5.75	41.50	41.50	41.50	41.50	41.50	41.50
22.00	22.000	2.88	6.75	4.25	6.00	45.00	45.00	45.00	45.00	45.00	45.00
24.00	24.000	3.00	6.88	4.50	6.25	48.50	48.50	48.50	48.50	48.50	48.50

VALVE AND FLANGE DATA (*CONTINUED*)

CLASS — 600

(A) WEIGHT (lb) - *Warning: Valve weight varies greatly from vendor to vendor*

SIZE (in.)	OD (in.)	BLIN	WNF	SLPN	LAPJ	FGTV	GTV	FGLV	GLV	FCKV	CKV
0.50	0.840	4	4	4	4	12	6	12	6	10	4
0.75	1.050	5	6	5	5	16	8	16	8	13	5
1.00	1.315	5	7	5	5	38	30	38	30	27	19
1.25	1.660	7	8	7	7	56	45	56	45	40	29
1.50	1.900	10	12	9	9	79	65	79	65	56	42
2.00	2.375	12	13	11	11	102	85	102	85	72	55
2.50	2.875	19	20	17	17	152	126	152	126	107	82
3.00	3.500	24	27	20	19	185	155	185	155	131	101
3.50	4.000	35	32	27	27	226	185	226	185	161	120
4.00	4.500	47	48	43	42	335	270	335	270	240	175
5.00	5.563	78	78	73	73	550	440	550	440	395	286
6.00	6.625	101	96	95	93	773	630	773	630	552	409
8.00	8.625	159	137	135	132	1,223	1,020	1,223	1,020	865	663
10.00	10.750	267	225	213	231	1,890	1,570	1,890	1,570	1,340	1,020
12.00	12.750	341	272	261	286	2,472	2,080	2,472	2,080	1,743	1,352
14.00	14.000	437	406	318	349	3,317	2,840	3,317	2,840	2,323	1,846
16.00	16.000	603	577	442	476	4,403	3,740	4,403	3,740	3,094	2,431
18.00	18.000	762	652	573	566	5,930	5,070	5,930	5,070	4,155	3,295
20.00	20.000	976	811	733	725	6,900	5,800	6,900	5,800	4,869	3,770
22.00	22.000	1,166	949	895	886	8,023	6,680	8,023	6,680	5,684	4,342
24.00	24.000	1,355	1,157	1,056	1,046	9,584	8,000	9,584	8,000	6,784	5,200

(B) LENGTH (in.)

SIZE (in.)	OD (in.)	BLIN	WNF	SLPN	LAPJ	FGTV	GTV	FGLV	GLV	FCKV	CKV
0.50	0.840	0.56	2.06	.88	.88	6.50	6.50	6.50	6.50	6.50	6.50
0.75	1.050	0.62	2.25	1.00	1.00	7.50	7.50	7.50	7.50	7.50	7.50
1.00	1.315	0.69	2.44	1.06	1.06	8.50	8.50	8.50	8.50	8.50	8.50
1.25	1.660	0.81	2.62	1.12	1.12	9.00	9.00	9.00	9.00	9.00	9.00
1.50	1.900	0.88	2.75	1.25	1.25	9.50	9.50	9.50	9.50	9.50	9.50
2.00	2.375	1.00	2.88	1.44	1.44	11.50	11.50	11.50	11.50	11.50	11.50
2.50	2.875	1.12	3.12	1.62	1.62	13.00	13.00	13.00	13.00	13.00	13.00
3.00	3.500	1.25	3.25	1.81	1.81	14.00	14.00	14.00	14.00	14.00	14.00
3.50	4.000	1.38	3.38	1.94	1.94	15.00	15.00	15.00	15.00	15.00	15.00
4.00	4.500	1.50	4.00	2.12	2.12	17.00	17.00	17.00	17.00	17.00	17.00
5.00	5.563	1.75	4.50	2.38	2.38	20.00	20.00	20.00	20.00	20.00	20.00
6.00	6.625	1.88	4.62	2.62	2.52	22.00	22.00	22.00	22.00	22.00	22.00
8.00	8.625	2.19	5.25	3.00	3.00	26.00	26.00	26.00	26.00	26.00	26.00
10.00	10.750	2.50	6.00	3.38	4.38	31.00	31.00	31.00	31.00	31.00	31.00
12.00	12.750	2.62	6.12	3.62	4.62	33.00	33.00	33.00	33.00	33.00	33.00
14.00	14.000	2.75	6.50	3.69	5.00	35.00	35.00	35.00	35.00	35.00	35.00
16.00	16.000	3.00	7.00	4.19	5.50	39.00	39.00	39.00	39.00	39.00	39.00
18.00	18.000	3.25	7.25	4.62	6.00	43.00	43.00	43.00	43.00	43.00	43.00
20.00	20.000	3.50	7.50	5.00	6.50	47.00	47.00	47.00	47.00	47.00	47.00
22.00	22.000	3.75	7.75	5.25	6.88	51.00	51.00	51.00	51.00	51.00	51.00
24.00	24.000	4.00	8.00	5.50	7.25	55.00	55.00	55.00	55.00	55.00	55.00

VALVE AND FLANGE DATA (*CONTINUED*)

CLASS — 900

(A) WEIGHT (lb) - *Warning: Valve weight varies greatly from vendor to vendor*

SIZE (in.)	OD (in.)	BLIN	WNF	SLPN	LAPJ	FGTV	GTV	FGLV	GLV	FCKV	CKV
0.50	0.840	8	9	8	8	22	10	22	10	18	6
0.75	1.050	9	10	9	9	26	12	26	12	21	8
1.00	1.315	12	12	12	12	83	65	83	65	60	42
1.25	1.660	13	13	13	13	100	80	100	80	71	52
1.50	1.900	19	19	19	19	124	95	124	95	90	62
2.00	2.375	32	31	32	32	188	140	188	140	139	91
2.50	2.875	45	46	46	46	259	190	259	190	192	123
3.00	3.500	38	38	37	36	296	240	296	240	211	156
3.50	4.000	−1	−1	−1	−1	−1	−1	−1	−1	−1	−1
4.00	4.500	67	64	66	64	499	400	499	400	359	260
5.00	5.563	104	103	100	100	710	560	710	560	514	364
6.00	6.625	133	130	128	125	1,037	845	1,037	845	741	549
8.00	8.625	232	222	207	223	1,531	1,220	1,531	1,220	1,103	793
10.00	10.750	338	316	293	325	2,600	2,160	2,600	2,160	1,843	1,404
12.00	12.750	475	434	388	433	3,482	2,900	3,482	2,900	2,467	1,885
14.00	14.000	574	642	460	477	4,830	4,140	4,830	4,140	3,381	2,691
16.00	16.000	719	785	559	588	6,059	5,220	6,059	5,220	4,231	3,393
18.00	18.000	1,030	1,074	797	820	8,296	7,100	8,296	7,100	5,810	4,615
20.00	20.000	1,287	1,344	972	1,048	9,308	7,850	9,308	7,850	6,560	5,102
22.00	22.000	1,865	1,786	1,398	1,525	10,607	8,510	10,607	8,510	7,628	5,531
24.00	24.000	2,442	2,450	1,823	2,002	12,235	9,500	12,235	9,500	8,909	6,175

(B) LENGTH (in.)

SIZE (in.)	OD (in.)	BLIN	WNF	SLPN	LAPJ	FGTV	GTV	FGLV	GLV	FCKV	CKV
0.50	0.840	.88	2.38	1.25	1.25	8.50	8.50	8.50	8.50	8.50	8.50
0.75	1.050	1.00	2.75	1.38	1.38	9.00	9.00	9.00	9.00	9.00	9.00
1.00	1.315	1.12	2.88	1.62	1.62	10.00	10.00	10.00	10.00	10.00	10.00
1.25	1.660	1.12	2.88	1.62	1.62	11.00	11.00	11.00	11.00	11.00	11.00
1.50	1.900	1.25	3.25	1.75	1.75	12.00	12.00	12.00	12.00	12.00	12.00
2.00	2.375	1.50	4.00	2.25	2.25	14.50	14.50	14.50	14.50	14.50	14.50
2.50	2.875	1.62	4.12	2.50	2.50	16.50	16.50	16.50	16.50	16.50	16.50
3.00	3.500	1.62	4.00	2.12	2.42	15.00	15.00	15.00	15.00	15.00	15.00
3.50	4.000	−1.00	−1.00	−1.00	−1.00	−1.00	−1.00	−1.00	−1.00	−1.00	−1.00
4.00	4.500	1.75	4.50	2.75	2.75	18.00	18.00	18.00	18.00	18.00	18.00
5.00	5.563	2.00	5.00	3.12	3.12	22.00	22.00	22.00	22.00	22.00	22.00
6.00	6.625	2.19	5.50	3.38	3.38	24.00	24.00	24.00	24.00	24.00	24.00
8.00	8.625	2.50	6.38	4.00	4.50	29.00	29.00	29.00	29.00	29.00	29.00
10.00	10.750	2.75	7.25	4.25	5.00	33.00	33.00	33.00	33.00	33.00	33.00
12.00	12.750	3.12	7.88	4.62	5.62	38.00	38.00	38.00	38.00	38.00	38.00
14.00	14.000	3.38	8.38	5.12	6.12	40.50	40.50	40.50	40.50	40.50	40.50
16.00	16.000	3.50	8.50	5.25	6.50	44.50	44.50	44.50	44.50	44.50	44.50
18.00	18.000	4.00	9.00	6.00	7.50	48.00	48.00	48.00	48.00	48.00	48.00
20.00	20.000	4.25	9.75	6.25	8.25	52.00	52.00	52.00	52.00	52.00	52.00
22.00	22.000	4.88	10.63	7.13	9.38	56.00	56.00	56.00	56.00	56.00	56.00
24.00	24.000	5.50	11.50	8.00	10.50	61.00	61.00	61.00	61.00	61.00	61.00

VALVE AND FLANGE DATA (*CONTINUED*)

CLASS — 1500

(A) WEIGHT (lb) - *Warning: Valve weight varies greatly from vendor to vendor*

SIZE (in.)	OD (in.)	BLIN	WNF	SLPN	LAPJ	FGTV	GTV	FGLV	GLV	FCKV	CKV
0.50	0.840	8	9	8	8	22	10	22	10	18	6
0.75	1.050	9	10	9	9	26	12	26	12	21	8
1.00	1.315	12	12	12	12	38	20	38	20	31	13
1.25	1.660	13	13	13	13	50	30	50	30	39	19
1.50	1.900	19	19	19	19	69	40	69	40	54	26
2.00	2.375	32	31	32	32	98	50	98	50	80	32
2.50	2.875	45	46	46	46	152	83	152	83	123	54
3.00	3.500	61	61	61	60	207	115	207	115	166	75
3.50	4.000	−1	−1	−1	−1	−1	−1	−1	−1	−1	−1
4.00	4.500	90	90	90	92	355	220	355	220	278	143
5.00	5.563	172	162	162	162	713	470	713	470	548	305
6.00	6.625	197	202	202	208	1,023	720	1,023	720	771	468
8.00	8.625	363	334	319	347	1,619	1,140	1,619	1,140	1,219	741
10.00	10.750	599	546	528	577	2,642	1,850	2,642	1,850	1,994	1,202
12.00	12.750	928	843	820	902	4,040	2,810	4,040	2,810	3,056	1,826
14.00	14.000	−1	1,241	1,016	1,076	4,644	3,120	4,644	3,120	3,552	2,028
16.00	16.000	−1	1,597	1,297	1,372	7,346	5,400	7,346	5,400	5,455	3,510
18.00	18.000	−1	2,069	1,694	1,769	8,521	5,980	8,521	5,980	6,428	3,887
20.00	20.000	−1	2,614	2,114	2,189	10,701	7,530	10,701	7,530	8,065	4,894
22.00	22.000	−1	3,230	2,746	2,834	13,699	9,580	13,699	9,580	10,346	6,227
24.00	24.000	−1	4,153	3,378	3,478	17,717	12,650	17,717	12,650	13,289	8,222

(B) LENGTH (in.)

SIZE (in.)	OD (in.)	BLIN	WNF	SLPN	LAPJ	FGTV	GTV	FGLV	GLV	FCKV	CKV
0.50	0.840	0.88	2.38	1.25	1.25	8.50	8.50	8.50	8.50	8.50	8.50
0.75	1.050	1.00	2.75	1.38	1.38	9.00	9.00	9.00	9.00	9.00	9.00
1.00	1.315	1.12	2.88	1.62	1.62	10.00	10.00	10.00	10.00	10.00	10.00
1.25	1.660	1.12	2.88	1.62	1.62	11.00	11.00	11.00	11.00	11.00	11.00
1.50	1.900	1.25	3.25	1.75	1.75	12.00	12.00	12.00	12.00	12.00	12.00
2.00	2.375	1.50	4.00	2.25	2.25	14.50	14.50	14.50	14.50	14.50	14.50
2.50	2.875	1.62	4.12	2.50	2.50	16.50	16.50	16.50	16.50	16.50	16.50
3.00	3.500	1.88	4.62	2.88	2.88	18.50	18.50	18.50	18.50	18.50	18.50
3.50	4.000	−1.00	−1.00	−1.00	−1.00	−1.00	−1.00	−1.00	−1.00	−1.00	−1.00
4.00	4.500	2.12	4.88	3.56	3.56	21.50	21.50	21.50	21.50	21.50	21.50
5.00	5.563	2.88	6.12	4.12	4.12	26.50	26.50	26.50	26.50	26.50	26.50
6.00	6.625	3.25	6.75	4.69	4.69	27.75	27.75	27.75	27.75	27.75	27.75
8.00	8.625	3.62	8.38	5.62	5.62	32.75	32.75	32.75	32.75	32.75	32.75
10.00	10.750	4.25	10.00	6.25	7.00	39.00	39.00	39.00	39.00	39.00	39.00
12.00	12.750	4.88	11.12	7.12	8.62	44.50	44.50	44.50	44.50	44.50	44.50
14.00	14.000	5.25	11.75	−1.00	9.50	49.50	49.50	49.50	49.50	49.50	49.50
16.00	16.000	5.75	12.25	−1.00	10.25	54.50	54.50	54.50	54.50	54.50	54.50
18.00	18.000	6.38	12.88	−1.00	10.88	60.50	60.50	60.50	60.50	60.50	60.50
20.00	20.000	7.00	14.00	−1.00	11.50	65.50	65.50	65.50	65.50	65.50	65.50
22.00	22.000	7.50	15.00	−1.00	12.25	71.00	71.00	71.00	71.00	71.00	71.00
24.00	24.000	8.00	16.00	−1.00	13.00	76.50	76.50	76.50	76.50	76.50	76.50

VALVE AND FLANGE DATA (*CONTINUED*)

CLASS — 2500

(A) WEIGHT (lb) - *Warning: Valve weight varies greatly from vendor to vendor*

SIZE (in.)	OD (in.)	BLIN	WNF	SLPN	LAPJ	FGTV	GTV	FGLV	GLV	FCKV	CKV
0.50	0.840	11	11	11	11	29	12	29	12	24	8
0.75	1.050	12	12	11	11	35	18	35	18	28	12
1.00	1.315	15	16	15	15	53	30	53	30	42	19
1.25	1.660	23	25	23	23	75	40	75	40	60	26
1.50	1.900	31	34	31	30	107	60	107	60	85	39
2.00	2.375	50	48	49	48	154	80	154	80	125	52
2.50	2.875	70	66	69	69	234	130	234	130	188	84
3.00	3.500	105	113	102	99	323	170	323	170	263	110
3.50	4.000	–1	–1	–1	–1	–1	–1	–1	–1	–1	–1
4.00	4.500	164	177	158	153	567	330	567	330	451	214
5.00	5.563	272	293	259	259	974	585	974	585	769	380
6.00	6.625	418	451	396	387	1,434	840	1,434	840	1,140	546
8.00	8.625	649	692	601	587	2,352	1,450	2,352	1,450	1,844	942
10.00	10.750	1,248	1,291	1,148	1,120	4,222	2,500	4,222	2,500	3,347	1,625
12.00	12.750	1,775	1,919	1,611	1,573	6,267	3,850	6,267	3,850	4,919	2,502
14.00	14.000	–1	–1	–1	–1	–1	–1	–1	–1	–1	–1
16.00	16.000	–1	–1	–1	–1	–1	–1	–1	–1	–1	–1
18.00	18.000	–1	–1	–1	–1	–1	–1	–1	–1	–1	–1
20.00	20.000	–1	–1	–1	–1	–1	–1	–1	–1	–1	–1
22.00	22.000	–1	–1	–1	–1	–1	–1	–1	–1	–1	–1
24.00	24.000	–1	–1	–1	–1	–1	–1	–1	–1	–1	–1

(B) LENGTH (in.)

SIZE (in.)	OD (in.)	BLIN	WNF	SLPN	LAPJ	FGTV	GTV	FGLV	GLV	FCKV	CKV
0.50	0.840	1.19	2.88	1.56	1.56	10.38	10.38	10.38	10.38	10.38	10.38
0.75	1.050	1.25	3.12	1.69	1.69	10.75	10.75	10.75	10.75	10.75	10.75
1.00	1.315	1.38	3.50	1.88	1.88	12.13	12.13	12.13	12.13	12.13	12.13
1.25	1.660	1.50	3.75	2.06	2.06	13.75	13.75	13.75	13.75	13.75	13.75
1.50	1.900	1.75	4.38	2.38	2.38	15.13	15.13	15.13	15.13	15.13	15.13
2.00	2.375	2.00	5.00	2.75	2.75	17.75	17.75	17.75	17.75	17.75	17.75
2.50	2.875	2.25	5.62	3.12	3.12	20.00	20.00	20.00	20.00	20.00	20.00
3.00	3.500	2.62	5.62	3.62	3.62	22.75	22.75	22.75	22.75	22.75	22.75
3.50	4.000	–1.00	–1.00	–1.00	–1.00	–1.00	–1.00	–1.00	–1.00	–1.00	–1.00
4.00	4.500	3.00	7.50	4.25	4.25	26.50	26.50	26.50	26.50	26.50	26.50
5.00	5.563	3.62	9.00	5.12	5.12	31.25	31.25	31.25	31.25	31.25	31.25
6.00	6.625	4.25	10.75	6.00	6.00	36.00	36.00	36.00	36.00	36.00	36.00
8.00	8.625	5.00	12.50	7.00	7.00	40.25	40.25	40.25	40.25	40.25	40.25
10.00	10.750	6.50	16.50	9.00	9.00	50.00	50.00	50.00	50.00	50.00	50.00
12.00	12.750	7.25	18.25	10.00	10.00	56.00	56.00	56.00	56.00	56.00	56.00
14.00	14.000	–1.00	–1.00	–1.00	–1.00	–1.00	–1.00	–1.00	–1.00	–1.00	–1.00
16.00	16.000	–1.00	–1.00	–1.00	–1.00	–1.00	–1.00	–1.00	–1.00	–1.00	–1.00
18.00	18.000	–1.00	–1.00	–1.00	–1.00	–1.00	–1.00	–1.00	–1.00	–1.00	–1.00
20.00	20.000	–1.00	–1.00	–1.00	–1.00	–1.00	–1.00	–1.00	–1.00	–1.00	–1.00
22.00	22.000	–1.00	–1.00	–1.00	–1.00	–1.00	–1.00	–1.00	–1.00	–1.00	–1.00
24.00	24.000	–1.00	–1.00	–1.00	–1.00	–1.00	–1.00	–1.00	–1.00	–1.00	–1.00

Note: –1.00 means not available.
(1 in. = 25.4 mm; 1 lb = 0.4536 kg.)

APPENDIX F

ASME B31.1 ALLOWABLE STRESS, SE, (ksi)
(NOTE: 1 ksi = 6.895 MPA)

MATERIAL-NAME	MAT GRP	(E) FACT	SU (ksi)	At Temperature (°F) 100 / 750 / 1150	200 / 800 / 1200	300 / 850 / 1250	400 / 900 / 1300	500 / 950 / 1350	600 / 1000 / 1400	650 / 1050 / 1450	700 / 1100 / 1500
A53/A/S	CS	1.00	48.0	13.7 / 10.7 / 0.0	13.7 / 9.0 / 0.0	13.7 / 0.0 / 0.0	13.7 / 0.0 / 0.0	13.7 / 0.0 / 0.0	13.7 / 0.0 / 0.0	13.7 / 0.0 / 0.0	12.5 / 0.0 / 0.0
A53/B/S	CS	1.00	60.0	17.1 / 13.0 / 0.0	17.1 / 10.8 / 0.0	17.1 / 0.0 / 0.0	17.1 / 0.0 / 0.0	17.1 / 0.0 / 0.0	17.1 / 0.0 / 0.0	17.1 / 0.0 / 0.0	15.6 / 0.0 / 0.0
A53/A/E	CS	0.85	48.0	11.7 / 9.1 / 0.0	11.7 / 7.7 / 0.0	11.7 / 0.0 / 0.0	11.7 / 0.0 / 0.0	11.7 / 0.0 / 0.0	11.7 / 0.0 / 0.0	11.7 / 0.0 / 0.0	10.6 / 0.0 / 0.0
A53/B/E	CS	0.85	60.0	14.6 / 11.1 / 0.0	14.6 / 9.2 / 0.0	14.6 / 0.0 / 0.0	14.6 / 0.0 / 0.0	14.6 / 0.0 / 0.0	14.6 / 0.0 / 0.0	14.6 / 0.0 / 0.0	13.3 / 0.0 / 0.0
A106/A	CS	1.00	48.0	13.7 / 10.7 / 0.0	13.7 / 9.0 / 0.0	13.7 / 0.0 / 0.0	13.7 / 0.0 / 0.0	13.7 / 0.0 / 0.0	13.7 / 0.0 / 0.0	13.7 / 0.0 / 0.0	12.5 / 0.0 / 0.0
A106/B	CS	1.00	60.0	17.1 / 13.0 / 0.0	17.1 / 10.8 / 0.0	17.1 / 0.0 / 0.0	17.1 / 0.0 / 0.0	17.1 / 0.0 / 0.0	17.1 / 0.0 / 0.0	17.1 / 0.0 / 0.0	15.6 / 0.0 / 0.0
A106/C	HS	1.00	70.0	20.0 / 14.8 / 0.0	20.0 / 12.0 / 0.0	20.0 / 0.0 / 0.0	20.0 / 0.0 / 0.0	20.0 / 0.0 / 0.0	20.0 / 0.0 / 0.0	19.8 / 0.0 / 0.0	18.3 / 0.0 / 0.0
API-5L/A/S	CS	1.00	48.0	13.7 / 10.7 / 0.0	13.7 / 9.0 / 0.0	13.7 / 0.0 / 0.0	13.7 / 0.0 / 0.0	13.7 / 0.0 / 0.0	13.7 / 0.0 / 0.0	13.7 / 0.0 / 0.0	12.5 / 0.0 / 0.0
API-5L/B/S	CS	1.00	60.0	17.1 / 13.0 / 0.0	17.1 / 10.8 / 0.0	17.1 / 0.0 / 0.0	17.1 / 0.0 / 0.0	17.1 / 0.0 / 0.0	17.1 / 0.0 / 0.0	17.1 / 0.0 / 0.0	15.6 / 0.0 / 0.0
API-5L/A/E	CS	0.85	48.0	11.7 / 9.1 / 0.0	11.7 / 7.7 / 0.0	11.7 / 0.0 / 0.0	11.7 / 0.0 / 0.0	11.7 / 0.0 / 0.0	11.7 / 0.0 / 0.0	11.7 / 0.0 / 0.0	10.6 / 0.0 / 0.0
API-5L/B/E	CS	0.85	60.0	14.6 / 11.0 / 0.0	14.6 / 9.2 / 0.0	14.6 / 0.0 / 0.0	14.6 / 0.0 / 0.0	14.6 / 0.0 / 0.0	14.6 / 0.0 / 0.0	14.6 / 0.0 / 0.0	13.4 / 0.0 / 0.0
A335/P1	CM	1.00	55.0	15.7 / 15.4 / 0.0	15.7 / 14.9 / 0.0	15.7 / 14.5 / 0.0	15.7 / 0.0 / 0.0	15.7 / 0.0 / 0.0	15.7 / 0.0 / 0.0	15.7 / 0.0 / 0.0	15.7 / 0.0 / 0.0

ASME B31.1 Allowable Stress, SE, (ksi)
(Note: 1 ksi = 6.895 MPa) (*CONTINUED*)

MATERIAL-NAME	MAT GRP	(E) FACT	SU (ksi)	100 / 750 / 1150	200 / 800 / 1200	300 / 850 / 1250	400 / 900 / 1300	500 / 950 / 1350	600 / 1000 / 1400	650 / 1050 / 1450	700 / 1100 / 1500
A335/P2	LC	1.00	55.0	15.7 / 15.4 / 0.0	15.7 / 14.9 / 0.0	15.7 / 14.5 / 0.0	15.7 / 13.9 / 0.0	15.7 / 9.2 / 0.0	15.7 / 5.9 / 0.0	15.7 / 0.0 / 0.0	15.7 / 0.0 / 0.0
A335/P11	LC	1.00	60.0	17.1 / 14.8 / 0.0	17.1 / 14.4 / 0.0	17.1 / 14.0 / 0.0	16.8 / 13.6 / 0.0	16.2 / 9.3 / 0.0	15.7 / 6.3 / 0.0	15.4 / 4.2 / 0.0	15.1 / 2.8 / 0.0
A335/P12	LC	1.00	60.0	17.1 / 15.5 / 0.0	16.8 / 15.3 / 0.0	16.5 / 14.9 / 0.0	16.5 / 14.5 / 0.0	16.5 / 11.3 / 0.0	16.3 / 7.2 / 0.0	16.0 / 4.5 / 0.0	15.8 / 2.8 / 0.0
A335/P21	2C	1.00	60.0	17.1 / 16.6 / 0.0	17.1 / 16.6 / 0.0	16.6 / 16.0 / 0.0	16.6 / 12.0 / 0.0	16.6 / 9.0 / 0.0	16.6 / 7.0 / 0.0	16.6 / 5.5 / 0.0	16.6 / 4.0 / 0.0
A335/P22	2C	1.00	60.0	17.1 / 16.6 / 0.0	17.1 / 16.6 / 0.0	16.6 / 16.6 / 0.0	16.6 / 13.6 / 0.0	16.6 / 10.8 / 0.0	16.6 / 8.0 / 0.0	16.6 / 5.7 / 0.0	16.6 / 3.8 / 0.0
A369/FP1	CM	1.00	55.0	15.7 / 15.4 / 0.0	15.7 / 14.9 / 0.0	15.7 / 14.5 / 0.0	15.7 / 0.0 / 0.0	15.7 / 0.0 / 0.0	15.7 / 0.0 / 0.0	15.7 / 0.0 / 0.0	15.7 / 0.0 / 0.0
A369/FP2	LC	1.00	55.0	15.7 / 15.4 / 0.0	15.7 / 14.9 / 0.0	15.7 / 14.5 / 0.0	15.7 / 13.9 / 0.0	15.7 / 9.2 / 0.0	15.7 / 5.9 / 0.0	15.7 / 0.0 / 0.0	15.7 / 0.0 / 0.0
A369/FP11	LC	1.00	60.0	17.1 / 14.8 / 0.0	17.1 / 14.4 / 0.0	17.1 / 14.0 / 0.0	16.8 / 13.6 / 0.0	16.2 / 9.3 / 0.0	15.7 / 6.3 / 0.0	15.4 / 4.2 / 0.0	15.1 / 2.8 / 0.0
A369/FP12	LC	1.00	60.0	17.1 / 15.5 / 0.0	16.8 / 15.3 / 0.0	16.5 / 14.9 / 0.0	16.5 / 14.5 / 0.0	16.5 / 11.3 / 0.0	16.3 / 7.2 / 0.0	16.0 / 4.5 / 0.0	15.8 / 2.8 / 0.0
A369/FP21	2C	1.00	60.0	17.1 / 16.6 / 0.0	17.1 / 16.6 / 0.0	16.6 / 16.0 / 0.0	16.6 / 12.0 / 0.0	16.6 / 9.0 / 0.0	16.6 / 7.0 / 0.0	16.6 / 5.5 / 0.0	16.6 / 4.0 / 0.0
A369/FP22	2C	1.00	60.0	17.1 / 16.6 / 0.0	17.1 / 16.6 / 0.0	16.6 / 16.6 / 0.0	16.6 / 13.6 / 0.0	16.6 / 10.8 / 0.0	16.6 / 8.0 / 0.0	16.6 / 5.7 / 0.0	16.6 / 3.8 / 0.0
A312/TP304L	SS	1.00	70.0	16.7 / 13.3 / 0.0	16.7 / 13.0 / 0.0	16.7 / 0.0 / 0.0	15.8 / 0.0 / 0.0	14.7 / 0.0 / 0.0	14.0 / 0.0 / 0.0	13.7 / 0.0 / 0.0	13.5 / 0.0 / 0.0
A312/TP304L/E	SS	0.85	70.0	14.2 / 11.3 / 0.0	14.2 / 11.1 / 0.0	14.2 / 0.0 / 0.0	13.4 / 0.0 / 0.0	12.5 / 0.0 / 0.0	11.9 / 0.0 / 0.0	11.7 / 0.0 / 0.0	11.4 / 0.0 / 0.0
A312/TP304	SS	1.00	75.0	20.0 / 15.5 / 7.7	20.0 / 15.2 / 6.1	18.9 / 14.9 / 0.0	18.3 / 14.6 / 0.0	17.5 / 14.3 / 0.0	16.6 / 14.0 / 0.0	16.2 / 12.4 / 0.0	15.8 / 9.8 / 0.0
A312/TP304/E	SS	0.85	75.0	17.0 / 13.2 / 6.6	17.0 / 12.9 / 5.2	16.1 / 12.6 / 0.0	15.5 / 12.4 / 0.0	14.8 / 12.1 / 0.0	14.1 / 11.9 / 0.0	13.8 / 10.5 / 0.0	13.5 / 8.3 / 0.0
A312/TP316L	SS	1.00	70.0	16.7 / 13.2 / 0.0	16.7 / 12.9 / 0.0	16.7 / 12.7 / 0.0	15.7 / 0.0 / 0.0	14.8 / 0.0 / 0.0	14.0 / 0.0 / 0.0	13.7 / 0.0 / 0.0	13.5 / 0.0 / 0.0
A312/TP316L/E	SS	0.85	70.0	14.2 / 11.2 / 0.0	14.2 / 11.0 / 0.0	14.2 / 10.8 / 0.0	13.4 / 0.0 / 0.0	12.5 / 0.0 / 0.0	11.9 / 0.0 / 0.0	11.7 / 0.0 / 0.0	11.4 / 0.0 / 0.0

(*continued on next page*)

ASME B31.1 ALLOWABLE STRESS, SE, (ksi)
(NOTE: 1 ksi = 6.895 MPA) (*CONTINUED*)

MATERIAL-NAME	MAT GRP	(E) FACT	SU (ksi)	At Temperature (°F)							
				100 / 750 / 1150	200 / 800 / 1200	300 / 850 / 1250	400 / 900 / 1300	500 / 950 / 1350	600 / 1000 / 1400	650 / 1050 / 1450	700 / 1100 / 1500
A312/TP316	SS	1.00	75.0	20.0	20.0	20.0	19.3	18.0	17.0	16.6	16.3
				16.1	15.9	15.7	15.6	15.4	15.3	15.1	12.4
				9.8	7.4	0.0	0.0	0.0	0.0	0.0	0.0
A312/TP316/E	SS	0.85	75.0	17.0	17.0	17.0	16.4	15.3	14.5	14.1	13.9
				13.7	13.5	13.4	13.2	13.1	13.0	12.9	10.5
				8.3	6.3	0.0	0.0	0.0	0.0	0.0	0.0
A312/TP321	SS	1.00	75.0	20.0	20.0	19.1	18.7	18.7	18.3	17.9	17.5
				17.2	16.9	16.7	16.5	16.4	16.2	9.6	6.9
				5.0	3.6	0.0	0.0	0.0	0.0	0.0	0.0
A312/TP321/E	SS	0.85	75.0	17.0	17.0	16.2	15.9	15.9	15.5	15.2	14.9
				14.6	14.4	14.2	14.1	13.9	13.8	8.2	5.9
				4.3	3.1	0.0	0.0	0.0	0.0	0.0	0.0
A321/TP321H	SS	1.00	75.0	20.0	20.0	19.1	18.7	18.7	18.3	17.9	17.5
				17.2	16.9	16.7	16.5	16.4	16.2	12.3	9.1
				6.9	5.4	0.0	0.0	0.0	0.0	0.0	0.0
A312/TP321H/E	SS	0.85	75.0	17.0	17.0	16.2	15.9	15.9	15.5	15.2	14.9
				14.6	14.4	14.2	14.1	13.9	13.8	10.5	7.7
				5.9	4.6	0.0	0.0	0.0	0.0	0.0	0.0
A312/TP347	SS	1.00	75.0	20.0	20.0	18.8	17.8	17.2	16.9	16.8	16.8
				16.8	16.8	16.8	16.7	16.6	16.0	12.1	9.1
				6.1	4.4	0.0	0.0	0.0	0.0	0.0	0.0
A312/TP347/E	SS	0.85	75.0	17.0	17.0	16.0	15.1	14.6	14.3	14.3	14.3
				14.3	14.3	14.3	14.2	14.1	13.6	10.3	7.8
				5.2	3.8	0.0	0.0	0.0	0.0	0.0	0.0
A312/TP347H	SS	1.00	75.0	20.0	20.0	18.8	17.8	17.1	16.9	16.8	16.8
				16.8	16.8	16.8	16.7	16.6	16.4	16.2	14.1
				10.5	7.9	0.0	0.0	0.0	0.0	0.0	0.0
A312/TP347H/E	SS	0.85	75.0	17.0	17.0	16.0	15.1	14.6	14.3	14.3	14.3
				14.3	14.3	14.3	14.2	14.1	14.0	13.7	12.0
				8.9	6.7	0.0	0.0	0.0	0.0	0.0	0.0

INDEX

A

ABS (absolute) earthquake combination, 419–22
acoustic pulsation, 97, 439
active anchor length. *See* virtual anchor length
AF (attenuation factor), 316
air-cooled heat exchanger connections, 268–70. *See also* stationary equipment interfaces
alignment guides, 228
all-in-one analysis, 291
allowable stresses
   at high temperature, 16
   austenitic stainless steels with two sets of, 35
   bases for establishing in Codes, 98–99
   corrosion and, 76
   Piping Code, 98–101, 331–35
   rotating equipment connections and, 76, 285
   table, 474–76
allowable vibration displacement and velocity, see steady-state vibration
alternating stress intensity, 129
analysis of piping assembly, 55
   data points and node points, 57
   finite element method, 56–57, 409
   mathematical assembly, 58–59
anchors, 151–52
   drag anchors, 152, 354
   ideal anchor evaluations, 61–62
   main anchors, 226
   pipelines, 335–36, 353–54
   stiffness, 26
   virtual, 336
angle of repose of soil, 340–41
ANSI B31, 20
AQ (attenuation quotient), 316
area replacement method, 110
ASME B31, 20, 91–92, 330–31
   B31.4 Liquid Petroleum Pipeline, 331–33
   B31.8 Gas Transmission Pipeline, 333–35
attenuation factor (AF), 316
attenuation quotient (AQ), 316
axial deformation, flexible connections, 215, 219
axial slip joints, 235–36

B

balanced spring load, 168
ball valves, 5
ball/ball-and-socket joints, 5, 8, 236–37
bandwidth optimization, 57
Barlow's formula, 103
beam formulas, 53–55
beam on elastic foundation, 133–35
   application to cylindrical shells, 139–41
   infinite beam with concentrated load, 135–37
   semi-infinite beam, 138–39
bellow expansion joints
   accommodating axial deformation, 225–30
   accommodating lateral and rotational movements, 230–32
   bellow elements, 213–18
   catalog data
      background, 218–19
      calculating operational movements, 223–24
      cold spring of expansion joint, 224–25
      using, 220–23
   See also gimbal bellow joints; hinge bellow joints
bend rotation, due to pressure, 25, 81–82
bending moments, stresses due to, 40–42, See also stress, bending
bends, 8, 46, 56, 66–70, 105
bladder-type dampeners, 318
Bourdon tube effect, 25, 81–82
bowing temperature, 94–95, 357
 internal thermal stresses generated by, 359–60
braces, 155
branch connections, 5, 8
   component stresses, calculation, 114
      nuclear piping, 130–32
   pressure design, 109–12
   reinforcement of, 110–11
   See also unmatched small branch connections
branch legs, stress calculation in Code, 116–17
brittle rupture, 10–11
butt welded fittings, 461. *See also* furnace butt welded pipes
butterfly valves, 5

C

cantilever beams, 55
centrifugal compressors, 308–12
centrifugal pumps
   API Standard 610 pumps, 304–8
   categories, 300

characteristics related to piping interface, 301–2
    non-API pumps, 303–4
    piping support schemes, 302–3
check valves, 5
choking force, ring girder supports, 180–82
choking model
    discontinuity stresses, 142–43
    See also Kellogg's choking model
choking phenomenon in gas flow, 392
circumferential weld. *See* girth welds
clamped shoes, 153
clamping stresses, 24
Class 1 nuclear piping
    Code differences from non-nuclear, 125–32
coarse sand, 340
Code (Piping Code), 19–22
    applicable codes, 91–92
    basic allowable stresses, 98–101, 331–35
    inclusion in owner's design specification, 27
    loadings to be considered, 92–98
    pressure design, 101–13
    rationales and intent, 91
    stress intensification factors (SIFS), 69–71
code stress compliance charts, 124–25
code stress compliance reports, 124–25
code table stresses. *See* Code (Piping Code), basic allowable stresses
cold, and brittleness, 11–12
cold balanced spring load, 168–69
cold spring, 76–77
    analysis of cold sprung systems, 79–80
    expansion joints, 224–25
    gap size and location, 77–78
    multi-branched systems, 79
    procedure, 78–79
column buckling, 17
combined bending stress, 115
complete quadratic combination (CQC), 419–21
component stresses
    calculations in Code, 113–17
    Code, 113–25
components, types of, 4–5
composite drawings, position in piping design process, 1–2
computer programs
    analysis of piping resting on supports, 161
    benefits of, 2, 38
    checking results of, 39
    field proven systems, 88–89
    load calculations for risers, 171
consistent mass matrix approach, 411–12
constant-effort supports, 156, 165–66
construction gaps, 26
contoured welded-on branch connections, 5, 8

corner pressure balanced expansion joints, 229–30
corrosion
    allowance, 18, 28, 102
    and allowable stress, 76
corrosion failure, 18–19
corrugated hose, 237–39
Coulomb damping, 382, 388
CQC (complete quadratic combination), 419–21
creep failure, and welds, 6
creep rupture, 14–17
critical pressure ratio in gas flow, 392–93
cumulative damage, 129
curved pipe, pressure design, 104–6
curved pipe beam elements, 68. *See also* ovalization
cyclic stress. *See* fatigue failure

D

dampeners for pulsation flow
    bladder-type, 318
    reciprocating compressors and pumps, 317–19
damping, 387–89
deflection shapes of common beams, 430–31
depth of soil cover, 341
Der Kiureghian correlation coefficient, 419
design pressure. *See* pressure, design
design specifications, 26–30
design temperature. *See* temperature, design
direct friction force approach, 174–75
discontinuity stresses
    beam deflection equations applied to cylindrical shells, 139–41
    choking model, 142–43
    definitions, 133
    effective widths, 141–42
    junctions between dissimilar materials, 143–44
        with a rigid section, 146–47
        with similar modulus of elasticity, 144–46
    vessel shell rotation, 147–49
displacement stress, 113, 121–24
    calculation in Code, 121–24
    weld strength reduction factor, 6
    See also thermal expansion
DLF (dynamic load factor), 374–75
double-acting stops, 155
down comers, 170. *See also* risers
drag anchors, 152, 354
Dresser Coupling, 235
ductile materials, 10
ductile rupture, 10
dynamic analysis
    accounting for uncertainties, 426–28
    impacts, 373–75

See also harmonic analyses, response spectra analyses, time-histpry analyses
See also MDOF systems; SDOF systems
dynamic fluid load, Code, 96–98
dynamic friction, 174
dynamic load factor (DLF), 374–75

E

earthquake load analysis
  Code, 96
  industry practice, 23
  nuclear piping, 126
  See also ground motion; occasional stress; response spectra method; sustained stress
effective stress, 53
EJMA standard, 211
elastic analysis, 37
elastic equivalent stress, 13, 67
elastic-plastic discontinuity analysis, 129–30
elasticity
  modulus of, 34, 465
  shear modulus of, 38
elbow bends, 8, 461
electric fusion welded pipes, 6
electric resistance welded pipes, 6
endurance limit, 13, 444–5
equivalent stress intensification factors, 117
erosion failure, 18–19
European code, 91–92
excessive flexibility, 86–88
  rotating equipment connections, 321–22
expansion, see thermal expansion
expansion joints
  cold spring, 224–25
  problems with, 241
    direction of the anchor force, 241–42
    improper installation, 322
    improperly installed anchors, 244–45
    with tie-rods and limit rods, 242–44
  See also bellow expansion joints
expansion loops, 63
  leg length estimation, 63–65
expansion stress, 115
  See also thermal expansion
external pressure. See pressure, external
extruded tee, 5, 8

F

failures, 18–19
  modes of, 9-19
  theories of, 52
  See also creep rupture; fatigue failure; stability failure; static stress rupture
fatigue failure, 12–14, 71, 76, 221, 444

field proven systems, 87–88. *See also* industry practice
finite element method, 56–57, 409
flanges
  bend flexibility and SIFs, 71
  classes, 7
  diagram, 5
  piping load evaluations
    Class 2 Nuclear Piping Rules, 254–55
    equivalent pressure method, 251–54
    flange stress calculation, 253–54
    leakage concerns, 247–48
    rating table lookup, 252–53
    standard design procedure, 248–50
    unofficial position of B31.3, 250–51
  usage, 6
  valve and flange data, 467–73
  See also un-insulated flange connections
flexibility factor, 69, 71
  pressure effect on, 82
flexible connections, 8, 209–10
  analysis, 211–12
  catalog data use, 220–21
    adjustment for operating cycles, 221–23
    bellow effective area, 220
    effect of rotation, 220
    non-concurrent movements, 221
    pressure rating, 220
    torsional moment, 223
  problems with, 210
  standards, 211
  suitable situations, 211
  See also bellow expansion joints
flexible hoses
  application and analysis, 238–41
  types, 237–38
  uses, 237
flow velocity, 392–95
  versus sonic velocity, 389–90
fluid flow, shaking forces from, 395–97
forged elbows, pressure design, 106
forged welding tee, 5, 8, 461
  pressure design, 109–10
free-body diagrams, 39
friction angle (soils), 340–41
furnace butt welded pipes, 6

G

g factor of earthquake loading, 413–14
gas constant, 390
gas-filled surge chambers, 316–18
gas pipelines
  Code, 21–22
  design factors, 334
  location classes, 331

See also ASME B31, B31.8 Gas Transmission Pipeline
gate valves, 5
gimbal bellow joints, 8, 232–33, 234
girth welds (circumferential welds), 5, 6
    weld strength reduction factor, 101
global coordinate systems, 58
globe valves, 5
gravel, 340
ground motion, 413–15. See also earthquake load analysis
guide gaps, 26
guided cantilever beams, 55
guides, 155

H

half-coupling welded-on branch connections, 5, 8
hangers, 153, 166, 168. See also variable spring hangers
harmonic analysis
    accounting for uncertainties, 427–28
    analysis procedure, 439–44
    anchor vibration displacement, 98, 439
    general form of harmonic forcing function, 437–42
    pulsation force, 439–42
    static equivalent solution of un-damped systems, 437
    vortex shedding forces, 437–39
heat exchanger connections, 265. See also air-cooled heat exchanger connections; stationary equipment interfaces
high-cycle fatigue, 13–14
high pressure pipe, 103
hinge bellow joints, 8, 232–34
Hook's law, 34, 38
hoop pressure stress, 45, 46–47, 332
hoses, see flexible hoses
hot balanced spring load, 168
hydrodynamic loads, inclusion in owner's design specification, 27
hydrogen attack, 19
hydrostatic testing
    and spring supports, 162–63
    compressor piping, 312

I

ideal anchor evaluations, 61–62
idling equipment, 294
impact loads. See also ductile materials
impact testing, and temperature, 11–12
impulse function, 398
impulse loading, accounting for uncertainties, 428
in-line pressure balanced expansion joints, 228–29
in-plane bending stress, 114
    SIF (stress intensification factor), 69
industry practice, 22–23
    small piping, 26

supports and restraints friction, 25–26
    See also field proven systems
inherent flexibility, 65
integrally reinforced welded-on branch connections, 5, 8
    pressure design, 109–10
    See also branch connections, reinforcement of
interlocked hose, 238
isometric drawings, position in piping design process, 2

J

job specification/requirements, 28. See also project specification

K

Kármán force coefficient, 438
Kellogg's choking model
    for bending on cylindrical shell nozzle, 259–61
    for integral support attachments, 198–202

L

lap-joint flanges, 5, 7
large pipes, definitions, 178
Larson-Miller parameter (LMP), 16–17
levels of services, nuclear piping, 126
limit-rods, problems with expansion joints, 243–44
limit stops, 155
line-by-line analysis, 291–92
line lists, 29
    position in piping design process, 1
line number, 29
line stops, 155
LMP (Larson-Miller parameter), 16–17
load cases, 23–24
local coordinate flexible joints, 212
local support stresses, 24
local thermal stresses, 24–25
long-radius elbow bends, 8
longitudinal joint efficiency, allowable stress and, 113
longitudinal pressure stress, 45–46, 118
longitudinal welded pipes, 5, 6
    weld strength reduction factor, 101
Lorenz factors, 47
low-cycle fatigue, 13–14
low-type tank connections
    allowable piping loads, 274–80
    definition, 270
    displacement and rotation, 271–72
    practical considerations, 280–82
    stiffness coefficients, 272–74

M

main anchors, 226
mass lumping, 410

material specification engineers, role in piping design process, 1
material specifications, 28–29
   position in piping design process, 1
maximum distortion energy theory, 52
maximum energy theory, 52
maximum shear theory, 52
maximum strain theory, 52
maximum stress theory, 52
MDOF systems (Multiple Degree of Freedom), 409–10
   consistent mass matrix approach, 411–12
   free vibration and modal superposition, 412–13
   mass lumping, 410
minimum bend radius, flexible hoses, 239–40
miter bends, 5, 8
   pressure design, 106–9
Mohr's Circle, 51
moment restraints. *See* rotational restraints
moments
   bending, 40–42
   of inertia, 42–43
   stresses due to, 40–44, 48
   summation of, 40
   torsion, 43–44
multi-dimensional stresses, 49

N

Napier's law, 394
natural circular frequency, 377
natural frequency, 376–77
natural period, 377
NB-3600, 92
NC-3600, 92
ND-3600, 92
needle valves, 5
Newmark mid-point constant acceleration, 449
nominal diameter, 6
nominal stress, nuclear piping, 127
nominal thickness, 6, 460. *See also* equivalent stress intensification factors
non-concurrent movements, use of catalog data for flexible connections, 221
notches, ductile ruptures, 10–11
NRC (Nuclear Regulatory Commission) directives, 125
nuclear piping
   branch connections
      component stresses calculation, 130–32
      flexibility, 261
   earthquake load analysis, 126
   levels of services, 126
   nominal stress, 127
   occasional stress, 126
   peak stress range, 128–29
   pressure design, 126
   wind load analysis, 126
   See also Class 1 nuclear piping
nuclear power plant piping codes, 20–22
Nuclear Regulatory Commission (NRC) directives, 125

O

occasional loads, inclusion in owner's design specification, 27
occasional stress, 113
   calculation in Code, 119–21, 332–33, 334
   nuclear piping, 126
   weld strength reduction factor, 6
offset method of determining yield point, 35
on-off interface approach, 419
one-way stops, 155
operating pressure. *See* pressure, operating
operating temperature. *See* temperature, operating
out-plane bending stress, 114
   SIF (stress intensification factor), 69
ovalization of bend cross-section, 68–69
owner's design specification, 26–28

P

P&IDs (piping and instrument diagrams), position in piping design process, 1
passive loads, 98
peak stress, 113
peak stress range, nuclear piping, 128–29
PGs (planar guides), 230
pipe materials, 6
pipe sizes, 6
pipe stress analysis
   neglect of, 2
   scope of, 2–4
pipe stress and support
   effort required, 2
   position in piping design process, 2
pipe support, terminology, 152
pipelines
   analytical model construction, 351–53
   anchors, 335–36, 353–54
   bends, 338–39, 348
      large, 349–51
   Codes, 21, 330–31
   example behavior calculations, 346–48
   extension to reduce stress, 353
   fully restrained sections, 337–38
   minimum bend radius, 339
   movement of free ends, 336–37, 348
   movement of restrained ends, 337
   pressure elongation, 25, 81, 335
   soil-pipe interactions, 344–46, 347–48
   soil resistance against axial pipe movement, 341–43
   soil resistance in lateral direction, 343–44
   soil resistance simulation, 348–49

support friction, 26, 172–73
See also gas pipelines
uses of, 329–30
piping, importance of, 1
Piping Code, 19–22
piping design process, 1–2
piping designers, role in piping design process, 1–2
piping flexibility, 63
   excessive, 86–88
   inherent, 65
piping flexibility analysis, 2, 61
   general procedure, 83–86
piping mechanical engineers, role in piping design process, 2
planar guides (PGs), 230
plant walk-down/through, 30
plug valves, 5
pneumatic test, inclusion in owner's design specification, 27
Poisson's ratio, 37–38
Poisson's strain, 38
pound classification of flanges, 7
power boiler attachment design formula, 194–98
power piping, Code, 21
power/steam boilers connections, 265–66. *See also* stationary equipment interfaces
pre-spring. *See* cold spring
pressure
   design, 29, 93, 102
   effect on bend flexibility and SIFs, 82
   effect on flexibility, 25, 80
   external, 17, 103
   operating, 29
   test, 93
   upset, 93
pressure balanced universal expansion joints, 232
pressure design
   branch connections, 109–12
   Code, 101–2
   curved pipe, 104–6
   miter bends, 106–9
   nuclear piping, 126
   other components, 113
   straight pipe, 102–4
pressure elongation
   as self-limiting load, 82
   effect on flexibility, 80–81
   pipelines, 25, 81, 335
pressure loadings, Code, 93
pressure rating, use of catalog data for flexible connections, 220
pressure-temperature rating, 7
pressure vessel connections

Kellogg's choking model for bending on cylindrical shell nozzle, 259–61
loadings imposed to piping from vessel, 257–58
nozzle flexibility at spherical shell, 263–64
nozzle flexibility on cylindrical vessel, 261–62
nuclear piping branch flexibility, 261
reinforcing pads, 264
vessel shell flexibility, 258–59
See also stationary equipment interfaces
pressure waves, 97–98
primary stage creep, 15
principal planes, 36
principal stresses, 36
process engineers, role in piping design process, 1
process flow diagrams, position in piping design process, 1
process heater connections, 266–68. *See also* stationary equipment interfaces
process piping, Code, 21
project specification, 28–30
proportional limit of stress-strain curve, 34
pulsation flow
   harmonic analysis, 439–42
   reciprocating compressors and pumps, 313–15
   See also acoustic pulsation
pumping circuits
   with excessive flexibility, 88
   See also centrifugal pumps

Q

quick check on piping flexibility
   formulas, 63–65
   caution, 65

R

real anchor evaluations, 62–63
reciprocating compressors and pumps, 313
   items to be addressed, 319–21
   pulsation dampeners, 317–19
   pulsation pressure, 315–16
   See also pulsation flow
rectangular pulse, 380
reference fatigue life of bellows, 218–19
refractory lined pipes, 365–69
   equivalent modulus of elasticity, 365–66
   hot-cold pipe junctions, 366–69
reinforced fabricated tee, 5, 8
   pressure design, 109–10
   See also branch connections, reinforcement of
relief displacement, 217
relief strain, 38
resilient friction restraint approach, 175–76
resilient supports. *See* spring supports
response spectra method, 415–17

absolute closely spaced modal combination, 421–22
combination of results, 417–19
comparison of modal combination methods, 419–21
compensation for the higher modes truncated, 422–23
design response spectra, 424–26
resting supports, 155, 159–62
restraints, 151, 153–54
   stiffness, 26
   See also supports and restraints
rigid body acceleration. *See* ZPA
rigidity, modulus of, 38
ring girders, 153
risers, 170
   analysis method, 172
   load calculation, 171–72
   support schemes, 170–71
Roark's saddle, 179–80
rollers, 177
Rosenblueth correlation coefficient, 419
rotating equipment connections, 9
   allowable piping load
      history, 286–88
      manufacturer's figures, 285–86
   allowable stress, 76
   analysis approaches, 290–92
   design procedure example, 324–26
   effect of piping loads, 288–89
   excessive flexibility, 321–22
   fit-up, 294–95
   improper expansion joint installations, 322
   multi-unit installations, 293–94
   nozzle movements, 289–90
   spring hanger selection to minimize weight load, 292–93
   theoretical restraints, 322–24
   See also centrifugal compressors; centrifugal pumps; reciprocating compressors and pumps; steam turbines
rotating equipment piping
   design considerations, 4
   support friction, 26, 172
rotational restraints, 157
rotational slip joints. *See* ball/ball-and-socket joints
rotational spring rate, 217
roundhouse stress-strain curve, 34, 35
routing study, position in piping design process, 1
rubber-type bellows, 213
rupture safety factor, 99

S

saddle supports
   Roark's saddle, 179–80
   Zick's saddle, 185–90

saddles, 153
safety-relief valves, 8
safety valve relieving forces
   closed discharge systems, 401–3
   open discharge systems, 397–401
schedule numbers, 6
schematic planning drawings, position in piping design process, 1
Schorer's ring girder, 180–85
scraper (pig) launching station analysis, 352
SDOF systems (Single Degree of Freedom), 386–87
   damped systems, 382
      alternative equation for SDOF damped vibration, 384
      forced vibration on damped system, 385–86
      free vibration on viscously damped system, 382–83
      free vibration on viscously damped system with initial displacement, 383–84
   definitions, 375–76
   un-damped systems, 376–82
      with constant force, 377–78
      with harmonic load, 378–79
      with impulse loads, 380–82
   working formula, 376
seamless pipes, 6
secondary stage creep, 15
self-limiting stress, 66–67
self-spring, 73
shaking forces from fluid flow, 395–97
shear distribution factor, 47, 178
shear strain, 38
shear stresses
   in principal planes, 36
   maximum, 36
   soils, 341
shell connections, 8–9. *See also* stationary equipment interfaces
shock waves, 97–98
shoes, 152–53
short-radius elbow bends, 8
SIF (stress intensification factor)
   basis, 69
   Code (Piping Code), 69–71
   curved pipes, 68–71, 105
   in various Codes, 123–24
   pressure effect on, 82
   sustained loads, 25, 70–71
   See also equivalent stress intensification factors; torsion stress intensification factor
silt, 340
single-acting stops, 155
single tied expansion joints, 230
sliding friction, 174
sliding plates, 177–78

sliding supports, 155
slip joints, 235–37
slip-on flanges, 5, 7
slug flow, 97
small piping, industry practice, 26
snubbers, 156, 388–89
socket-welded connections, 370–71
soil mechanics, level of knowledge required, 340
soils
  friction angle, 340–41
  friction coefficients, 343
  resistance against axial pipe movement, 341–43
  resistance in lateral direction, 343–44
  resistance simulation, 348–49
  shearing stress, 341
  soil-pipe interactions, 344–46, 347–48
  types, 340
sonic velocity, 390–92
  versus flow velocity, 389–90
space maintenance stops, 155
specialty items, position in piping design process, 2
specific heat ratio, 391
spiral welded pipes, 5, 6
  weld strength reduction factor, 101
spring constant. See anchors, stiffness
spring rates, 151
  use of catalog data for flexible connections, 220
spring supports
  deciding when to use, 162–63
  terminology, 156
  See also constant-effort supports; variable spring hangers
SRSS (square root of the sum of the squares) method, 23, 115, 121, 418–21
stability failure, 17–18
stainless steel pipes, schedule numbers, 6
standards, 20. See also Piping Code
static equilibrium, 40
static friction, 174
static stress rupture, 9–12
stationary equipment interfaces
  allowable piping load, 264–65
  flanges
    Class 2 Nuclear Piping Rules, 254–55
    equivalent pressure method, 251–54
    flange stress calculation, 253–54
    leakage concerns, 247–48
    rating table lookup, 252–53
    standard design procedure, 248–50
    unofficial position of B31.3, 250–51
  interface effects, 247
  sensitive valves, 255–56
  See also heat exchanger connections; power/steam boilers connections; pressure vessel connections; process heater connections

steady-state vibration, 428
  allowable vibration displacement and velocity, 428–37
  basic vibration patterns, 428–29
  stress evaluation, 444–45
  See also harmonic analysis
steam hammer effects, 24, 97–98
  steam turbine trip load, 403–6
  See also occasional stress; sustained stress
steam, superheated, critical pressure ratio, 393
steam turbines
  mechanical drives, 296
    allowable for combined resultant loads, 297–98
    allowable loads at individual connections, 296–97
    piping layout strategy, 298–300
  power plants, 295–96
  trip load, 403–6
stiffening rings, stresses at, 184–85, 187–88, 189–90
stiffness coefficient, 53
stop gaps, 26
stops, 155
straight pipe, pressure design, 102–4
straight pipe beam elements, 68
strain, 33–34
  elastic relationship with stress, 36–37
stratification of flow, 361–63
stress amplitude, 13
stress corrosion failures, 19
stress indices, 119, 126
stress intensification factor. See SIF
stress intensity, 52, 126
  alternating, 129
  calculation in Code, 115
stress reduction provisions, for cold temperature, 12
stress–strain curve, 34
stresses
  at skewed plane, 35–36
  calculation of, 33–34
  due to forces, 47–48
  due to internal pressure, 45–47
  due to moments, 40–44, 48
  elastic relationship with strain, 36–37
  multi-dimensiona, 49–53
  of piping components, 113–17
Strouhal number, 438
structural analysis, 113
struts, 154
stub-in connection. See un-reinforced fabricated tee
suction standpipes, 319
suddenly applied load, 374–75
support spring constant. See spring rates
supports, 152–53
  terminology, 151, 166

supports and restraints
　details in project specification, 29–30
　device terminology, 151–57
　friction
　　centrifugal compressors, 312
　　effects, 172–74
　　including in analysis, 174–76
　　inclusion in owner's design specification, 28
　　industry practice, 25–26
　　methods of reducing, 176–78
　　types, 172
　integral support attachments
　　analysis methods, 194
　　Kellogg's choking model, 198–202
　　power boiler formula, 194–98
　　WRC-107, 202–5
　large pipes, 178–79
　　choice of support type, 190–92
　　Roark's saddle, 179–80
　　Schorer's ring girder, 180–85
　　Zick's saddle, 185–90
　spacing, 152–59
　stiffness and displacement, 205–6
surge bottles, 315–16. *See also* gas-filled surge chambers
sustained loads
　inclusion in owner's design specification, 28
　stress intensification factors (SIFS), 25, 70–71
sustained stress, 113, 117–19
system engineers, role in piping design process, 1

T
tank settlement, 281
Teflon, sliding plates, 177
temperature
　design, 29, 93
　flexibility, 29, 93
　operating, 29
temperature, and brittleness, 11–12
temperature gradient, across pipe wall, 95
temperature loadings, Code, 93–95
tensile strength, 33–34
terminal connections, 8
tertiary stage creep, 15
test pressure, Code, 93
test stress, Code, 93
thermal bowing
　created by a tiny line, 364–65
　displacement and stress produced, 357–59
　occurrences of, 357, 360–64
　See also bowing temperature
thermal discontinuity, 95
thermal expansion
　allowable stress range, 13, 71–76, 129, 332, 334
　as self-limiting stress, 66–67
　calculation in Code, 121–24

　forces and stresses, 61
　　straight pipes, 61–63
　　See also local thermal stresses
　rates for, 463
　support friction, 172
　wall thickness and, 65
thermal gradient stresses, 24
through-run moment components, 131–32
tie-rods, problems with expansion joints, 242–43
time-history analysis, 446
　damping, 446–48
　example analysis, 451–56
　integration schemes, 448–50
　time step, stability, and accuracy, 450–51
torsion moments, stresses due to, 43–44
torsion stress, 115
torsion stress intensification factor, 115
torsional moment, use of catalog data for flexible connections, 223
torsional spring rate, 219
total end force. *See* impulse function
Tresca Stress. *See* stress intensity
tripartite log-log graphs, 424
trunnion shoes, 153
tube bundle header connections, 9. *See also* stationary equipment interfaces
two-dimensional stresses, 49–51

U
ultimate strength, 9, 35
un-insulated flange connections, 369
un-intensified stress, 431
un-reinforced fabricated tee (stub-in connection), 5, 8
　pressure design, 109, 111–12
under-tolerance, 6
universal expansion joints, 230–32
universal gas constant, 390
unmatched small branch connections, 369–70

V
valves, 5, 8
　valve and flange data, 467–73
　See also safety valve relieving forces
variable spring hangers
　load setting, 168–69
　selection procedure, 163–65
　types and installation, 166–68
variable spring supports. *See* spring supports
vessel connections. *See* stationary equipment interfaces
vibration lines, 23
vibration velocity
　allowable, 436–37

measuring, 436
See also steady-state vibration, allowable vibration displacement and velocity
virtual anchor length, 336
virtual anchor point, 336
viscous damping, 382, 387–88
von Mises Stress. *See* effective stress
vortex shedding, 96–97

W

water hammer effects, 24, 97–98. *See also* occasional stress; reciprocating compressors and pumps, pulsation dampeners; sustained stress
weather effects, 95
weight effects, 95
weight grades, 6
weld strength reduction factor, 6–7, 100–101
welded connections, 6. *See also* contoured welded-on branch connections; socket-welded connections
welded-on branch connections, 5
welded pipes, 6
welding neck flanges, 5, 7
welding tee. *See* forged welding tee
Wilson θ method, 449
wind load analysis
  Code, 95
  industry practice, 23
  nuclear piping, 126
  vortex shedding, 439
  See also occasional stress; sustained stress
working spring rate of below, 214–15
WRC-107 stress evaluation, 202–5

Y

yield point, 35
yield strength, 35
yield stress, 35
Young's modulus, 34

Z

Zick's saddle, 185–90
ZPA (zero period acceleration), 423